Columbia Crew Survival Investigation Report

NASA/SP-2008-565

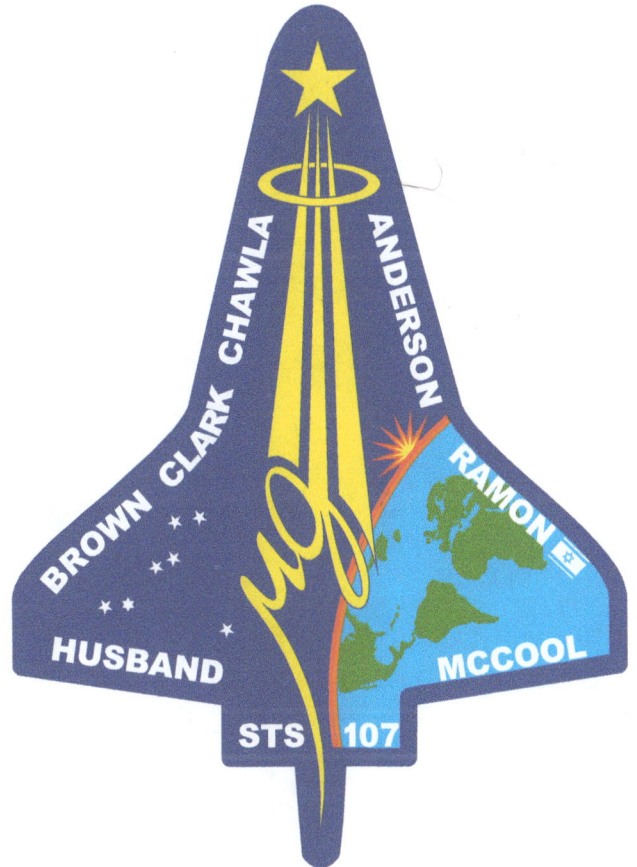

Contents

Future Work

Appendix: Ballistic Tutorial

Figures

Tables

Acronyms and Abbreviations

A/G	air-to-ground
A/G1	air-to-ground 1
A/N PRC	Army/Navy personal radio communications
AC	alternating current
AC3	alternating current Bus 3
ACES	advanced crew escape suit
AFIP	Air Force Institute of Pathology
AOD	automatic opening device
AOS	acquisition of signal
APU	auxiliary power unit
ASCAN	astronaut candidate
BFS	backup flight software
BIP	bio-instrument pass-through
BLIMP-K	Boundary Layer Integral Matrix Procedure-Kinetic
BN	ballistic number
BRC	Biodynamic Research Corporation
Btu	British thermal unit
Btu/lbm	Btu per pound of mass
c.g.	center of gravity
CAIB	*Columbia* Accident Investigation Board
CAPCOM	capsule communicator
CCA	communications carrier assembly
C_D	coefficient of drag
CDR	Commander
CE	Catastrophic Event
CEE	crew escape equipment
CEL	Concept Exploration Laboratory
CES	Crew Escape System
CFD	computational fluid dynamics
CH4	Channel 4
CM	crew module
CMCE	Crew Module Catastrophic Event
CO_2	carbon dioxide
COMM	communications
CRD	*Columbia* Reconstruction Database
CRP	*Columbia* Research and Preservation
CRT	cathode ray tube
CSS	control stick steering
CSWG	Crew Survival Working Group
CTB	cargo transfer bag
CWE	crew worn equipment
D/O PREP	deorbit preparation
DAP	digital autopilot
DD.ddddd	degrees and decimal degrees
DD MM.mmm	degrees, minutes, and decimal minutes
DD MM SS.sss	degrees, minutes, seconds, and decimal seconds
DNA	deoxyribonucleic acid
DoD	Department of Defense

DP/dt	pressure change rate
DSC	dual suit controller
ECLSS	Environmental Control and Life Support System
EI	entry interface
EOS	Emergency Oxygen System
ET	external tank
EVA	extravehicular activity
FAA	Federal Aviation Administration
FBI	Federal Bureau of Investigation
FCS	Flight Control System
FF	forward fuselage
FOV	field of view
fps	feet per second
FREESTAR	Fast Reaction Experiments Enabling Science, Technology, Applications, and Research
FSP	Fault Summary Page
GMT	Greenwich Mean Time
GPC	general purpose computer
GPS	Global Positioning System
hh:mm:ss	hours:minutes:seconds
ICU	individual cooling unit
ISAG	Image Science and Analysis Group
ISS	International Space Station
JSC	Johnson Space Center
KEAS	knots equivalent air speed
KSC	Kennedy Space Center
LCAT	Large Core Arc Tunnel
LCG	liquid cooling garment
LED	light emitting diode
LIB	left inboard
LiOH	lithium hydroxide
LOB	left outboard
LOC	loss of control
LOS	loss of signal
LPU	life preserver unit
LRWG	Late Re-entry Working Group
MADS	Module Auxiliary Data System
MAGR	Miniaturized Airborne Global Receiver
MAP	middeck access panel
MAR	middeck accommodation rack
MCC	Mission Control Center
μm	micron
MMACS	Mechanical, Maintenance, Arm, and Crew Systems
MPS	Main Propulsion System
MS	Mission Specialist
MSID	measurement and stimulus indication
N_2	nitrogen
NBC	National Broadcasting Corporation
NET	no earlier than
NLT	no later than
NO	nitric oxide
NTSB	National Transportation Safety Board

O$_2$	oxygen
OEX	orbiter experiment
OMS	Orbital Maneuvering System
ORSAT	Object Reentry Survival Analysis Tool
OSL	off-scale low
PASS	Primary Avionics Software System
PGSC	payload general support computer
PLBD	payload bay door
PLT	Pilot
PPA	personal parachute assembly
ppCO$_2$	partial pressure of carbon dioxide
ppO$_2$	partial pressure of oxygen
PS	Payload Specialist
psf	pounds per square foot
psi	pounds per square inch
psia	pounds per square inch absolute
psid	pounds per square inch differential
QD	quick disconnect
RCC	reinforced carbon-carbon
RCS	Reaction Control System
RGPC	reconstructed general purpose computer
RHC	rotational hand controller
RPTA	rudder pedal transducer assembly
SARSAT	search and rescue satellite-aided tracking
SCSIIT	Spacecraft Crew Survival Integrated Investigation Team
SEAWARS	Seawater Activated Release System
SES	Shuttle Engineering Simulator
SFRM	space flight resource management
SINDA	Systems Improved Numerical Differencing Analyzer
SMPTE	Society of Motion Picture and Television Engineers
SMS	shuttle mission simulator
SORT	Simulation and Optimization of Rocket Trajectories
SSME	space shuttle main engine
SUPA	shuttle urine pretreat assembly
3-D	three-dimensional
TAA	tunnel adapter assembly
TCDT	terminal countdown demonstration test
TD	Total Dispersal
	thickness dimension
TDRS-W	Telemetry, Tracking, and Data Relay Satellite-West
TELCU	thermal electric liquid cooling unit
TET	tape elapsed time
TIG	time of ignition
TIM	Technical Interchange Meeting
TiO$_2$	titanium oxide
TPS	Thermal Protection System
TSUB-A	Telonics Satellite Uplink Beacon-A
TUC	time of useful consciousness
WCS	Waste Collection System
WSB	water spray boiler

Executive Summary

Background

NASA commissioned the Columbia Accident Investigation Board (CAIB) to conduct a thorough review of both the technical and the organizational causes of the loss of the Space Shuttle *Columbia* and her crew on February 1, 2003. The accident investigation that followed determined that a large piece of insulating foam from *Columbia*'s external tank (ET) had come off during ascent and struck the leading edge of the left wing, causing critical damage. The damage was undetected during the mission. The CAIB's findings and recommendations were published in 2003 and are available on the web at http://caib.nasa.gov/. NASA responded to the CAIB findings and recommendations with the Space Shuttle Return to Flight Implementation Plan.[1] Significant enhancements were made to NASA's organizational structure, technical rigor, and understanding of the flight environment. The ET was redesigned to reduce foam shedding and eliminate critical debris. In 2005, NASA succeeded in returning the space shuttle to flight. In 2010, the space shuttle will complete its mission of assembling the International Space Station and will be retired to make way for the next generation of human space flight vehicles: the Constellation Program.

The Space Shuttle Program recognized the importance of capturing the lessons learned from the loss of *Columbia* and her crew to benefit future human exploration, particularly future vehicle design. The program commissioned the Spacecraft Crew Survival Integrated Investigation Team (SCSIIT). The SCSIIT was asked to perform a comprehensive analysis of the accident, focusing on factors and events affecting crew survival, and to develop recommendations for improving crew survival for all future human space flight vehicles. To do this, the SCSIIT investigated all elements of crew survival, including the design features, equipment, training, and procedures intended to protect the crew. This report documents the SCSIIT findings, conclusions, and recommendations.

Results

One of the more difficult problems facing the SCSIIT was how to characterize events that occurred in an operating regime that was far outside the collective experience of aircraft accident investigation and without significant applicable test data. The investigation relied on data in the form of video, recovered debris, and medical findings, each supplemented with modeling and analyses when needed. The SCSIIT used these data to identify all events with lethal potential (even those that occurred after the crew was deceased) during entry so that threats to crew survival could be described and methodically approached in future designs. In the course of the investigation, five events with lethal potential were identified.

1. **Depressurization of the crew module at or shortly after orbiter breakup.**
   The pressure suit used by space shuttle crews on ascent and entry was not a part of the initial design of the orbiter. It was introduced in response to the *Challenger* accident. While it protects the crew from many contingency scenarios, there are several areas where integration difficulties diminish the capability of the suit to protect the crew. The *Columbia* depressurization event occurred so rapidly that the crew members were incapacitated within seconds, before they could configure the suit for full protection from loss of cabin pressure. Although circulatory systems functioned for a brief time, the effects of the depressurization were severe enough that the crew could not have regained consciousness. This event was lethal to the crew.

---

[1]NASA's Implementation Plan for Space Shuttle Return to Flight and Beyond, 10th Edition, June 3, 2005.

Key Recommendations

Space shuttle crew training should include greater emphasis on the transition between problem-solving and survival operations.

Future spacecraft must fully integrate suit operations into the design of the vehicle and provide features that will protect the crew without hindering normal operations.

2. **Exposure of unconscious or deceased crew members to a dynamic rotating load environment with a lack of upper body restraint and nonconformal helmets.**
    When the orbiter lost control, the resultant motion was not lethal but did require crew members to brace against the motion. The forebody, which is made up of the crew module and forward fuselage, separated at orbiter breakup. The forebody continued to rotate. After the crew lost consciousness due to the loss of cabin pressure, the seat inertial reel mechanisms on the crews' shoulder harnesses did not lock. As a result, the unconscious or deceased crew was exposed to cyclical rotational motion while restrained only at the lower body. Crew helmets do not conform to the head. Consequently, lethal trauma occurred to the unconscious or deceased crew due to the lack of upper body support and restraint.

Key Recommendations

Crew procedures must be re-evaluated in light of the findings regarding the motion of the intact orbiter and the forebody after separation.

Future spacecraft should be evaluated for loss of control motion and dynamics for adequate integration into development, design, and crew training.

Future spacecraft seats and suits should be integrated to ensure proper restraint of the crew in off-nominal situations while not affecting operational performance. Future crewed spacecraft vehicle design should account for vehicle loss of control to maximize the probability of crew survival.

3. **Separation of the crew from the crew module and the seat with associated forces, material interactions, and thermal consequences.**
    The breakup of the crew module and the crew's subsequent exposure to hypersonic entry conditions was not survivable by any currently existing capability. It was an extremely significant event, but it was very difficult to characterize because many events appeared to happen in a short period of time. The actual maximum survivable altitude for the crew module following a breakup of the orbiter is too complex to compute because it depends on the altitude and velocity at release as well as rotational dynamics that are understood only in a general way. The lethal-type consequences of exposure to entry conditions included traumatic injury due to seat restraints, high loads associated with deceleration due to a change in ballistic number, aerodynamic loads, and thermal events. Crew circulatory functions ceased shortly before or during this event. The ascent and entry suit had no performance requirements for occupant protection from thermal events. The only known complete protection from this event would be to prevent its occurrence.

Key Recommendation

Future vehicle design should incorporate an analysis for loss of control/breakup to optimize for the most graceful degradation of vehicle systems and structure to enhance chances for crew survival. Operational procedures can then integrate the most likely scenarios into survival strategies.

4. **Exposure to near vacuum, aerodynamic accelerations, and cold temperatures.**
    The ascent and entry suit system is certified to a maximum altitude of 100,000 feet and velocity of 560 knots equivalent air speed. It is uncertain whether it can protect a crew member at higher altitudes and air speeds.

Key Recommendation

Crew survival suits should be evaluated as an integrated system to determine the various weak points (thermal, pressure, windblast, chemical exposure, etc.). Once identified, alternatives should be explored to strengthen the weak areas.

5. **Ground impact.**
The ascent and entry suit system provides protection from ground impact with a parachute system. The current parachute system requires manual action by a crew member to activate the opening sequence.

Key Recommendation

Future spacecraft crew survival systems should not rely on manual activation to protect the crew.

Improving Crew Survival Investigations

The SCSIIT also identified recommendations regarding crew survival investigations. These include that:

- Crew survival investigations should be given high priority for all future spacecraft mishaps. Medically sensitive data should always be protected to preserve the privacy of the victims and their families. Because there is a limited database of information, each accident provides crucial understanding of the environment and expanding the envelope for survival.

- Data management proved to be critical to the investigation in many areas, such as equipment identification marking, debris recovery ground coordinates, database documentation, and tracking versions of reports, briefings, and analyses. Preservation of debris and data that may be of value in future investigations should be standardized and continued.

The SCSIIT investigation was performed with the belief that a comprehensive, respectful investigation could provide knowledge that would improve the safety of future space flight crews and explorers. By learning these lessons and ensuring that we continue the journey begun by the crews of Apollo 1, *Challenger*, and *Columbia*, we help to give meaning to their sacrifice and the sacrifice of their families. It is for them, and for the future generations of explorers, that we strive to be better and go farther.

Introduction

Human space flight is still in its infancy; spacecraft navigate narrow tracks of carefully computed ascent and entry trajectories with little allowable deviation. Until recently, it remained the province of a few governments. As private industry and more countries join in this great enterprise, we must share findings that may help protect those who venture into space. In the history of NASA, this approach has resulted in many improvements in crew survival. After the Apollo 1 fire, sweeping changes were made to spacecraft design and to the way crew rescue equipment was positioned and available at the launch pad. After the *Challenger* accident, a jettisonable hatch, personal oxygen systems, parachutes, rafts, and pressure suits were added to ascent and entry operations of the space shuttle.

As we move toward a time when human space flight will be commonplace, there is an obligation to make this inherently risky endeavor as safe as feasible. Design features, equipment, training, and procedures all play a role in improving crew safety and survival in contingencies. In aviation, continual improvement in oxygen systems, pressure suits, parachutes, ejection seats, and other equipment and systems has been made. It is a core value in the aviation world to evaluate these systems in every accident and pool the data to understand how design improvements may improve the chances that a crew will survive in a future accident.

The *Columbia* accident was not survivable. After the *Columbia* Accident Investigation Board (CAIB) investigation regarding the cause of the accident was completed, further consideration produced the question of whether there were lessons to be learned about how to improve crew survival in the future.

This investigation was performed with the belief that a comprehensive, respectful investigation could provide knowledge that can protect future crews in the worldwide community of human space flight. Additionally, in the course of the investigation, several areas of research were identified that could improve our understanding of both nominal space flight and future spacecraft accidents.

This report is the first comprehensive, publicly available accident investigation report addressing crew survival for a human spacecraft mishap, and it provides key information for future crew survival investigations. The results of this investigation are intended to add meaning to the sacrifice of the crew's lives by making space flight safer for all future generations.

The Columbia *Accident Investigation Board Report*

The CAIB completed its investigation into the *Columbia* mishap and published Volume I of its report in August 2003. Five supporting volumes were subsequently completed and published. The CAIB Report provides a thorough study of the accident and its causes. Since the crew had no role in causing the accident, the CAIB Report contained limited discussion of crew-related events. Although the CAIB Report included no formal recommendations concerning crew survival, it did contain the following relevant observation:[2]

> Observation 10.2-1 *Future crewed-vehicle requirements should incorporate the knowledge gained from the* Challenger *and* Columbia *accidents in assessing the feasibility of vehicles that could ensure crew survival even if the vehicle is destroyed.*

---

[2]*Columbia* Accident Investigation Board Report, Volume I, Section 10.2, Crew Escape and Survival, August 2003.

Additionally, Appendix G12, Crew Survivability, page 355, Volume V, October 2003 added:

> *To enhance the likelihood of crew survivability, NASA must evaluate the feasibility of improvements to protect the crew cabin in existing orbiters.*
>
> *NASA should investigate techniques that will prevent the structural failure of the CM [crew module] due to thermal degradation of structural properties and determine their feasibility for application.*
>
> *Future crewed vehicles should incorporate the knowledge gained from the 51-L [Challenger] and STS-107 mishaps in assessing the feasibility of designing vehicles that will provide for crew survival even in the face of a mishap that results in the loss of vehicle.*
>
> *Crew procedures and techniques for use of CWE [crew worn equipment] should be standardized and complied with by all crewmembers.*

To address post-*Columbia* Return to Flight actions, the Space Shuttle Program approved the formation of a multidisciplinary Spacecraft Crew Survival Integrated Investigation Team (SCSIIT) in July 2004. The team's primary objective was to combine engineering and medical analyses to determine what happened to the crew module and the *Columbia* crew to enhance crew safety and survival for future human space flights. This effort built upon and extended the activities of the Crew Survival Working Group (CSWG), which was formed at the time of the CAIB investigation.

In many regards this investigation presented several challenges. First, space flight is a relatively new and rare experience and there have been only a few fatal mishaps. Consequently, there is no integrated or widely available body of information regarding the analysis of spacecraft accidents for crew survival. The environment of atmospheric entry is also unique when compared to aviation. The SCSIIT had to break new ground in conducting the investigation of a singular event in such a complex environment. The team had to modify existing models and tools, normally used for specific nominal situations in a predictive manner, to understand the mishap environment. Many of the technical tools and concepts used will be of great assistance to a future spacecraft accident investigator. With the proliferation of commercial and international human space flight activities, it is crucial that all participants begin to develop a more comprehensive process and database of information regarding spacecraft accident investigation.

Because of the nature of the *Columbia* accident, there are many unknowns associated with it. The SCSIIT attempted to address these unknowns through calculated judgment and some speculation. In the end, there were varying degrees of certainty and confidence. The word "probable" refers to events that the team was very confident occurred. "Likely" refers to events that the team is somewhat confident occurred, although supporting evidence may be less definite. When an event is described as "possible," it generally reflects a lack of data to confirm or refute the scenario but is still considered valuable to mention. The reconstruction of this accident relied on a wide array of data. This report reflects the final consensus reached by the investigators.

Summary of Conclusions and Recommendations

Lethal events

The SCSIIT framed its analysis by attempting to identify all of the potentially lethal events that occurred during the mishap, including those that occurred after the crew was deceased. This allowed the team to identify specific threats to crew survival at different phases of entry and address those threats in recommendations for future vehicles. In the course of the investigation, the SCSIIT identified five events with lethal potential. These events are summarized below, along with the findings and recommendations that accompany each one. Each event is discussed in detail in the body of the report.

1. **The first event with lethal potential was depressurization of the crew module, which started at or shortly after orbiter breakup.**

   The majority of the SCSIIT findings related to the first lethal event were connected to the operational incompatibilities of the advanced crew escape suit (ACES) with the orbiter. The launch and entry suit was added in response to the *Challenger* accident, rather than as a part of the original vehicle design. The ACES was the successor to that suit. The suit protects the crew in many scenarios; however, there are several areas where integration difficulties diminish the capability of the suit to protect the crew. Integration issues include: the crew cannot keep their visors down throughout entry because doing so results in high oxygen concentrations in the cabin; gloves can inhibit the performance of nominal tasks; and the cabin stow/deorbit preparation timeframe is so busy that sometimes crew members do not have enough time to complete suit-related steps prior to atmospheric entry.

   As *Columbia* entered the atmosphere, one crew member was not yet wearing the ACES helmet and three crew members were not wearing gloves. Per nominal procedures, the crew wearing helmets had visors up. There was a period of about 40 seconds after the orbiter loss of control (LOC) but prior to depressurization when the crew was conscious and capable of action. Part of this short timeframe was undoubtedly employed in recognizing that a problem existed, as the indications of LOC developed gradually. The crew members could have closed their visors in this timeframe but did not. The SCSIIT attributed this to the training regimen, which separates vehicle systems training from emergency egress training and does not emphasize the *transition* between problem resolution and a survival situation. Once the cabin depressurization began, the rate of depressurization incapacitated the crew so quickly that even those crew members who had fully donned the ACES did not have time to lower their visors. Although circulatory systems functioned for a brief time, the crew could not have regained consciousness upon descent to lower altitudes due to the effects of the depressurization.

   ### Key Recommendations

   - Crew survival systems and procedures should be incorporated early into future spacecraft designs to ensure that they are compatible with nominal operations and that sufficient time exists to ensure all safety-critical equipment can be configured prior to entry interface.

   - The training program should be evaluated to determine how to best incorporate the transition from problem-solving to survival.

   - Future spacecraft crew survival systems should not rely solely on manual activation to protect the crew.

2. **The second event with lethal potential was unconscious or deceased crew members exposed to a dynamic rotating load environment with nonconformal helmets and a lack of upper body restraint.**

   The orbiter lost control, probably when the hydraulic systems failed due to hot gas intrusion in the left wing. The resulting motion was not lethal but did require bracing by the crew. The forebody (crew module and forward fuselage) eventually separated and the crew module lost pressure at orbiter break-up. When it separated, the forebody began a multi-axis rotation at approximately 0.1 revolution/second. Loads due to deceleration significantly decreased at the moment of breakup due to the change in ballistic number, but began to climb as the forebody continued to decelerate.

   After the crew module depressurized and the crew lost consciousness, the seat inertial reel mechanisms failed to lock despite the off-nominal motion. The reels were not defective; they were simply not designed to lock under the conditions the forebody experienced. The upper harness straps failed at some point prior to the forebody breakup, causing the straps to recoil back into the inertial reel mechanism. Because the reel mechanisms did not lock, the unconscious or deceased crew members were exposed to cyclical rotational motion while their upper bodies were inadequately restrained. Helmets that did not conform to the head and the lack of upper body restraint resulted in injuries and lethal trauma.

Current emergency egress procedures for a vehicle LOC or breakup assume that the crew module will eventually stop rotating and will stabilize in a specific attitude. Aerodynamic analysis completed during this investigation shows that this is extremely unlikely. Further, the procedures are based on ascent conditions only.

Key Recommendations

- Future spacecraft suits and seat restraints should use state-of-the-art technology in an integrated solution to minimize crew injury and maximize crew survival in off-nominal acceleration environments. Inertial reels should be evaluated for appropriateness of design for off-nominal scenarios.

- Helmets should provide head and neck protection in off-nominal dynamic load conditions. The current space shuttle inertial reels should be manually locked at the first sign of an off-nominal situation.

- A team of crew escape instructors, flight directors, and astronauts should be assembled to assess orbiter procedures in the context of ascent, deorbit, and entry contingencies.

- Future spacecraft should be evaluated while still in the design phase for dynamics and entry thermal and aerodynamic loads during a vehicle LOC for adequate integration into development, design, and crew training.

- Future crewed spacecraft vehicle design should account for vehicle LOC contingencies to maximize the probability of crew survival.

3. **The third event with lethal potential was separation from the crew module and the seats with associated forces, material interactions, and thermal consequences. This event is the least understood due to limitations in current knowledge of mechanisms at this Mach number and altitude. Seat restraints played a role in the lethality of this event.**
The breakup of the crew module and resultant exposure of the crew to entry conditions was an extremely significant event but was very difficult to characterize since many related events occurred in a short period of time. The consequences of exposure to entry conditions included traumatic injury related to seat restraints, high loads associated with deceleration due to a change in ballistic number, aerodynamic loads, and thermal events. All crew were deceased before, or by the end of, this event. The ACES has no performance requirements for occupant protection from thermal events and may not provide adequate protection even for egress scenarios involving heat and flames. There is no known complete protection from the breakup event except to prevent its occurrence.

The actual maximum survivable altitude for the crew module following a breakup of the orbiter is too complex to compute because it depends on the altitude and velocity at release as well as rotational dynamics, which are understood only in a general way.

Key Recommendations

- Future vehicle design should incorporate analysis for LOC/breakup to optimize for the most graceful degradation to vehicle systems and structure to enhance chances for crew survival. Operational procedures can then integrate the most likely scenarios into survival strategies.

- Future spacecraft suits, seats, and seat restraints should use state-of-the-art technology in an integrated solution to minimize crew injury and maximize crew survival in off-nominal acceleration environments.

- Crew survival systems should be evaluated as an integrated system that includes boots, helmet, seat restraints, etc. to determine the various weak points (thermal, pressure, windblast, chemical exposure, etc.). Once identified, alternatives should be explored to strengthen the weak areas. Materials with low resistance to chemicals, heat, and flames should not be used on equipment that is intended to protect the wearer from such hostile environments.

4. **The fourth event with lethal potential was exposure to near vacuum, aerodynamic accelerations, and cold temperatures.**
The ACES system is certified to operate at a maximum altitude of 100,000 feet, and certified to survive exposure to a maximum velocity of 560 knots equivalent air speed. The operating envelope of the orbiter is much greater than this. The actual maximum protection environment for the ACES is not known.

The recommendation to strengthen the weak areas of the suit system will increase the probability of survival through this type of event as well.

5. **The final event with lethal potential was ground impact.**
The ACES system provides protection from ground impact with a parachute system. The current parachute system requires manual action by a crew member to activate the opening sequence.

The earlier recommendation that future survival systems should not rely on manual activation will address this lethal event as well.

Crew survival accident investigation

Although this investigation was a follow-up to the actual mishap investigation, there were many findings, conclusions, and recommendations that apply to spacecraft accident investigations in general and crew survival investigations in particular. The recommendations address both NASA processes and investigation processes in general.

This crew survival investigation was difficult to do because of both the technical complexity and the sensitivity of the topic. Other Return to Flight activities took priority over the crew survival follow-up investigation, leading to resource issues for the SCSIIT.

Key Recommendations

- In the event of a future fatal spacecraft mishap, NASA should place a high priority on the performance of crew survival investigations.

- Medically sensitive and personal effects data should always be protected to preserve the privacy of the victims and their families. Issues surrounding public release of this type of sensitive information during a NASA accident investigation should be resolved and policies documented throughout the agency to ensure future crew survival investigations are performed.

- Stress debriefings and other counseling services should be available to those experiencing ongoing stress as a result of participating in the debris recovery and investigation.

- Data management proved to be critical to the investigation in many areas. Specifically, location of and access to debris recovery ground coordinates, database documentation, and configuration management of versions of reports, briefings, and analyses were all important. Many elements were highly successful, but improvements could be made. Global Positioning System coordinates for recovered items should be standardized.[3] Configuration control for documents was not initiated as early as it could have been. Additionally, *Challenger* supporting data were generally not cataloged by references to crew survival or the crew module. It was extremely difficult to find relevant data. *Challenger* debris is unpreserved and inaccessible for analysis. Report generation should start early in the investigation process to help provide consistency and documentation. Preservation of debris and data that may be of value in future investigations should be continued using the approach of the *Columbia* Research and Preservation Team. Accident investigation teams should develop standard templates across all areas of data management for the types of investigations performed for *Columbia*.

---

[3]STS-107 *Columbia* Reconstruction Report, NSTS-60501, June 30, 2003, p. 142.

- Configuration management documentation of seat and suit components had a significant impact on the investigation – positive when done well and negative when inadequate. Serialization and quality marking requirements and policies for space flight hardware should be developed to the lowest component level practical to aid in accident investigation.

- There were many findings relative to ground-based video of the entry and mishap. The video was a vital source of data for understanding the accident, especially after telemetry was no longer available. Video was used to timeline key events and to help understand the motion and trajectory of the objects of interest. One video that proved important had been mis-categorized initially, when the timeline of the accident was not yet understood, and was not used until very late in the investigation. After a mishap timeline has been established, videos should be re-reviewed to ensure relevant data are being used.

Other

Thermal analysis of some titanium components showed that entry heating alone was insufficient to cause the damage seen. Shock wave interactions can account for the damage to some extent, but arc jet testing showed that titanium combustion may also have played a role. In many other cases, evaluation of material performance in the low-pressure, high-temperature environment did not exist. Studies should be performed to further characterize the material behavior of titanium in entry environments to better understand optimal space applications of this material.

Summary

In summary, many findings, conclusions, and recommendations have resulted from this investigation that will be valuable both to spacecraft designers and accident investigators. This report provides the reader an expert level of knowledge regarding the sequence of events that contributed to the loss of *Columbia*'s crew on February 1, 2003 and what can be learned to improve the safety of human space flight for all future crews. It is the team's expectation that readers will approach the report with the respect and integrity that the subject and the crew of *Columbia* deserve.

Report Format

The *Executive Summary* highlights the intention of the investigation and key results.

Introduction explains the purpose and scope of the report and summarizes key results.

Conclusions and Recommendations contains a brief summary of all conclusions and recommendations.

Chapter 1 Integrated Story provides the sequence of events of the mishap that related to the crew.

1.1 Integrated Investigation Results brings together into one integrated story the results from the various aspects of the investigation.

1.2 Master Timeline provides a reference timeline of key events including vehicle configuration and status, crew activities, and changing circumstances.

Chapters 2 and 3 provide detailed insight into the examination, analysis, and understanding of evidence and results. Findings, conclusions, and recommendations are embedded in these chapters. The different sections address the accident from the individual subject matter perspective. By intent, these sections contain highly interrelated data and information that are shared or duplicated. This repetition provides mutually supportive information from the different aspects of the investigation. It is expected that future technical readers may only have interest in a few of these sections, so repetition was accepted for completeness so that each section could stand on its own. To aid future investigations, this report contains substantially more technical data and information than is normally contained in an aviation mishap report.

Chapter 2 Vehicle Failure Assessment describes the vehicle analyses. These were used to understand what happened to the orbiter and the crew module structures.

2.1 Motion and Thermal Analyses describes the analyses performed to understand the motion of the vehicle and crew module and the resultant loads acting upon the crew and structure. Thermal analyses rely heavily on trajectory assessments and, therefore, are also covered in this section.

2.2 Orbiter Breakup Sequence describes the analyses performed to understand the sequence in which the orbiter breakup occurred.

2.3 Crew Cabin Pressure Environment Analysis describes the integrated analyses used to understand the cabin pressure conditions of the crew module during the mishap.

2.4 Forebody Breakup Sequence describes the analyses performed to understand the sequence of the breakup of the forebody (crew module and forward fuselage) of the orbiter.

Chapter 3 Occupant Protection addresses crew and crew equipment special assessments.

3.1 Crew Seats and *3.2 Crew Worn Equipment* address the function and performance of the equipment intended to protect the crew in the experienced motion, load, and thermal environment.

3.3 Crew Training addresses procedures and preparations associated with examined events and activities.

3.4 Crew Analysis encompasses the awareness the crew had of events, crew actions in response to the events, and the events of lethal potential to which the crew was exposed.

Chapter 4 Investigative Methods and Processes explains the structure and makeup of the team, the approach taken to conduct the investigation, a description of the tools used, and the collection and management of data and information.

Future Work addresses suggestions for forward work.

Appendix A contains a tutorial on ballistic trajectories, providing more conceptual insight into a critical topic discussed extensively in the report.

Conclusions and Recommendations

The first event with lethal potential was depressurization of the crew module, which started at or shortly after orbiter breakup.

Conclusion L1-1. After loss of control at GMT[4] 13:59:37 and prior to orbiter breakup at GMT 14:00:18, the *Columbia* cabin pressure was nominal and the crew was capable of conscious actions. (p. 2-89, p. 3-82)

Conclusion L1-2. The depressurization was due to relatively small cabin breaches above and below the middeck floor and was not a result of a major loss of cabin structural integrity. (p. 2-93)

Conclusion L1-3. The crew was exposed to a pressure altitude above 63,500 feet, indicating that the cabin depressurization event occurred above this altitude. (p. 2-91, p. 3-83)

Conclusion L1-4. The crew was not exposed to a cabin fire or thermal injury prior to depressurization, cessation of breathing, and loss of consciousness. (p. 3-89)

Conclusion L1-5. The depressurization incapacitated the crew members so rapidly that they were not able to lower their helmet visors. (p. 2-90, p. 3-84)

Recommendation L1-1. Incorporate objectives in the astronaut training program that emphasize understanding the transition from recoverable systems problems to impending survival situations. (p. 3-66)

Recommendation L1-2. Future spacecraft and crew survival systems should be designed such that the equipment and procedures provided to protect the crew in emergency situations are compatible with nominal operations. Future spacecraft vehicles, equipment, and mission timelines should be designed such that a suited crew member can perform all operations without compromising the configuration of the survival suit during critical phases of flight. (p. 3-38, p. 3-86)

Recommendation L1-3/L5-1. Future spacecraft crew survival systems should not rely on manual activation to protect the crew. (p. 3-20, p. 3-44, p. 3-84)

Recommendation L1-4. Future suit design should incorporate the ability for crew members to communicate visors-down without relying on spacecraft power. (p. 3-82)

The second event with lethal potential was unconscious or deceased crew members exposed to a dynamic rotating load environment with nonconformal helmets and a lack of upper body restraint.

Conclusion L2-1. Between orbiter breakup and the forebody[5] breakup, the free-flying forebody was rotating about all three axes at approximately 0.1 rev/sec and did not trim into a specific attitude. (p. 2-23)

---

[4]Greenwich Mean Time.
[5]The orbiter forebody consists of the crew module, forward fuselage, forward Reaction Control System, nose cap, and nose landing gear.

Conclusion L2-2. The seat inertial reels did not lock. (p. 3-20)

Conclusion L2-3. Lethal injuries resulted from inadequate upper body restraint and protection during rotational motion. (p. 3-20, p. 3-87)

Recommendation L2-1. Assemble a team of crew escape instructors, flight directors, and astronauts to assess orbiter procedures in the context of ascent, deorbit, and entry contingencies. Revise the procedures with consideration to time constraints and the interplay among the thermal environment, expected crew module dynamics, and crew and crew equipment capabilities. (p. 3-67)

Recommendation L2-2. Prior to operational deployment of future crewed spacecraft, determine the vehicle dynamics, entry thermal and aerodynamic loads, and crew survival envelopes during a vehicle loss of control so that they may be adequately integrated into training programs. (p. 2-10, p. 2-29, p. 3-67)

Recommendation L2-3. Future crewed spacecraft vehicle design should account for vehicle loss of control contingencies to maximize the probability of crew survival. (p. 3-67)

Recommendation L2-4/L3-4. Future spacecraft suits and seat restraints should use state-of-the-art technology in an integrated solution to minimize crew injury and maximize crew survival in off-nominal acceleration environments. (p. 3-20, p. 3-53, p. 3-87, p. 3-88)

Recommendation L2-5. Incorporate features into the pass-through slots on the seats such that the slot will not damage the strap. (p. 3-24)

Recommendation L2-6. Perform dynamic testing of straps and testing of straps at elevated temperatures to determine load-carrying capabilities under these conditions. Perform testing of strap materials in high-temperature/low-oxygen/low-pressure environments to determine materials properties under these conditions. (p. 3-27)

Recommendation L2-7. Design suit helmets with head protection as a functional requirement, not just as a portion of the pressure garment. Suits should incorporate conformal helmets with head and neck restraint devices, similar to helmet/head restraint techniques used in professional automobile racing. (p. 3-53, p. 3-87)

Recommendation L2-8. The current shuttle inertial reels should be manually locked at the first sign of an off-nominal situation. (p. 3-21, p. 3-88)

Recommendation L2-9. The use of inertial reels in future restraint systems should be evaluated to ensure that they are capable of protecting the crew during nominal and off-nominal situations without active crew intervention. (p. 3-88)

The third event with lethal potential was separation from the crew module and the seats with associated forces, material interactions, and thermal consequences. This event is the least understood due to limitations in current knowledge of mechanisms at this Mach number and altitude. Seat restraints played a role in the lethality of this event.

Conclusion L3-1. Complete loss of hydraulic pressure to the aerosurfaces resulting from the breach in the left wing was the probable proximal cause for the vehicle loss of control. (p. 2-6)

Conclusion L3-2. The breakup of both *Challenger* and *Columbia* resulted in most of the X_o 582[6] ring frame bulkhead remaining with the crew module or forebody. (p. 2-84)

---

[6]X_o 582 refers to the location of the bulkhead in the orbiter coordinate frame. This bulkhead is immediately aft of the crew module.

Conclusion L3-3. The actual maximum survivable altitude for a breakup of the space shuttle is not known. (p. 2-29)

Conclusion L3-4. The seat restraint system caused lethal-level injuries to the unconscious or deceased crew members when they separated from the seat. (p. 3-88)

Recommendation L3-1. Future vehicles should incorporate a design analysis for breakup to help guide design toward the most graceful degradation of the integrated vehicle systems and structure to maximize crew survival. (p. 2-87, p. 2-139, p. 3-88)

Recommendation L3-2. Future vehicles should be designed with a separation of critical functions to the maximum extent possible and robust protection for individual functional components when separation is not practical. (p. 2-6)

Recommendation L3-3. Future spacecraft design should incorporate crashworthy, locatable data recorders for accident/incident flight reconstruction. (p. 2-36)

Recommendation L2-4/L3-4. Future spacecraft suits and seat restraints should use state-of-the-art technology in an integrated solution to minimize crew injury and maximize crew survival in off-nominal acceleration environments. (p. 3-53)

Recommendation L3-5/L4-1. Evaluate crew survival suits as an integrated system that includes boots, helmet, and other elements to determine the weak points, such as thermal, pressure, windblast, or chemical exposure. Once identified, alternatives should be explored to strengthen the weak areas. Materials with low resistance to chemicals, heat, and flames should not be used on equipment that is intended to protect the wearer from such hostile environments. (p. 3-46, p. 3-63)

The fourth event with lethal potential was exposure to near vacuum, aerodynamic accelerations, and cold temperatures.

Conclusion L4-1. Although the advanced crew escape suit (ACES) system is certified to operate at a maximum altitude of 100,000 feet and to survive exposure to a maximum velocity of 560 knots equivalent air speed, the actual maximum protection environment for the ACES is not known. (p. 3-46)

See **Recommendation L3-5/L4-1** above, which also addresses this event.

The final event with lethal potential was ground impact.

Conclusion L5-1. The current parachute system requires manual action by a crew member to activate the opening sequence. (p. 3-44)

See **Recommendation L1-3/L5-1** above, which also addresses this event.

Recommendation A1. In the event of a future fatal human space flight mishap, NASA should place high priority on the crew survival aspects of the mishap both during the investigation as well as in its follow-up actions using dedicated individuals who are appropriately qualified in this specialized work. (p. 4-5, p. 4-9)

Recommendation A2. Medically sensitive and personal debris and data should always be available to designated investigators but protected from release to preserve the privacy of the victims and their families. (p. 4-11)

Recommendation A3. Resolve issues and document policies surrounding public release of sensitive information relative to the crew during a NASA accident investigation to ensure that all levels of the agency understand how future crew survival investigations should be performed. (p. 4-11)

Recommendation A4. Due to the complexity of the operating environment, in addition to traditional accident investigation techniques, spacecraft accident investigators must evaluate multiple sources of information including ballistics, video analysis, aerodynamic trajectories, and thermal and material analyses. (p. 4-9)

Recommendation A5. Develop equipment failure investigation marking ("fingerprinting") requirements and policies for space flight programs. Equipment fingerprinting requires three aspects to be effective: component serialization, marking, and tracking to the lowest assembly level practical. (p. 3-35, p. 3-63)

Recommendation A6. Standard templates for accident investigation data (document, presentation, data spreadsheet, etc.) should be used. All reports, presentations, spreadsheets, and other documents should include the following data on every page: title, date the file was created, date the file was updated, version (if applicable), person creating the file, and person editing the file (if different from author). (p. 4-10)

Recommendation A7. To aid in configuration control and ensure data are properly documented, report generation must begin early in the investigation process. (p. 4-10)

Conclusion A8-1. Spacecraft accidents are rare, and each event adds critical knowledge and understanding to the database of experience. (p. 3-84, p. 4-11)

Recommendation A8. As was executed with *Columbia*, spacecraft accident investigation plans must include provisions for debris and data preservation and security. All debris and data should be cataloged, stored, and preserved so they will be available for future investigations or studies. (p. 3-85, p. 4-11)

Recommendation A9. Post-traumatic stress debriefings and other counseling services should be available to those experiencing ongoing stress as a result of participating in the debris recovery and investigation. Designated personnel should follow up on a regular basis to ensure that individual needs are being met. (p. 4-12)

Recommendation A10. Global Positioning System receivers used for recording the latitude/longitude of recovered debris must all be calibrated the same way (i.e., using the same reference system), and the latitude/longitude data should be recorded in a standardized format.[7] (p. 4-25)

Recommendation A11. All video segments within a compilation should be categorized and summarized. All videos should be re-reviewed once the investigation has progressed to the point that a timeline has been established to verify that all relevant video data are being used. (p. 2-49, p. 4-23)

---

[7]STS-107 *Columbia* Reconstruction Report, NSTS-60501, June 30, 2003, p. 142.

Conclusion A13-1. Titanium may oxidize and combust in entry heating conditions dependent on enthalpy, pressure, and geometry. (p. 2-45)

Conclusion A13-2. The heating from a Type IV shock-shock[8] impingement and titanium combustion (in some combination) likely resulted in the damage seen by the forward payload bay door rollers and the x-links.[9] (p. 2-45)

Recommendation A13. Studies should be performed to further characterize the material behavior of titanium in entry environments to better understand optimal space applications of this material. (p. 2-46)

---

[8]This refers to a specific type of intersecting hypersonic shock waves and is discussed in Section 2.1.
[9]The x-links are fittings that attach the crew module to the forward fuselage of the orbiter (see Section 2.1).

Chapter 1 – Integrated Story

1.1 Integrated Investigation Results

1.1.1 Events with lethal potential

There were five events identified with lethal potential to the crew.

The first event with lethal potential was depressurization of the crew module, which started at or shortly after orbiter breakup.

The second event with lethal potential was unconscious or deceased crew members exposed to a dynamic rotating load environment with nonconformal helmets and a lack of upper body restraint.

The third event with lethal potential was separation from the crew module and the seats with associated forces, material interactions, and thermal consequences. This event is the least understood due to limitations in current knowledge of mechanisms at this Mach number and altitude. Seat restraints played a role in the lethality of this event.

The fourth event with lethal potential was exposure to near vacuum, aerodynamic accelerations, and cold temperatures.

The final event with lethal potential was ground impact.

1.1.2 Integrated summary of events

This section provides an integrated summary of key events during the *Columbia* mishap as they relate to the crew and orbiter forebody (figures 1.1-1 and 1.1-2).[1] Figures 1.1-3 and 1.1-4 show depictions of the flight deck and middeck seats.

---

[1] The orbiter forebody consists of the crew module, forward fuselage, forward Reaction Control System (RCS), nose cap, and nose landing gear.

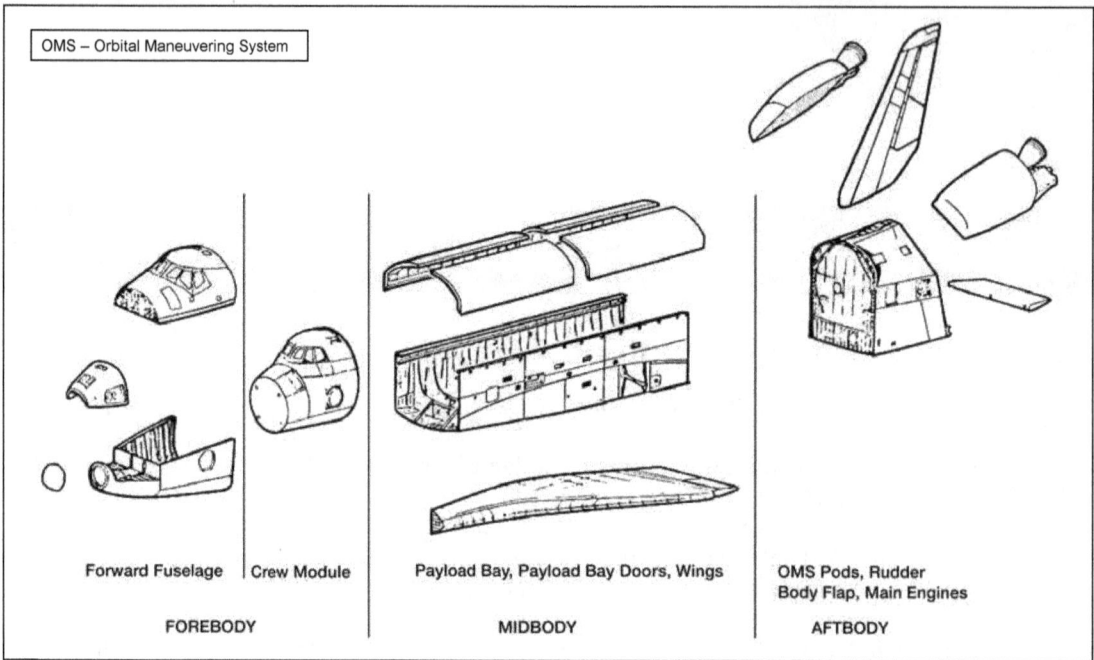

Figure 1.1-1. *Depiction of the orbiter forebody, midbody, and aftbody elements.*

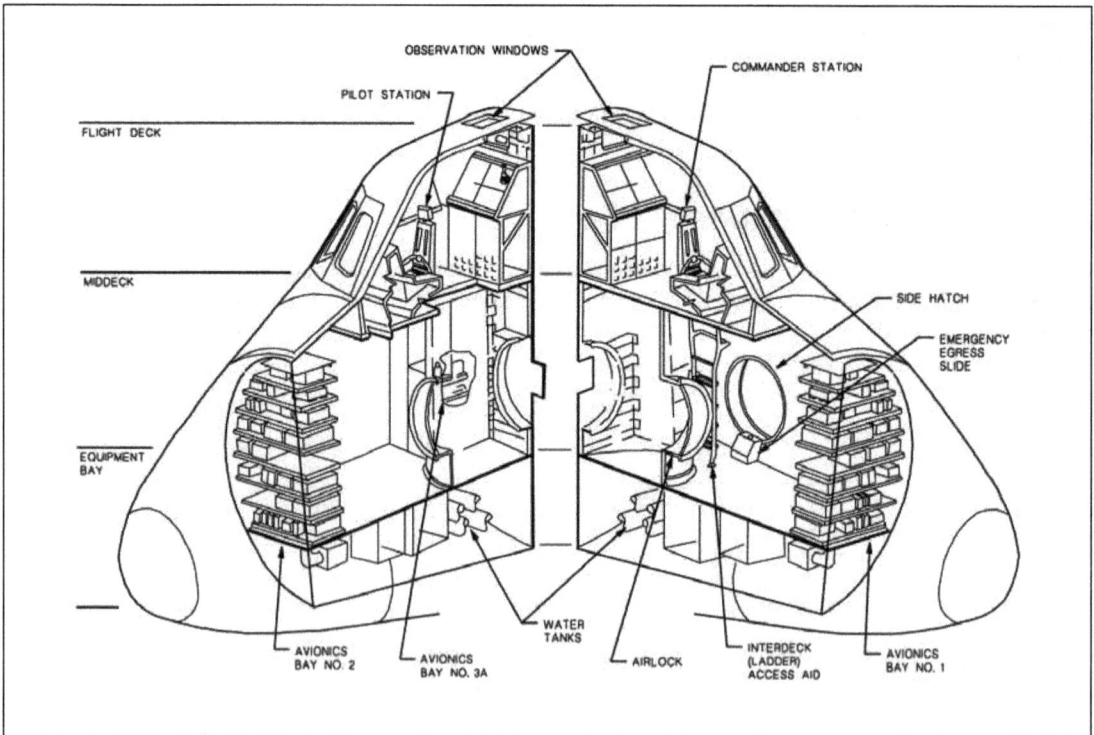

Figure 1.1-2. *Depiction of the crew module flight deck and middeck within the forebody.* [Adapted from the Shuttle Operations Data Book]

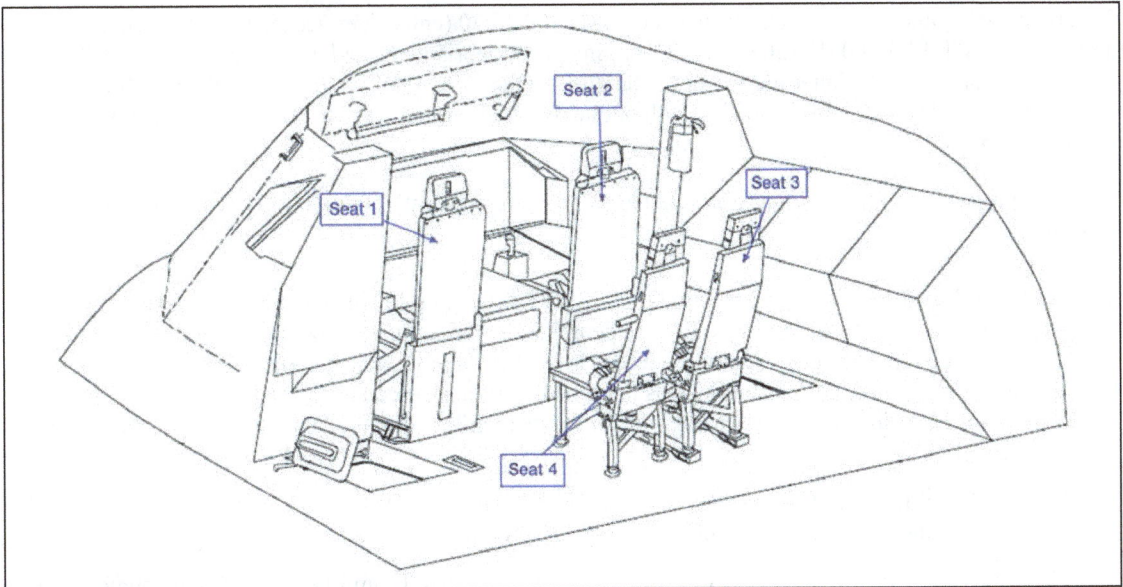

Figure 1.1-3. *Depiction of the flight deck seats.*

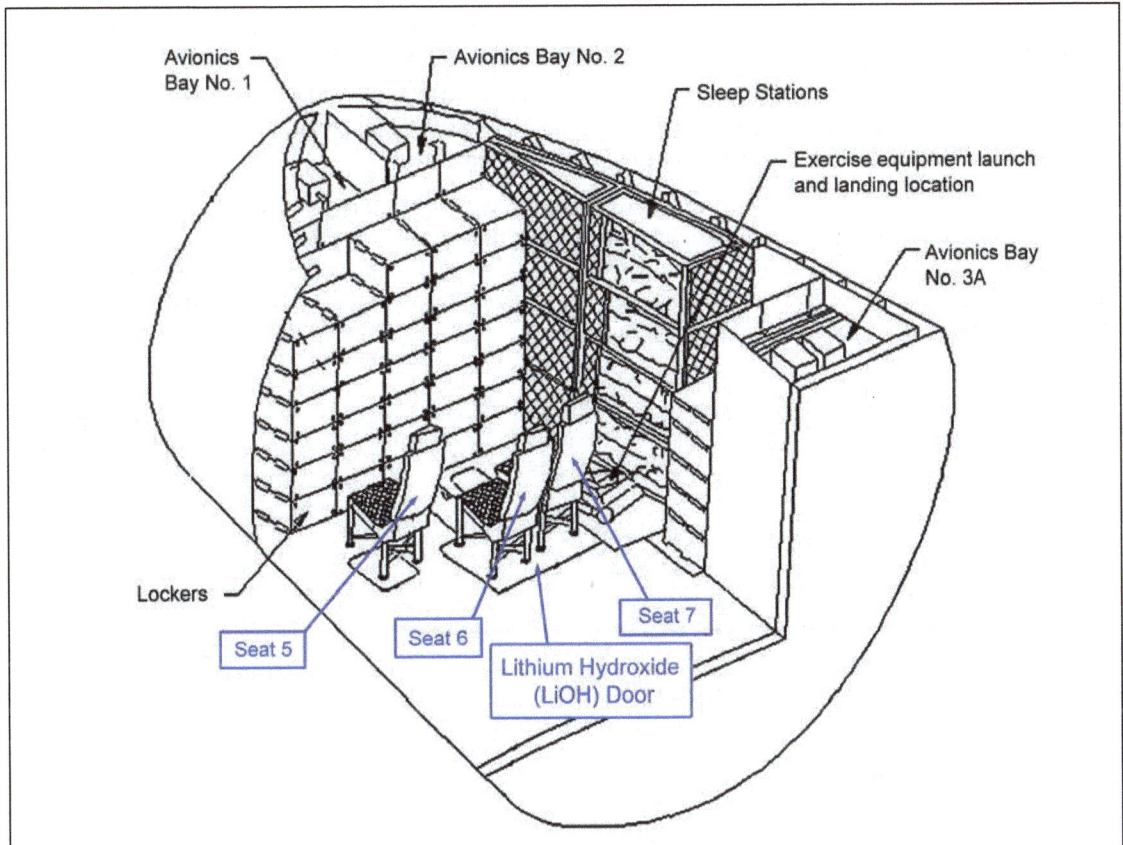

Figure 1.1-4. *Depiction of the middeck seats.* [Adapted from the Shuttle Crew Operations Manual]

This timeline begins at Greenwich Mean Time (GMT) 09:15:30 (entry interface (EI)–16119 seconds) and ends at GMT 14:35:00 (EI+3051) (by which time most debris items had impacted the ground). This timeline overlaps with the latter portion of the timeline of the *Columbia* Accident Investigation Board (CAIB) Report.[2] The cause of the mishap will not be discussed because it was fully covered in the CAIB Report.

This timeline is divided into six phases, based on key events. Each phase of the timeline is addressed in sequence.

- Phase 1 [GMT 09:15:30 (EI–16119) to GMT 13:44:09 (EI)]: From the beginning of the deorbit preparation portion of the mission to EI. The deorbit preparation timeline begins 4 hours prior to the deorbit burn. After the burn, the orbiter descends in altitude until atmospheric drag effects become noticeable, roughly at EI. EI is defined as the time the orbiter descends through an altitude of 400,000 feet. At EI, *Columbia* was approximately 4,300 nautical miles from the landing site, traveling in excess of Mach 24.

- Phase 2 [GMT 13:44:09 (EI) to GMT 13:59:32 (EI+923)]: From EI to loss of signal (LOS). LOS is the loss of voice and real-time data transmissions from *Columbia*.

- Phase 3 [GMT 13:59:32 (EI+923) to GMT 14:00:18 (EI+969)]: From LOS to the Catastrophic Event (CE). The CE is defined as the initiation of the orbiter breakup into the primary subcomponents of the forebody, midbody,[3] and aftbody.[4] The CAIB timeline ends with the CE.

- Phase 4 [GMT 14:00:18 (EI+969) to GMT 14:00:53 (EI+1004)]: From the CE to the Crew Module Catastrophic Event (CMCE). The CMCE is defined as the initiation of the forebody breakup.

- Phase 5 [GMT 14:00:53 (EI+1004) to GMT 14:01:10 (EI+1021)]: From the CMCE to Total Dispersal (TD). TD is defined as the time when the crew module was substantively broken down into subcomponents and was no longer visible on ground-based videos.

- Phase 6 [GMT 14:01:10 (EI+1021) to approximately GMT 14:35:00 (EI+3051)]: From the TD to ground impact of the crew remains and the majority of the crew module debris.

Figure 1.1-5 shows the overall graphical timeline. This timeline represents the best fit to known and inferred data, but it is subject to some inherent uncertainty.

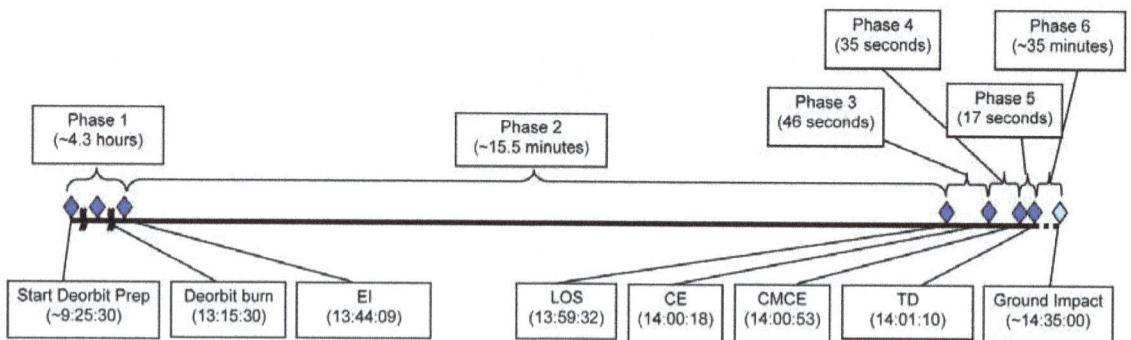

Figure 1.1-5. *Overall timeline.*

---

[2]*Columbia* Accident Investigation Board Report, Volume I, August 2003, pp. 40-41.
[3]The midbody consists of the payload bay and wings.
[4]The aftbody consists of the aft fuselage structure and internal components, the main engines, the left and right OMS pods (including the aft RCS), and the vertical tail.

The early portion of the timeline was developed from objective data such as on-board and downlinked vehicle instrumentation data, on-board video data recovered from the debris, and air-to-ground crew communications. As the timeline progresses into the later phases, the analysis increasingly relied on derived data such as results from ballistic analyses, thermal analyses, ground-based video, and recovered debris. All available data sets were integrated for this analysis.

The environmental conditions experienced by the crew are the focus for each phase. The conditions of interest are atmospheric pressure, thermal situation, and acceleration.[5] The data and methods used to understand the environment are described individually for each phase, and known and inferred events are integrated with the environmental information into a time-based sequence.

1.1.2.1 *Phase 1: Deorbit preparation to entry interface*
[GMT 09:15:30 (EI–16119) to GMT 13:44:09 (EI)]
4 hours, 28 minutes, 39 seconds in duration

This section discusses events affecting the crew from the deorbit preparation timeframe (4 hours prior to the deorbit burn) until EI, about 4 hours and 20 minutes in total. Figure 1.1-6 shows a timeline of this phase. All times are in GMT.

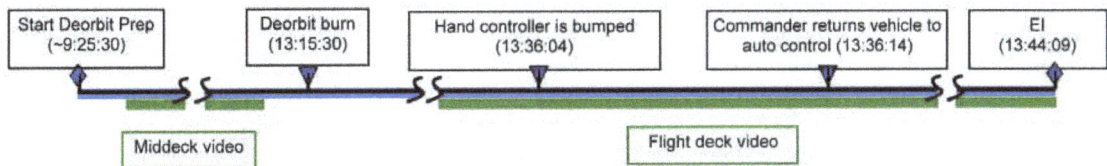

Figure 1.1-6. *Phase 1 timeline with key events*. Green bars represent times when video data are available. Blue bar represents when voice and telemetry transmissions are available (throughout this phase).

The principal sources of data for this phase are orbiter transmissions and recovered on-board video of middeck and flight deck activities. Transmissions include vehicle general purpose computer (GPC)-generated telemetry and audio transmissions made by the crew. The recovered middeck video shows the crew involved in deorbit preparation checklist activities about 2 hours prior to EI from approximately GMT 11:40:00 (EI–7449) to GMT 12:10:00 (EI–5649). The recovered flight deck video shows the flight deck crew seated and preparing for entry. This video was time-synchronized with audio transmissions and crew keystrokes recorded on the ground. The time duration of the video spanned across EI, showing activities from GMT 13:35:34 (EI–515) to GMT 13:48:45 (EI+276).

The crew performed cabin stow activities the day prior to entry and on the morning of entry day. Items were stowed and secured to prevent articles from coming loose in the cabin during entry. Objects aboard the orbiter are stowed in lockers, or are bagged and strapped down in the airlock and in the SPACEHAB module. The only "loose" items on the middeck or flight deck are clips, kneeboards, checklists, timers, writing instruments, drink bags used for fluid loading, and some crew escape equipment (CEE) prior to donning.[6] Any items that are not worn are restrained with VELCRO® or tethers.

The recovered middeck and flight deck videos show that all seats in the crew module were installed and the escape pole was in the process of being installed (figure 1.1-7). Recovered debris analysis shows that the pole installation was completed.

---

[5]Discussions of acceleration will use the Aerospace Medical Association convention. References to the standard gravitational acceleration of the Earth will use the lowercase "g." References to acceleration acting on objects and crew members in multiples of the Earth's gravitational acceleration will use the uppercase "G."

[6]CEE consists of things such as the g-suit, advanced crew escape suit (ACES), parachute, etc. See Section 3.2 Crew Worn Equipment for details.

The cabin stow and deorbit preparation portion of a shuttle mission is a busy period; according to many experienced crew members, shuttle crews often struggle to complete all actions in the time allotted, giving priority to time-critical orbiter systems activities. It is an accepted operational practice for a crew to reorder the tasks as necessary. The middeck video, which ended approximately an hour before the deorbit burn, indicates that the *Columbia* crew members were using their discretion to order their tasks. As a result, the middeck video cannot be precisely time-synchronized using either the published checklist or the crew-specific plan.

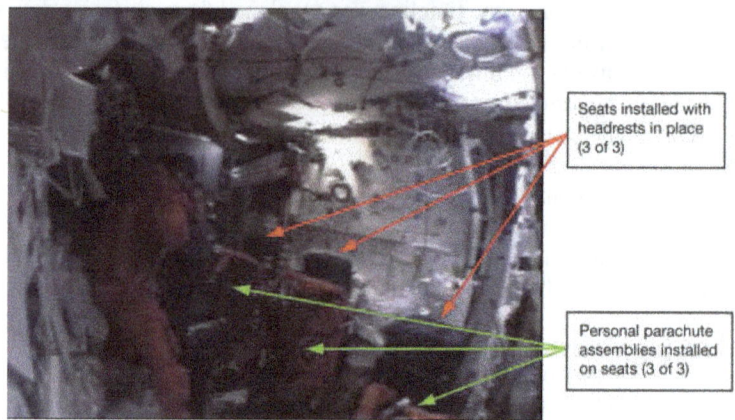

Seats installed with headrests in place (3 of 3)

Personal parachute assemblies installed on seats (3 of 3)

Figure 1.1-7. *This early image from the recovered middeck video shows progress of the crew in deorbit preparation tasks.*

At 45 minutes prior to time of ignition (TIG)–45 minutes), the Commander (CDR) and Pilot (PLT) began working tasks in the entry checklist. By TIG–30 minutes, the rest of the crew should have completed items in the deorbit preparation checklist and transitioned to the entry checklist.

Deorbit burn occurred at GMT 13:15:30 (EI–1719/TIG+0). The burn was nominal, and *Columbia* began entry into the Earth's atmosphere. Per the checklist, a few tasks remain to be completed after the burn, including stowing the last laptop computer, which requires a crew member to be out of the seat. Crew equipment configuration items on the entry checklist (all crew members seated and strapped in, helmets and gloves donned, and suit pressure checked) were not entirely completed prior to EI. At least one crew member was not wearing the helmet and several were not wearing gloves.

The flight deck video shows that conditions on the flight deck were nominal during the entire time of the video recording. The video shows the flight deck crew finishing most checklist tasks close to the planned times. However, one flight deck crew member did not yet have gloves in place in time for the ACES pressure check. One event of note occurred at GMT 13:36:04 (EI–485/TIG+1234) when the CDR bumped the rotational hand controller (RHC) accidentally. Movement of the RHC out of the centered position caused the digital autopilot (DAP) to "downmode" from the "Auto" mode to "Inertial" mode. When this occurred, a "DAP DOWNMODE RHC" caution and warning message was displayed, the INRTL button on the C3 panel was illuminated, and a tone, which can be heard in the recovered flight deck video, was annunciated. An immediate reactivation of the autopilot was performed by the CDR. The capsule communicator (CAPCOM) in the Mission Control Center (MCC) then requested the CDR to enter "another Item 27," which is a command to fully recover the vehicle attitude from the bumped RHC.

Bumping of the RHC is a relatively common occurrence by either the PLT or the CDR because the ACES is bulky and the area near the controls is confined. Such RHC bumps with prompt recovery represent a very low hazard to the crew. The original design specifications of the orbiters were for a shirtsleeve environment (i.e., no special clothing needed to be worn). Although pressure suits have been worn during launch and entry since the *Challenger* accident, no modifications were made to displays and controls to accommodate the ACES.

1.1.2.2 *Phase 2: Entry interface to loss of signal*
[GMT 13:44:09 (EI) to GMT 13:59:32 (EI+923)]
15 minutes, 23 seconds in duration

This section discusses events affecting the crew from EI (GMT 13:44:09) to LOS (the last audio and real-time telemetry transmission received from *Columbia*) at GMT 13:59:32 (EI+923).[7] This section discusses *Columbia*'s entry and the minimal indications available to *Columbia*'s crew and the MCC that *Columbia*'s structure was compromised. Ground-based video of the orbiter's flight is first available in this phase. This phase was approximately 15 minutes long. Figure 1.1-8 shows the timeline and key events for Phase 2.

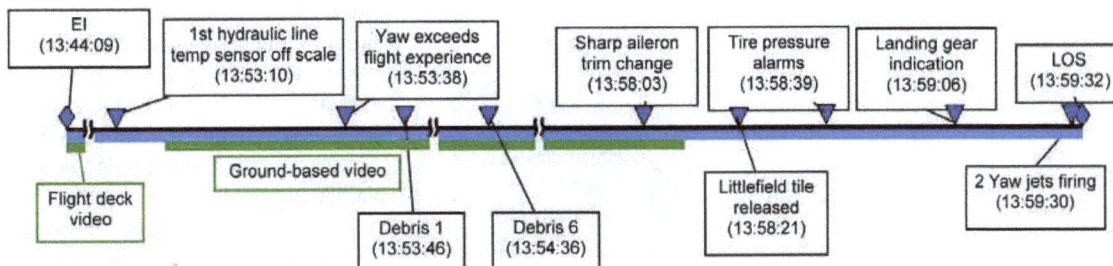

Figure 1.1-8. *Phase 2 timeline with key events.* Green bars represent times when video data are available. Blue bar represents when the Modular Auxiliary Data System/orbiter experiment (MADS/OEX) recorder data, and voice and telemetry transmissions are available (throughout this phase).

The sources of information that provided data for the investigation of Phase 2 are: orbiter transmissions; the MADS/OEX recorder; recovered flight deck video; ground-based video; and recovered debris. The MADS/OEX recorder was an on-board data collection recording system located in the crew module. Recorded data parameters consisted of structural temperature, strain, and accelerations from sensors located throughout the orbiter and concentrated in the left wing. This system was unique to *Columbia*. The MADS/OEX system did not display data to the crew or transmit telemetry to the MCC. Consequently, no MADS/OEX data were available in real time. The MADS/OEX recorder was recovered intact in the debris field and the data were recovered. The recovered flight deck video, beginning in Phase 1 at GMT 13:35:34 (EI–515), contains data through GMT 13:48:45 (EI+276), 4 minutes into this phase. Ground-based imagery recorded the entry of *Columbia* from the coast of California to the final breakup over Texas, with a gap in coverage from eastern New Mexico to western Texas.

Crew cabin pressure and the thermal environment inside the cockpit were nominal throughout this phase. At EI, atmospheric drag on the orbiter began to gradually increase. The initial roll and subsequent roll reversals caused accelerations of up to 0.8 G during entry. During this period of a shuttle mission, crew members typically experience heaviness, dizziness, and sometimes stomach awareness or mild nausea. Based upon telemetry, accelerations were nominal despite the damage-induced aerodynamic changes to the orbiter. The orbiter was shedding debris during at least part of this phase, although changes in mass properties were small and no detectable load spikes were noted by the MCC, reported by the crew, or found in post-mishap analysis data. No anomalous orbiter systems conditions were displayed to the crew until the end of this phase.

At EI, the vehicle also began to experience the thermal effects of the Earth's atmosphere. Shock waves due to the vehicle's hypersonic velocities and the frictional effect of the atmosphere began to heat the orbiter's surface. Temperatures on the surface of the orbiter during entry vary by location, with the nose and leading edge of the wings experiencing temperatures greater than 2,800°F (1,538°C). The Thermal Protection

---

[7]Although LOS is defined here as the loss of audio and real-time data transmissions, some instrument-based telemetry was received shortly into Phase 3; this will be discussed in the next section.

System (TPS) of the orbiter consists of reinforced carbon-carbon (RCC) panels, tile, and thermal blankets, and is designed to protect the orbiter's structure from this nominal entry heating.

The CAIB concluded that the breach in the TPS on the leading edge of the left wing allowed hot gas to penetrate the wing and to work its way aft and inboard, toward the midbody fuselage and left main landing gear wheel well (figure 1.1-9). CAIB analysis showed that this eventually caused significant internal damage to *Columbia*'s left wing, changing the wing's aerodynamic properties. There were no indications to the crew of this ongoing damage. Throughout phase 2, the orbiter Flight Control System (FCS) corrected for the damage-induced yaw and roll moments, and control of the orbiter was maintained.[8]

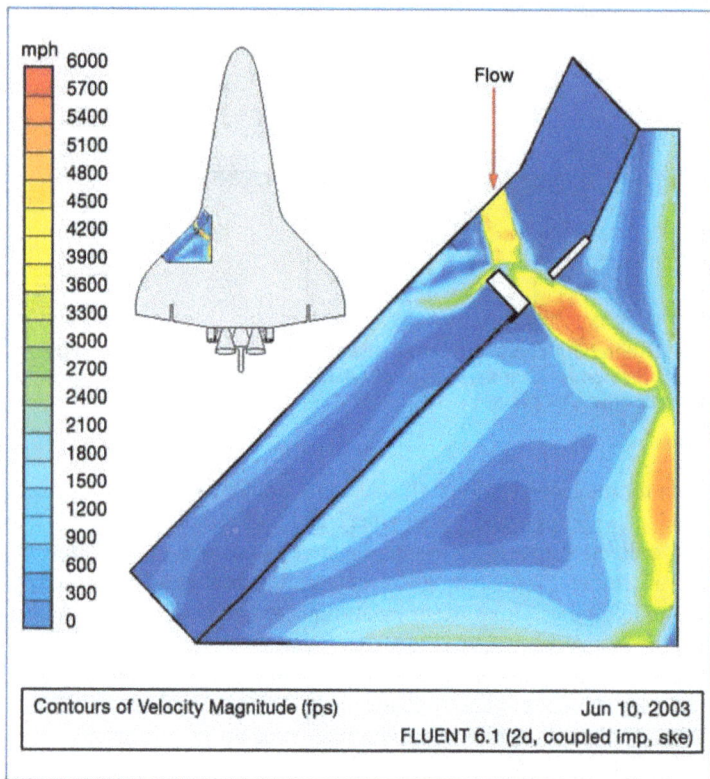

Figure 1.1-9. *The effects of wing Thermal Protection System damage as reported by the* Columbia *Accident Investigation Board.*

The recovered flight deck video shows the CDR requesting a suit pressure integrity check from the other crew members. Suit pressurization checks for the CDR, PLT, and Mission Specialist (MS) 4 were observed in the video and validated with telemetry of the oxygen (O_2) system. After the ACES pressure checks, the crew members turned off the flow of O_2 to the ACES and opened their visors. The O_2 must be turned off because the suit vents O_2-enriched air into the cabin with the visor down and O_2 flowing. Venting O_2-enriched air eventually creates an increased concentration of O_2, leading to an increased fire hazard. To prevent carbon dioxide buildup inside the helmet, visors are returned to the open position. Open visors have the added benefit of improved crew comfort and communication. Telemetry data were consistent with one or two more suit pressure integrity checks occurring after the end of the flight deck video. It could not be determined which crew member(s) performed these checks. The video, which ended at GMT 13:48:45 (EI+276), shows the four flight deck crew members suited, seated, and strapped in with helmets donned. All except one had fully donned and connected gloves. The recovered flight deck video indicates that the crew was not aware of any problems.

At GMT 13:49:32 (EI+323), a nominal roll to the right was completed for energy management. At GMT 13:50:53 (EI+404), *Columbia* started the expected 10-minute window of nominal peak heating.[9]

At GMT 13:51:46 (EI+457), the inertial sideslip (yaw) angle began a negative trend (yaw to the left), although the angle remained within previous flight experience for almost 2 minutes (figure 1.1-10). At GMT 13:52:05 (EI+476), the yaw moment changed due to increased drag from the damaged left wing; the orbiter's FCS commanded the aileron trim to compensate (figure 1.1-11). Neither the yaw moment change nor the aileron trim change was obvious to either the MCC or the crew as an off-nominal condition, although post-accident analysis concluded that this was the first indication of the orbiter's response to the

---

[8]*Columbia* Accident Investigation Board Report, Volume I, August 2003, p. 78.
[9]*Columbia* Accident Investigation Board Report, Volume I, August 2003, p. 38.

changing aerodynamic properties brought about by the left wing damage. Otherr post-mishap analysis determined that damage inside the wing began no later than GMT 13:52:17 (EI+488).[10]

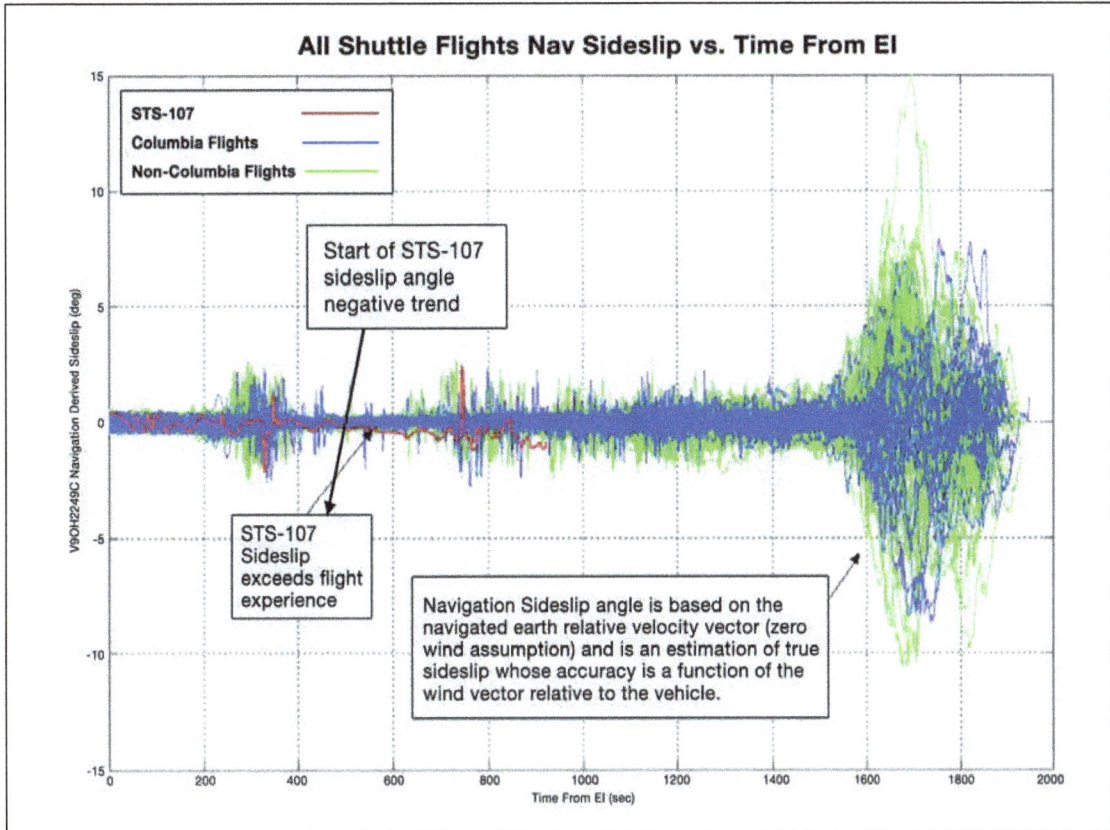

Figure 1.1-10. *Negative trend starts at GMT 13:51:46 (EI+457) and exceeds flight experience at GMT 13:53:38 (EI+569).*[11]

---

[10]*Columbia* Accident Investigation Board Report, Volume I, August 2003, pp. 68 and 71.
[11]Integrated Entry Environment Team Final Report, May 30, 2003, Figure 6.6-2, p. 30; taken verbatim from document.

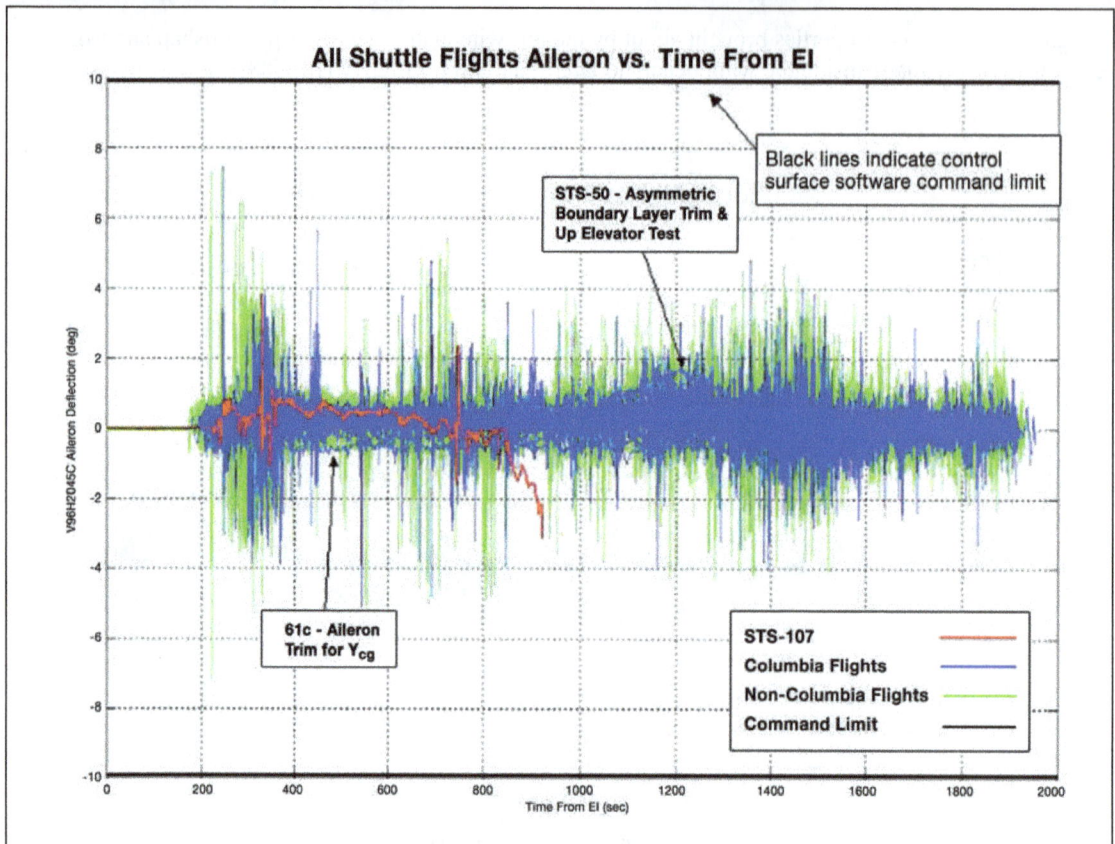

Figure 1.1-11. *The sharp divergence from previous flight experience starts at approximately GMT 13:58:19 (EI+850).*[12]

The left main landing gear brake line in the inboard sidewall of the wheel well began to show an off-nominal temperature rise rate at GMT 13:52:17 (EI+488).

MCC personnel became aware of an off-nominal flight condition when four hydraulic return line temperature sensors in the left wing went off-scale low from GMT 13:53:10 (EI+541)[13] to GMT 13:53:36 (EI+567). A sensor suddenly going off-scale usually indicates a failure of the sensor or the wiring. Loss of an individual sensor for various reasons has occurred in previous missions. However, the simultaneous failure of multiple sensors from separate redundant systems was an event outside previous flight experience. These temperature data were not available to the crew and the crew was not notified of these indications. The loss of sensors generated concern in the MCC, and investigation by the flight control team began immediately.

Ground-based video coverage of *Columbia* was acquired by videographers unassociated with NASA at GMT 13:53:15 (EI+546).[14] Figure 1.1-12 is the first ground-based image of *Columbia* acquired. The bright spot circled is the "orbiter envelope" and is the nominally produced hot gas and plasma that surround the orbiter during entry. The actual shape of the orbiter (and most debris) cannot be seen in any non-telescopic-based video; only the surrounding hot gas/plasma envelope is visible.

---

[12]Integrated Entry Environment Team Final Report, May 30, 2003, Figure 6.6-1, p. 30; taken verbatim from document.
[13]*Columbia* Accident Investigation Board Report, Volume I, August 2003, description of these sensor failures.
[14]All video frames of the orbiter prior to GMT 13:59:32 (EI+923) were processed by the STS-107 Image Analysis Team. Details of video processing can be found in the *Columbia* Accident Investigation Board Report, Volume III, Appendix E.2, STS-107 Image Analysis Team Final Report, October 2003. All videos are assumed to have an approximately 1-second error.

Columbia crossed the California coastline at GMT 13:53:26 (EI+557).

At GMT 13:53:38 (EI+569), the sideslip angle (left yaw) exceeded all previous flight experience (figure 1.1-10).

The first known debris shedding event on entry was identified as Debris 1 and most likely originated from the left wing. Debris 1 becomes visible on ground-based video at GMT 13:53:46 (EI+577). Later ballistic analysis estimated a release time of GMT 13:53:44.8 (EI+575).[15] Luminosity measurements and calculated rates of deceleration were used to determine that the mass was < 8 lbs.[16] Figure 1.1-13 shows the orbiter and the debris.

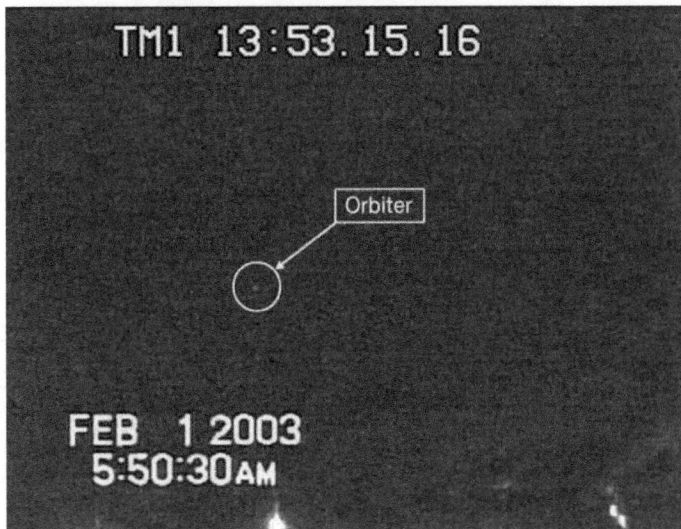

Figure 1.1-12. *This is the first frame of ground-based video. Columbia is circled. Time 1 (TM1) shows the Greenwich Mean Time. The Pacific Standard Time displayed on the lower portion of the image is inexact.*

The Mechanical, Maintenance, Arm, and Crew Systems (MMACS) officer in the MCC notified the Flight Director of the off-scale low hydraulic line temperature sensors at GMT 13:54:24 (EI+615).

The brightest debris shedding event that occurred in this phase, Debris 6, is first visible on video at GMT 13:54:36 (EI+627) (figure 1.1-14). Ballistic estimates determined that the actual release time was 4 seconds earlier. Luminosity measurements and calculated rates of deceleration were used to determine that the mass was probably a few hundred pounds.[17] There were no data from sensors, instrumental indications, or apparent crew recognition of this debris loss.

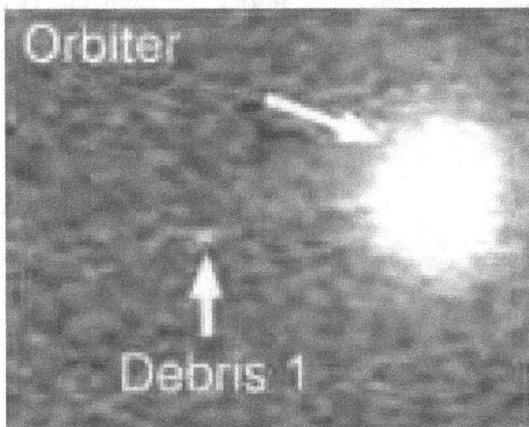

Figure 1.1-13. *Video capture of the first observed incident of debris being shed.* The orbiter is traveling from left to right in this image.

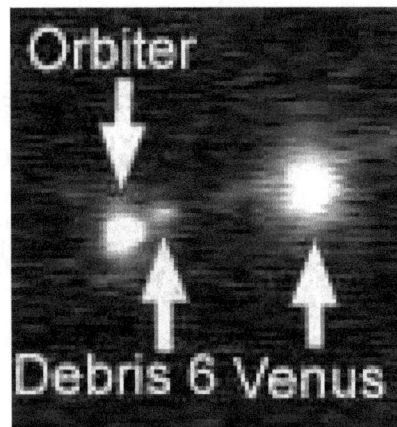

Figure 1.1-14. *The brightest debris event known to have occurred prior to loss of signal.* The orbiter is traveling from right to left.

---

[15]All ballistic analysis release times have a ±5-second uncertainty.
[16]*Columbia* Accident Investigation Board Report, Volume III, Appendix E.2, STS-107 Image Analysis Team Final Report, October 2003, p. 110.
[17]*Columbia* Accident Investigation Board Report, Volume III, Appendix E.2, STS-107 Image Analysis Team Final Report, October 2003, p. 110.

At approximately GMT 13:54:30 (EI+621), cabin O_2 partial pressure and cabin pressure telemetry indicated signatures consistent with additional ACES pressurization events, indicating that the crew was continuing suit activities.

The first planned roll reversal was initiated from right wing low to left wing low at GMT 13:56:30 (EI+741).

At approximately GMT 13:58:03 (EI+834), the aileron trim begins to diverge sharply from the expected values to counteract the increasing adverse moments due to the left wing damage (figure 1.1-15).

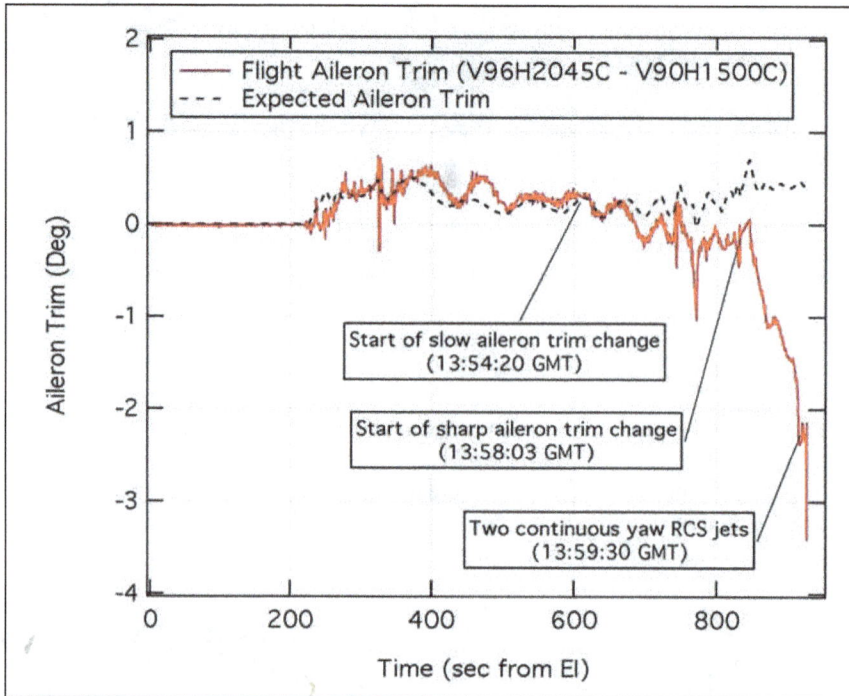

Figure 1.1-15. *Aileron trim discrepancy*.[18]

Western ground-based video coverage ends at GMT 13:58:12 (EI+843).

Ballistic analysis indicates that the westernmost piece of recovered *Columbia* debris was released at GMT 13:58:21 (EI+852). This debris, a tile from the left wing upper surface located just inboard of RCC panels 8 and 9, was found in Littlefield, Texas[19] (figure 1.1-16).

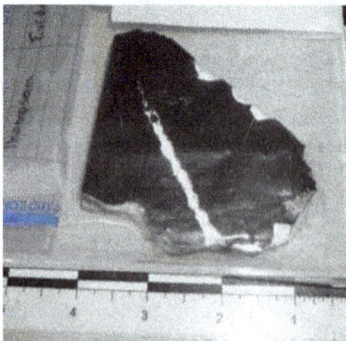

Figure 1.1-16. *The Littlefield tile*. [Picture from the *Columbia* Reconstruction Database, debris item no. 14768]

---

[18]Integrated Entry Environment Team Final Report, May 30, 2003, Figure 6.3-4, p. 20.
[19]STS-107 *Columbia* Reconstruction Report, NSTS-60501, June 2003, pp. 21 and 121.

Columbia's crew received the first indication of a problem at GMT 13:58:39 (EI+870) when the first of four fault messages was annunciated on the on-board Backup Flight Software monitor. These messages were accompanied by an audible tone. The fault messages indicated a loss of pressure on the left main landing gear tires. These indications also were presented to the flight control team in the MCC. The CDR and PLT called up the fault page for these messages and reviewed the information. The failure the crew saw would be familiar, although slightly different from what they saw in training. One of the failure scenarios the crew practiced during training was a circuit breaker trip that resulted in one-half of the tire pressure sensors being disabled. A circuit breaker trip would disable some sensors for all of the tires (left main gear, right main gear, and nose gear), but the failure signature during the accident involved all the tire pressure sensors on the left main gear only. At GMT 13:58:48 (EI+879), the crew began a call to the MCC but the call was broken and not repeated. Brief interruptions of communications often occur due to the tracking and data relay satellite antenna pointing angles changing relative to the orbiter's transceivers. This specific dropout of communication was not unexpected.

At GMT 13:59:06 (EI+897), 10 seconds after the fourth of four tire pressure fault messages, telemetry indicated the "LEFT MAIN GEAR DOWN" lock sensor transferred to "ON." Other sensors indicated that the landing gear door was still closed and the landing gear was locked in the up (stowed) position. These mixed signals caused the left landing gear position indicator to display a "barber pole" (figure 1.1-17), which indicates an indeterminate landing gear position. Post-accident analysis of the data and recovered debris indicates that the left landing gear was locked in the "up" position and the landing gear door was closed. The signal indicating that the gear was down was a false signal that was likely triggered by damage to the sensor system (sensor, wiring harness, etc.). Based on training experience, the crew was probably attempting to diagnose the situation given that it involved the same landing gear as the tire pressure messages and indicated a potential landing gear deployment problem.

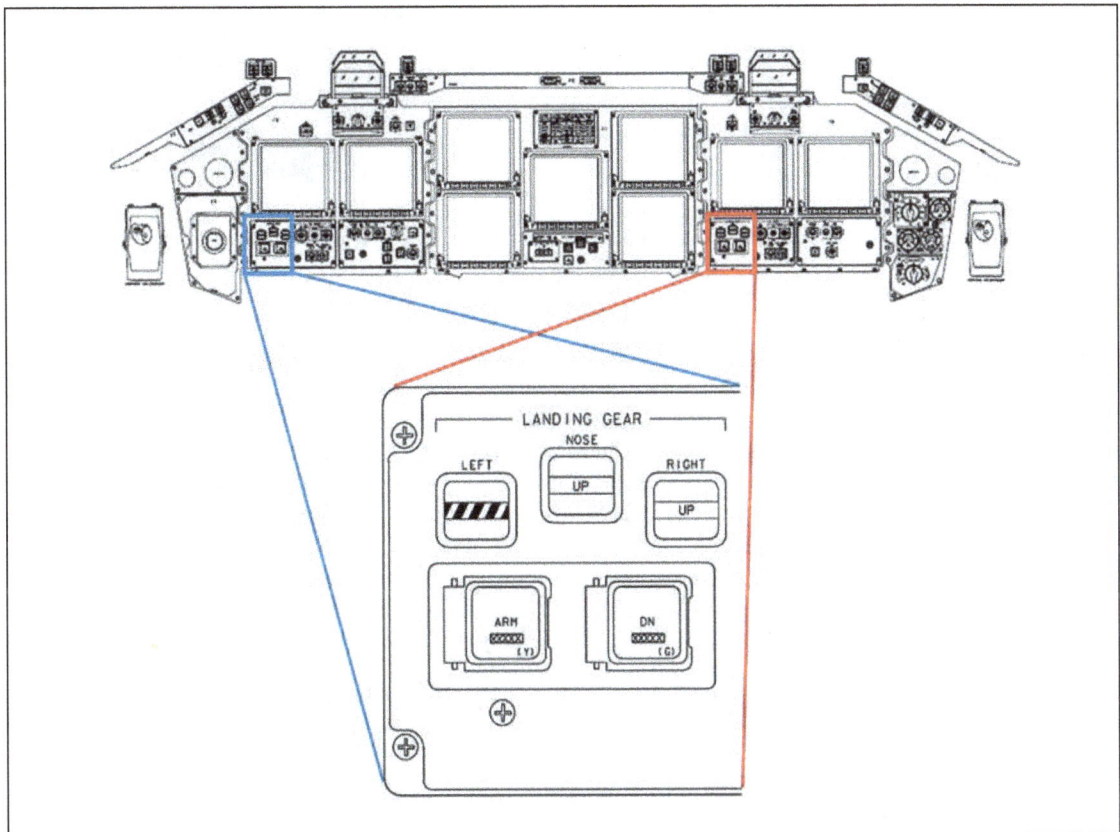

Figure 1.1-17. *Landing gear indicator panel, identical on both sides of the flight deck forward display panels.* **Left indicator showing "barber pole" (indeterminate position). [Adapted from the Space Shuttle Systems Handbook]**

At GMT 13:59:29 (EI+920), the orbiter yaw and roll rates exceeded the ability of the aileron trim to compensate for the changing drag of the deformed left wing. One second later, the R2R and R3R RCS jets[20] activated. Typically, RCS jets pulse throughout entry, adjusting the orbiter's flight path as needed. RCS jets had been pulsing nominally until this time when R2R and R3R began firing continuously as the orbiter attempted to counteract the increased left wing drag and resulting yaw moment. A small light on a panel in front of the CDR would have become illuminated continuously (figure 1.1-18). Experience shows that this jet fire light is not easily noticed.

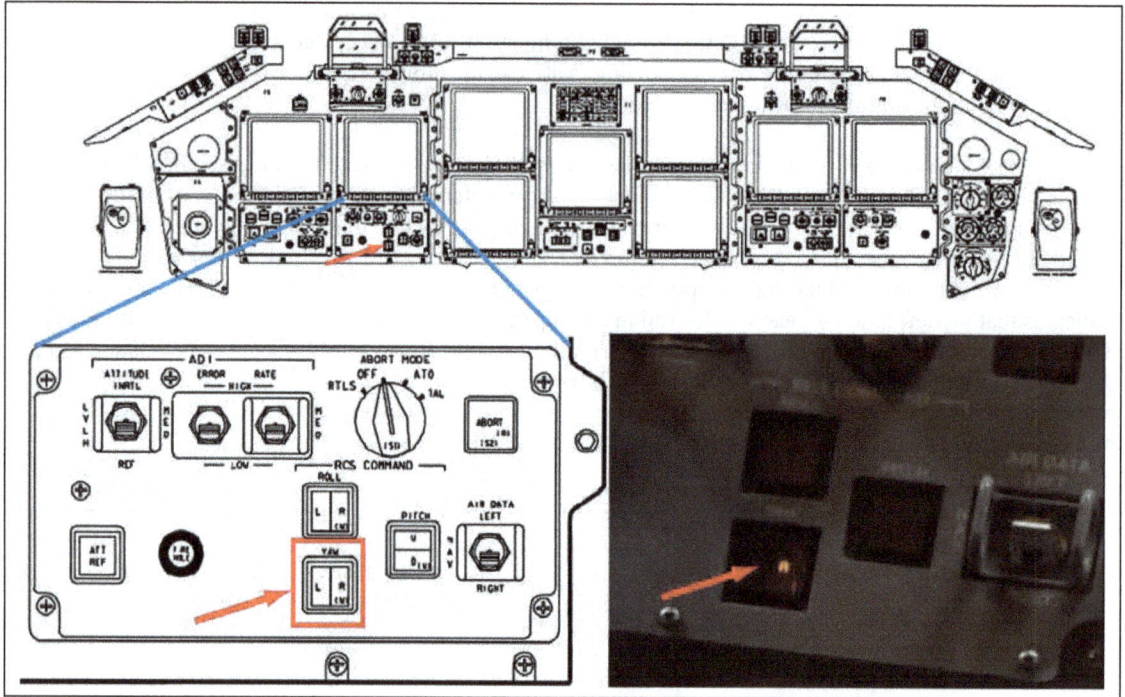

Figure 1.1-18. *Reaction Control System thruster status display.* [Pictures (top and bottom left) adapted from the Space Shuttle Systems Handbook; picture (bottom right) from the Shuttle Engineering Simulator]

At GMT 13:59:32 (EI+923), the crew acknowledged a call from the MCC but the crew's response was interrupted in mid-sentence ("Roger, uh …"[21]). This was the final call heard from *Columbia*. This is also the time of LOS, when all audio and real-time data to the MCC from *Columbia* was lost. A short dropout (seconds) was expected at this time based on pre-mission analysis as the orbiter switched from one communication satellite to another. The CAPCOM replied to the partial transmission to let the crew know that the flight controllers saw the tire pressure fault messages and that the MCC did not understand the last transmission.[22] The MCC personnel recognized that there were problems occurring with *Columbia*, but the telemetry signatures were such that these personnel were unable to complete analysis of the wide-ranging (and seemingly unrelated) problems before contact was lost.

There were no indications to the crew and the MCC that the loss of audio communications and real-time data was more than a brief condition. To all on-board appearances, *Columbia* only had a potential issue with landing gear deployment; a non-trivial event, but the crew had time to troubleshoot the problem. Changing drag on the left wing was just beginning to develop into a potentially recognizable problem.

---

[20]Right-firing RCS jets on the right OMS pod.
[21]*Columbia* Accident Investigation Board Report, Volume I, August 2003, p. 43.
[22]The CAPCOM continued to attempt to contact the crew on different radio frequencies to re-establish voice communications.

Based on seat debris and medical findings, at the end of this phase one middeck crew member had not fully ingressed the seat yet, although the action may have been in progress. This crew member was responsible for completing post-deorbit burn tasks and was assigned to be the last to ingress the seat.

1.1.2.3 *Phase 3: Loss of signal to Catastrophic Event*
[GMT 13:59:32 (EI+923) to GMT 14:00:18 (EI+969)]
46 seconds in duration

This section discusses key events that affected the crew from LOS at GMT 13:59:32 (EI+923) to the CE, which began at 14:00:18 (EI+969) (figure 1.1-19). During this period of time at about GMT 13:59:37 (EI+928), loss of control (LOC) of *Columbia* occurred. LOC marks the beginning of the transition from controlled flight to an uncontrolled ballistic entry. This phase is 46 seconds long.

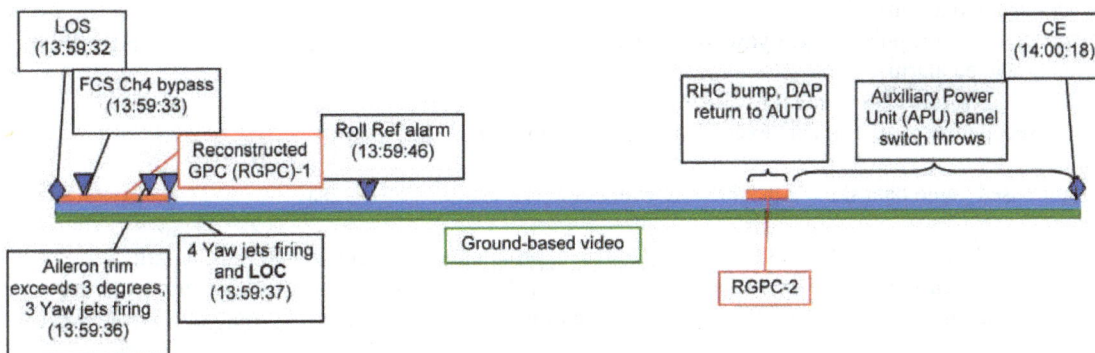

Figure 1.1-19. *Phase 3 timeline with key events.* **Real-time voice and telemetry transmissions were not available in this phase or subsequent phases. The green bar represents time when video data are available; the blue bar represents when the Modular Auxiliary Data System/orbiter experiment recorder data are available (both available throughout this phase). Red bars represent times when reconstructed general purpose computer data were available.**

Available instrumentation data (recorded and recovered) become scarce in Phase 3 (and nonexistent in subsequent phases). The sources of data used to reconstruct conditions in Phase 3 are: reconstructed telemetry; MADS/OEX recorder; ground-based videos; recovered debris; and aerodynamic and ballistics analyses. Reconstructed general purpose computer (RGPC) data were data that were recorded at the telemetry receiving ground station at White Sands, New Mexico, but not transmitted to the MCC in real time due to quality filtering. After the accident, the data were retrieved and manually reconstructed.[23] The RGPC data include real-time parameter data (such as pressures, temperatures, and switch positions), time-stamped alert messages, and non-time-stamped alert messages. Ground-based video was re-established starting at approximately GMT 13:59:32.5 (EI+923), at about LOS. For the first 16 seconds of this phase, a single video supplies coverage. Additional video coverage begins at GMT 13:59:48 (EI+939). Recovered *Columbia* debris was used for reconstruction via visual inspection, material sampling, and ballistic analysis. This led to conclusions regarding thermal events and material loads. Ballistic analysis was performed on select debris items to help understand the events and their sequence.

At GMT 13:59:33 (EI+924), data from RGPC-1 showed the primary software system annunciated that FCS Channel 4 had been automatically bypassed out of the control loop. This bypass occurred because of a failed wire bundle and resulted in a Master Alarm. The Master Alarm was annunciated visually and aurally. While there is no crew action associated with this frequently trained FCS fault message other than to perform a message reset, the crew may have called up a display to analyze the failure. Crews are trained to troubleshoot systems errors, and this crew would have been evaluating this new message along with the previous tire pressure and landing gear down-lock indications to assess whether there was a common

---

[23]*Columbia* Accident Investigation Board Report, Volume II, Appendix D.9, Data Review and Timeline Reconstruction Report, October 2003.

system fault that could account for all of these messages. It is unknown whether the increasing aileron trim and thruster firings were noticed by the flight deck crew members.

At GMT 13:59:36 (EI+927), the third RCS yaw jet, R4R, began firing continuously and aileron trim exceeded 3 degrees. There is no alarm associated with a deviating trim condition, and the crew is not expected to monitor the trim during this period of entry. At GMT 13:59:37 (EI+928), the fourth and last RCS yaw jet, R1R, began firing continuously.

To summarize, in the minute prior to LOC, the crew received several indications of various vehicle systems problems:

1. 58 seconds prior: the first of four tire pressure alert messages was displayed.
2. 31 seconds prior: left main landing gear indicator transitioned to an indeterminate state (no annunciated alarm).
3. 7 seconds prior: pulsing RCS yaw light became constant as two RCS jets began firing continuously (no annunciated alarm).
4. 4 seconds prior: FCS channel bypass message and Master Alarm.
5. 0.6 second prior: aileron trim exceeded 3 degrees (no annunciated alarm).

Ground-based video data show a brightening event at GMT 13:59:37 (EI+928).

RGPC-1 ends[24] at GMT 13:59:37.4 (EI+928) with an approximately 25-second gap in data until RGPC-2 data begins at GMT 14:00:02.660 (EI+953). RGPC-2 data include messages generated during the 25-second gap; some of the messages do not have time tags, and some message time tags are corrupted.

Vehicle LOC probably occurred at GMT 13:59:37 (EI+928).[25] The CAIB Report concluded that "During re-entry this breach in the TPS allowed superheated air to penetrate through the leading edge insulation and progressively melt the aluminum structure of the left wing, resulting in a weakening of the structure until increasing aerodynamic forces caused loss of control, failure of the wing, and breakup of the orbiter."[26] An in-depth review of the data by the Integrated Entry Environment team provided further insight into the probable sequence of events. The RGPC-2 data showed that a "ROLL REF" alarm was recorded at GMT 13:59:46 (EI+937), only 9 seconds after the end of RGPC-1. A ROLL REF alarm generally indicates that the drag of the orbiter has exceeded the entry drag profile. The Integrated Entry Environment team concluded that the most credible scenario that could cause this message within 9 seconds would be from a pitch deviation rather than a roll deviation (which would be expected with increasing drag on the wing).[27] A complete loss of hydraulics would cause the elevons and body flap to move to a floating position, resulting in an uncontrolled pitch-up. RGPC-2 data (approximately 25 seconds later) showed that the hydraulics systems failed, but no time signature was available to confirm when the loss occurred. Video data supported this time for LOC. The Spacecraft Crew Survival Integration Investigation Team (SCSIIT) concluded that the LOC occurred as a result of the loss of hydraulics at GMT 13:59:37 (EI+928). The loss of hydraulics likely occurred when all three redundant hydraulic systems lost pressure due to breaches in the hydraulic lines from thermal damage in the left wing. A visual simulation of the pitch-up associated with this type of LOC is shown in figure 1.1-20.

---

[24]Although data were received after GMT 13:59:37 (EI+928), none could be reconstructed until RGPC-2. See *Columbia* Accident Investigation Board Report, Volume II, Appendix D.19, Qualification and Interpretation of Sensor Data from STS-107, October 2003.

[25]This LOC time (GMT 13:59:37 (EI+928)) occurs 42 seconds earlier than that concluded in the *Columbia* Accident Investigation Board Report, Volume I, August 2003, p. 73.

[26]*Columbia* Accident Investigation Board Report, Volume I, Executive Summary, August 2003, p. 9.

[27] EG-DIV-08-32- Integrated Entry Team Report, Appendix G - Post-LOS Analysis.

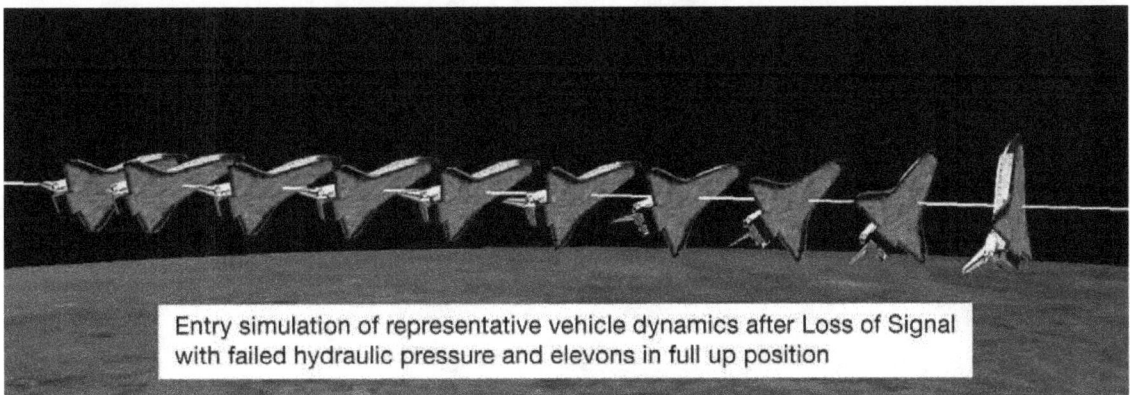

Entry simulation of representative vehicle dynamics after Loss of Signal with failed hydraulic pressure and elevons in full up position

Figure 1.1-20. *Sequence (1-second intervals) showing a simulation of orbiter loss of control pitch-up from GMT 13:59:37 (EI+928) to GMT 13:59:46 (EI+937).* White line indicates vehicle trajectory relative to the ground.

The LOC event marked the beginning of the transition from a controlled gliding trajectory into an uncontrolled ballistic trajectory. Once attitude control was lost, orbiter heating, lift, and drag were dictated predominantly by ballistic number.[28] The out-of-control orbiter configuration had significantly more drag than the nominal entry configuration, and the trajectory became steeper. Changes in the flight profile are recognizable in the ground-based video as changes in the orbiter's trail and in the brightness of the visual signal ("brightening events").

Video imagery shows a dynamically changing orbiter trail after GMT 13:59:37 (EI+928) with a braided or corkscrew appearance, implying motion of the orbiter. However, the specific attitude of the orbiter cannot be derived from ground-based imagery. Brightening events, objects separating, "puffs," and splitting of the trail are all seen in the video during this timeframe. Ballistic analysis of debris could not positively correlate a specific orbiter source to shedding events seen in the video. However, it is known that the left wing and the left OMS pod were being compromised.[29] Figure 1.1-21 shows video frames from GMT 13:59:35.5 (EI+925) through GMT 13:59:43.5 (EI+934) and displays some of these dynamic changes, although they are much more clearly seen in the video.

---

[28]Ballistic number is affected by the coefficient of drag of an object (which changes with its velocity), its weight, and the area presented to the velocity vector. A low ballistic number indicates high drag. The initial trajectory of an object with a low ballistic number is steeper than the trajectory of an object with a high ballistic number. See Section 2.1.
[29]*Columbia* Accident Investigation Board Report, Volume I, August 2003, p. 68.

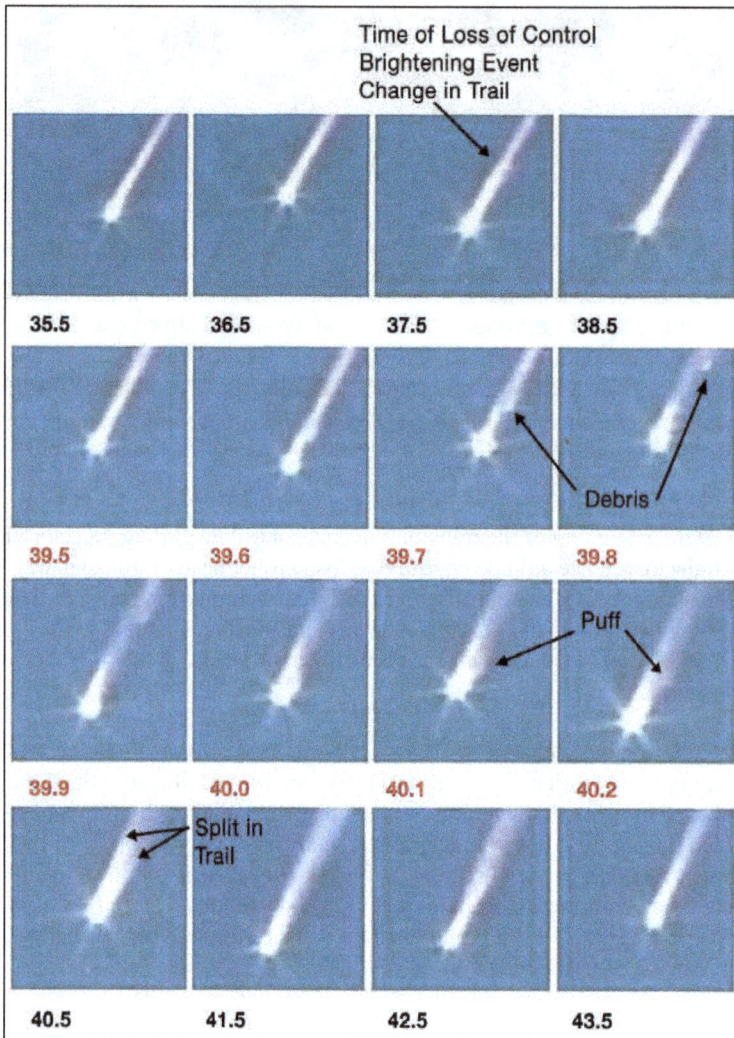

Figure 1.1-21. *Video frame captures from ground-based video, beginning at GMT 13:59:35.5 (EI+926) and ending at GMT 13:59:43.5 (EI+937). The numbers below each frame indicate the seconds after GMT 13:59:00. The frames in the first and last rows are 1 second apart. The frames in the second and third rows are 0.1 second apart.*

For the crew, the first strong indications of the LOC would be lighting and horizon changes seen through the windows and changes on the vehicle attitude displays. Additionally, the forces experienced by the crew changed significantly and began to differ from the nominal, expected accelerations. The accelerations were translational (due to aerodynamic drag) and angular (due to rotation of the orbiter). The translational acceleration due to drag was dominant, and the direction was changing as the orbiter attitude changed relative to the velocity vector (along the direction of flight).

Results of a shuttle LOC simulation show that the motion of the orbiter in this timeframe is best described as a highly oscillatory slow (30 to 40 degrees per second) flat spin, with the orbiter's belly generally facing into the velocity vector. It is important to note that the velocity vector was still nearly parallel to the ground as the vehicle was moving along its trajectory in excess of Mach 15. The crew experienced a swaying motion to the left and right (Y-axis) combined with a pull forward (X-axis) away from the seatback. The Z-axis accelerations pushed the crew members down into their seats. These motions might induce nausea, dizziness, and disorientation in crew members, but they were not incapacitating. The total acceleration experienced by the crew increased from approximately 0.8 G at LOC to slightly more than 3 G by the CE (figure 1.1-22).

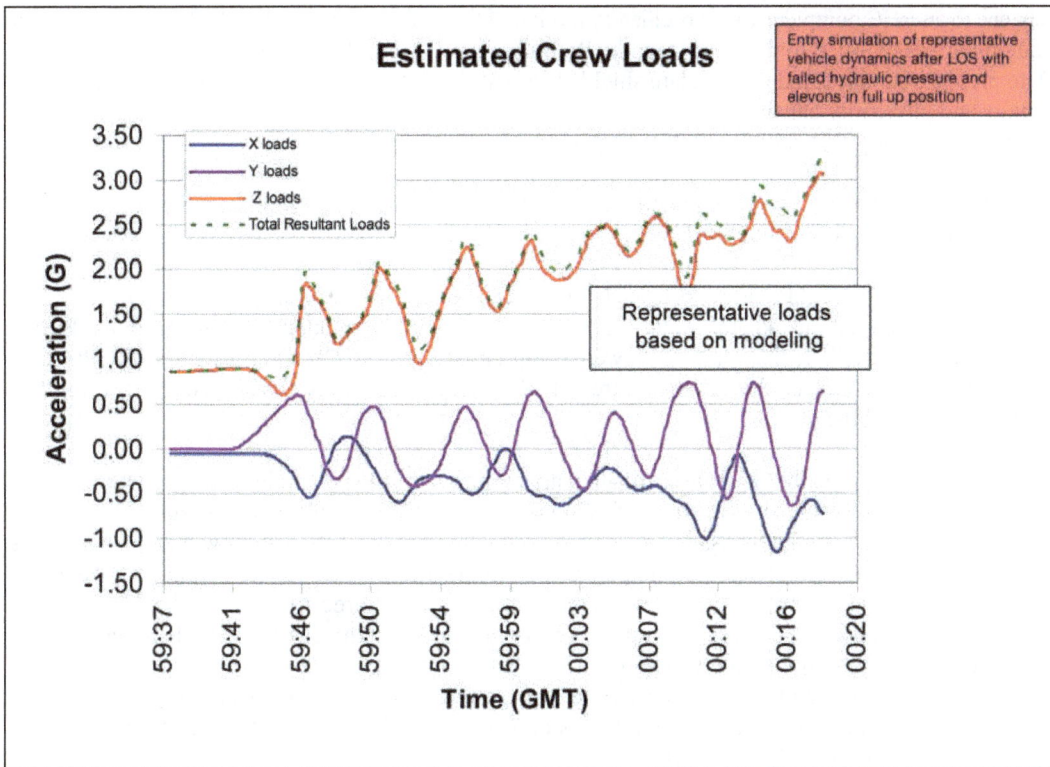

Figure 1.1-22. *Resultant acceleration (G) vs. time prior to Catastrophic Event in the crew coordinate frame from the aerodynamic simulation model.*

The onset of this highly oscillatory flat spin likely resulted in the need for crew members to brace as they attempted to diagnose and correct the orbiter systems. As mentioned in the previous phase discussion, one middeck crew member had not completed seat ingress and strap-in at the beginning of this phase. Seat debris and medical analyses indicate that this crew member was not fully restrained before loss of consciousness. Only the shoulder and crotch straps appear to have been connected. The normal sequence for strap-in is to attach the lap belts to the crotch strap first, followed by the shoulder straps. Analysis of the seven recovered helmets indicated that this same crew member was the only one not wearing a helmet. Additionally, this crew member was tasked with post-deorbit burn duties. This suggests that this crew member was preparing to become seated and restrained when the LOC dynamics began. During a dynamic flight condition, the lap belts hanging down between the closely space seats would be difficult to grasp due to the motion of the orbiter, which may be why only the shoulder straps were connected.

At GMT 13:59:46 (EI+937), ground-based video indicates that a bright piece of debris was released followed by a second piece 2 seconds later. This second piece of debris[30] separated from the orbiter's trail and decelerated slowly, remaining visible for more than 37 seconds before dispersing into significantly fainter pieces. Ballistic analyses of ground debris indicate that pieces of the left OMS pod were being shed starting at about GMT 13:59:49 (EI+940).

RGPC-2 data show a message reset sometime between GMT 13:59:37.4 (EI+928) and GMT 14:00:05 (EI+956). This action is a nominal crew response to a fault message and requires a crew member to man-ually acknowledge the message by keyboard entry on the center panel. RGPC-2 data indicate that the RHC was moved beyond neutral sometime between GMT 14:00:01.7 (EI+952) and GMT 14:00:03.6 (EI+954), triggering a "DAP DOWNMODE RHC" message at GMT 14:00:03.637 (EI+954). This message, which is identical to the DAP DOWNMODE message that occurred at GMT 13:36:04 (EI+485) in the first phase,

---

[30]Identified as Debris D in the CAIB timeline.

was likely due to an RHC bump due to the oscillatory motion of the orbiter. At GMT 14:00:03.678 (EI+954), the orbiter autopilot was returned to the AUTO mode. Returning the DAP to AUTO requires either the CDR or the PLT to press a button located on the glare shield. These actions indicate that the CDR or the PLT was still mentally and physically capable of processing display information and executing commands, and that the orbiter dynamics were still within human performance limitations.

RGPC-2 data show normal crew module temperature and pressure through the end of the period of reconstructed data [GMT 14:00:04.826 (EI+956)].

The RGPC-2 data show normal Freon flow through the radiators on the inside of the payload bay doors (PLBDs). This indicates that the radiators and PLBDs also retained structural integrity up to this point. Based on structural analysis, it is likely that the PLBDs were compromised prior to CE, after the end of RGPC-2 data. Loss of the PLBDs reduced the structural strength of the orbiter midbody and allowed hot gas to impinge upon the sills in the payload bay.

The RGPC-2 data also indicate that while all three auxiliary power units (APUs)[31] were running, all three hydraulic systems had zero pressure and zero quantities in the reservoirs. With the loss of hydraulic pressures and the vehicle LOC, the crew likely assumed a generic problem with the APUs. A crew module panel was recovered with switch configurations indicating an attempt by the PLT to recover the hydraulic systems and hydraulic pressure by performing steps to initiate a restart of two of the three APUs. Switches for the same two of the three system hydraulic circulation pumps were also in the "On" position. While turning on the hydraulic circulation pump is not on the emergency checklist, it nonetheless can provide some limited hydraulic pressure and shows good systems knowledge by the crew members as they worked to attempt to restore orbiter control. These switch positions were not reflected in RPGC-2 data and, therefore, must have occurred after GMT 14:00:05 (EI+956).

Although the orbiter continued to shed debris, ground-based video from GMT 14:00:09 (EI+960) to GMT 14:00:18 (EI+969) shows a thin, relatively consistent trail, suggesting that the conditions remained steady for a short period of time (~9 seconds). Aerodynamic modeling indicates that this was a time of growing stresses on the orbiter and increasing Gs on the crew.

1.1.2.4 Phase 4: Catastrophic Event to Crew Module Catastrophic Event
[GMT 14:00:18 (EI+969) to GMT 14:00:53 (EI+1004)]
35 seconds in duration

This section discusses events affecting the crew from the CE at GMT 14:00:18 (EI+969) to the CMCE at GMT 14:00:53 (EI+1004) (figure 1.1-23). Separation of the forebody from the midbody and aftbody occurred at or just after CE. This phase lasted 35 seconds.

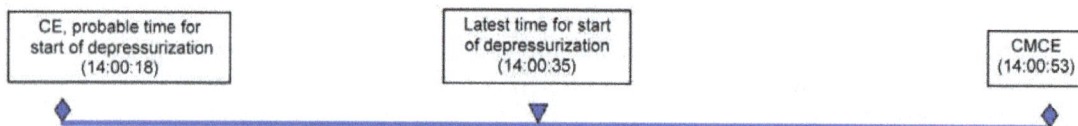

Figure 1.1-23. *Phase 4 timeline with key events.*

No telemetry or orbiter systems data are available during this phase. An on-board Global Positioning System (GPS) data recorder stopped at GMT 14:00:18.7 (EI+969) and the MADS/OEX recorder tape spool stopped at GMT 14:00:19 (EI+970) when the forebody lost power.[32] The sources of data that are available for reconstruction of events include ground-based video, recovered debris, medical findings, and modeling.

[31]The APUs drive the hydraulic pumps.
[32]The fuel cells, which provide all orbiter electrical power, are located in the midbody. Separation of the midbody from the forebody resulted in a total loss of power in the forebody.

Analysis of ground-based video identified specific events such as debris shedding and luminosity changes. A major luminosity event and orbiter trail characteristic change occurred at GMT 14:00:18.3 (EI+969) and is identified as the initiation of a major structural breakup called the CE. The aftbody and the forebody were identifiable as separate objects by GMT 14:00:25 (EI+976). Triangulating the video data provided relative motion, which was analyzed for an estimate of the deceleration and rotation rates of the forebody.

Columbia debris analysis consisted of five different methods: cluster analysis (plots of the ground location of recovered debris sorted by origination point on the orbiter), visual observations of debris, material analyses of melted deposits on select debris, and ballistic and thermal analysis of select debris. See Chapter 2.

Modeling was performed for various properties of the forebody. Aero thermal modeling provided estimated heat exposure. Aerodynamic modeling was used to evaluate the possible stable modes of the forebody. Aerodynamic modeling also provided estimates of G-loads, which increased as the forebody decelerated. The effects of the changing G-loads on both the forebody and the crew were analyzed. A combined environmental and structural analysis was performed to understand the effects of depressurization due to a single hole (or multiple holes equivalent to the same cross-sectional area) and the subsequent delta-pressure effects on the crew module structure.

At GMT 14:00:18 (EI+969), video showed that a significant event (the CE) occurred to the orbiter (figure 1.1-24). The GPS Miniaturized Airborne Global Receiver (MAGR) experiment, which was located in the middeck and powered by a fuel cell in the payload bay, experienced a loss of power at the CE. Less than 1 second later, the MADS/OEX recorder, which was also located in the crew module and similarly powered from the payload bay, also experienced a total power loss. The conclusion was that the forward and midbody orbiter segments separated at the CE. The CE is actually the start of a period of several seconds in which the orbiter underwent a major structural breakup. At GMT 14:00:25 (EI+976), there were visual indications that the orbiter was in multiple pieces. Ballistics analysis and structural debris analysis supports this period as the breakup event.

GMT 14:00:18.23   GMT 14:00:18.26   GMT 14:00:18.30

Figure 1.1-24. *The Catastrophic Event is depicted in these three frames of video that cover 0.1 second.* There is no change in the magnification/zoom factor. The third frame represents GMT 14:00:18.3 (EI+969).

The CAIB Report, based on data provided by the Crew Survival Working Group (CSWG), concluded that "Separation of the crew module/forward fuselage assembly from the rest of the orbiter likely occurred immediately in front of the payload bay (between X_0 576[33] and X_0 582 ring frame bulkheads)."[34] However, the SCSIIT's subsequent in-depth review of the debris field showed that significant portions of the X_0 582 ring frame bulkhead were found intermingled with the crew module debris field. Structural analysis led to the conclusion that the forebody separated from the midbody aft of the X_0 582 ring frame bulkhead. An aerodynamic simulation indicates that the structural operating load limits for the orbiter were not exceeded, indicating thermal degradation likely played a role in the failure.

The simulation also indicates that the crew module attachment fittings' (the x-links, y-links, and z-link) load limits were not exceeded. Although the exact sequence is not known, structural debris analysis suggests that the initial failure occurred immediately aft of the starboard x-link, where it attached to the payload bay

---

[33]The X_0 terminology refers to the X position in inches in the orbiter coordinate frame, where the X-axis runs the length of the orbiter from fore to aft. See Section 2.1 for a visual graphic of the orbiter coordinate frame.

[34]*Columbia* Accident Investigation Board Report, Volume 1, August, 2003, p. 77.

sill, and was likely from a combination of thermal degradation and structural loads. The forebody probably separated from the midbody from starboard to port. Based on structural evidence and the debris field, the crew module remained with the forward fuselage indicating that the links attaching the two structures remained intact. The remaining orbiter structure separated into aft and midbody/right wing segments. See Chapter 2. Figure 1.1-25 shows the recovered x-links.

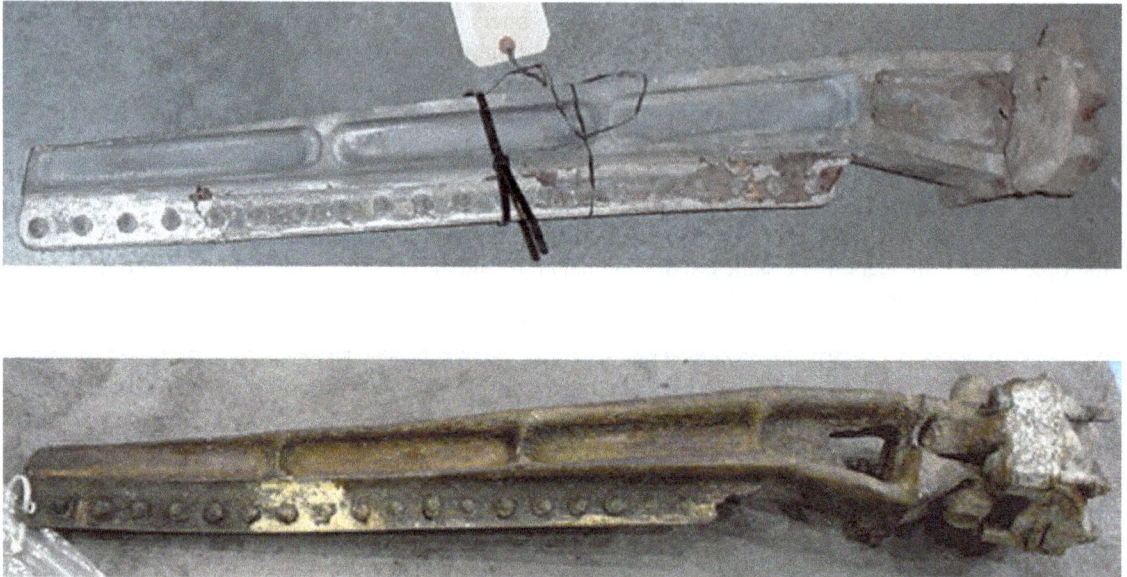

Figure 1.1-25. *Port x-link, debris item no. 1678* (top), *and starboard x-link, debris item no. 1765* (bottom), *from the* Columbia *Reconstruction Database.*

The resulting jerk acting on the crew module attach fittings as the forebody separated from midbody structure caused motion of the crew module within the forward fuselage shell. The crew module pressure vessel impacted the forward fuselage, which apparently resulted in damage to the crew module pressure vessel, internal crew module structure, and forward fuselage structure. Debris evidence shows that internal damage occurred to some volumes and lockers on the middeck in close proximity to the pressure vessel shell. At least one crew module pressure vessel breach occurred in the lower equipment bay or middeck area (figure 1.1-26), probably at or near the time of the CE, but definitely not later than GMT 14:00:35 (EI+986) ±5 sec. This time is based on the ballistic analysis of a recovered mission patch confirmed to have come from inside the crew module from one of the volumes (Volume E) that suffered damage.

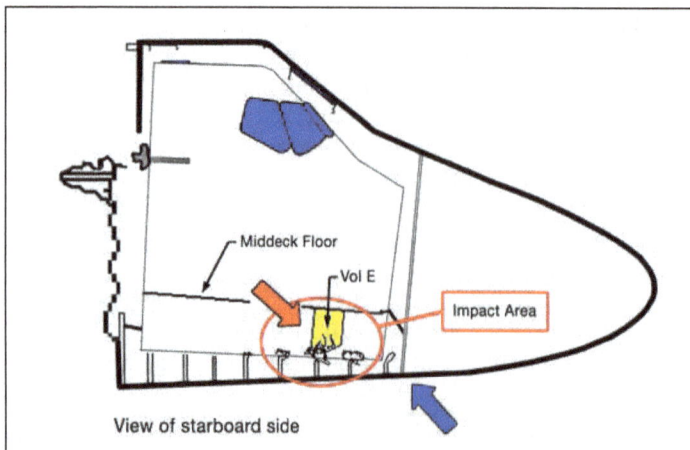

Figure 1.1-26. *Scenario showing how the crew module pressure vessel could impact the forward fuselage, and the middeck Volume E could impact the crew module pressure vessel, with resultant damage.*

The start of crew cabin depressurization can be narrowed to a range of 17 seconds, from between GMT 14:00:18 (EI+969) to GMT 14:00:35 (EI+986) ±5 sec (see Section 2.3). Crew module debris items recovered west of the main crew module debris field were 8 in. in diameter or smaller, were not comprised of crew module primary structure, and originated from areas above and below the middeck floor. This indicates that the crew module depressurization was due to multiple breaches (above and below the floor), and that these breaches were initially small. Another crew module breach possibly occurred at the starboard x-link area, but no significant flight deck debris is seen west of the CMCE-related debris, suggesting that this breach occurred later rather than earlier in the timeline.

When the forebody separated from the midbody, the crew members experienced three dramatic changes in their environment:

1. all power was lost,
2. the motion and acceleration environment changed; and
3. crew cabin depressurization began within 0 to 17 seconds.

With the loss of power, all of the lights and displays went dark (although each astronaut already had individual chem-lights activated). The intercom system was no longer functional and the orbiter O_2 system was no longer available for use, although individual, crew worn Emergency Oxygen System (EOS) bottles were still available.

As the forebody broke free from the rest of the orbiter, its ballistic number underwent a sharp change from an average ballistic number of 41.7 pounds per square foot (psf) (out of control intact orbiter) to 122 psf (free-flying forebody). The aerodynamic drag of the forebody instantaneously decreased, resulting in a reduction in the translational deceleration from approximately 3.5 G to about 1 G (figure 1.1-27).

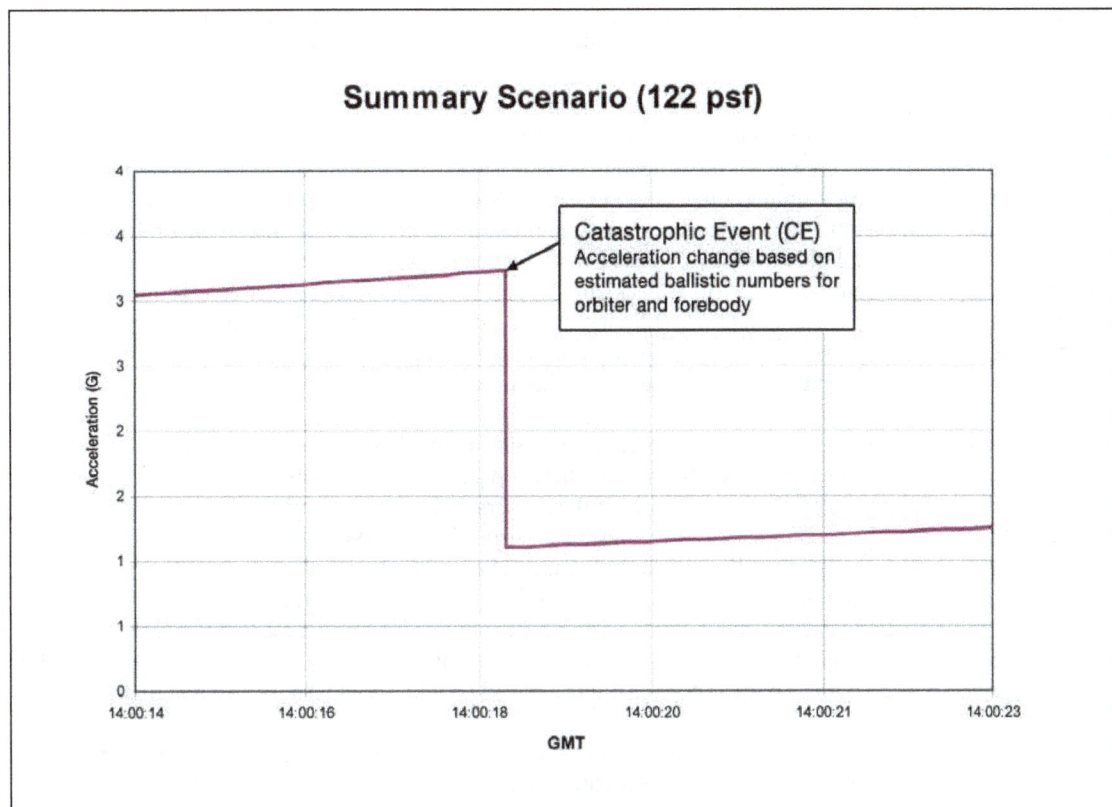

Figure 1.1-27. *Estimated change in total G experienced by the forebody due to a change in ballistic number from the orbiter breakup.*

The asymmetrical starboard to port separation of the forebody from the midbody would have induced rotation in the forebody, introducing new angular accelerations. The angular accelerations acting on the forebody at this time are impossible to accurately characterize due to inadequate data, but they likely changed significantly due to the abrupt change in the center of axis of rotation when the center of gravity (c.g.) changed from X_o 1075.5 to X_o 470.8 (figure 1.1-28).

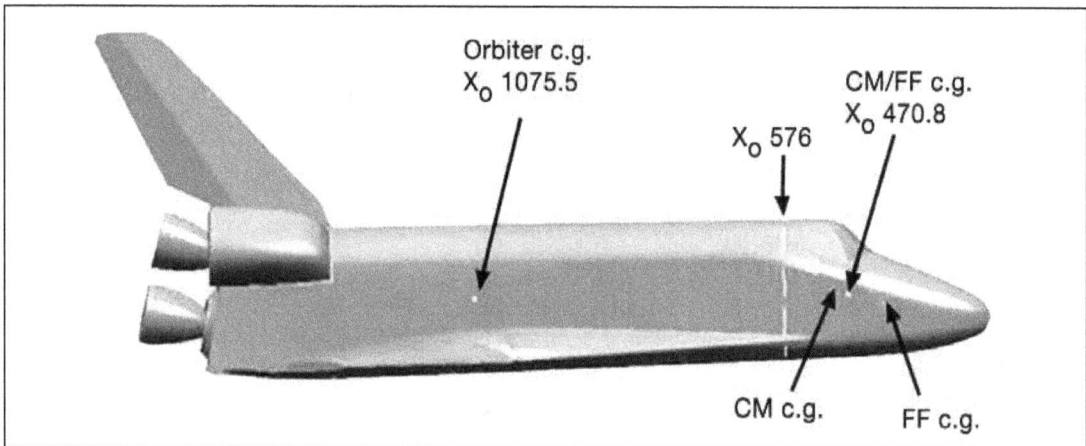

Figure 1.1-28. *X-axis center-of-gravity locations for the intact orbiter, the crew module, the forward fuselage, and the forebody (crew module plus forward fuselage). The X_o 576 is the aft bulkhead of the crew module.*

Also, as the forebody broke from the vehicle, the crew module moved within the forward fuselage shell resulting in transient rates of change (described earlier). Although it is probable that momentary sharp changes in acceleration caused high instantaneous G-loads, medical evidence indicates that the crew cabin pressure and load environment at the CE were still within human limitations for survival.

Effects of cabin depressurization on the crew would depend on the rate of depressurization. Existing CEE is capable of protecting the crew from rapid decompression via pressure suit, helmet, and either the orbiter O_2 or an individual EOS for a limited time. However, recovered crew equipment shows crew visors were in the nominal (up) position rather than emergency configuration (down and locked). Inspection of the wrist and glove rings showed that the glove wrist rings were not attached to the suit for two crew members on the middeck and one crew member on the flight deck, and one crew member had not yet donned the helmet. The change (from the crew's vantage point) from a nominal entry profile to the LOC and subsequent separation of the forebody from the orbiter all occurred in approximately 40 seconds. Experience shows that this is not sufficient time to don gloves and helmets.

Histological (tissue) examination of all crew member remains showed the effects of depressurization. Neither the effects of CE nor the accelerations immediately post-CE would preclude the crew members who were wearing helmets from closing and locking their visors at the first indication of a cabin depressurization. This action can be accomplished in seconds. This strongly suggests that the depressurization rate was rapid enough to be nearly immediately incapacitating. The exact rate of cabin depressurization could not be determined, but based on video evidence complete loss of pressure was reached no later than (NLT) GMT 14:00:59 (EI+1010), and was likely much earlier. The medical findings show that the crew could not have regained consciousness after this event. Additionally, respiration ceased after the depressurization, but circulatory functions could still have existed for a short period of time for at least some crew members.

The first event with lethal potential was depressurization of the crew module, which started at or shortly after orbiter breakup. Existing crew equipment protects for this type of lethal event, but inadequate time existed to configure the equipment for the environment encountered.

After the CE, the forebody was exposed to a high thermal environment as it decelerated and descended into an increasingly dense atmosphere. TPS tile and blankets on the forward fuselage protected the crew module, but the aft bulkhead was unprotected. However, debris field analysis indicates that the aft bulkhead remained intact until crew module breakup. The volume between the forward fuselage and the crew module had openings to the environment, which could result in the entry of heated gas. Crew module breaches could allow the entry of this gas into the crew module after the dynamic pressure outside the crew module exceeded atmospheric pressure inside the crew module. A few molten globules of metal were found on recovered seat harness straps, indicating the presence of heated metal inside the crew module while the unconscious or deceased crew members were still restrained in their seats. Although the timing of the deposition cannot be determined precisely, it may have been very close to the crew module breakup. There is no evidence to suggest that the overall crew module internal structure temperature was severe, but local hot spots may have existed near breaches.

The orbiter had substantial rates of rotation in all axes of rotation when RGPC-2 data ended at GMT 14:00:04 (EI+955). The orbiter breakup at the CE imparted motion to the forebody, and the forebody began rotating after it broke free from the vehicle. Aerodynamic modeling indicates that the free-flying forebody would not achieve a stable attitude. Videos of the forebody show brightening and dimming, implying rotational motion. Triangulation analysis of the forebody in video showed a slow wobble motion in all three axes, also supporting rotation or tumbling. Thermal damage seen on external portions of the forebody indicates intermittent exposure to heat. Based on the wobble motion, rotation rates gradually increased with an estimated initial average rate of 0.1 revolution per second (36 degrees per second) around a changing body axis (see Section 2.1). This rate is not extreme, and even peak rates toward the end of this phase result in angular accelerations of less than 2 G. Translational deceleration due to aerodynamic drag also increased, up to approximately 3 G at the CMCE. The loss and redistribution of mass as forward fuselage structure failed and separated would affect the rotation rate. Modeling suggests that the rate could continue to build, up to 0.5 revolution per second, although this was not verified with video data. The crew members seated farthest from the crew module c.g. experienced the highest angular accelerations due to the greater distance (moment arm) from the center of rotation. This acceleration was in addition to the translational accelerations and, depending on the attitude of the rotating forebody, the accelerations experienced by the crew members could vary from about –1 G to +5 G.

Under rapidly changing accelerations, the design intention is that inertial reels on the seat restraint shoulder harnesses will lock, and remain locked, until manually disengaged by the crew member. To lock, the inertial reel mechanism used on orbiter seats requires 1.78 G to 2 G of strap acceleration, in a direction orthogonal to the mechanism (straight out of the seatback). The abrupt dynamics associated with the CE would be expected to have locked the inertial reel. In the subsequent multi-axial rotating environment experienced during this phase, it is expected that the unconscious or deceased crew periodically would arrive at a posture allowing harness retraction. The harness would then remain retracted if the inertial reels had locked. However, seat analysis shows that several of the shoulder harness restraints failed with the inertial reel straps partly or fully extended, and other inertial reel straps were extended at some point during this phase (see Section 3.1). Either the acceleration on the straps was insufficient to lock the harness, the loading was not orthogonal (preventing harness retraction), ACES equipment blocked the strap retraction slot, or some combination of all three effects occurred. The net effect was that the crew members had no upper body restraint and were restrained solely by their lap belts.

The combination of the lack of upper body restraint and a helmet that, by design, does not internally conform to the head while exposed to cyclical motion resulted in lethal mechanical injuries for some of the unconscious or deceased crew members. The circulatory system of most of the unconscious or deceased crew was still functioning at the time of these lethal injuries. If the harnesses had been locked or the crew had been conscious and able to brace, the injuries likely would not have been lethal.

The second event with lethal potential was unconscious or deceased crew members exposed to a dynamic rotating load environment with nonconformal helmets and a lack of upper body restraint.

Existing seat and helmet design did not protect the crew from this lethal event.

Crew module structure temperature increased during this phase, resulting in a corresponding reduction in structural strength. The increase in loads due to the increasing deceleration and increasing rotational rate and thermal degradation resulted in eventual structural failure.

In summary, in the 35 seconds from the CE to the CMCE, the forebody detached, the crew module breached and depressurized, and the forebody experienced increased heating and began to structurally degrade.

Figure 1.1-29 shows an overall summary of accelerations, heat rates, and trajectory. *Information for post-CMCE (GMT 14:00:53 (EI+1004) and beyond) is predicted data for an* **intact crew module** *and* **is not representative** *of actual trajectory, accelerations, and thermal environment for the crew or for individual components of the crew module.*

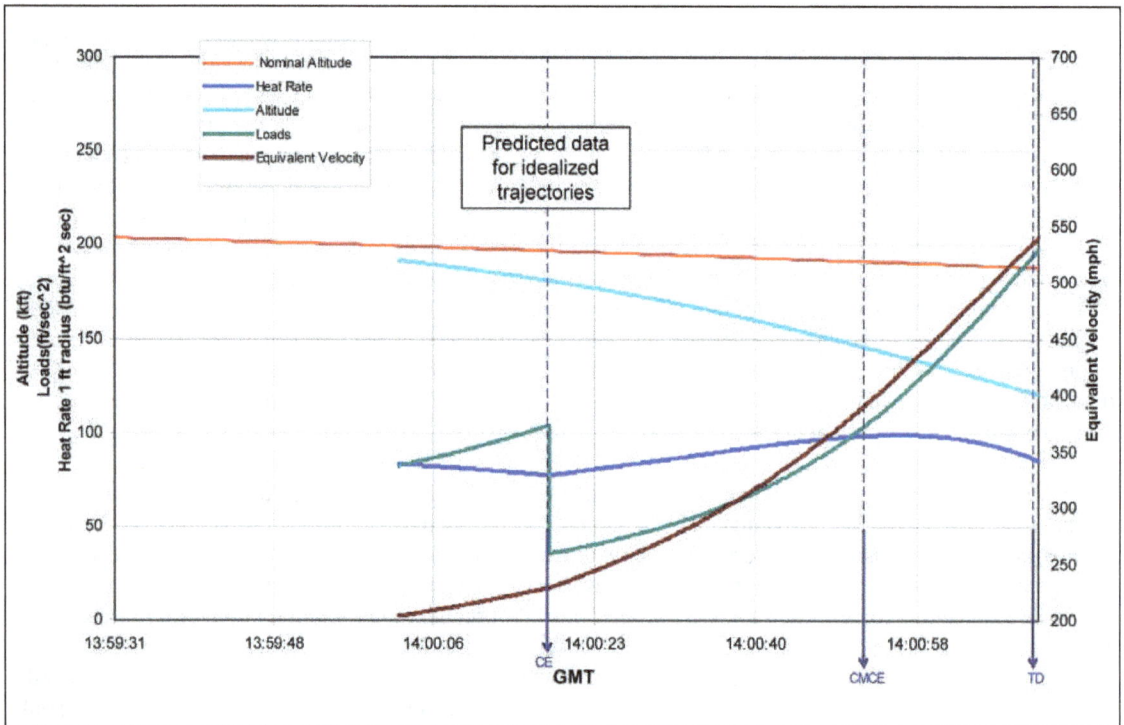

Figure 1.1-29. *Predicted acceleration and thermal data for an idealized trajectory of the intact orbiter with a ballistic number of 42 pounds per square foot, and post-Catastrophic Event free-flying forebody with a ballistic number of 122 pounds per square foot.*

1.1.2.5 *Phase 5: Crew Module Catastrophic Event to Total Dispersal*
[GMT 14:00:53 (EI+1004) to GMT 14:01:10 (EI+1021)]
17 seconds in duration

This section discusses events affecting the crew from the CMCE at GMT 14:00:53 (EI+1004) to TD at GMT 14:01:10 (EI+1021). This phase lasted 17 seconds.

Three sources of data were available for analysis of Phase 5: video, debris, and medical findings.

Video data show that a significant forebody brightening event began at GMT 14:00:53 (EI+1004). By GMT 14:01:10 (EI+1021), the forward fuselage, crew module, and trailing debris clearly originating from the forebody are no longer visible in the video. In all videos, the last of this debris disappears from sight while in the middle of the field of view (FOV) rather than leaving the FOV, indicating either speed and/or size decreased such that its brightness, which was created by frictional heating from drag, was below the sensitivity of the video camera.

Debris analysis consisted of five techniques: cluster analysis, visual observations, materials analyses, ballistics, and thermal modeling predictions (see Section 2.4). Materials analysis was performed on select items to evaluate thermal exposure (see Sections 2.1 and 3.2). Materials evaluations were performed on the seats, some crew equipment, and some forebody structure. Additionally, some thermal modeling was done to estimate peak thermal temperatures after separation for various items. The model results were then compared to the actual debris appearance.

Based on video and ballistic evidence, at GMT 14:00:53 (EI+1004) a significant breakup event began. This event is designated as the CMCE.

The CMCE started with the separation of the forward fuselage from the crew module, exposing the entire crew module to the thermal effects of entry. The main forebody debris field included all recovered crew module pressure vessel structure, almost 90 percent of recovered forward fuselage structure, and around 90 percent of the crew module contents. This indicates that the failures of the forward fuselage and crew module were closely associated. Ballistic analysis confirmed this assessment.

The video recorded from an Apache helicopter operating in the area of Ft. Hood, Texas shows a significant event of two objects with similar luminosity and ballistic number separating simultaneously from the forebody (figure 1.1-30). The remaining central object maintained integrity for several more seconds in the video. Shortly after these items peeled away, the remaining object began to lose large pieces of structure.

The conclusion was drawn that these two objects were most likely the upper and lower forward fuselage sections, leaving the crew module (the central object) intact but no longer protected. Within seconds, the crew module began to lose structural integrity as well.

Forward fuselage debris shows localized thermal damage and very little evidence of debris-debris interaction. Large portions of structure were recovered intact. Material deposition on the interior of the forward fuselage debris was not significant. Reconstruction of the forward fuselage debris supports a structural failure from starboard-to-port and forward-to-aft.

The crew module breakup was rapid (<15 seconds). The range of ballistic numbers of the debris items resulted in quickly diverging individual trajectories such that very little debris-debris interaction occurred. Cluster analysis of the debris field shows that the crew module forward X_{cm}[35] 200 bulkhead debris is farther west than the crew module aft X_o 576 bulkhead debris. This indicates that the failure

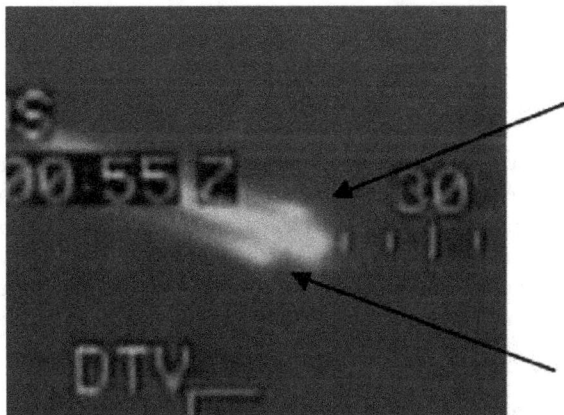

Figure 1.1-30. *Video frame from the Apache video at GMT 14:00:55 (EI+1006) showing two similar luminosity objects separating from the forebody.*

of the forward bulkhead happened prior to the failure of the aft bulkhead. The middeck floor debris field begins at the same longitude as the forward bulkhead, suggesting that the failures were nearly simultaneous. Debris evidence from the crew module structure suggests a starboard-to-port breakup of the middeck area, which probably included the forward bulkhead. Cluster analysis and evidence of significant heating of the flight deck floor and the flight deck seats indicates that the flight deck was intact for a short period of time (probably less than 5 seconds) after separation from the middeck. Cluster analysis indicates that the airlock stayed with the flight deck, possibly connected by the aft bulkhead.

---

[35]The X_{cm} terminology refers to the X position in inches in the crew module coordinate frame, where the X-axis runs the length of the crew module from fore to aft. The crew module coordinate frame axes are coincident with the orbiter frame axes, but with a different X-axis origin.

There is no evidence of an explosion or a fire. Analysis of thermal vectors on numerous debris items showed multiple independent heat vectors across the structure. For example, many recovered middeck floor panels were nearly pristine with paint still visible, while floor panels from immediately adjacent locations had melted materials deposited on them and other signs of high thermal exposure (figure 1.1-31). After breakup, individual items experienced their own trajectories and heat exposure. This heat exposure can vary enormously with ballistic number and other effects such as shadowing from other debris items and orientation of the item into the heat vector. The lack of consistent directional heating vectors on crew module debris suggests heating was due to individual item trajectories and random exposure during breakup rather than a major breach resulting in directional heating.

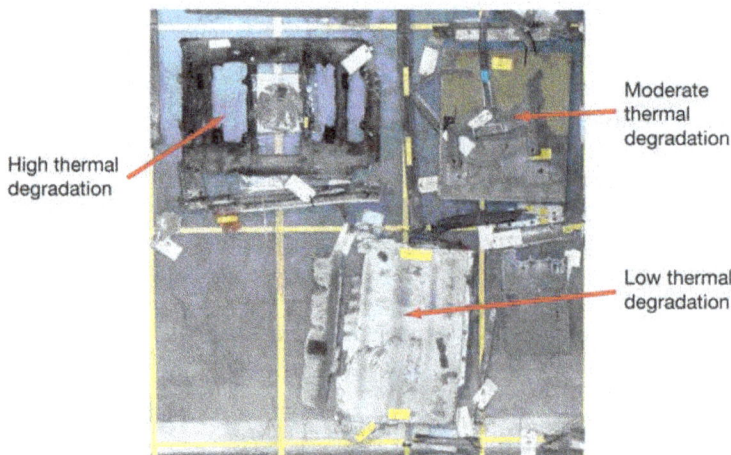

Figure 1.1-31. *Middeck floor debris in original relative orientation showing varying thermal exposure.*

The exact time and sequence that the crew and seats separated from the crew module is unknown. A comprehensive evaluation of ballistic analysis of debris, crew member remains, and crew worn equipment indicates that the middeck crew remains were separated from the crew module prior to the flight deck crew remains, supporting the conclusion that the flight deck stayed intact a few seconds longer than the middeck.

The dynamic pressure environment exposure caused the mechanical failure of the crew suits (common to high-speed accidents, but somewhat unexpected given the aerodynamic pressure of only 450 to 550 psf). The suit is designed to maintain structural integrity when exposed to a windblast that is up to 560 knots equivalent air speed (KEAS) (806 psf). This assumes that the helmet visor is down. The helmet visors being in the up position is the most likely explanation for the hastened disruption of the suits. Although suit disruption was primarily due to aerodynamic (mechanical) loads, the thermal environment and atomic oxygen in the atmosphere may have been a contributing factor.

The third event with lethal potential was separation from the crew module and the seats with associated forces, material interactions, and thermal consequences. This event is the least understood due to limitations in current knowledge of mechanisms at this Mach number and altitude. Seat restraints played a role in the lethality of this event. Although the seat restraints played a significant role in the lethal-level mechanical injuries, there is currently no full range of equipment to protect for this event. This event was not survivable by any means currently known to the investigative team. All circulatory functions had ceased by the end of this phase.

Whether an item separated from the crew module or the crew module lost significant mass, an instantaneous change in ballistic number occurred and resulted in varying deceleration and thermal profiles. The accelerations varied from over 30 G for a short duration (less than 5 seconds) to over 10 G for up to 20 seconds. The range of ballistic numbers of debris generated a range of thermal conditions. Objects with higher ballistic numbers take longer to decelerate, and experience longer periods of heating and lower G-spikes (see Section 2.1). Video shows a large deceleration of the crew module relative to the main engines at GMT 14:01:08 (EI+1019). The visual object actually represents a cloud of multiple objects that experienced deceleration at varying rates. Several seconds elapsed before objects of varying ballistic numbers separated visually from each other, creating the impression of a solid object in the video for a few seconds. This is consistent with a gradually expanding breakup caused by items having a wide range of ballistic numbers and deceleration trajectories, resulting in a widely spread debris cloud.

TD was complete by GMT 14:01:10 (EI+1021).

1.1.2.6 *Phase 6: Total Dispersal to ground impact*
[GMT 14:01:10 (EI+1021) to approximately 14:35:00 (EI+3051)]
approximately 34 minutes in duration

This section discusses events from TD (GMT 14:01:10 (EI+1021)) to the ground impact of debris. Heavy items impacted the ground much sooner than lighter objects but traveled much farther from the point of separation. It was calculated that a small fragment of cloth with a ballistic number of .5 psf would impact the ground around 33 minutes after separation. Based on this calculation, all crew module debris was likely on the ground (including very lightweight objects) by GMT 14:35:00 (EI+3051).

Three sources of data were available for Phase 6: video, debris analysis, and medical findings.

Very limited video data were available as the crew module rapidly disappeared from the FOV as it dispersed into smaller and smaller debris. The smaller size and loss of heating as the debris decelerated reduced the ability to detect items in the video. Debris analysis consisted of four techniques: visual observations, ballistics analysis, cluster analysis, and materials analysis (see Chapter 4).

The forebody breakup event occurred between 145,000 feet and 105,000 feet at an ambient pressure of approximately 0.03 pound per square inch (psi). After the deceleration peak, the overall deceleration would stabilize to 1 G at terminal velocity. At ground level, the ambient absolute pressure condition was approximately 14.7 psi and the temperature was 59°F (15°C).

The fourth event with lethal potential was exposure to near vacuum, aerodynamic accelerations, and cold temperatures. Current crew survival equipment is not certified to protect the crew above 100,000 feet, although it may potentially be capable of protecting the crew.

At the altitude the deceased crew departed from the crew module, the environmental risks include lack of O_2, low air pressure, high thermal exposure as a result of deceleration from high Mach numbers, and exposure to cold temperatures. Existing shuttle CEE is certified to protect a crew member exposed to an atmospheric/altitude environment up to 100,000 feet. Anecdotal evidence from the survival of the pilot of an SR-71 mishap[36] suggests that an intact, pressurized suit similar to the ACES can also protect a crew member at least up to speeds of Mach 3.

Shuttle crew members carry a personal O_2 supply that provides O_2 independent of the orbiter supply. This system can provide enough O_2 for a crew member to reach the ground from altitudes much greater than 100,000 feet, so it is not the limiting factor in the system.

The ground impact without parachute protection generated a very large instantaneous G event.

The final event with lethal potential was ground impact. Existing shuttle CEE protects for ground impact with a parachute. However, the crew must manually initiate the parachute opening sequence, or the parachute must be used in conjunction with the crew escape pole of the shuttle to initiate the parachute automatic opening sequence.

---

[36]Department of the Air Force, SR-71 Aircraft Mishap Report, January 25, 1966.

1.2 Master Timeline

The SCSIIT Master Timeline was developed as a tool to aid the SCSIIT with the investigation of what happened to the crew of STS-107. The SCSIIT Master Timeline began with the tailoring of the CAIB Master Timeline, Revision 15 to highlight crew-related events. Additional events were added from various sources, including:

- recorded telemetry
- RGPC data
- MADS/OEX recorder data
- MAGR data
- recovered on-board videos
- ground-based videos
- air-to-ground audio
- forensic analysis of medical findings
- engineering forensic analysis of vehicle and CEE debris
- ballistic analysis of vehicle and CEE debris

The timeline is divided into six phases:

- Phase 1: From the deorbit preparation checklist timeline initiation to EI. The deorbit preparation checklist timeline begins 4 hours prior to the deorbit burn. EI is defined as the time at which an altitude of 400,000 feet was reached. [GMT 09:15:30 (EI–16119 seconds) – GMT 13:44:09 (EI)]

- Phase 2: From EI to LOS. LOS is the time of the loss of voice and real-time data from *Columbia*. [GMT 13:44:09 (EI) – GMT 13:59:32 (EI+923)]

- Phase 3: From LOS to the CE. The CE is defined as the initiation of the orbiter breakup into the primary subcomponents of the forebody, midbody and aftbody. The CAIB timeline ends with the CE. [GMT 13:59:32 (EI+923) – GMT 14:00:18 (EI+969)]

- Phase 4: From the CE to the CMCE. The CMCE is defined as the initiation of the forebody breakup. [GMT 14:00:18 (EI+969) – GMT 14:00:53 (EI+1004)]

- Phase 5: From the CMCE to TD. TD is defined as the time at which the crew module was broken down into its subcomponents. [GMT 14:00:53 (EI+1004) – GMT 14:01:10 (EI+1021)]

- Phase 6: From the TD to ground impact of the crew and the bulk of the crew module debris. [GMT 14:01:10 (EI+1021) – approximately GMT 14:35:00 (EI+3051)]

All events are presented in GMT. In addition, events prior to the TIG of the deorbit burn also include the time prior to TIG. After TIG, the events include the time from TIG and the time to EI. After EI is reached, all events are presented in just GMT and EI.

Five symbols are used in the timeline to aid the reader in scanning for events of a certain category. The symbols are:

- ⚷ – indicates a crew-related event.
- ▦ – indicates an event that is based/observed on video footage.
- ⚶ – indicates a vehicle-related event.
- ⚸ – indicates a vehicle-related event that occurred at the time of the separation of the forebody (crew module and forward fuselage) from the midbody.
- ⚹ – indicates a vehicle-related event that occurred after the separation of the forebody from the midbody.

The timeline does not include every event. For a comprehensive listing of events, the reader should consult sources such as:

- *Columbia* Accident Investigation Board Master Timeline, Revision 15
- *Columbia* Accident Investigation Board Report, Volume II, Appendix D.19, Qualification and Interpretation of Sensor Data from STS-107
- OEX Data Evaluation of End of Mission Data for STS-107, Vehicle Data Mapping Team, presentation, 8-22-2003, Rev B
- STS-107 Investigation Action Response: OVE-204 Crew Inputs After Loss of COMM (Voice); CAIB-MRT-00099, 3-10-2003

1.2.1   Phase 1: Deorbit preparation to entry interface
GMT 09:15:30 (EI–16119) through GMT 13:44:09 (EI)

TIME		EVENT
09:15:30 (TIG[1]–04:00:00[1])	⚷	**Deorbit Preparation.** The crew begins working items on the deorbit preparation (D/O PREP) checklist at TIG–4 hours per pre-mission planning.
11:11:18 (TIG–02:04:12)	⚷	**OPS 301.** The crew, per nominal procedures, manually enters the OPS 301 command to initiate the Pre-Deorbit Coast Major Mode software. This is the first entry-phase software sequence in preparation for entry. [*Telemetry, Tracking, and Data Relay Satellite-West (TDRS-W) data*]
~11:40:00 (TIG–01:35:30)	⚷ ▦	**Recovered Middeck Video Begins.** Approximately 30 minutes of video (without audio) were recovered from a camera in the middeck. On the video, the crew is shown working through D/O PREP checklist items. Times are approximate due to the lack of audio and visual cues to synchronize activities seen on the video with GMT (figure 1.2-1).

---

[1]TIG is time of ignition and refers to the start time of the deorbit burn. TIG–hh:mm:ss is the time before the burn begins in hours (hh), minutes (mm), and seconds (ss). TIG+hh:mm:ss is the time after the burn begins.

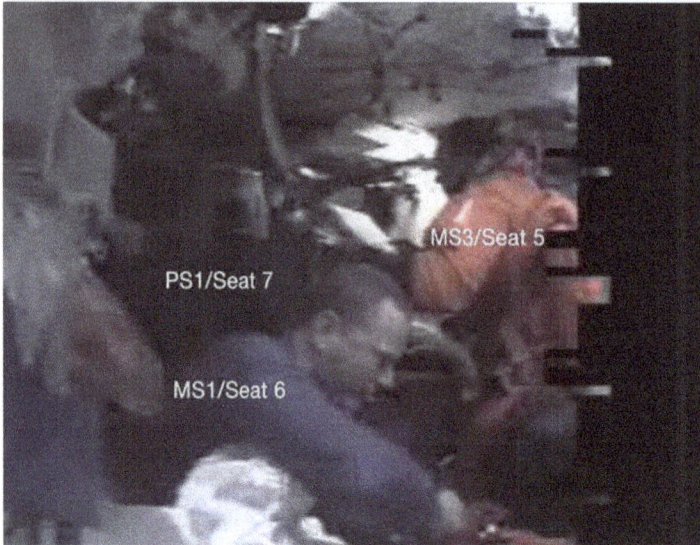

Figure 1.2-1. *Start of the recovered middeck video.*

The following was observed at the start of the video (figure 1.2-2):

- All middeck seats (seat 5, seat 6, and seat 7) are installed on the middeck floor.
- Personal parachute assemblies for MS3/Seat 5, MS1/Seat 6, and Payload Specialist 1 (PS1)/Seat 7 are positioned on seatbacks.
- MS3/Seat 5 has already donned the ACES (excluding gloves and helmet) and parachute harness.
- MS1/Seat 6 and PS1/Seat 7 have yet to don the ACESs.
- The escape pole is still stowed on the middeck ceiling (on-orbit location).

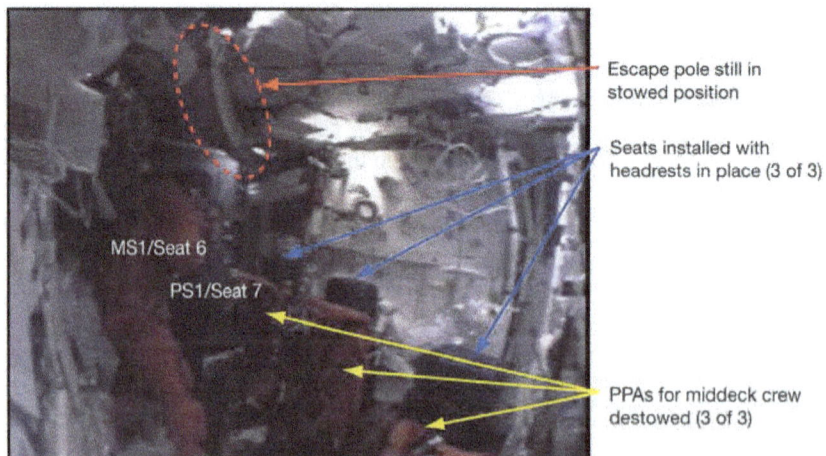

Figure 1.2-2. *Middeck configuration during duration of the recovered video.*

TIME		EVENT

~11:48:56 (TIG–01:26:34)   📽 🧍   CDR/Seat 1 dons ACES (excluding gloves and helmet) and parachute harness and goes to flight deck.

📽 🧍   MS3/Seat 5 and PS1/Seat 7 are observed fluid-loading in preparation for return to 1 G.

📽 🧍   MS1/Seat 6 dons ACES (excluding gloves and helmet) and parachute harness.

📽 🧍   PS1/Seat 7 and MS1/Seat 6 pass helmets in helmet bags to flight deck crew members.

12:10:00 (TIG–01:05:30)   📽 🧍🚀   **Recovered Middeck Video Ends.** The video ends with PS1/Seat 7, MS3/Seat 5, and MS1/Seat 6 beginning to re-move the escape pole from the ceiling to install it. The D/O PREP checklist calls for the pole to be installed as one of many activities in the Entry Cabin Configuration block, which starts at TIG–03:50:00. NOTE: Evaluation of the debris reveals that the pole was installed in the launch/entry position, so the crew completed the installation (figure 1.2-3).

Figure 1.2-3. *End of the recovered middeck video.* [Mission Specialist 3/Seat 5 in foreground]

TIME		EVENT
12:53:04 (TIG–00:22:26)		**OPS 302.** Per nominal procedures, the crew manually enters the OPS 302 command to initiate the *Deorbit Execute* Major Mode software. [*Telemetry, TDRS-W data*]
13:10:00 (TIG–00:05:00)		The MCC gives the "GO" for deorbit burn.[2]
13:10:30 (TIG–00:05:00)		**TIG–5.** TIG refers to the time of the planned ignition of the OMS engines (referred to as the deorbit burn) to reduce the orbiter's velocity enough to result in entry into the atmosphere. This is a benchmark time to make sure the crew starts the APU in time for the deorbit burn and is also the time at which the last MS is to be seated. [*Based on TIG event in Master Timeline, Rev. 15 Baseline*]
13:10:39 (TIG–00:04:51)		**APU 2 Start – Low Press.** Three APUs provide pressure to the orbiter hydraulic systems (engine gimbals, elevons, and body flap). Only one APU is used to support the deorbit burn. [*Master Timeline, Rev. 15 Baseline*]
13:15:30 (TIG–00:00:00)		**TIG: Deorbit Burn Begins.** This is the beginning of the deorbit burn using the OMS engines. [*Master Timeline, Rev. 15 Baseline*]
13:18:08 (EI–1561)		**Deorbit Burn Ends.** This is the end of the deorbit burn. [*Master Timeline, Rev. 15 Baseline*]
13:20:21 (EI–1428)		**OPS 303.** Per nominal procedures, the crew manually enters the OPS 303 command to initiate the *Pre-entry Monitor* Major Mode software.
13:26:09 (EI–1080)		**Forward RCS Dump Start.** This is a nominal operation to deplete the forward RCS fuel and oxidizer tanks in preparation for entry. [*Master Timeline, Rev. 15 Baseline*]
13:27:12 (EI–1017)		**Forward RCS Dump Complete.** [*Master Timeline, Rev. 15 Baseline*]
13:31:25 (EI–764)		**APU 1 Start.** APU 1 is started per the deorbit procedures. [*Master Timeline, Rev. 15 Baseline*]
13:31:29 (EI–760)		**APU 3 Start.** APU 3 is started per the deorbit procedures. [*Master Timeline, Rev. 15 Baseline*]
13:31:57 (EI–732)		APU 1 is performing nominally. [*Master Timeline, Rev. 15 Baseline*]

---

[2]*Columbia* Accident Investigation Board Report, Volume I, August 2003, p. 38.

TIME		EVENT
13:31:59 (EI–730)	⛴	APU 2 is performing nominally. [*Master Timeline, Rev. 15 Baseline*]
13:32:01 (EI–728)	⛴	APU 3 is performing nominally. [*Master Timeline, Rev. 15 Baseline*]
13:32:11 (EI–718)	🧍	PLT/Seat 2: "Houston, here comes SSME HYD repress." This is part of the procedure to use the hydraulic system to move the space shuttle main engines (SSMEs) to the desired position for entry and landing. [*SCSIIT air-to-ground 1 (A/G1) Tape Elapsed Time (TET)*[3] *02:20:26*]
13:32:18 (EI–711)	🧍	CAPCOM: "And we're ready, Willie. No deltas." This message informs the crew that there were no changes to the planned procedure. [*SCSIIT A/G1 TET 02:20:33*]
13:32:22 (EI–707)	🧍	PLT/Seat 2: "Copy, no deltas." [*SCSIIT A/G1 TET 02:20:37*]
13:35:16 (EI–533)	🧍	CAPCOM: "And *Columbia*, Houston. The HYD fluid thermal conditioning will not be required today. We'll meet you on the cards." [*SCSIIT A/G1 TET 02:23:31*]
13:35:26 (EI–523)	🧍	CDR/Seat 1: "And we copy, Houston. HYD fluid thermal conditioning not required, and we copy going to the cards." [*SCSIIT A/G1 TET 02:23:41*]
13:35:32 (EI–517)	🧍	CAPCOM: "And, Rick, don't want to lead you astray, and don't forget about the stuff on page 3-44." [*SCSIIT A/G1 TET 02:23:47*] Page 3-44 is part of the entry checklist; it contains the steps for enabling the RHCs by turning on the flight controller power and activating the entry video camera system. The last step has the crew go to the Entry Maneuvers cue card.
13:35:34 (EI–515)	🧍🎞	**Recovered Flight Deck Video Begins.** The first audio on the tape is of the CAPCOM completing the sentence recorded on the A/G audio at approximately GMT 13:35:32 (EI-517): "…and don't forget about the stuff on page 3-44." [*SCSIIT A/G1 TET 02:23:51*] The PLT/Seat 2 is shown adjusting g-suit setting (figure 1.2-4).

---

[3] TET is the tape elapsed time from the start of the audio recording file.

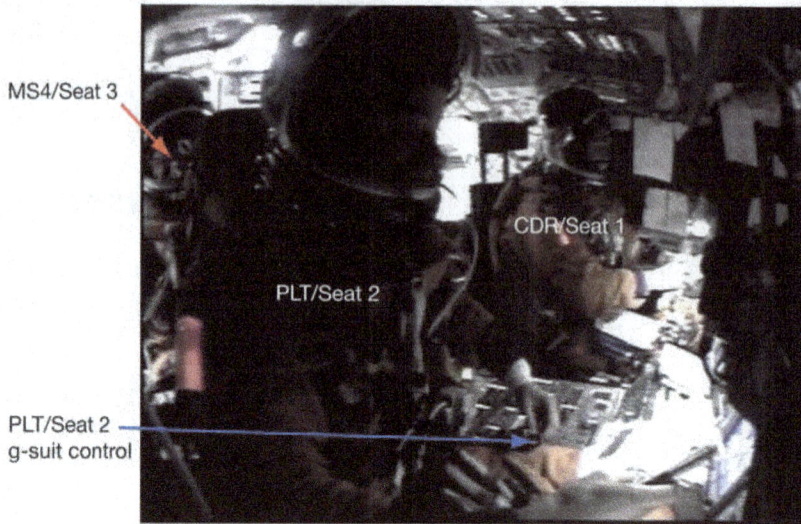

Figure 1.2-4. *Pilot/Seat 2 adjusting g-suit setting.*

All visible crew members (CDR/Seat 1, PLT/Seat 2, and MS4/Seat 3) are fully suited except for gloves and are strapped in. All helmet visors are OPEN per nominal procedure.

CDR/Seat 1: "Right, we're checking that. We've got the flight controller power on. We're working through the rest of it as well. Thanks." [*SCSIIT A/G1 TET 02:23:57*]

CAPCOM: "Sounds good."

13:36:02 (EI–487)

MS4/Seat 3 is shown starting to don gloves (figure 1.2-5).

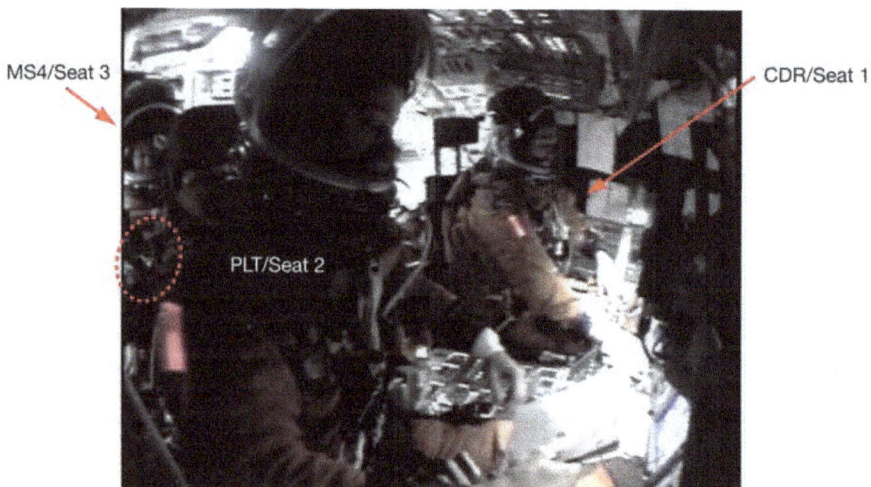

Figure 1.2-5. *Mission Specialist 4/Seat 3 donning glove.*
[Glove is within dotted circle]

13:36:04 (EI–485)   👤 🎞   Between GMT 13:36:04 (EI-485) and GMT 13:36:06 (EI-483), the CDR/Seat 1 is performing entry preparation actions that lead to the RHC being bumped (figure 1.2-6).

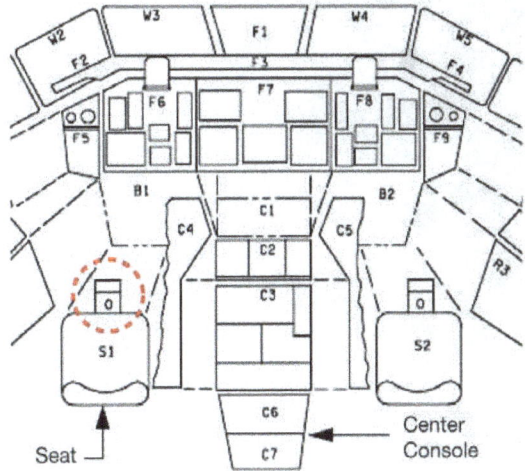

Figure 1.2-6. *Location of Commander's seat and rotational hand controller.* [Left picture from the Shuttle Training Simulator looking from starboard to port; right picture adapted from the Space Shuttle Systems Handbook looking from aft to forward]

13:36:07 (EI–482)   👤 🎞   **DAP DNMODE RHC.** Primary Avionics Software System (PASS) DAP DOWNMODE RHC (time from telemetry analysis). The message indicates that the RHC movement in the previous event was sufficient to mode the DAP out of AUTO into Inertial mode. When this occurs, a "DAP DOWNMODE RHC" caution and warning message is displayed, the INRTL button on the C3 panel is illuminated (see arrow in figure 1.2-7), and a tone, which can be heard in the recovered video tape, is annunciated (figure 1.2-7).

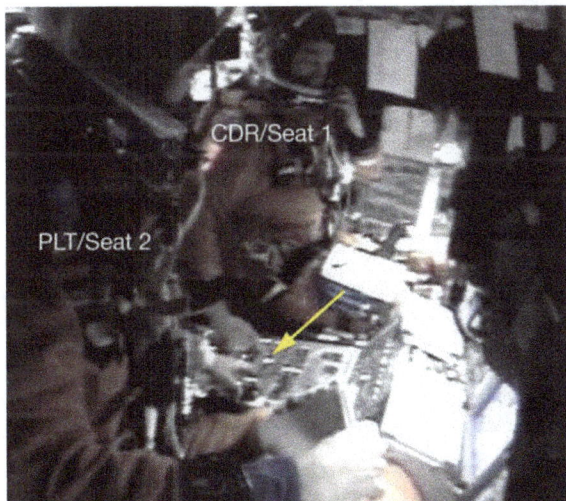

Figure 1.2-7. *Digital autopilot button on the C3 panel is illuminated.*

13:36:14 (EI–475)   CDR/Seat 1 responds to the DAP DOWNMODE RHC message by pressing the illuminated AUTO button on the C3 panel to restore the DAP to AUTO (figures 1.2-8 and 1.2-9).

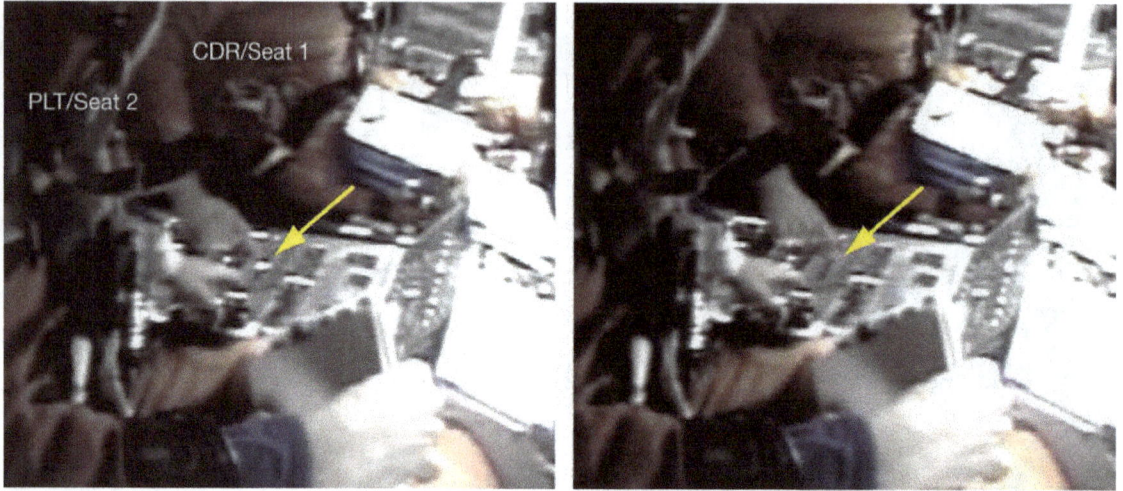

Figure 1.2-8. *Digital autopilot is no longer in AUTO, INRTL light is ON* (left figure); *Commander/Seat 1 restores digital autopilot to AUTO, INRTL light is OFF* (right figure).

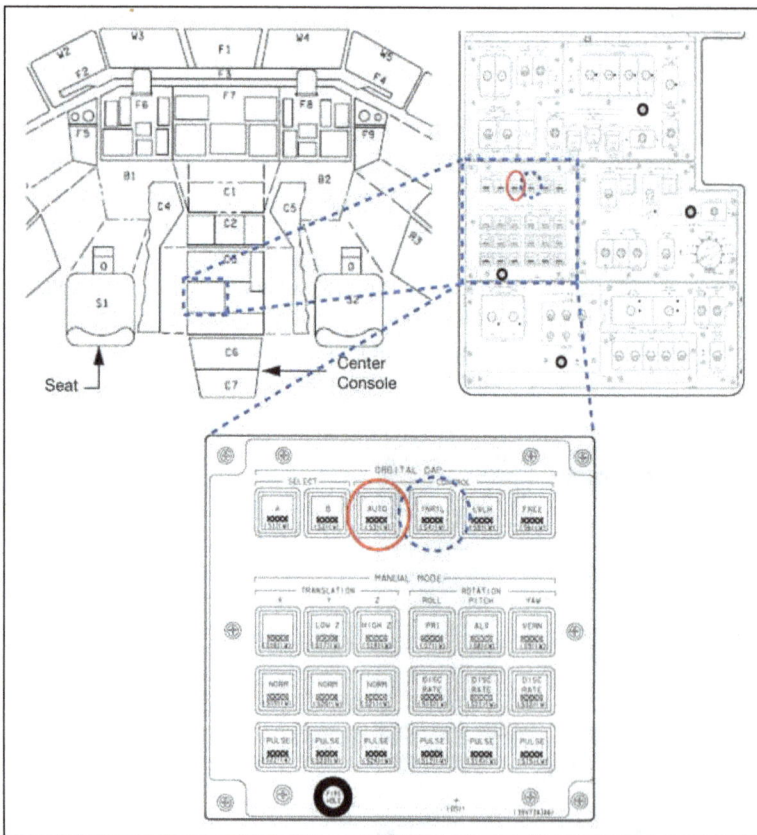

Figure 1.2-9. *Location of the "INRTL" button* (blue circle) *that illuminated when the digital autopilot moded out of AUTO, and the "AUTO" button* (red circle) *that the Commander pressed to restore the digital autopilot to AUTO.* [Adapted from the Space Shuttle Systems Handbook]

TIME	EVENT

13:37:31 (EI–398)   👤 🎞   CAPCOM: "*Columbia*, Houston for Rick. We'll take another ITEM 27 please." This is required to resume the maneuver (to the EI–5-minute attitude) that was interrupted by the bumped RHC. The CDR/Seat 1 acknowledges the request. [*A/G recording and recovered flight deck video*]

13:37:39 (EI–390)   👤 🎞   ITEM 27. CDR/Seat 1 manually inputs the ITEM 27 command using the keypad on the C2 panel. This fully recovers the vehicle from the bumped RHC. [*From video and telemetry*] (figures 1.2-10 and 1.2-11)

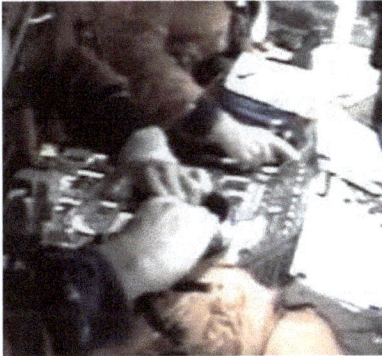

Figure 1.2-10. *Commander/Seat 1 inputting an ITEM 27.*

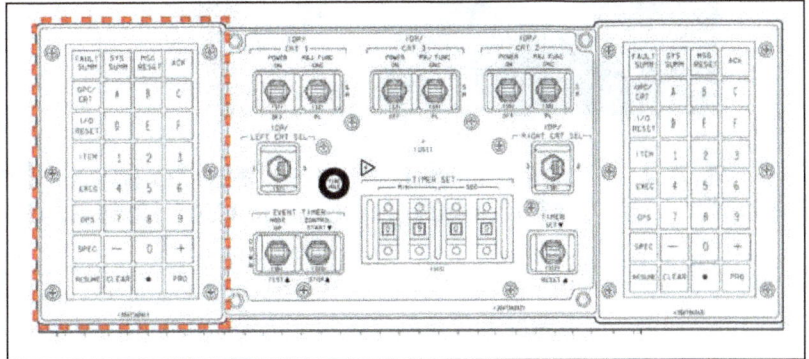

Figure 1.2-11. *Entry keypad* (red dashes) *on the C2 panel.* [Adapted from the Space Shuttle Systems Handbook]

13:37:44 (EI–385)   👤 🎞   CDR/Seat 1: "And thanks for that, Houston. We gave you an ITEM 27. We bumped the stick earlier."

           👤 🎞   CAPCOM: "Not a problem, Rick."

13:38:50 (EI–319)   👤 🎞   CDR/Seat 1 enters the OPS 304 command into the queue. OPS 304 is the *Entry* Major Mode software.

13:38:56 (EI–313)   👤 🎞   CDR/Seat 1: "And, Houston, we'll get the 304 at 5 minutes."

13:39:04 (EI–305)   👤 🎞   CAPCOM: "Rick, we're ready for OPS 304."

13:39:09 (EI–300)   👤 🎞   **OPS 304.** The CDR/Seat 1 executes OPS 304. [*Flight deck video, Telemetry, TDRS-W data – cathode ray tube (CRT) 1, Master Timeline, Rev. 15*]

13:40:12 (EI–237)   👤 🎞   PLT/Seat 2 gloves are observed ON and MATED (figure 1.2-12).

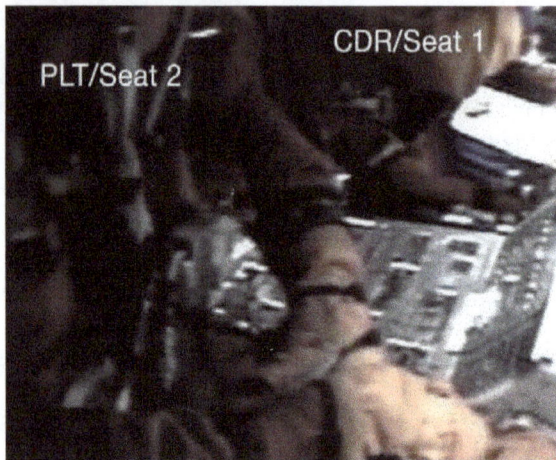

Figure 1.2-12. *Pilot/Seat 2 observed with gloves ON and MATED.*

1.2.2   Phase 2: Entry interface to loss of signal
[GMT 13:44:09 (EI) through GMT 13:59:32 (EI+923)]

13:44:09 (EI)   **EI.** This is the point where the orbiter is considered to be first encountering the atmosphere. (GPS derived) [*Master Timeline, Rev. 15 Baseline*]

                                       Alt   = 400,000 feet [per definition of EI]
                                       Mach = 24.57 (*Master Timeline, Rev. 15 Baseline*]
                                       Qbar = ~0.01 psf [modeling]

13:44:15 (EI+006)   CDR/Seat 1 states, "Just past EI."

13:44:58 (EI+049)   CDR/Seat 1 request for everyone to check suit pressure integrity.

13:45:13 (EI+064)   CDR/Seat 1 observed with helmet visor down and latched in preparation for the suit pressure check and the communication check. MS2/Seat 4 visor observed OPEN and both gloves OFF (figure 1.2-13).

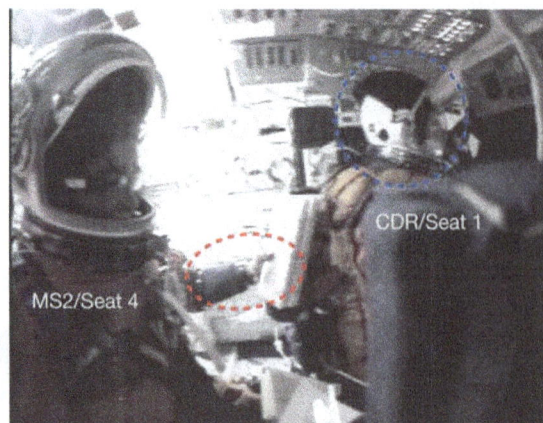

Figure 1.2-13. *Commander/Seat 1 visor down and latched* (blue dashed circle) *and Mission Specialist 2/Seat 4 with helmet on and left glove off* (red dashed circle).

Time		Event
13:45:24 (EI+075)	👤 🎞	**Suit Pressure/Communications (COMM) Check.** CDR/ Seat 1, PLT/Seat 2, and MS4/Seat 3 complete COMM check. Visors are DOWN and locked during this check. Analysis of the O_2 supply pressure telemetry identified the O_2 supply pressure drop from the CDR/Seat 1, PLT/Seat 2, and MS4/Seat 3 suit pressure check.
	👤 🎞	During the suit pressure integrity check, the CDR/Seat 1, PLT/Seat 2, and/or MS4/Seat 3 microphone activated the intercom system. Breathing is heard for about 20 seconds. CDR/Seat 1 comments, "It's noisy in there, isn't it?"
13:45:50 (EI+101)	👤 🎞	MS4/Seat 3 states going visor UP after suit pressure check.
13:46:18 (EI+129)	👤 🎞	PLT/Seat 2 is observed with visor OPEN.
13:46:48 (EI+159)	🛰	Mach = 24.66 [*Master Timeline, Rev. 15 Baseline*] Qbar = 0.5 psf [*Master Timeline, Rev. 15 Baseline*]
13:47:13 (EI+184)	👤 🎞	**Accelerometer Bit Flip.** PLT/Seat 2 states that he observed a bit-flip on the accelerometer, indicating that the entry deceleration loads are starting to build as expected and were finally large enough to be registered by the vehicle accelerometers.
13:47:20 (EI+191)	👤 🎞	CDR/Seat 1 states to crew, "We're at a hundredth of a G."
13:47:38 (EI+209)	👤 🎞	CDR/Seat 1 and PLT/Seat 2 are observed with gloves still ON and MATED (red circles in figure below), visors OPEN. MS2/Seat 4 is observed starting to don gloves (yellow circle in figure) and is also observed with left glove ON but NOT MATED and right glove OFF (figure 1.2-14).

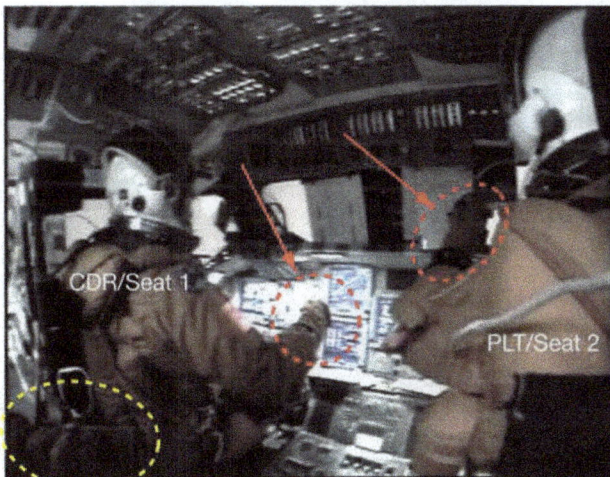

Figure 1.2-14. *Mission Specialist 2/Seat 4 donning left glove; Commander/Seat 1 and Pilot/Seat 2 with gloves ON and MATED.*

13:47:51 (EI+222)   🔺📽 MS4/Seat 3 with right glove still ON and MATED (figure 1.2-15).

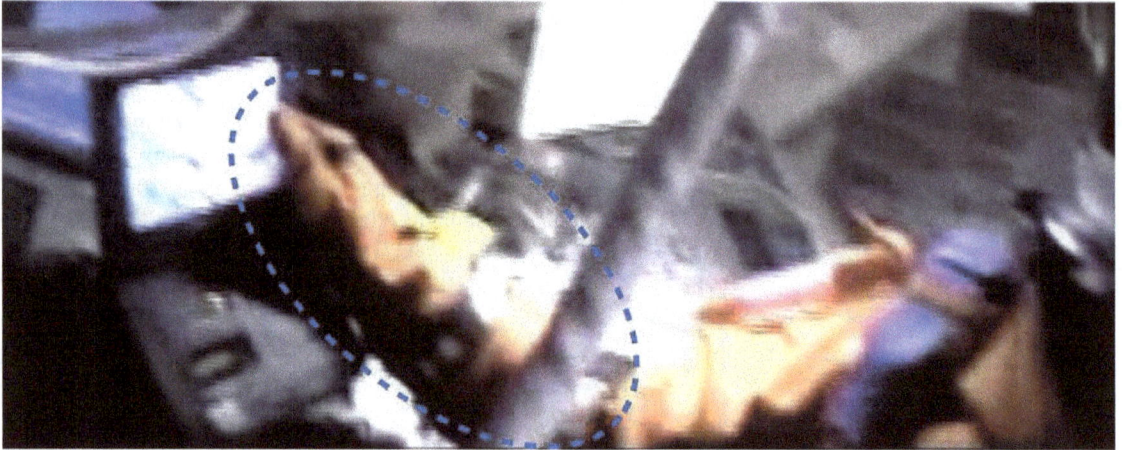

Figure 1.2-15. *Mission Specialist 4/Seat 3 right glove still ON and MATED.*

13:47:52 (EI+223)   🔺 Mach = 24.66 [*Master Timeline, Rev. 15 Baseline*]
Qbar = 2.0 psf [*Master Timeline, Rev. 15 Baseline*]

🔺 **Elevon and Body Flap Active.** When the Qbar increases to 2 psf, the elevons and body flap aerodynamic control surfaces become effective for controlling the vehicle and are added as active effectors to the vehicle control logic.

13:48:45 (EI+276)   🔺📽 **End of Recovered Flight Deck Video.** This is the last frame with a discernable image (figure 1.2-16).

PLT/Seat 2
right window
(window 5)

Figure 1.2-16. *Last discernable image from recovered flight deck video.* [Window 5 and the right side of the crew module visible]

TIME		EVENT

13:49:16 (EI+307) ⚕ Mach = 24.57 [*Master Timeline, Rev. 15 Baseline*]
Qbar = ~10.0 psf [*Master Timeline, Rev. 15 Baseline*]

⚕ **Roll Jets Deactivated.** When the Qbar increases to 10 psf, the roll jets are removed from the control logic. [*Master Timeline, Rev. 15 Baseline*]

13:49:32 (EI+323) ⚕ **Start of First Planned Roll to the Right for Energy Management.** [*Master Timeline, Rev. 15 Baseline*] (figure 1.2-17).

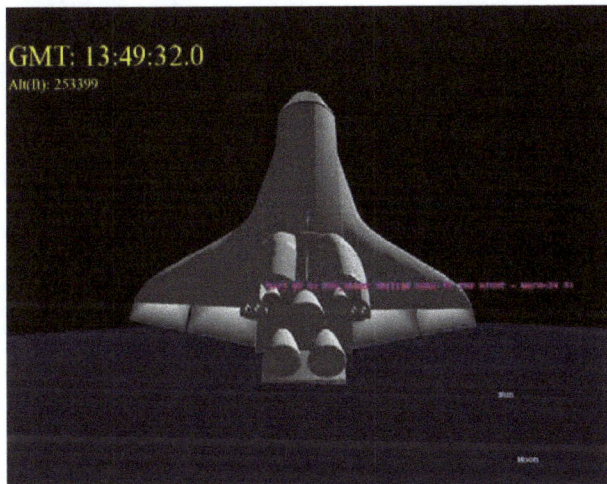

Figure 1.2-17. *Attitude of* Columbia *at the start of the first planned roll.*

⚕ Mach = 24.51 [*Master Timeline, Rev. 15 Baseline*]
Qbar = ~15 psf [modeling]

13:50:00 ⚕ **Completion of First Planned Roll to the Right for Energy Management.** [*Master Timeline, Rev. 15 Baseline*] (figure 1.2-18).

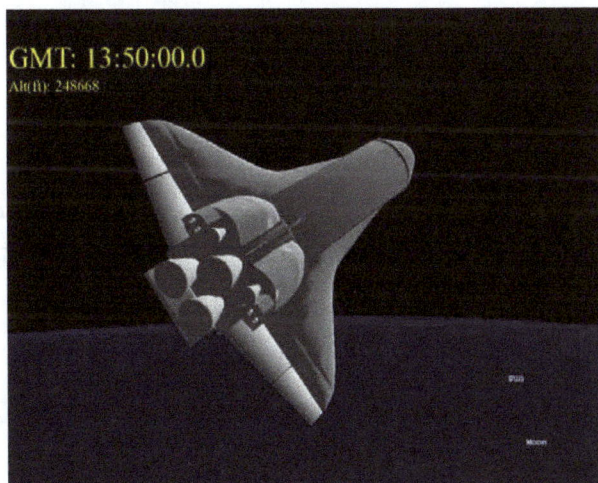

Figure 1.2-18. *Attitude of* Columbia *at the end of the first planned roll.*

TIME		EVENT

13:50:30 (EI+381)   ⚛   First indication of entry heating. A thermal sensor measurement and stimulus indication (MSID) V09T1702A in the aft fuselage center bottom bond line registers a normal rise in temperature due to entry heating. [*Master Timeline, Rev. 15 Baseline*]

13:50:53 (EI+404)   ⚛   Start of peak heating. [*Master Timeline, Rev. 15 Baseline*]

                                ⚛   Alt   = 243,048 feet [*Master Timeline, Rev. 15 Baseline*]
                                              Mach   = 24.12 [*Master Timeline, Rev. 15 Baseline*]
                                              Qbar   = ~22 psf [modeling]

13:51:19 -13:52:49   ⚛   Nominal yaw jet firings were occurring during this time (GMT 13:51:19 (EI+430) to GMT 13:52:49 (EI+520)). When the yaw jet(s) fires, an indicator on the F6 panel will illuminate while the jet(s) is on (see arrows on figure 1.2-19). [*Master Timeline, Rev. 15 Baseline*]

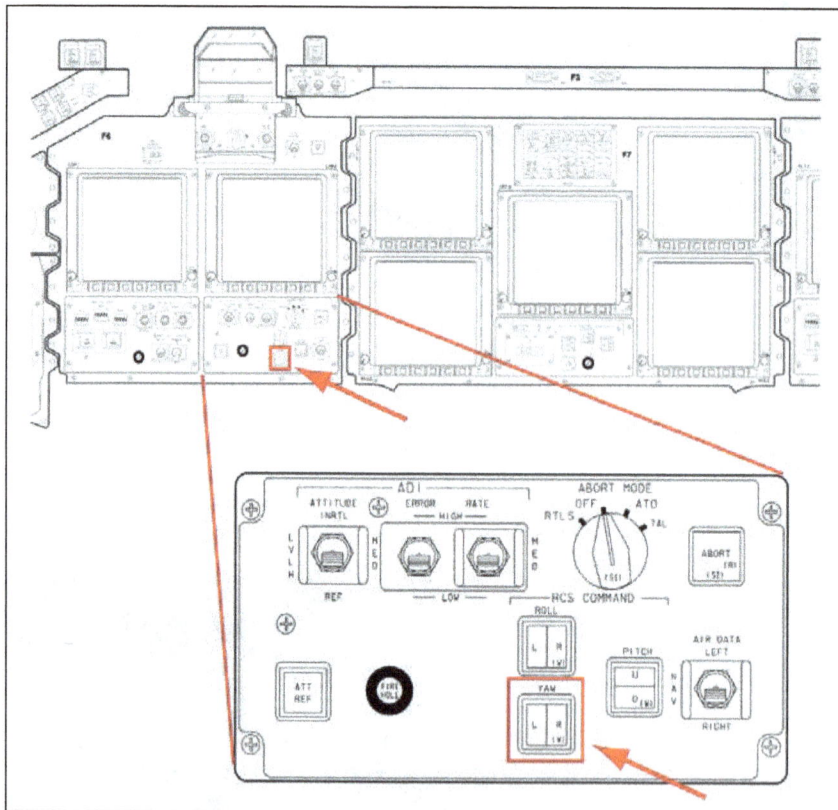

Figure 1.2-19. *Location of the yaw jet indicator light.* [Adapted from the Space Shuttle Systems Handbook]

TIME		EVENT

13:51:46 (EI+457) ⚠ Inertial sideslip angle (beta) goes negative (yaw to port/ left) and stays negative until the LOS from *Columbia* at GMT 13:59:32 (EI+923) (figure 1.2-20). [*Master Timeline, Rev. 15 Baseline*]

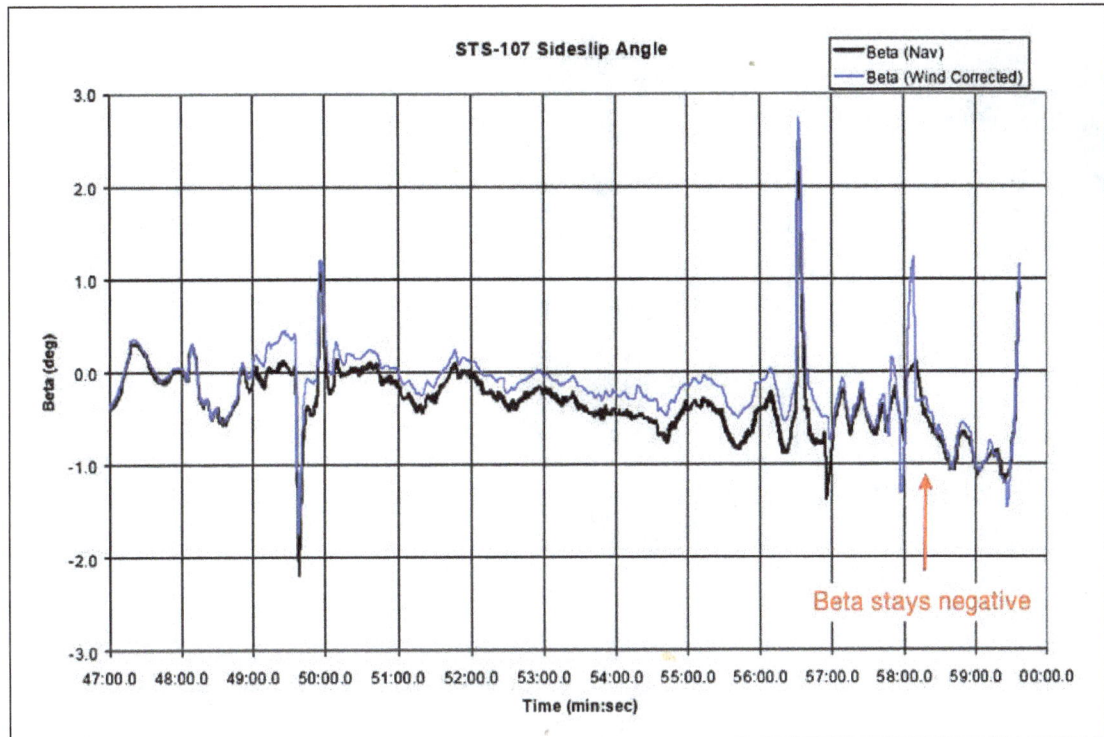

Figure 1.2-20. *STS-107 sideslip angle vs. time.*[4]

13:52:05 (EI+476) ⚠ **Yaw-Moment Changed.** Post-accident analysis determined that this was the first clear indication of off-nominal aerody-namics. This information was not available/visible to the crew or controllers in real time. [*Master Timeline, Rev. 15 Baseline*]

13:52:17 (EI+488) ⚠ Left main landing gear brake line temperature sensor (MSID V58T1703A), which was located on the inboard sidewall of the wheel well, starts to indicate an off-nominal temperature rise rate. This is the first indication of off-nominal system readings in the left wing. The information is not visible to the crew. [*Master Timeline, Rev. 15 Baseline*]

⚠ **Damage to Left Wing Begins – NLT Time.** Post-accident analysis determined that damage inside the left wing began NLT this time.[5]

[4]Integrated Entry Environment Team Final Report, May 30, 2003, Figure I-21, p. I-12.
[5]*Columbia* Accident Investigation Board Report, Volume I, August 2003, pp. 68 and 71.

⚶ Approximately 300 miles west of the California coast (figure 1.2-21)

Alt = 236,791 feet [*Master Timeline, Rev. 15 Baseline*]
Mach = 23.58 [*Master Timeline, Rev. 15 Baseline*]
Qbar = ~27 psf [modeling]

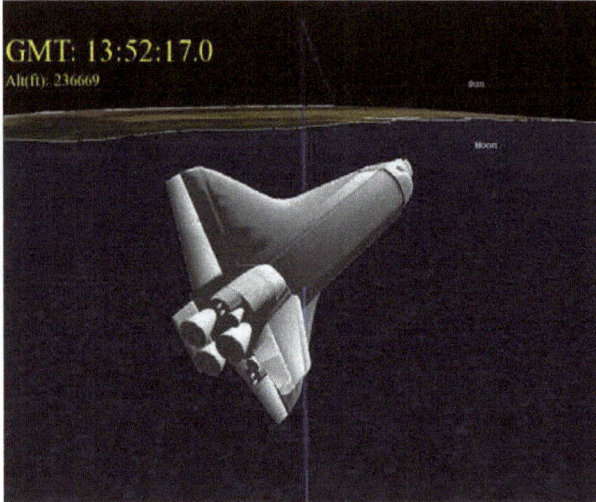

GMT: 13:52:17.0
Alt(ft): 236669

Figure 1.2-21. Columbia *approaching the California coastline.*

13:53:00 (EI+531) ⚶ Qbar = ~29 psf

13:53:01 (EI+532) ⚶ Off-nominal rolling moment. Post-accident analysis identifies the first clear indication of off-nominal rolling moment. Start of steady growth in roll moment. [*Master Timeline, Rev. 15 Baseline*]

13:53:10 (EI+541) ⚶ **First Indication in MCC of Off-nominal Readings.** Four hydraulic return line temperature sensors in the left wing went off-scale low (OSL) between GMT 13:53:10 (EI+541) and GMT 13:53:36 (EI+557). OSL refers to a reading that is below the lower display limit. When several sensors go OSL it usually indicates a suite of sensors has failed. These sensor failures were described in the CAIB Report. This information was not available to the crew.

13:53:15 (EI+546) ±2 sec ▤ At GMT 13:53:15 (EI+546) ±2 sec, ground-based video coverage of *Columbia* is acquired by videographers who are unassociated with NASA (figure 1.2-22).

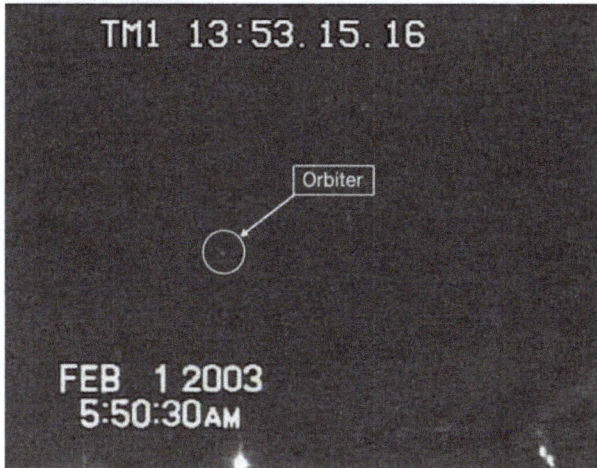

Figure 1.2-22. *This is the first frame of ground-based video.* Columbia *is circled. Time 1 (TM1) shows the Greenwich Mean Time. The Pacific Standard Time displayed on the lower portion of the image is inexact.*

13:53:26 (EI+557)   ⚓   *Columbia* crosses the California coastline west of Sacramento. [*Master Timeline, Rev. 15 Baseline*] (figure 1.2-23).

Figure 1.2-23. Columbia *crossing the California coastline.*

⚓   Alt   = 231,600 feet [*Master Timeline, Rev. 15 Baseline*]
Mach = 23.0 [*Master Timeline, Rev. 15 Baseline*]
Qbar = ~30 psf [modeling]

13:53:38 (EI+569)   ⚓   Sideslip angle exceeded all previous flight experience.[6]

13:53:46 (EI+577) ±2 sec   ⚓ 📖   **First Observed Incident of Debris Being Shed.** This is most likely a piece of the left wing. Debris 1 is seen just aft of the orbiter envelope 1 second after a trail anomaly that consisted of a noticeably luminescent section of the plasma trail. There were no reported entry observations while *Columbia* was over the Pacific Ocean prior to GMT 13:53:15 (EI+546), so it is

---

[6]Integrated Entry Environment Team Final Report, May 30, 2003, p. 15.

not possible to know whether debris was being shed before this time. Image of Debris 1 is 0.6 second after first visual detection of it. Analysis indicates that the mass of the debris was probably less than 8 lbs. Neither the crew nor MCC personnel were aware of the shedding events (blue arrow shows direction of flight) (figure 1.2-24).

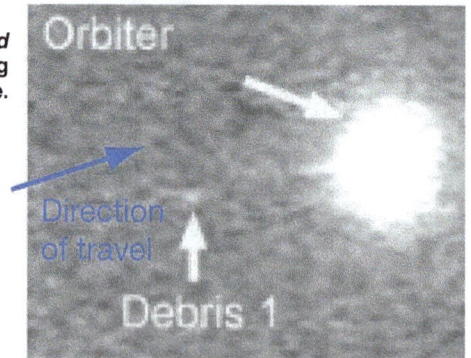

Figure 1.2-24. *Video capture of the first observed incident of debris being shed.* The orbiter is traveling from left to right in this image.

13:54:20 (EI+611) ±10 sec   The beginning of the slow elevon trim change starts at GMT 13:54:20 (EI+611) ±10 sec. While there is a display that shows the aileron trim movement (figure 1.2-25), the initial change was so small that it would not be detectable by the crew. [*Master Timeline, Rev. 15 Baseline*]

Figure 1.2-25. *Elevon trim display.* [Picture of the cockpit and display from the Shuttle Engineering Simulator]

13:54:24 (EI+615)   **MCC Team Made Aware of Off-nominal Readings.** The MMACS flight controller makes a call over the MCC voice loops that the four hydraulic return line temperature sensors in the left wing went OSL at GMT 13:53:10 (EI+541).

13:54:25 (EI+616)   *Columbia* crosses the California-Nevada border. [*Master Timeline, Rev. 15 Baseline*] (figure 1.2-26).

Figure 1.2-26. Columbia *crossing the California-Nevada border.*

       ⚙   Alt   = 227,400 feet [*Master Timeline, Rev. 15 Baseline*]
                     Mach = 22.5 [*Master Timeline, Rev. 15 Baseline*]
                     Qbar   = ~34 psf [modeling]

13:54:30 (EI+621)   ⚙   Indication of a second suit pressure check by three to five crew members (figure 1.2-27).

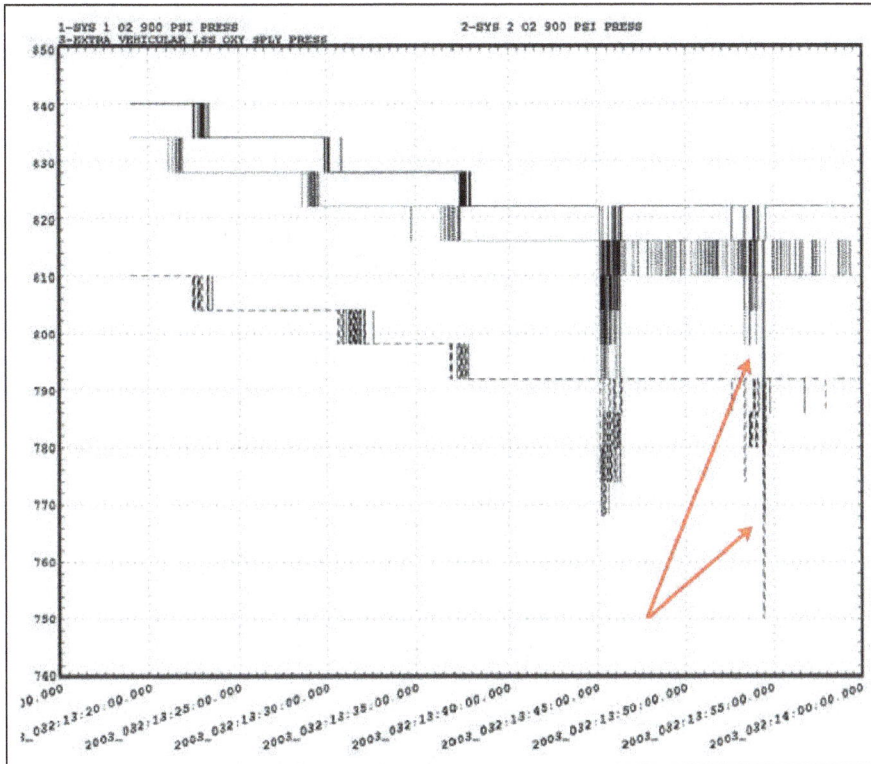

Figure 1.2-27. *Plot of main oxygen supply pressure showing second series of suit pressure checks* (red arrows).

TIME		EVENT

13:54:32 (EI+623) ⚠ 🎞 Debris 6 released. The brightest debris-shedding event occurring in this phase, Debris 6, is first visible on video at GMT 13:54:36 (EI+627) (figure 1.2-28). Ballistic estimates determined that the actual release time was 4 seconds earlier. Luminosity measurements and calculated rates of deceleration were used to determine that the mass was probably a few hundred pounds.[7] There were no data from sensors, instrumental indications, or apparent crew recognition of this debris loss.

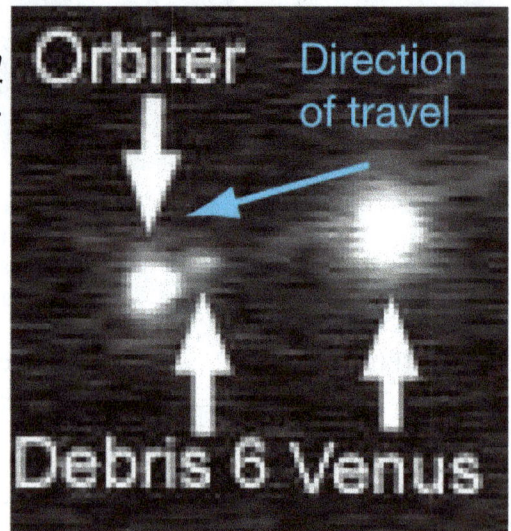

Figure 1.2-28. *The brightest debris event known to have occurred prior to loss of signal.* The orbiter is traveling from right to left.

13:54:33 (EI+624) ⚠ At GMT 13:54:33.52 (EI+624.52), RCS yaw jet R3R fired followed by RCS yaw jet R2R at GMT 13:54:33.54 (EI+624.54). It is unknown whether the jet firings were in response to the debris-shedding event. [*Master Timeline, Rev. 15 Baseline*]

13:54:36 (EI+627) ⚠ 🎞 Debris 6 first visible on ground-based video.

13:55:21 (EI+672) ⚠ Atmospheric drag on the orbiter was producing the nominal deceleration load on the crew of approximately 0.3 G.

13:56:02 (EI+713) ⚠ Aft RCS pitch jets deactivated when Qbar reached 40 psf. At 40 psf, the elevons and body flap have sufficient control authority to control the pitch of the orbiter.

13:56:30 (EI+741) ⚠ **First Roll Reversal Initiated.** *Columbia* initiated a roll from right wing low to left wing low (figure 1.2-29).

---

[7]*Columbia* Accident Investigation Board Report, Volume III, Appendix E.2, STS-107 Image Analysis Team Final Report, October 2003.

Figure 1.2-29. *Attitude at the start of the roll reversal.*

13:56:55 (EI+766) 🛩 **First Roll Reversal Completed.** Left wing low (figure 1.2-30).

Figure 1.2-30. *Attitude at the completion of the roll reversal.*

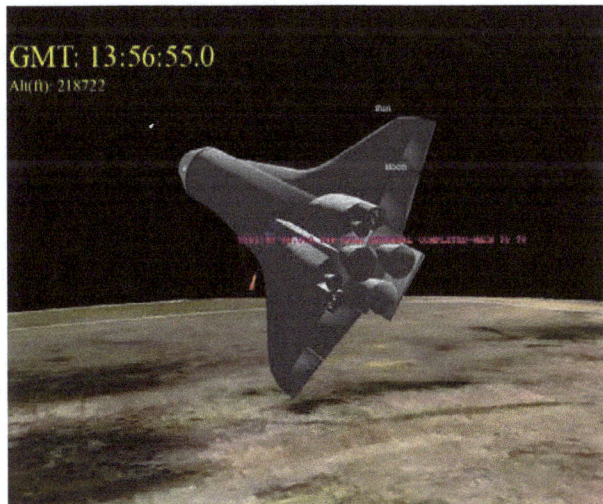

🛩 Alt = 218,817 feet [*Master Timeline, Rev. 15 Baseline*]
Mach = 20.76 [*Master Timeline, Rev. 15 Baseline*]
Qbar = ~43 psf [modeling]

13:57:14 (EI+785) ±1 sec 🛩 🎞 **Starfire Image.** An image of the shuttle was taken while it was over the Starfire Optical Range located at the Kirtland Air Force Base, New Mexico. The photograph has been enhanced, and a wireframe representation of the orbiter has been overlaid. The bright areas behind the shuttle could be traced to orbiter sources and, thus, were generally considered to be nominal, as was the asymmetry that was seen in the bright gas around the nose area. The bulges that were seen (as marked) at the left wing could not be explained by a nominal condition (figure 1.2-31).

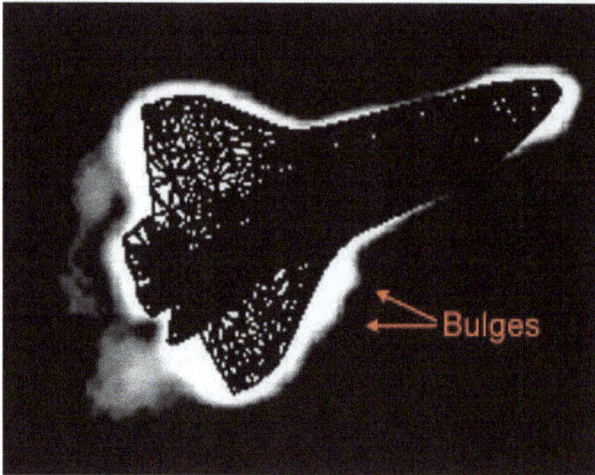

Figure 1.2-31. *Infrared photograph of the shuttle during entry.*[8]

TIME		EVENT
13:57:54 (EI+825) ±1 sec	🛦 🎞	Flare 1 is an asymmetrical brightening of the orbiter shape at GMT 13:57:54.7 (EI+825) ±1 sec. This brightening was detected in images taken by a charge coupled device camera on a telescope, so its image is well magnified. Another brightening event was detected 6 seconds later (see GMT 13:58:00 (EI+831)). The two small images are the raw images from the data. The left image is interpreted to be the orbiter in a nominal condition. The image on the right is 0.4 second later. The larger image has been rotated to its correct viewing orientation and enhanced; a wireframe model of the orbiter at approximately the correct scale and orientation has been overlaid (figure 1.2-32).

---

[8]*Columbia* Accident Investigation Board Report, Volume V, Appendix G.7, Starfire Team Final Report, June 3, 2003, Figure 3, p. 360.

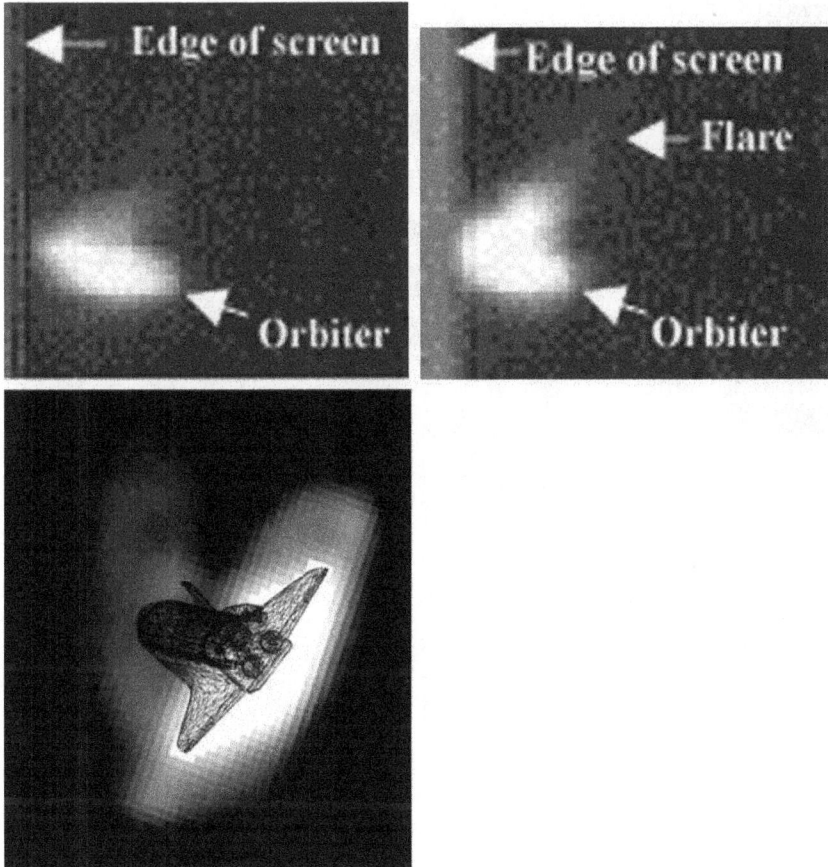

Figure 1.2-32. *Enhanced video captures showing the first observed flare in the trail.*[9]

13:58:00 (EI+831) ±1 sec   🔺🎞   Flare 2 is an asymmetrical brightening of the orbiter shape at GMT 13:58:00.5 (EI+831.5) ±1 sec (figure 1.2-33).

[9]*Columbia* Accident Investigation Board Report, Volume V, Appendix G.7, Starfire Team Final Report, June 3, 2003, Figure 6, p. 361 and Figure 6A, p. 362.

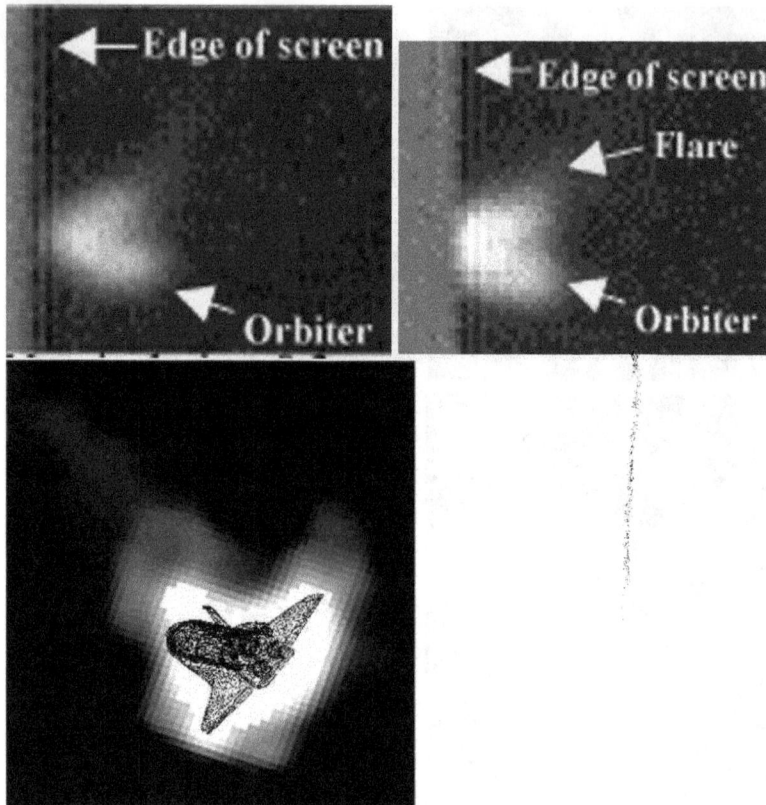

Figure 1.2-33. *Enhanced video captures showing the second observed flare in the trail.*

	Mach	= 19.8 [modeling]
	Qbar	= ~53 psf [modeling]

13:58:03 (EI+834)   ⚛   **Start of "Sharp" Elevon Trim Increase.** The FCS is now compensating for increasingly asymmetric aerodynamic loading and is commanding the elevon trim at a much higher than nominal rate. Time is uncertain (±10 sec) (figure 1.2-34). [*Master Timeline Rev. 15*]

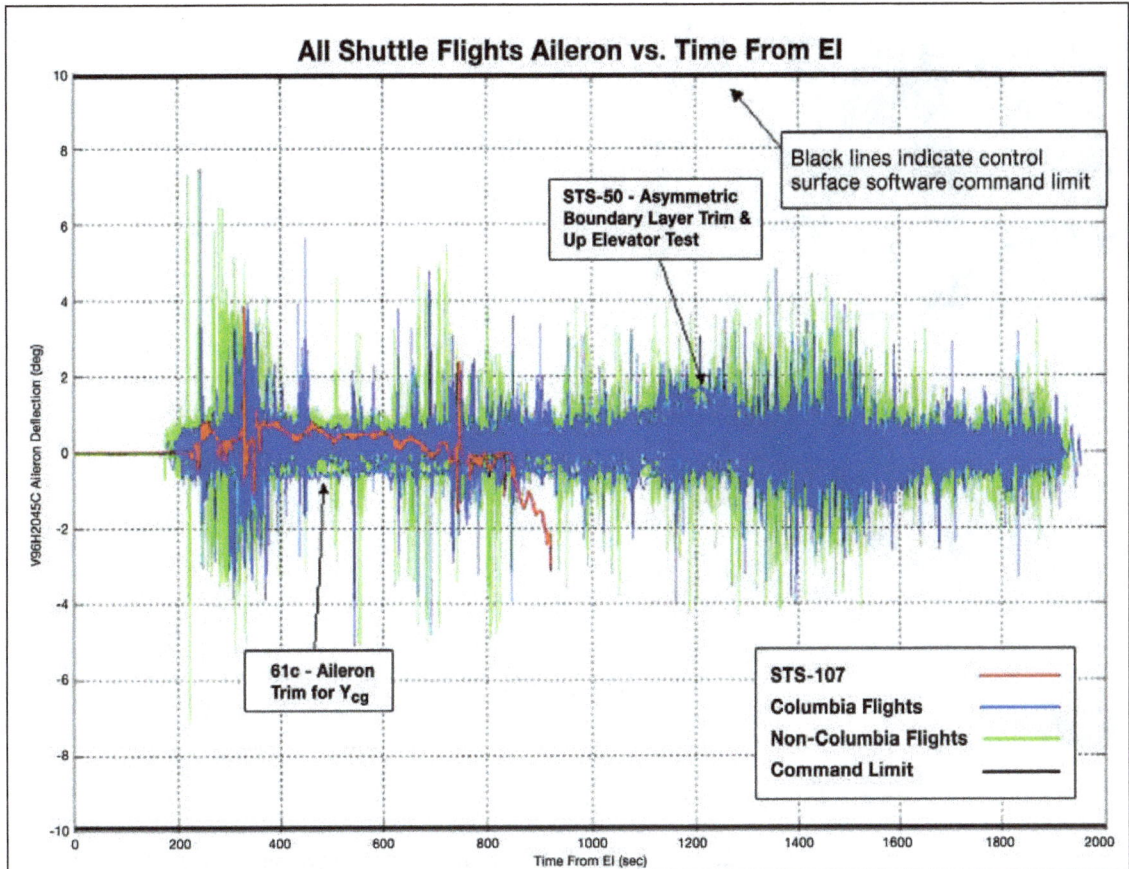

Figure 1.2-34. *The sharp divergence from previous flight experience starts at approximately GMT 13:58:19 (EI+850).*[10]

	⚠	Alt = 212,007 feet [*Master Timeline, Rev. 15 Baseline*] Mach = 19.77 [*Master Timeline, Rev. 15 Baseline*] Qbar = ~54 psf [modeling]
13:58:12 (EI+843)	⚠ 🎞	End of ground-based video coverage for western portion of entry.
13:58:21(EI+852) ±5 sec	⚠	**Littlefield Tile Released**. This piece of tile was recovered in Littlefield, Texas; it is the westernmost piece of recovered debris (figure 1.2-35).

---

[10]Integrated Entry Environment Team Final Report, May 30, 2003, Figure 6.6-1, p. 30.

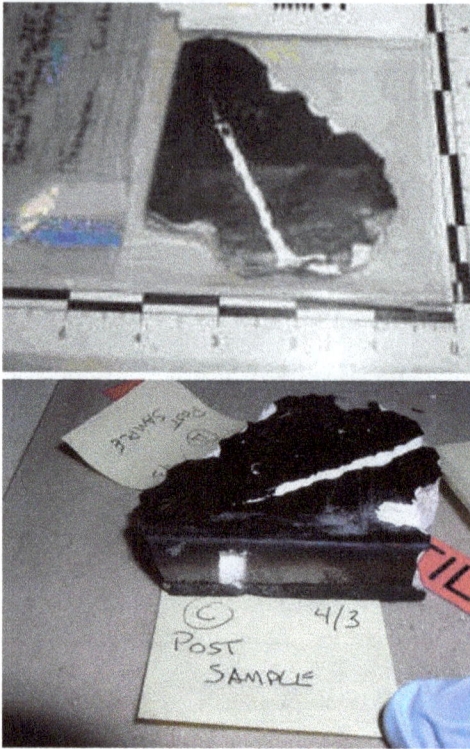

Figure 1.2-35. *The Littlefield tile.* [Picture from the *Columbia* Reconstruction Database, debris item no. 14768]

13:58:39 (EI+870)   ⚠ 👤   TIRE PRESS LOB. This message, indicating a left outboard (LOB) tire pressure fault, was recorded by the backup flight software (BFS) in the downlink stack and had a time of GMT 13:58:39.94 (EI+870.94). The fault message was annunciated on the crew displays and with an audio tone. [*TDRS-W data; Master Timeline, Rev. 15 Baseline*]

13:58:40 (EI+871)   ⚠ 👤   Main landing gear left inboard (LIB) tire pressure reading went OSL. Downlink telemetry from the GPC recorded that the BFS was indicating a tire pressure fault message. The fault message was annunciated on the crew displays and with an audio tone. [*TDRS-W data; Master Timeline, Rev. 15 Baseline*]

13:58:41 (EI+872)   ⚠ 👤   TIRE PRESS LIB. This message, indicating an LIB tire pressure fault, was recorded by the BFS in the downlink stack and had a time of GMT 13:58:41.84 (EI+872.84). The fault message was annunciated on the crew displays and with an audio tone. [*TDRS-W data; Master Timeline, Rev. 15 Baseline*]

TIME		EVENT

3:58:48 (EI+879) — A partial voice transmission from *Columbia* is received over the A/G: "And, uh, Hou…" The vehicle and crew were still performing nominally.[11]

— CRT3: FAULT SUMM. The crew called up the fault summary display to look at the messages. [*TDRS-W data*]

13:58:49 (EI+880) — BFS FSM: SM0 TIRE PRESS LIB. This message, indicating an LIB tire pressure fault, was recorded by the BFS in the down-link stack and had a time of GMT 13:58:49.54 (EI+880.54). The fault message was annunciated on the crew displays and with an audio tone. [*Master Timeline, Rev. 15 Baseline*]

13:58:56 (EI+887) — BFS FSM: SM0 TIRE PRESS LOB. This message, indicating an LOB tire pressure fault, was recorded by the BFS in the downlink stack and had a time of GMT 13:58:56.26 (EI+887.26). The fault message was annunciated on the crew displays and with an audio tone. [*Master Timeline, Rev. 15 Baseline*]

13:59:06 (EI+897) — Telemetry records the LEFT MAIN GEAR DOWN-lock sensor transferred to ON. This indicated that the left main landing gear was down and locked in the deployed position. Other sensors indicated that the landing gear door was still closed and the landing gear was locked in the stowed position. The mixed signals would result in the landing gear position indicator for the left gear displaying a "barber pole," which would indicate an indeterminate gear position. Analysis of the data and recovered debris indicates that the landing gear was locked in the stowed position and the landing gear door was closed. The signal, indicating that the gear was down, was a false signal that was likely triggered by damage to the sensor system (sensor, wiring harness, etc.) (figure 1.2-36).

---

[11]*Columbia* Accident Investigation Board Report, Volume I, August 2003, p. 42.

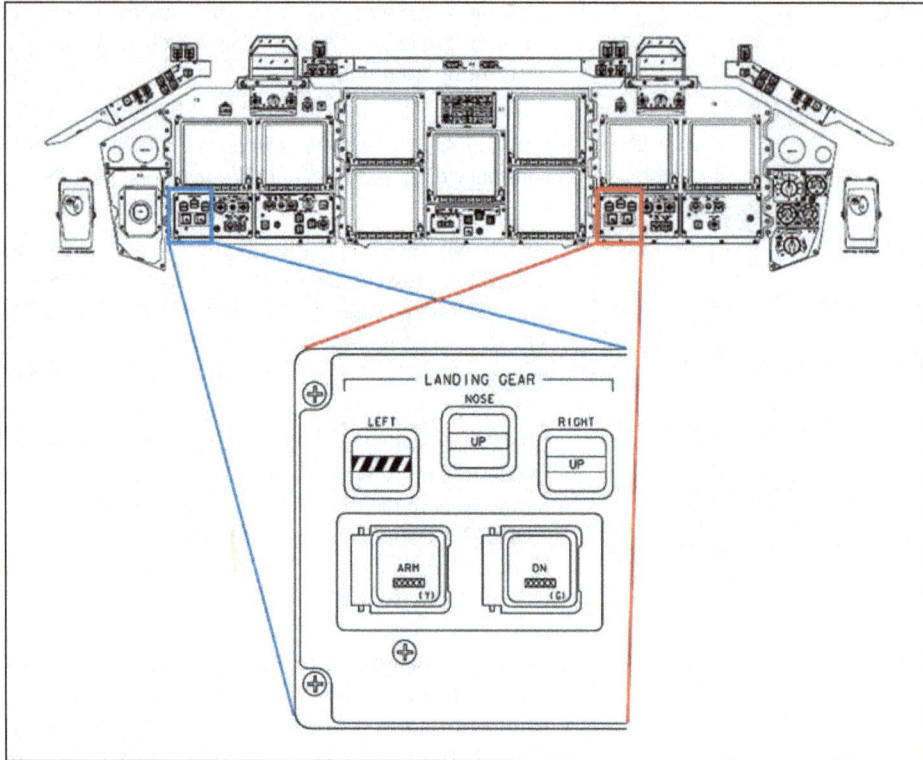

Figure 1.2-36. *Landing gear indicator panel, identical on both sides of the flight deck forward display panels.* Left indicator showing "barber pole" (indeterminate position). [Adapted from the Space Shuttle Systems Handbook]

13:59:24 (EI+915)	🚶	The MCC calls *Columbia* regarding the tire pressure fault message, "And, *Columbia*, Houston, we see your tire pressure message and we did not copy your last call."[12] [*STS-107 A/G recording*]
13:59:29 (EI+920)	🛰	Aerodynamic control authority is exceeded. Damage to the left wing exceeds the elevon control surface ability to compensate. The aileron trim deflection derived from flight data shows that the aileron trim rate did reach the 4.2 deg/sec rate limit. This indicates that the amount of lateral control (aileron and yaw RCS jets) that was required to trim the vehicle was constantly increasing and by LOS+5 seconds was quickly approaching the limits of the FCS.[13]

Change in vehicle attitude from GMT 13:57:08 (EI+779) to GMT 13:59:29 (EI+920). |
| 13:59:30 (EI+921) | 🛰 | **RCS Yaw JETS FIRING. RCS Yaw Jets (R2R and R3R) Begin Firing Continuously.** Aft right RCS yaw jets R2R and R3R started firing at GMT 13:59:30.66 (EI+921.66) and GMT 13:59:30.68 (EI+921.68), respectively, to correct an |

---

[12]*Columbia* Accident Investigation Board Report, Volume I, August 2003, p. 43.
[13]Integrated Entry Environment Team Final Report, May 30, 2003, p. 20.

increasing yaw to the left, and were still firing when all data were lost at GMT 13:59:37.4 (EI+928.4).

The only indication the crew would have that the jets were firing is from status lights on the F6 panel underneath the CDR2 display and the fuel quantity display on the overhead panel decrementing (figure 1.2-37).

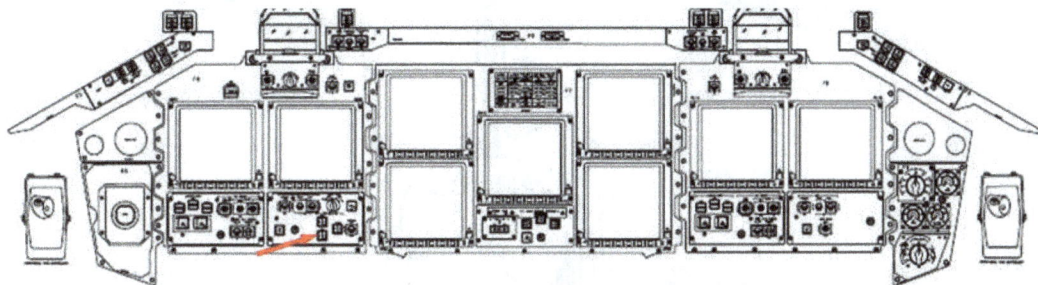

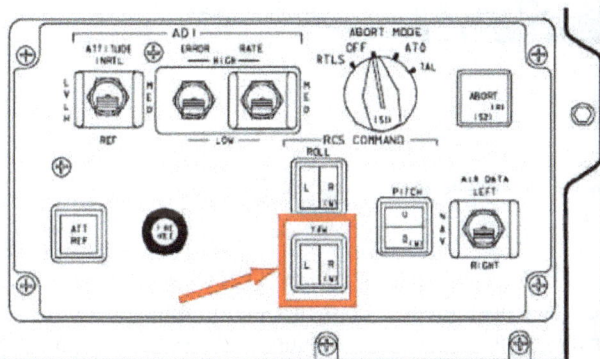

Figure 1.2-37. *Reaction Control System thruster status display.* [Pictures [top and bottom left) adapted from the Space Shuttle Systems Handbook; picture (bottom right) from the Shuttle Engineering Simulator]

13:59:31 (EI+922)   Fault Summary Page (FSP) Message Downlink Stack (last five messages) (figure 1.2-39):

```
FSP1: SM0    TIRE PRESS   L  OB   32/13:58:56.26
FSP2: SM0    TIRE PRESS   L  IB   32/13:58:49.54
FSP3: SM0    TIRE PRESS   L  IB   32/13:58:41.48
FSP4: SM0    TIRE PRESS   L  OB   32/13:58:39.94
FSP5: MPS[14] PNEU           REG   32/13:58:04.42
```

Last observed elevon deflections (figure 1.2-38):[15]

---

[14]MPS = Main Propulsion System.
[15]Integrated Entry Environment Team Final Report, May 30, 2003, p. 15.

Left: −8.11 deg (deflected up)
Right: −1.15 deg (deflected up)

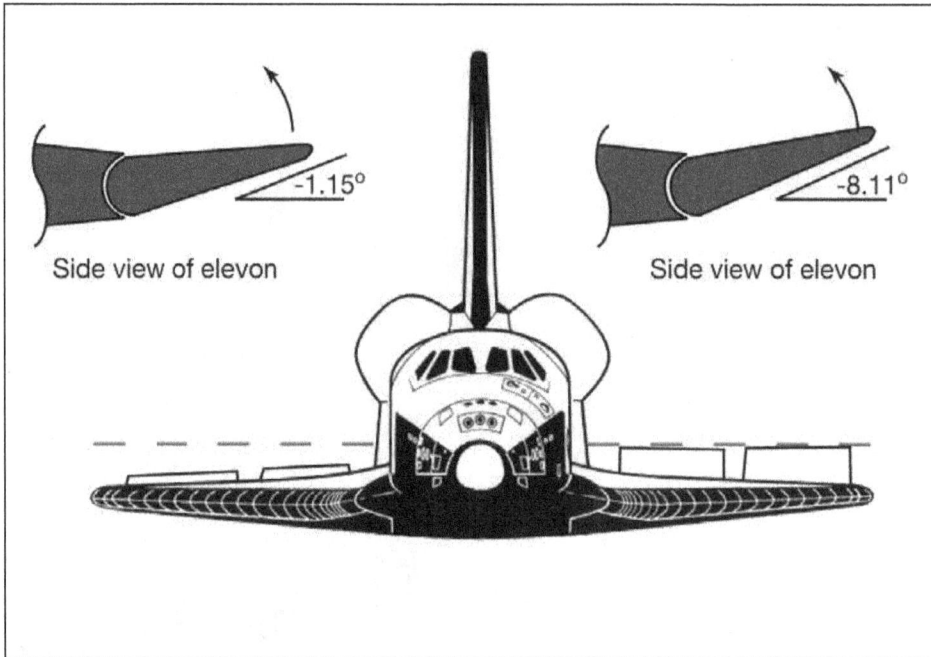

Figure 1.2-38. *Illustration of the last observed elevon deflections.*

At GMT 13:59:31.4 (EI+922.4) the FCS Channel 4 aerosurface position measurements start trending towards their null values, indicating a failure of the sensor due to a wiring short. This is the first indication of the eventual bypass of FCS Channel 4. [*Master Timeline, Rev. 15 Baseline*]

At GMT 13:59:31.478 (EI+922.478) all of the FCS Channel 4 bypass valves close (i.e., bypassed condition). This is a leading indicator of an aeroservo actuator failure. [*Master Timeline, Rev. 15 Baseline*]

1.2.3   Phase 3: Loss of signal to Catastrophic Event
[GMT 13:59:32 (EI+923) – GMT 14:00:18 (EI+969)]

13:59:32 (EI+923)   Near Dallas, Texas
                                     Alt   = ~200,700 feet [*Master Timeline, Rev. 15 Baseline*]
                                     Mach   = ~18.1 [*Master Timeline, Rev. 15 Baseline*]
                                     Qbar   = ~70 psf [modeling]

TIME	EVENT

Loss of Voice Communications. The last audio transmission from *Columbia*, "Roger, [truncated mid-word]…"[16] was cut off at GMT 13:59:32.136 (EI+923.136). A short-duration loss of voice communications was not unexpected because it coincided with the approximate time at which the on-board communications systems was to switch from the West TDRS to either the East TDRS or the ground station at Kennedy Space Center. [*STS-107 A/G recording*][17]

LOS. The last valid downlink frame was accepted by the Orbiter Data Reduction Complex at GMT 13:59:32.136 (EI+923.136).[18] The following measurements were recorded in the MCC at the time of LOS:

- Cabin pressure = 14.64 pounds per square inch absolute (psia)
- Cabin temperature =71.6°F (22°C)
- Humidity = 37.9%
- ppO_2 levels (three sensors)
    3.14 psia (sensor A),
    3.14 psia (sensor B),
    3.16 psia (sensor C)
- Pressure change rate (DP/dt) = 0.004 psi/minute (within sensor bias limit for 0)
- Partial pressure of carbon dioxide ($ppCO_2$) = 1.96 mmHg
- Cabin temperature setting = full cool
- Nitrogen (N_2) supply pressures
    1011 psia (System 1),
    1067 psia (System 2)
- O_2 supply pressures
    822 psia (System 1),
    809 psia (System 2)

Start of First Period of RGPC-1 Data. Data reconstructed from GMT 13:59:32.136 (EI+923.136) to GMT 13:59:37.396 (EI+928.396).

It is possible that the crew noticed the aileron trim increasing. Per procedures, no crew action would be required until the trim reached 3 degrees.

There is an increase in the roll error as the orbiter uses roll control to correct the yaw error and rate. [*RGPC*]

---

[16]*Columbia* Accident Investigation Board Report, Volume I, August 2003, p. 43.
[17]*Columbia* Accident Investigation Board Report, Volume I, August 2003, p. 43.
[18]STS-107 Investigation Action Response: OVE-204 Crew Inputs After Loss of COMM (Voice); CAIB-MRT-00099, 03/10/2003.

TIME		EVENT

The FCS Channel 4 problems that were developing at GMT 13:59:31 (EI+922) had progressed to the point that the FCS Channel 4 fail flags tripped (1 Hz) on all aerosurface actuators. [*Master Timeline, Rev. 15 Baseline; RGPC*]

During the RGPC-1 period, the vehicle remains in AUTO guidance and control.

Ground-based video coverage is regained.

13:59:33 (EI+924)

PASS FSM: **FCS CH 4.** The fault message is associated with the removal of FCS Channel 4 from the control loop and would have been annunciated on the crew displays. [*TDRS-E data*]

Master Alarm.[19] The Master Alarm associated with the FCS Channel 4 fault message would have been annunciated to the crew (figure 1.2-39).

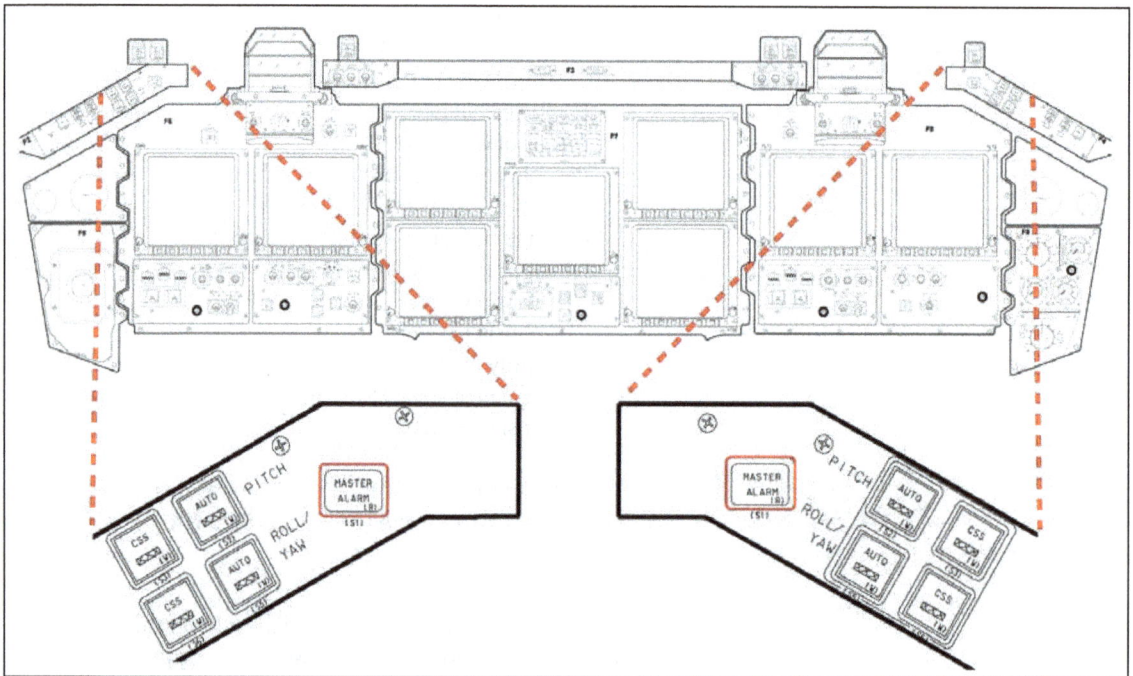

Figure 1.2-39. *Location of the master alarm light.* [Adapted from the Space Shuttle Systems Handbook]

---

[19]STS-107 Investigation Action Response: OVE-204 Crew Inputs After Loss of COMM (Voice); CAIB-MRT-00099, p. 4, 03/10/2003.

TIME		EVENT

13:59:34 (EI+925)   ⚠   FSP Message Downlink Stack (five deep):

FSP1: FCS	CH 4			32/13:59:33.68
FSP2: SM0	TIRE PRESS	L	OB	32/13:58:56.26
FSP3: SM0	TIRE PRESS	L	IB	32/13:58:49.54
FSP4: SM0	TIRE PRESS	L	IB	32/13:58:41.84
FSP5: SM0	TIRE PRESS	L	OB	32/13:58:39.94

13:59:36 (EI+927)   ⚠   **Third RCS Yaw Jet (R4R) Begins Firing Continuously.** Reconstructed data indicate that the DAP commanded a third RCS yaw jet (R4R) to fire at GMT 13:59:36.8 (EI+927.8) and that it fired continuously until end of data at GMT 13:59:37.4 (EI+928.4). [*Master Timeline, Rev. 15 Baseline*]

   ⚠   Aileron trim exceeds 3 degrees.

13:59:37 (EI+928)   ⚠   **LOSS OF CONTROL NO-EARLIER-THAN (NET) TIME**

   ⚠   **Beginning of the Orbiter Pitch-up.** Based on the time of the ROLL REF message (GMT 13:59:46 (EI+937)) and the divergence from the drag profile required to generate the ROLL REF alarm, this is the probable time that control was lost due to the probable loss of hydraulic pressure to the control surfaces. The loss of hydraulic pressure would have resulted in a Master Alarm being annunciated and the orbiter pitching up. While the drag had yet to diverge out of bounds of the drag profile, the orbiter was no longer in controlled flight. This marks the beginning of the transition from a controlled glide to an uncontrolled ballistic entry. Orbiter heating, lift, and drag would no longer be controlled, the ballistic number would be constantly changing with the changing attitudes, and downrange control would be lost (figure 1.2-40).

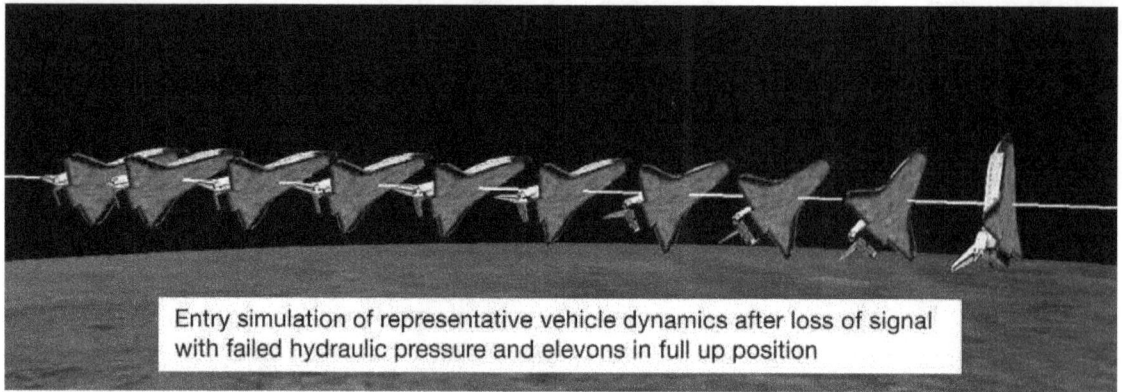

Entry simulation of representative vehicle dynamics after loss of signal with failed hydraulic pressure and elevons in full up position

Figure 1.2-40. *Sequence (1-second intervals) showing a simulation of orbiter loss of control pitch-up from GMT 13:59:37 (EI+928) to GMT 13:59:46 (EI+937).* White line indicates vehicle trajectory relative to the ground.

Video imagery shows a dynamically changing orbiter trail after GMT 13:59:37 (EI+928) with a braided or corkscrew appearance, implying motion of the orbiter. However, the specific attitude of the orbiter cannot be derived from ground-based imagery. Brightening events, objects separating, "puffs," and splitting of the trail are all seen in the video during this timeframe. Ballistic analysis of debris could not positively correlate shedding events seen in the video to a specific orbiter source. However, it is known that the left wing and left OMS pod were being compromised.[20] Figure 1.2-41 shows video frames from GMT 13:59:35.5 (EI+925.5) through GMT 13:59:43.5 (EI+934.5) and displays some of these dynamic changes, although they are much more clearly seen in the video.

---

[20]*Columbia* Accident Investigation Board Report, Volume I, August 2003, p. 68.

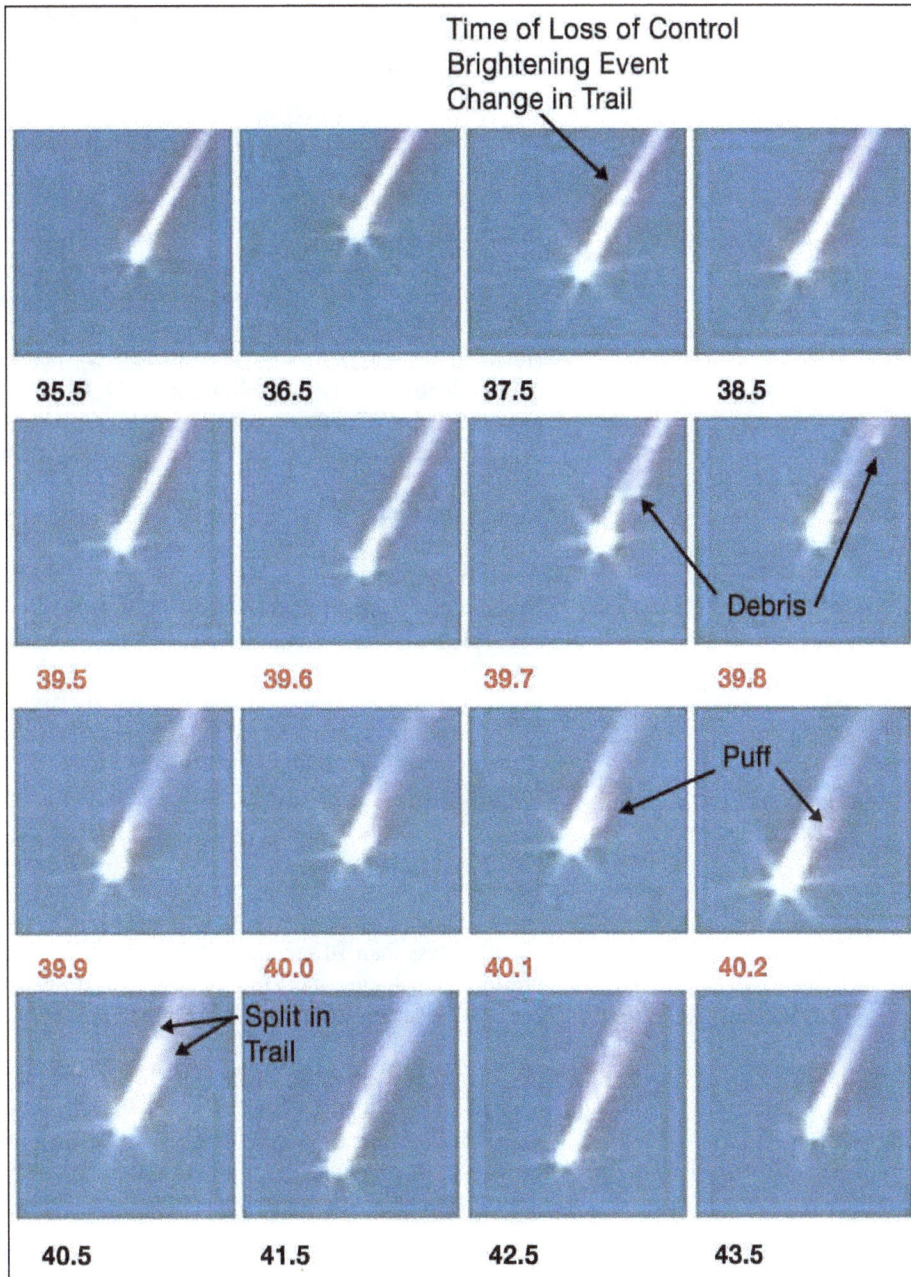

Figure 1.2-41. *Video frame captures from ground-based video, beginning at GMT 13:59:35.5 (EI+926.5) and ending at GMT 13:59:43.5 (EI+937.5). The numbers below each frame indicate the seconds after GMT 13:59:00. The frames in the first and last rows are 1 second apart. The frames in the second and third rows are 0.1 second apart.*

Beginning of Off-nominal Attitude and Loads. The CDR/Seat 1 and PLT/Seat 2 would have been aware of the off-nominal movement of the orbiter based on information from the flight displays and from the changing view in the forward windows. MS2/Seat 4 and MS4/Seat 3 may also have been aware of this information.

TIME		EVENT

	⚛	The rate of change of the elevon trim had reached the maximum allowed by the DAP while in AUTO. [*Master Timeline, Rev. 15 Baseline*]
	⚕	RGPC data recorded the left rudder pedal transducer assembly (RPTA) was near null at GMT 13:59:37.118 (EI+928.118),[21] indicating that there were no crew inputs on the rudder pedals.
	⚛	**Fourth RCS Yaw Jet (R1R) Begins Firing.** Reconstructed data indicate that the DAP commanded a fourth RCS yaw jet (R1R) to fire at GMT 13:59:37.3 (EI+928.3). The R1R fired continuously until the end of data at 13:59:37.396 (EI+928.396). With four jets firing continuously, the fuel supply would be rapidly depleted (within 30 seconds).
	⚛	**End of RGPC-1.** This is the end of the ~5-second period of RGPC data at GMT 13:59:37.396 (EI+928.396).[22] There were no confirmed indications of changes in positions of any of the switches on panel R2.[23] No GPC data were recoverable for the next 25 seconds.[24]
13:59:44 (EI+935)	⚛	Fuselage upper surface canopy thermocouple BP0439T (above window 2) starts to indicate an off-nominal sharp temperature increase followed by a drop to loss of sensor signal.[25]
13:59:46 (EI+937)	⚛	**ROLL REF.** BFS recorded the PASS generating a Roll Alarm fault message. At this time, the drag of the vehicle had exceeded the limits of the entry drag profile. This message occurred less than 10 seconds after the fourth yaw jet (R1R) began firing, suggesting a rapid change in orbiter aerodynamics. [*Master Time Line, Rev. 15 Baseline; BFS, RGPC-2*]
	▤	**Debris A (possible Left OMS Pod Cover).**[26] Video analysis identifies GMT 13:59:46.67 (EI+937.67) as the earliest time at which Debris A is observed. Debris A was identified as possibly being the cover of the left OMS pod. Debris A fades from view on the video at GMT 14:00:22.97 (EI+973.97).

---

[21]STS-107 Investigation Action Response: OVE-204 Crew Inputs After Loss of COMM (Voice); CAIB-MRT-00099, p. 4, 03/10/2003.

[22]STS-107 Investigation Action Response: OVE-204 Crew Inputs After Loss of COMM (Voice); CAIB-MRT-00099, p. 4, 03/10/2003.

[23]STS-107 Investigation Action Response: OVE-204 Crew Inputs After Loss of COMM (Voice); CAIB-MRT-00099, p. 2, 03/10/2003.

[24]Data in RGPC-2 had time-stamped data for some events that occurred during this 25-second period between RGPC-1 and RGPC-2.

[25]OEX Data Evaluation of End of Mission Data for STS-107, Vehicle Data Mapping Team, OV-102 Investigation, 08/22/2003, Rev 2.

[26]See Section 2.2, Figure 2.2-9 Debris Tree.

TIME		EVENT

13:59:52 (EI+943)   ⚜   At GMT 13:59:52.114 (EI+943.114) the BFS recorded the PASS generating an "L RCS LEAK" fault message. [*RGPC-2*] [*Master Timeline, Rev. 15 Baseline*].

🎞   **Debris D (possible Left OMS Pod).** Video analysis identifies this as the earliest time at which Debris D is observed. Debris D was identified as possibly being the left OMS pod. (figure 1.2-42).

Figure 1.2-42. *Debris D is depicted in this frame.*

14:00:01 (EI+952)   ⚜   At GMT 14:00:01.54 (EI+952.54) the BFS recorded an "L RCS LEAK" fault message. [*Master Timeline, Rev. 15 Baseline*]

🕴   **RHC Moved.** The RHC moved beyond neutral between approximately GMT 14:00:01.7 (EI+952.7) and GMT 14:00:03.6 (EI+954.6). The uncertainty in the time of the event is due to the processing rates of signal detection, signal processing, and message annunciation. This event resulted in the DAP downmode RHC Fault message that was recorded at GMT 14:00:03.637 (EI+954.637). By GMT 14:00:3.678 (EI+954.678) the crew returned the DAP back to AUTO, indicating that this RHC movement was accidental.

⚜   At GMT 14:00:01.90 (EI+952.90) the BFS recorded another "L RCS LEAK" fault message. [*Master Timeline, Rev. 15 Baseline*]

14:00:02 (EI+953)   ⚜   **Final Recovered GPS State Vector at GMT 14:00:02.12 (EI+953.12).** These data were recovered from the MAGR.[27]

---

[27]*Columbia* Investigation: GPS Receiver (MAGR) Memory Extraction Summary and Disposition presentation, May 12, 2003.

PASS FSM: LJET L RCS. This Class 2 alarm message may not be the result of the actual vehicle condition since the message was generated based on data from a sensor path that contained noise. If valid, the message is used to indicate when an RCS jet has failed on or failed off, or is leaking. The failed off condition could be an indication that the fuel was exhausted.

Beginning of Final Period of RGPC Data at GMT 14:00:02.660 (EI+953.660). Reconstructed GPC data (RGPC-2), lasting from GMT 14:00:02.660 (EI+953.660) to GMT 14:00:04.826[28] (EI+955.826), contained multiple bit errors.

Vehicle rates during this period were (Note: there is some uncertainty with the NAV-derived parameters since the high rates may have resulted in corruption of the inertial measurement unit state):

- roll rate command from the DAP was at –5.0 deg/sec during the entire RGPC-2 period, indicating that the DAP was in AUTO the entire time (maximum control rates allowed: AUTO is –5.0 deg/sec, control stick steering (CSS) is –6.0 deg/sec)
- Roll rate transitioned from +7 deg/sec (right roll) to –23 deg/sec (left roll)
- Yaw rate was at the sensor maximum of 20 deg/sec (nose right)
- Pitch rate was 5 deg/sec

Debris B (Portion of Left Wing). Video analysis identifies this as the earliest time at which Debris B is observed.

Debris C (Portion of Left Wing). Video analysis identifies this as the earliest time at which Debris C is observed.

14:00:03 (EI+954)

At GMT 14:00:03.47 (EI+954.47) BFS recorded an "L OMS TK P" fault message. This message is annunciated when the left OMS oxidizer ullage pressure (V43P4221C) or fuel tank ullage pressure (V43P4321C) is out of limits either high or low.

DAP DNMODE RHC Fault Message. At GMT 14:00:03.637 (EI+954.637) the DAP Downmode RHC Fault message was recorded in the fault message buffer due to the movement of the RHC beyond neutral between approximately GMT 14:00:01.7 (EI+952.7) and GMT 14:00:03.6 (EI+954.6). This fault message was corroborated by an initialization flag for the aerojet DAP roll stick function. During the final 2-second RGPC period, all available data indicate that the RHC remained

---

[28]STS-107 Investigation Action Response: OVE-204 Crew Inputs After Loss of COMM (Voice); CAIB-MRT-00099, p. 1, 03/10/2003.

in the detent position and that the DAP was in AUTO. This supports the RHC being bumped and the crew immediately returning the DAP to AUTO.

DAP in AUTO. At GMT 14:00:03.678 (EI+954.678) the RPGC data recorded the first indication from PASS that the DAP was in the AUTO mode.[29] This requires a manual crew input by either the CDR/Seat 1 or the PLT/Seat 2. This manual command was in response to the RHC movement at GMT 14:00:01 (EI+952) moding the DAP out of AUTO. Because moding the DAP from CSS to AUTO requires either the CDR/ Seat 1 or the PLT/Seat 2 to press two buttons located on both panels F6 (CDR/Seat 1) and F8 (PLT/Seat 2), at least one crew member was conscious and able to respond to events that were occurring on board and that the vehicle dynamics were within human performance capabilities for this action.

14:00:04 (EI+955)

RHC in Detent. At GMT 14:00:04.179 (EI+955.179) the RGPC data recorded the first indication from PASS or BFS that the RHC was in detent.[30] This indicates that there were no manual inputs at this time on the RHC.

MSG Reset on CRT 1. BFS recorded an MSG reset on CRT 1 some time after GMT 13:59:37 (EI+928) and before GMT 14:00:05 (EI+956).[31] This event is part of the RGPC-2 data and is considered a valid input. This action would be a nominal crew response to a fault message and requires a crew member to manually acknowledge the message by keyboard entry on the C2 panel.

End of RGPC-2. End of the final 2-second period of RGPC-2 data at GMT 14:00:04.826 (EI+955.826). At this time, the forward/mid/aft fuselage, PLBDs, right wing, and right OMS pod were still intact based on the following data: [*Master Timeline, Rev. 15 Baseline*]

- APUs were running and the panel R2 switches were ON. The APUs are located in the aftbody, indicating that this portion of the vehicle was still intact.
- Water spray boiler (WSB) cooling was evident. The WSB is located in the aftbody, indicating that this portion of the vehicle was still intact and the data lines from it to the crew module were still intact.

---

[29]STS-107 Investigation Action Response: OVE-204 Crew Inputs After Loss of COMM (Voice); CAIB-MRT-00099, 03/10/2003.
[30]STS-107 Investigation Action Response: OVE-204 Crew Inputs After Loss of COMM (Voice); CAIB-MRT-00099, 03/10/2003.
[31]STS-107 Investigation Action Response: OVE-204 Crew Inputs After Loss of COMM (Voice); CAIB-MRT-00099, 03/10/2003.

- Panel R2 switches in nominal positions. Since panel R2 was recovered with switches in positions consistent with an attempted APU restart, this indicates that the switch throws occurred after this time.
- Fuel cells were generating power, indicating that electrical power was still being produced.
- Power reactant, storage, and distribution tanks and lines were intact, indicating that this portion of the vehicle was still relatively intact.
- Main Propulsion System still intact indicated that the aft portion of the vehicle was still relatively intact.
- Helium tanks and lines were intact.
- Freon loops and radiators (located in the PLBDs) were intact; quantities, flow measurements, and pressure measurements were nominal. This indicates that the PLBDs were still intact at this time.
- RHCs were in detent during the entire period of RGPC-2, indicating that the crew had not tried to manually control the vehicle during this time period.
- Left RPTA indicated a small left rudder input/offset that remained nearly constant for the duration of RGPC-2, indicating that there may have been some pressure being applied on the pedal by one of the crew members. It should be noted that the RPTA signals are not transmitted to the flight control software at this phase of flight. In addition, the lack of hydraulic pressure would prevent control inputs to the control surfaces.
- Communications and navigation systems in the forebody were performing nominally, indicating that this part of the vehicle was relatively intact.
- Environmental Control and Life Support System (ECLSS) performance was nominal, indicating that the cabin environment was nominal.

The following systems were indicating off-nominal conditions:

- Hydraulic supply pressures were reading 0 psi and the reservoir quantities were at 0% on all three systems; the lack of hydraulic pressure results in the aerosurfaces (elevons, body flap, and rudder) were free to move in the wind stream.
- LIB and LOB elevon actuator temperatures were either OSL or no data existed.
- APU lube oil was possibly being overcooled by the WSB.
- The Flash Evaporator System appears to have shut down.
- Most of the left OMS pod sensors were OSL or off-scale high, or had no data available.

- The electrical power distribution circuit showed a general upward shift in main bus amps and a downward shift in main bus volts.
- Alternating current (AC) Bus 3 (AC3) Phase A inverter was off-line; this would not impair crew or vehicle performance.
- Elevated temperature readings were recorded at the bottom bondline centerline skin forward and aft of the wheel wells and at the portside structure over the left wing.

Another left wing piece separates based on ballistics analysis. This analysis was performed on a piece of recovered structure that was identified as being from the left wing.

14:00:05 (EI+956)   **End of GPC signal.** The last GPC signal, which contained no recoverable data, was recorded at GMT 14:00:05.121 (EI+956.121).

PLT/Seat 2 takes actions in apparent attempt to restart APUs 2 and 3 . With the loss of hydraulic pressures and the vehicle LOC, the crew likely assumed a generic problem with the APUs. A crew module panel was recovered with switch configurations indicating an attempt by the PLT/Seat 2 to recover the hydraulic systems and hydraulic pressure by performing steps to initiate the restart of APUs 2 and 3. Switches for APU 1 were in the nominal position. Switches for hydraulic circulation pumps 2 and 3 were also in the "On" position. While turning on the hydraulic circulation pump is not on the emergency checklist, it nonetheless can provide some limited hydraulic pressure and shows good systems knowledge by the crew members as they worked to attempt to restore orbiter control. These actions took place after the end of RGPC-2 (after GMT 14:00:05 (EI+956)) and prior to loss of consciousness.

14:00:11 (EI+962)   Video analysis identified a color change in the plume that is likely the result of an OMS tank rupturing.

14:00:13 (EI+964)   The last MADS/OEX recorder data value (left wing spar cap sensor, V12G909A), marked GMT 14:00:13.439 (EI+964.439), is recorded.[32]

---

[32]OEX Data Evaluation of End of Mission Data for STS-107, Vehicle Data Mapping Team, OV-102 Investigation, 08/22/2003, Rev 2.

1.2.4   Phase 4: Catastrophic Event to Crew Module Catastrophic Event
[GMT 14:00:18 (EI+969) through GMT 14:00:53 (EI+1004)]

14:00:18 (EI+969)

CATASTROPHIC EVENT. The CE is a period of time during which the orbiter vehicle is undergoing a major structural breakup. At this time, the accelerations on the forebody were estimated to be 3.5 Gs. The breakup sequence progressed over several seconds. Analysis of ground-based video of the event established the first detectable signs at GMT 14:00:18 (EI+969). Based on engineering analysis, the CE is thought to have started with the compromise of the PLBDs, exposing the payload bay longeron sill to entry heating. The skin splice between the midbody and the X_o 582 ring frame bulkhead area, aft of the starboard x-link, failed due to a combination of mechanical and thermal loads. The forebody rotated away from starboard to port, causing the port x-link to fail due to bending loads. As the forebody separated from the midbody, various power, data, and ECLSS lines failed and the crew module was free to move forward and strike the inside of the forward fuselage. At GMT 14:00:25 (EI+976), the separated forebody is marginally visible. The CE is further supported by a collection of evidence that includes ground-based video of a large debris-generating event, the loss of the MADS/OEX recorder function, and the debris footprint (figures 1.2-43 and 1.2-44).

Figure 1.2-43. *The Catastrophic Event is depicted in these three frames of video that cover 0.1 second.* There is no change in the magnification/zoom factor. The third frame represents GMT 14:00:18.3 (EI+969.3).

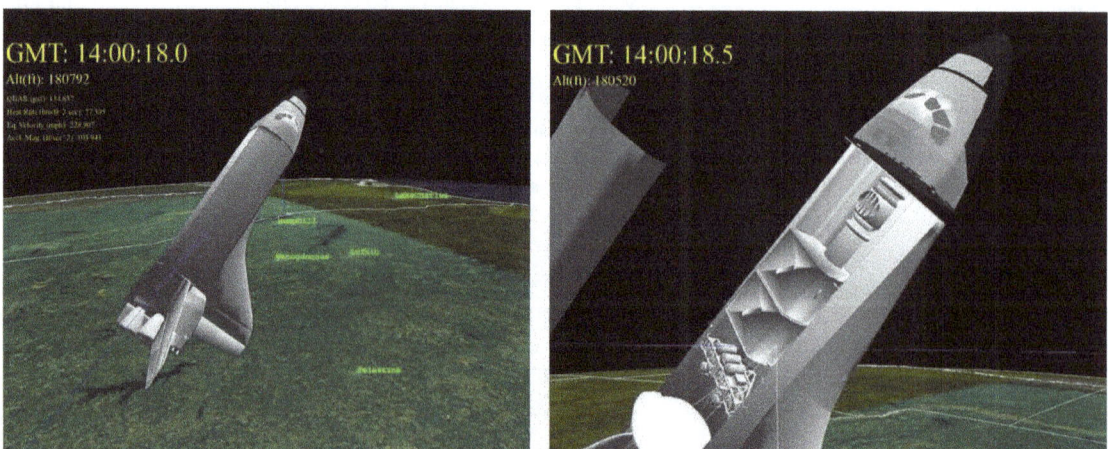

Figure 1.2-44. *Simulation of the Catastrophic Event.*

The separation of the forebody from the midbody resulted in:

- The crew module and forward fuselage becoming independent from the rest of the vehicle, moving farther downrange and remaining higher in altitude than the rest of the vehicle. Since there was no other debris ahead of the forebody, all material depositions and shock wave damage on the recovered forebody debris was generated by the forebody.

- Loss of electrical power – Electrical power for the orbiter systems is supplied from fuel cells located in Bays 1 and 2 of the midbody of the vehicle. Separation of the crew module and forward fuselage from the midbody caused a loss of power to the crew module. Loss of electrical power resulted in the loss of all powered systems, including:

  o Displays – The loss of displays resulted in a loss of all situational awareness from instrumentation.
  o Intercom System – The loss of the intercom system meant that crew-to-crew communications could only be performed by shouting.
  o Lighting – With the cabin lights lost, the only lighting was from the windows and the Cyalume chemical lights that are pre-positioned within the crew module and on the upper arms of the ACES of each of the crew members.
  o Ventilation – Electrically powered fans for circulating air through the cabin and the CO_2 scrubber shut down, resulting in a loss of cabin ventilation.
  o O_2 supply –Valves in the orbiter O_2 system are designed to close when power is lost, resulting in a loss of O_2 supply to the cabin and the suit O_2 hoses. Since none of the helmet visors were lowered and locked, it is unlikely that the crew members activated their emergency O_2 supplies since that step is performed after visors are lowered and locked.
  o MAGR – Last time value was recorded at GMT 14:00:18.6875 (EI+969.6875). No state vector data were recorded at this time.

- Sudden change in aerodynamic characteristics – The ballistic number of the forebody by itself is higher than the ballistic number of the rotating intact orbiter. This resulted in a reduction of drag with a corresponding reduction in translational force on the crew, estimated to have dropped from 3.5 Gs to 1 G.

- Unknown dynamic changes – In addition to the change in translational forces acting on the crew, the separation of the forebody resulted in a change to rotational loads.

Prior to the CE, the crew was experiencing rotational loads with the center of rotation located aft of the crew module at the center of mass of the orbiter. As a result of the crew module and forward fuselage separating from the rest of the orbiter, the rotational arm decreased (figure 1.2-45).

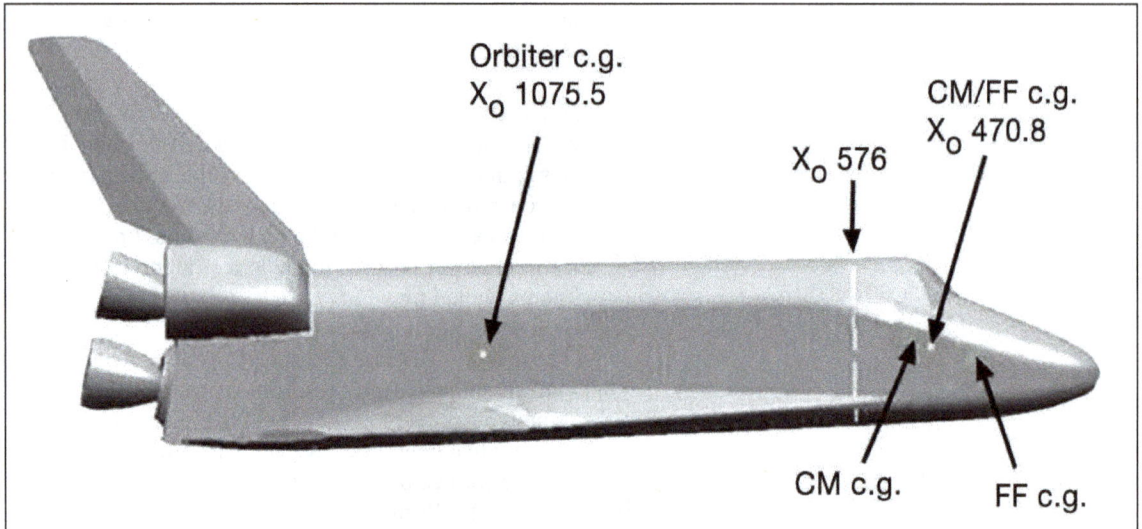

Figure 1.2-45. *X-axis center-of-gravity locations for the intact orbiter, the crew module, the forward fuselage, and the forebody (crew module plus forward fuselage). The X_o 576 is the aft bulkhead of the crew module.*

- Based on video analysis, triangulation data, and aerodynamic modeling, the initial rotation rate of the separated forebody was low, the rates built over time, and the forebody never trimmed.

The estimated vehicle state at GMT 14:00:18 (EI+969):

Alt = 181,000 feet [modeling]
Mach = 15 [modeling]
KEAS = 228 [modeling]
Qbar = 83 psf [modeling]

DEPRESSURIZATION BEGINS – NET Time. This is the first event of lethal potential.[33]

Loss of consciousness and cessation of respiration.

14:00:19 (EI+970)

The MADS/OEX recorder stopped recording at GMT 14:00:19.44 (EI+970.44).[34]

---

[33] This symbol is used to indicate a vehicle-related event that occurred after the separation of the crew module and forward fuselage from the midbody.

[34] *Columbia* Accident Investigation Board Report, Volume I, August 2003, p. 73.

14:00:25 (EI+976)   📽 🏛 First visual indication that the orbiter had broken into multiple pieces (figure 1.2-46).

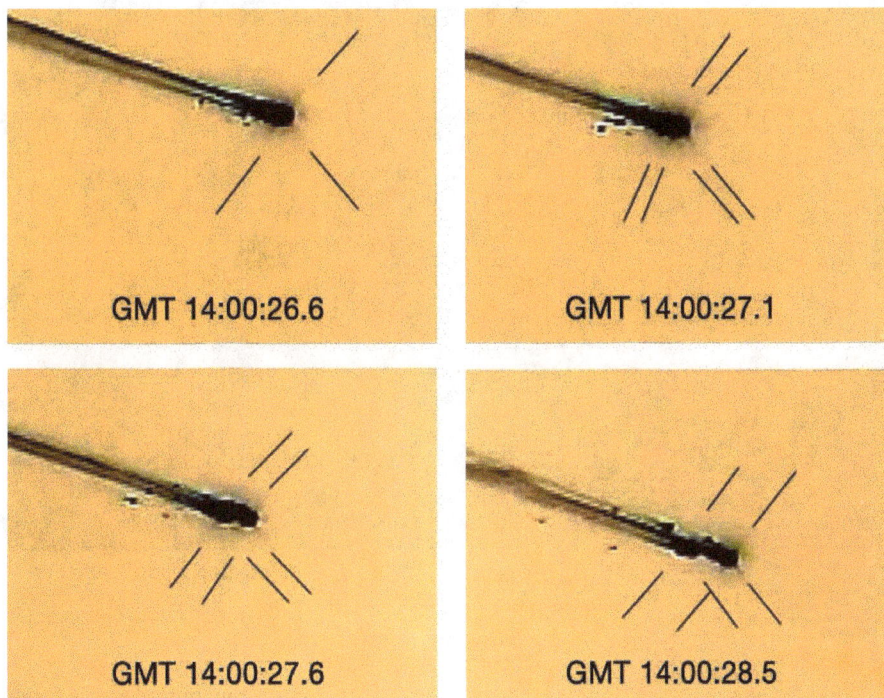

Figure 1.2-46. *Color-inverted video images of the start of the double star event, GMT 14:00:26.6 (EI+977.6).* **Black lines have been added to more clearly identify the two separate objects.**

14:00:35 (EI+986) ±5 sec   🏛 **DEPRESSURIZATION BEGINS – NLT Time.** This time is based on the release times for items from a storage compartment within the crew module.

1.2.5 Phase 5: Crew Module Catastrophic Event to Total Dispersal
[GMT 14:00:53 (EI+1004) through GMT 14:01:10 (EI+1021)]

14:00:53 (EI+1004)   📽 🏛 **CREW MODULE CATASTROPHIC EVENT Begins.** The CMCE is the initiation of a period of time during which the forebody was undergoing a major structural breakup. The breakup sequence progressed over several seconds and ends at approximately GMT 14:01:10 (EI+1021). Analysis of ground-based video of the event established the first detectable signs at GMT 14:00:53 (EI+1004). Based on engineering analysis, CMCE is thought to have started with the failure of the forward fuselage. Once the forward fuselage began to break away, the exposed crew module rapidly failed due to the combined effects of the high G-loads, aerodynamic forces, and thermal loads. The flight deck maintained structural integrity longer than the middeck (figure 1.2-47).

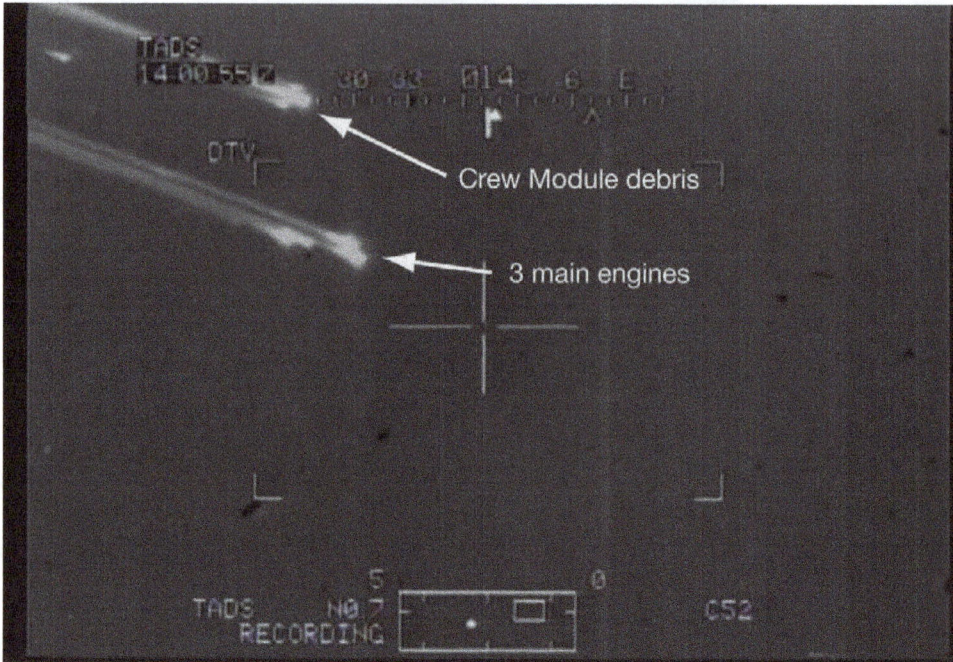

Figure 1.2-47. *Image from Apache[35] video showing crew module debris* (top cluster) *and the three main engines* (cluster to left of cross hairs).

	♦ ⚱	Thermal intrusion into the crew module – NLT time.
	♦ ⚱	Inertial reel straps fail – Metallic deposits on the inertial reel straps indicate that globules of molten metal were present in the crew module prior to the straps failing and retracting into the inertial reel housing.
14:00:57 (EI+1008)±5 sec	⚱	Based on ballistic analysis, a recovered piece of the forward RCS separated at this time.
14:00:59 (EI+1010)	⚱	**DEPRESSURIZATION COMPLETE – NLT Time.** Based on video evidence, the crew module no longer had sufficient structural integrity to maintain cabin pressure. However, the cabin depressurization was probably complete well before this time.
14:01:01 (EI+1012)±5 sec	⚱	**Middeck CEE Released.** Based on ballistic analysis, the earliest piece of recovered CEE from the middeck crew members separated at this time.
14:01:07 (EI+1018)±5 sec	⚱	**Middeck Accommodation Rack (MAR) Separates.** Based on ballistic analysis, the MAR, which was located on the port side of the middeck, separated at this time.

---

[35]The "Apache" video was filmed from an Apache helicopter.

TIME	EVENT

14:01:08 (EI+1019)±5 sec   🛆   **Flight Deck CEE Released.** Based on ballistic analysis, the earliest piece of recovered CEE from the flight deck crew members separated at this time.

1.2.6 Phase 6: Total Dispersal to ground impact
[GMT 14:01:10 (EI+1021) through approximately GMT 14:35:00 (EI+3051)]

14:01:10 (EI+1021)   ▤ 🛆   **TD** – Forward fuselage and crew module have fragmented into pieces too small to be detected on any of the ground-based videos (figure 1.2-48).

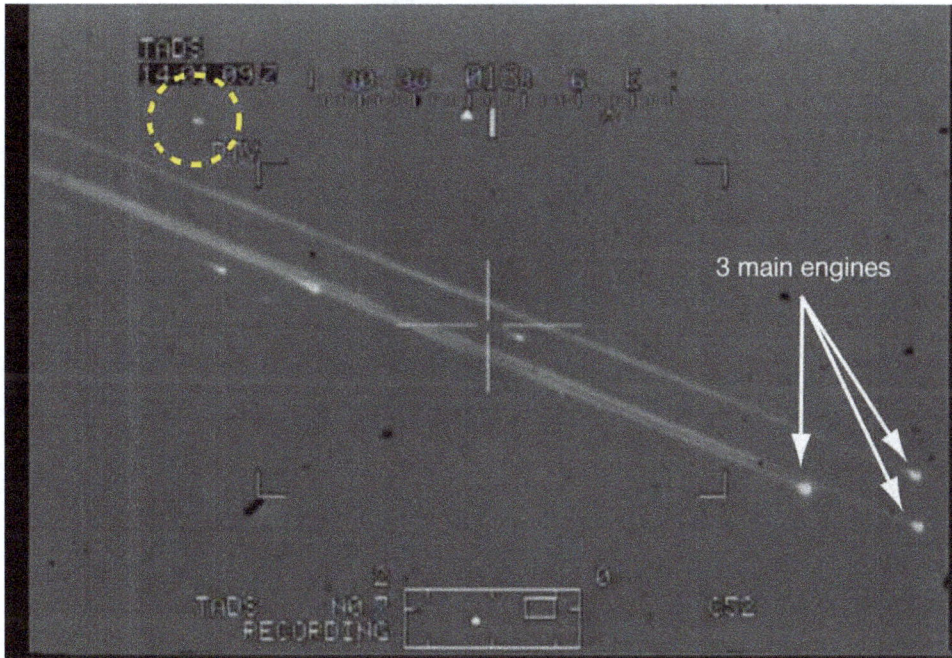

Figure 1.2-48. *Image is from the Apache video.* Dotted circle indicates area where crew module debris is last visible. The three points in the lower right are the three main engines.

14:35:00 (EI+3051)   🧍 🛆   Approximate time at which the crew remains and the majority of the crew module debris completed the free fall to the ground.

Chapter 2 – Vehicle Failure Assessment

2.1 Motion and Thermal Analyses

This section discusses the trajectory, dynamic attitude, and thermal analyses of the *Columbia* accident performed in support of the Spacecraft Crew Survival Integration Investigation Team (SCSIIT). The motion analyses provided information about the loads experienced by the crew and the vehicle through the evolving conditions. Because the thermal environment for atmospheric entry is highly dependent on the aerodynamic properties of an object as well as the aerodynamic conditions such as altitude and velocity, the aerodynamic analyses also provided reference data for thermal analyses. The purpose of the thermal analyses was to aid in understanding the probable failure sequence by comparing the physical condition of certain key items of debris with predicted entry heating.

"Trajectory" refers to the translational motion of an object relative to the Earth, and provides information about the deceleration of an object due to atmospheric drag. The trajectory of the intact orbiter after communications were lost was estimated from available aerodynamic data. However, this type of data did not exist for the forebody as a separate object following orbiter breakup at the Catastrophic Event (CE). An average ballistic number was approximated[1] for the forebody. A 3 degree-of-freedom trajectory simulation used this approximate ballistic number to estimate the trajectory of the forebody. The "reference trajectory" was a continuous trajectory used for ballistic and thermal analyses. The reference trajectory was a sequential combination of the trajectory of the intact orbiter followed by the trajectory of the forebody after separation from the orbiter.

"Dynamic attitude motion" refers to the rotational motion of an object relative to the trajectory. This information was necessary since aerodynamic drag was the predominant load on the vehicle. Because drag always acts in a direction opposite to a vehicle's forward motion, a rotating vehicle will experience the drag force in varying directions relative to the body axes. Simulations were developed to characterize the dynamic attitude motion of the intact orbiter and the forebody. The first simulation modeled estimated dynamic attitude for the intact orbiter from loss of signal (LOS) to the CE. This simulation incorporated telemetry and the Modular Auxiliary Data System/orbiter experiment (MADS/OEX) recorder data into an existing orbiter aerodynamic model. The second simulation modeled the estimated dynamic attitude of the free-flying forebody from the CE to the Crew Module Catastrophic Event (CMCE). This simulation had no direct data available. The aerodynamic characteristics of the forebody were estimated and incorporated into an aerodynamic model. The rotational loads and attitudes were then combined with the drag loads identified through the trajectory analysis to estimate loads in all axes for both the intact orbiter and the forebody.

The data from these model-based analyses should be considered only as *representative* of the type of motion the vehicle and crew experienced. The results of these analyses were compared to a relative motion analysis obtained from ground-based video. The analyses were found to be in general agreement.

To better quantify the limits of crew survival for future flights, the survivability of an exposed crew module (CM) following orbiter breakup for ascent and entry conditions were examined from a thermal perspective. The results were compared to the estimates currently used in crew procedures and to the conditions experienced for both the *Challenger* and the *Columbia* accidents. Thermal analyses were also completed on individual recovered items to predict the amount of heating that they would have experienced given a specific

---

[1]Ballistic number is an indicator of the performance of an un-powered object flying through the atmosphere. It is characterized by the ratio of the object's weight over its aerodynamic drag.

release time. These results were compared to the thermal damage observed on debris items to confirm ballistic release times and to help sequence events.

The following is a summary of findings, conclusions and recommendations in this section:

Conclusion L3-1. Complete loss of hydraulic pressure to the aerosurfaces resulting from the breach in the left wing was the probable proximal cause for the vehicle loss of control.

Recommendation L3-2. Future vehicles should be designed with a separation of critical functions to the maximum extent possible and robust protection for individual functional components when separation is not practical.

Finding. The forebody and the CM of the orbiter are not aerodynamically stable in attitude with any initial rates or lateral center of gravity (c.g.) other than zero.

Finding. Triangulation results suggest that the free-flying forebody rotated at approximately 0.1 rev/sec in a multi-axis motion.

Conclusion L2-1. Between orbiter breakup and the forebody breakup, the free-flying forebody was rotating about all three axes at approximately 0.1 rev/sec and did not trim into a specific attitude.

Finding. The estimate for maximum thermal survivability on ascent of 280,000 feet is a reasonable estimate.

Finding. The maximum thermal survivable breakup altitude for the CM on entry is approximately 150,000 feet.

Conclusion L3-3. The actual maximum survivable altitude for a breakup of the space shuttle is not known.

Recommendation L2-2. Prior to operational deployment of future crewed spacecraft, determine the vehicle dynamics, entry thermal and aerodynamic loads, and crew survival envelopes during a vehicle loss of control so that they may be adequately integrated into training programs.

Recommendation L3-3. Future spacecraft design should incorporate crashworthy, locatable data recorders for accident/incident flight reconstruction.

Finding. Thermal analyses predicted that entry aeroheating alone was insufficient by an order of magnitude to produce the observed thermal damage on the x-links. Therefore, the x-links must have experienced other heating mechanism(s) in addition to normal entry heating.

Conclusion A13-1. Titanium may oxidize and combust in entry heating conditions dependent on enthalpy, pressure, and geometry.

Conclusion A13-2. The heating from a Type IV shock-shock impingement and titanium combustion (in some combination) likely resulted in the damage seen by the forward payload bay door rollers and the x-links.

Recommendation A13. Studies should be performed to further characterize the material behavior of titanium in entry environments to better understand optimal space applications of this material.

2.1.1 Intact orbiter trajectory

Figures 2.1-1 through 2.1-3 define the body axis coordinate system and the attitude angles that will be discussed in this section. Figure 2.1-1 defines the body axis coordinate system (X_{BY}, Y_{BY}, Z_{BY}) and the pitch, yaw, and roll angles. Figure 2.1-2 defines the angle of attack (α), and figure 2.1-3 defines sideslip (β). These angles are defined relative to the X_{BY} axis and the velocity vector (V). Proj(V) is the projection of the velocity vector (line of sight along the trajectory).

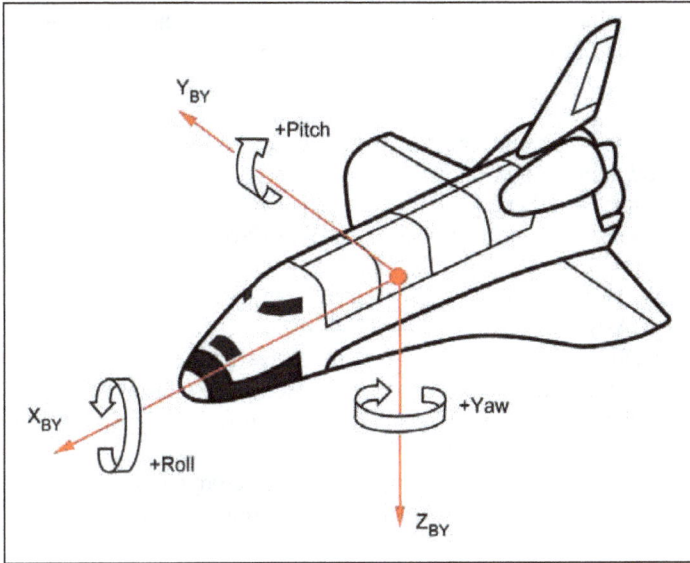

Figure 2.1-1. *Depiction of the body axis coordinate system and pitch, yaw, and roll definitions.*

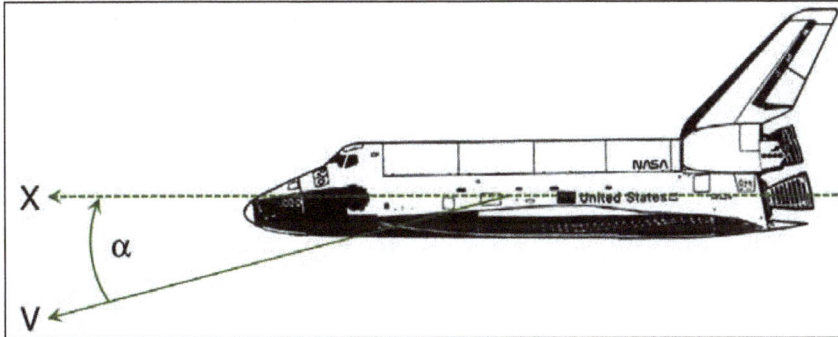

Figure 2.1-2. *Depiction of angle of attack.*

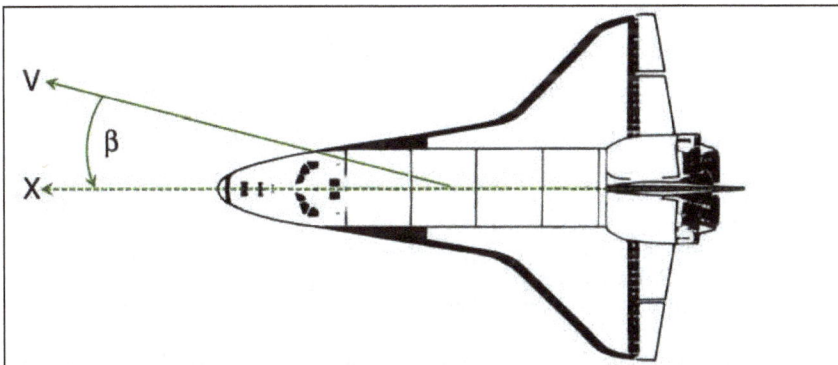

Figure 2.1-3. *Depiction of sideslip.*

Nominal dynamic, 3-axis coupled orbiter aerodynamic stability has been discussed in many documents and will not be discussed here. There are many documents available that discuss the orbiter's dynamic stability.[2,3,4] The aerodynamic data used in this aerodynamic analysis were obtained from the Orbiter Aerodynamic Databook.[5] The data in the Databook are limited to the operational range of the orbiter. After loss of control (LOC), the orbiter was probably out of that range in some parameters. However, no other data exists, and the Databook was the best source available. Engineering judgment was used to determine which data applied best to the conditions.

Some assumptions were necessary to provide a continuous reference trajectory for ballistic and thermal analyses. Four different sets of condition assumptions were used sequentially to determine the reference trajectory. To help clarify the differences for each of these phases, some discussion follows of these conditions and how they affected the trajectory.

An uncontrolled object entering the Earth's atmosphere is on a ballistic trajectory (see Ballistic Tutorial, Appendix). However, a vehicle may use aerodynamic controls and vehicle properties to generate lift, which modulates drag, to alter the ballistic number and, hence, the trajectory. The orbiter can generate lift when in the appropriate attitude. A simplified, static, 3-axis uncoupled analysis of aerodynamic stability shows that when in the high-altitude hypersonic flight regime, the orbiter is not statically stable in attitude. Active control is necessary to maintain orbiter stability. This active control is provided by the general purpose computers (GPCs) through the digital autopilot (DAP). The DAP uses the Flight Control System (FCS), a blend of Reaction Control System (RCS) jets and the elevons, body flap, and (at lower Mach numbers) rudder/speed brake aerosurfaces to control the orbiter attitude. As the orbiter descends into the atmosphere, the aerosurfaces have more control authority and the RCS jets play a lesser role in control.

The orbiter uses a drag-velocity profile to reduce its velocity, maintain heating and aerodynamic loads within orbiter limits, and reach the landing site. Figure 2.1-4 shows the entry drag-velocity profile and boundaries. The lower lift boundary is called the equilibrium glide boundary. This is typically defined as having 20% of vehicle lift remaining to adjust the trajectory, which is accomplished by modulating the angle of attack and bank angle. Bank angle also changes the trajectory laterally, so periodic roll reversals are accomplished to keep the orbiter from drifting too far crossrange.

---

[2]William T. Suit, "Summary of Longitudinal Stability and Control Parameters as Determined from Space Shuttle Columbia Flight Test Data," NASA Technical Memorandum 87768, August 1986.
[3]Robert Blanchard, Kevin Larman, Christina Moats, "Rarefied-Flow Shuttle Aerodynamics Flight Model," NASA Technical Memorandum 107698, February 1993.
[4]Robert Day, "Coupling Dynamics in Aircraft: A Historical Perspective," NASA Special Publication 532, 1997.
[5]Orbiter Aerodynamic Databook, STS85-0118, Volumes 1, 2, and 5, August 2001, Volume 3, February 1996, Volume 4, January 1994, Volume 6, December 2000.

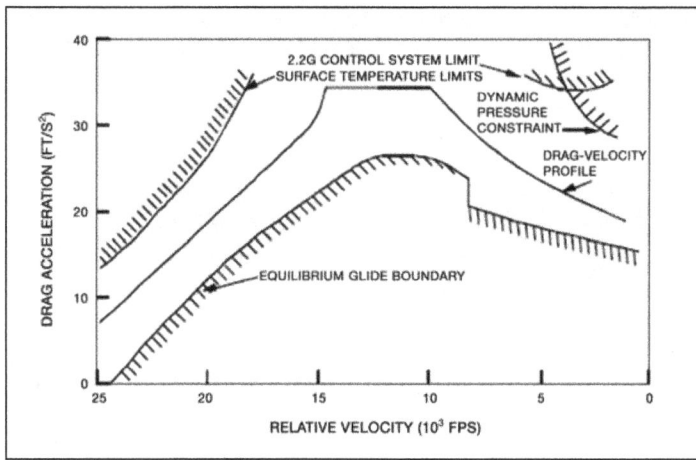

Figure 2.1-4. *Entry guidance drag-velocity profile and limits.*[6]

Existing orbiter aerodynamic data were used to estimate the ballistic number for the first phase. This phase of the trajectory was the "Nominal Orbiter Phase" with an intact, controlled orbiter ballistic number of 108 pounds per square foot (psf). This reflected conditions when signal was lost at Greenwich Mean Time (GMT) 13:59:32.

At some point after LOS, the orbiter lost attitude control. The degradation of the left wing appeared to be the most obvious cause. The Integrated Entry Environment Team pursued a more detailed investigation.

As mentioned above, the drag-velocity profile is controlled by the orbiter's angle of attack and bank angle. Increasing the bank angle increases the drag slowly, while increasing the angle of attack increases the drag quickly. When the reference drag profile computed by the on-board guidance is below the equilibrium glide boundary, a "ROLL REF" alarm is annunciated to the crew. During STS-107, this alarm was annunciated at GMT 13:59:46, only 9 seconds after the LOS.

To determine which parameter (bank angle or angle of attack) caused the ROLL REF alarm, the Integrated Entry Environment Team ran simulations varying each parameter.[7] The thermal degradation was causing increased drag on the left wing, which would induce increasing yaw and/or roll. However, the parameters driving the ROLL REF message proved to be much more sensitive to angle-of-attack deviations than bank angle deviations. Since LOS was at GMT 13:59:37, only 9 seconds passed before the message was annunciated. Based on these simulations, the Integrated Entry Environment Team concluded that it was most likely that a large angle-of-attack change (rather than bank angle) triggered the ROLL REF alarm, although the actual rate of change of the angle of attack could not be determined.

Reconstructed general purpose computer (RGPC) data from GMT 14:00:02.6 (referred to as RGPC-2) show all three hydraulic system pressures at zero, and the hinge moments of the aerosurfaces in the "up" position. Loss of hydraulic power would cause the pitch, roll, and yaw aerosurfaces to float. Pitch and roll RCS jets are no longer used for control at this Mach number and altitude, while yaw jets are still incorporated into control logic and could continue to provide some control in this axis.

Significantly, ground-based video shows a marked change in the appearance of the orbiter's trail at GMT 13:59:37 (immediately after the end of the first RGPC data set, RGPC-1) (figure 2.1-5). The width of the trail increases at this time, which likely indicates a change in the orbiter's flight condition. In addition to the change in the width of the trail, the trail appears to pulse or "corkscrew" over a period of less than 1 second. It is possible that a large debris event (such as loss of a major portion of the wing) may have caused

---

[6]Space Shuttle Orbiter Operational Level C Functional Subsystem Software Requirements; Guidance, Navigation and Control; Part A; Entry Through Landing Guidance.
[7]EG-DIV-08-32 – Integrated Entry Team Report Appendix G - Post-LOS Analysis.

the LOC. However, the debris event closest to the LOS that was apparent in the video occurred 2 seconds, or more than 60 frames (figure 2.1-6), after the change in the trail's appearance. This suggests that it was subsequent to the LOC and was not the cause. Video time errors are up to 1 second, but the relative sequence of the events (brightening, followed by a debris event) is not changed.

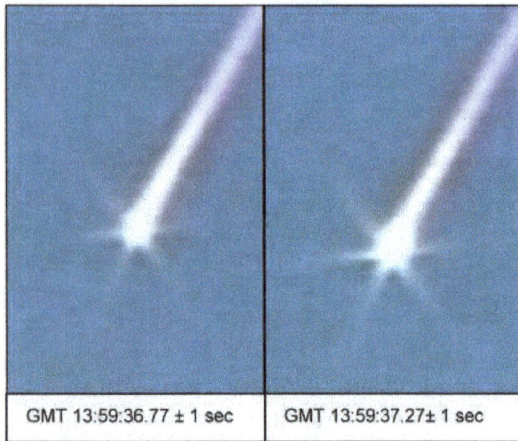

| GMT 13:59:36.77 ± 1 sec | GMT 13:59:37.27± 1 sec |

Figure 2.1-5. *Noticeable change in brightness at GMT 13:59:37±1 second.*

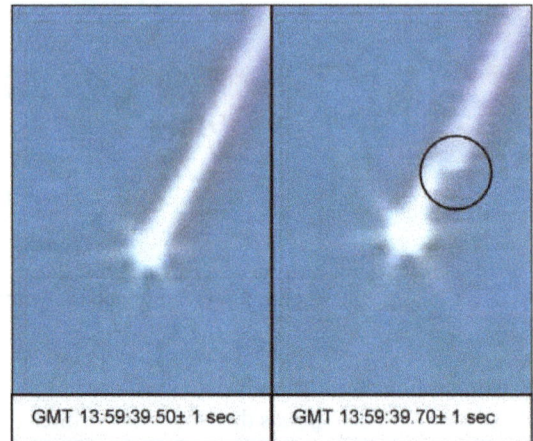

| GMT 13:59:39.50± 1 sec | GMT 13:59:39.70± 1 sec |

Figure 2.1-6. *Noticeable debris shedding event (circled at right) at GMT 13:59:39±1 second.*

It was concluded that the most credible scenario for LOC was the loss of all hydraulic systems, causing the aerosurfaces to float and resulting in an uncontrolled pitch-up. Since the left wing contains locations where all three hydraulic systems have lines that are in close proximity (including the left gear well), it is probable that the loss of hydraulics was due to thermal damage as a result of the breach in the wing. Consequently, the SCSIIT defined the vehicle LOC to begin at GMT 13:59:37, immediately after the LOS, when it was determined by inference that hydraulic power to the aerosurfaces was lost. Prior to this time, *Columbia*'s DAP was still in control of the orbiter. The FCS was commanding the elevons and the RCS yaw jets to counteract the increasing drag on the damaged left wing. The LOC marks the time when the *ability* to control the vehicle was lost, not when *Columbia* departed known attitudes (which occurred shortly thereafter).

Separation of redundant features is an important element in survivability design.[8] It should be noted, however, that with the RCS jets continually firing, propellant would have been rapidly exhausted, causing an inevitable LOC regardless of the condition of the hydraulic systems.

> **Conclusion L3-1.** Complete loss of hydraulic pressure to the aerosurfaces resulting from the breach in the left wing was the probable proximal cause for the vehicle loss of control.

> **Recommendation L3-2.** Future vehicles should be designed with a separation of critical functions to the maximum extent possible and robust protection for individual functional components when separation is not practical.

Figure 2.1-7 shows the increasing angle-of-attack simulation. The figure shows the orbiter attitude at 1-second intervals, beginning at GMT 13:59:37 and ending at GMT 13:59:46. The white line in the figure is the reference trajectory.

---

[8]Ball, Robert E., *The Fundamentals of Aircraft Combat Survivability Analysis and Design*, Second Edition, AIAA, Reston, VA, 2003 (The AIAA Textbook); http://www.aircraft-survivability.com/pages/books_frame.html.

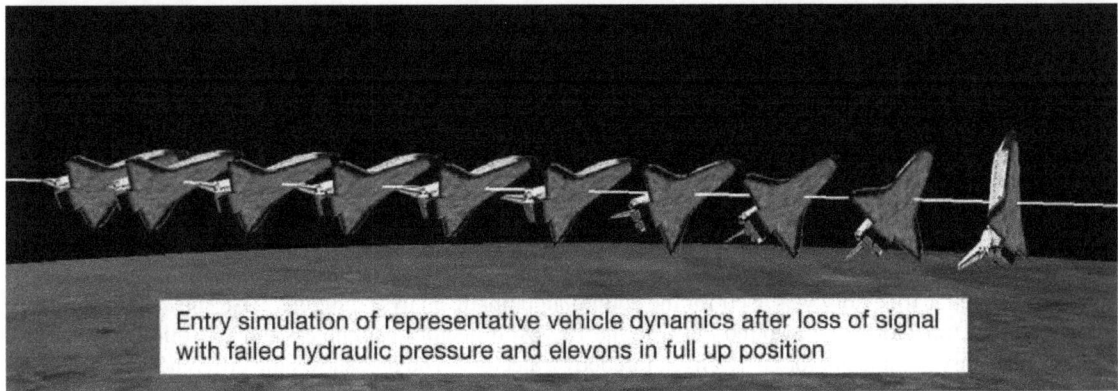

Entry simulation of representative vehicle dynamics after loss of signal with failed hydraulic pressure and elevons in full up position

Figure 2.1-7. *Sequence (1-second intervals) showing a simulation of orbiter loss of control pitch-up from GMT 13:59:37 to GMT 13:59:46.* White line indicates vehicle trajectory relative to the ground.

Following LOC, the ability of the orbiter to generate lift was greatly reduced due to the uncontrolled changes in attitude. These changes in attitude also affected ballistic number, which is dependent on the cross-section of an object presented to the drag vector. Consequently, the ballistic number for the orbiter after LOC had to be estimated. This estimate was obtained by averaging ballistic numbers from a variety of attitudes.

The periodic loss of structure as the orbiter shed debris also affected the mass properties and the cross-section of the orbiter. Debris shedding events were not understood enough to model, so for the purposes of this assessment mass properties were held constant.

This second phase of the trajectory was termed the "High Drag Orbiter Phase." The average ballistic number for the conditions between LOC and the last available GPC data (start of RGPC- 2, GMT 14:00:02.6) was estimated to be 41.7 psf. The motion of the orbiter was assumed to be benign enough that some lift was still generated, and this lift was incorporated into the trajectory. The drop in ballistic number as the orbiter changed attitude from the nominal orbiter ballistic number of 108 psf to a high drag condition of 41.7 psf would increase the deceleration of the orbiter from the nominal deceleration profile.

At the beginning of RGPC-2 and before the CE (orbiter breakup), the average ballistic number for this phase was still estimated to be 41.7 psf. However, the orbiter was assumed to be generating no lift, and only drag was incorporated into the trajectory. For this "No-lift Orbiter Phase," it was predicted from simulations that the orbiter's angle of attack was varying from 30 to 120 degrees and the damaged left wing was not producing lift. The angle of attack assumed for this phase of the simulation was 72 degrees because it was a mid-value angle of attack and closely corresponded with the last angle of attack data value that was recovered.

2.1.2 Forebody trajectory

The forebody trajectory analysis was initially performed by the Crew Survival Working Group (CSWG) and later updated by the SCSIIT. Once again, the assumption was made that mass property changes were negligible and the ballistic number was averaged across a range of attitudes.

The "Forebody Phase" begins at GMT 14:00:23, which was an early assessment time of orbiter breakup during the initial investigation. The average ballistic number of 150 psf for the detached forebody was estimated by Johnson Space Center (JSC) Engineering immediately after the *Columbia* accident. This estimate was used in a ballistic analysis to generate debris search areas. This analysis was provided to the *Columbia* Accident Investigation Board (CAIB) and published in the CAIB Report.

Using more detailed information from ground-based video, the Image Science and Analysis Group (ISAG) determined that the orbiter breakup occurred at GMT 14:00:18.3.[9] Later, higher-fidelity mass properties of the CM based on flight data were combined with a refined estimate of the geometry of the forebody to provide a more precise average ballistic number of 122 psf. Because ballistic analysis on many objects had already been conducted using the 150-psf reference trajectory, it was preferable not to change to the updated CE time and 122-psf reference trajectory for the additional ballistic analyses. A sensitivity study was conducted to determine how changing the reference trajectory average ballistic number from 150 psf to 122 psf would affect the release times. It was found that the release times only changed by 1 second. Since the overall error in ballistic release times was ±5 seconds, this is not significant. Therefore, the original 150-psf forebody reference trajectory was used for the additional ballistic analyses. Figure 2.1-8 shows the two forebody reference trajectories. The dotted lines are the 122-psf trajectory parameters and the solid lines are the 150-psf trajectory parameters for an idealized forebody trajectory. The plot shows the differences in the CE times, loads, heat rates, altitudes, and equivalent velocity between the two trajectories. The trajectories were propagated out even past the CMCE, the time at which the forebody broke up, to enhance the evaluation of the differences.

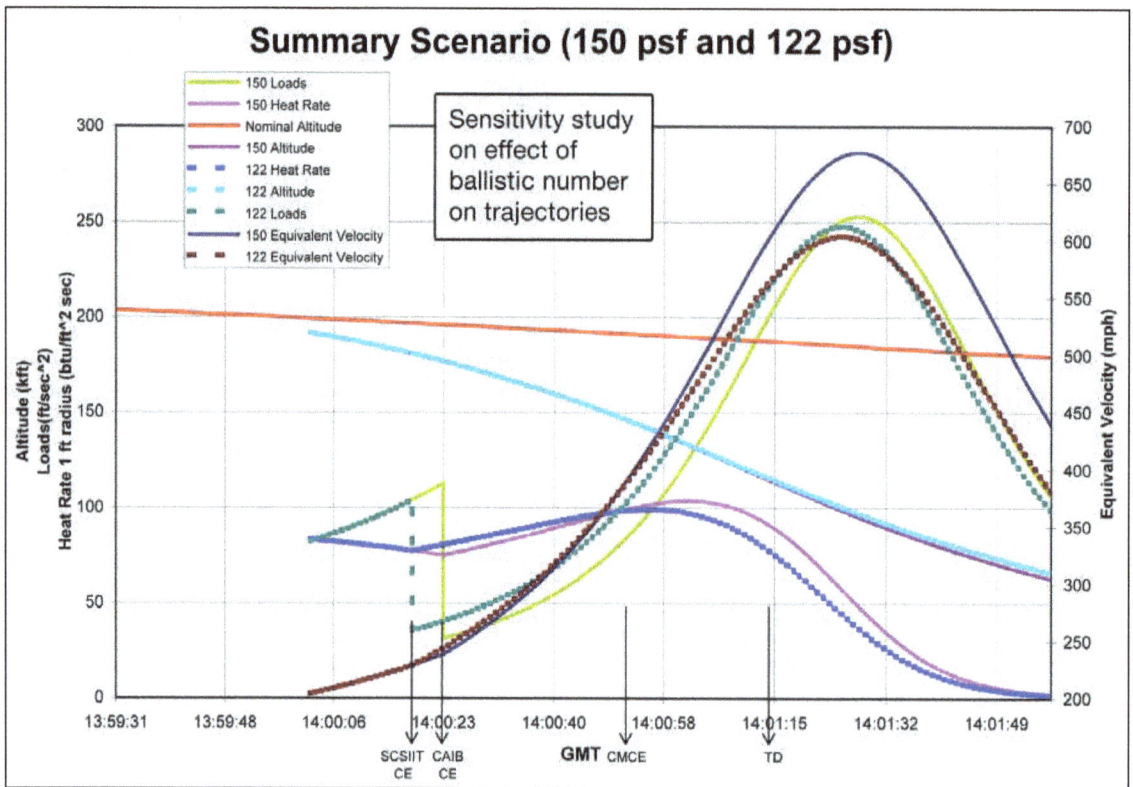

Figure 2.1-8. Comparison of 150-pounds-per-square-foot and 122-pounds-per-square-foot forebody projected trajectories.

The significance of this sudden change in ballistic number can also be seen in figure 2.1-8. As the ballistic number went from 41.7 psf to 150 psf, the deceleration dropped from approximately 3.3 G to approximately 1 G as experienced by the forebody. Figure 2.1-9 shows a more expanded view of this change in deceleration.

[9]*Columbia* Accident Investigation Board Report, Volume III, Appendix E.2, STS-107 Image Analysis Team Final Report, October 2003.

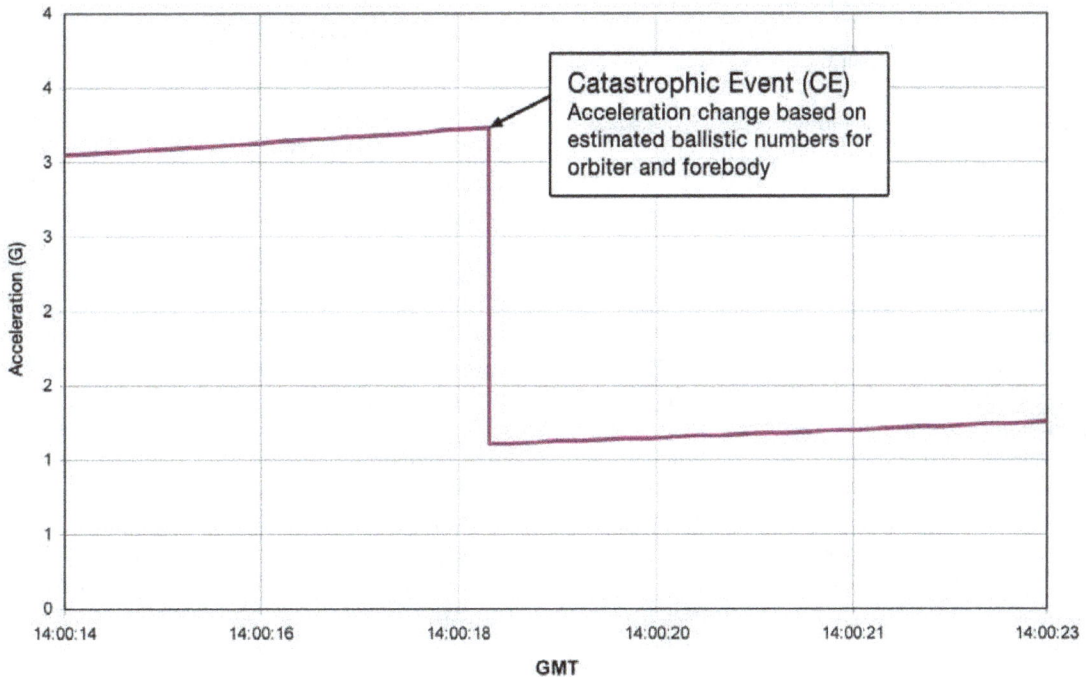

Figure 2.1-9. *Estimated change in total G experienced by the forebody due to a change in ballistic number from the orbiter breakup.*

In summary, the reference trajectory is divided into the following four phases:

1. "Nominal Orbiter Phase," with an average ballistic number of 108 psf and aerodynamic data re-covered from the vehicle. This phase starts 10 seconds before entry interface at 400,000 feet and lasts until LOC (GMT 13:43:59 to GMT 13:59:37).

2. "High Drag Orbiter Phase," with an average ballistic number of 41.7 psf and incorporating some lift generation. This phase starts at LOC and lasts until the beginning of RGPC-2 (GMT 13:59:37 to GMT 14:00:02).

3. "No-lift Orbiter Phase," with an average ballistic number of 41.7 psf and no lift generation (72-degree angle of attack assumed). This phase begins at the start of RGPC-2 and ends shortly after the beginning of the orbiter breakup at the CE (GMT 14:00:02 to GMT 14:00:23).

4. "Forebody Phase," with an average ballistic number of 150 psf. This phase begins shortly after the CE and ends at the forebody breakup (GMT 14:00:23 to GMT 14:00:53).

2.1.3 Intact orbiter attitude dynamics through the Catastrophic Event

An entry simulation was developed to estimate the attitude dynamics of the orbiter and the resulting accelerations on the vehicle structure and crew from LOS to the CE. When the orbiter was designed, there were no requirements to assess catastrophic scenarios to understand how the orbiter might behave in an out-of-control scenario. Test data were available to verify aerodynamic coefficients for the orbiter in nominal conditions, but for this investigation existing models had to be extrapolated beyond flight experience to understand the probable motion of the orbiter.

> **Recommendation L2-2.** Prior to operational deployment of future crewed spacecraft, determine the vehicle dynamics, entry thermal and aerodynamic loads, and crew survival envelopes during a vehicle loss of control so that they may be adequately integrated into training programs.

The simulation used the vehicle's preflight predicted mass properties. Changes in the c.g., moments of inertia, and mass due to the thermal damage were unknown and, therefore, not included. The aerodynamic characteristics determined for attitudes that were outside of the vehicle database were also based on the "un-damaged" geometry. The initial conditions of the simulation were based on the downlinked GPC data at LOS including position, velocity, attitude, any alarm/warning-related data; and the recovered MADS/OEX recorder sensor data. The first set of RGPC data (RGPC-1) and the second set of RGPC data (RGPC-2) were used in an attempt to synchronize the simulation with the actual flight conditions. These RGPC data indicated that all hydraulic systems had failed by this point, which would result in the elevons free-floating, probably in the full-up position. The yaw rates during the RGPC-2 period exceeded the maximum value of the data scale ("pegged out"). All of the rates were high enough that they may have affected the inertial measurement units' ability to maintain a reference, so the quality of acceleration, rate, or attitude data is unclear. However, the period of excessive rates was relatively short and it was the only data available. The MADS/OEX recorder pressure data from the lower right wing surface pressure transducers indicate large oscillations in the angle of attack 5 seconds after LOS. These MADS/OEX recorder pressure data were matched to the simulation and showed good agreement in terms of rates and attitude excursions. Even though this entry simulation used all the available data, damage to the wing and aerodynamic properties outside the database of experience could not be accounted for in this analysis. However, the results are representative of the attitudes, rates, and characteristic motion the vehicle probably experienced.

Figures 2.1-10 through 2.1-12 show the plots of the estimated orbiter angle of attack, roll angle, and sideslip angle from the start of the LOC period until the CE based on the simulation. The representative motion seen in these plots shows that the orbiter was oscillating around the velocity vector in all three axes.

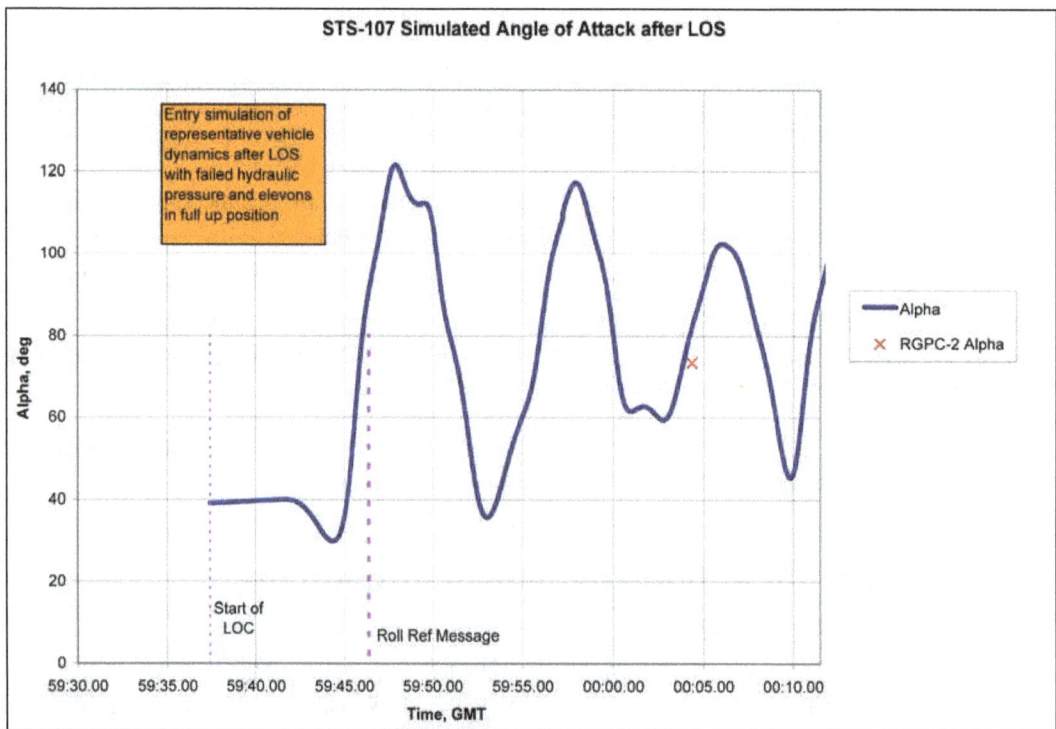

Figure 2.1-10. *STS-107 simulated angle of attack after loss of signal.* Red data indicate alpha condition recorded in the reconstructed general purpose computer-2 data.

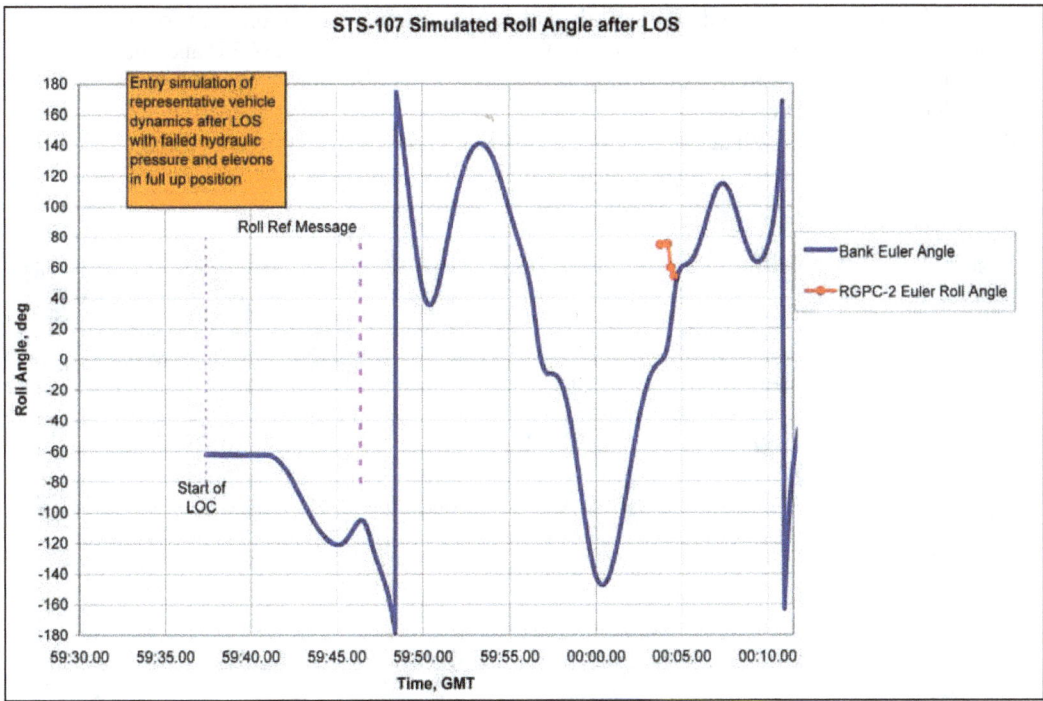

Figure 2.1-11. *STS-107 simulated roll angle after loss of signal*. Red data indicate roll conditions recorded in the reconstructed general purpose computer-2 data.

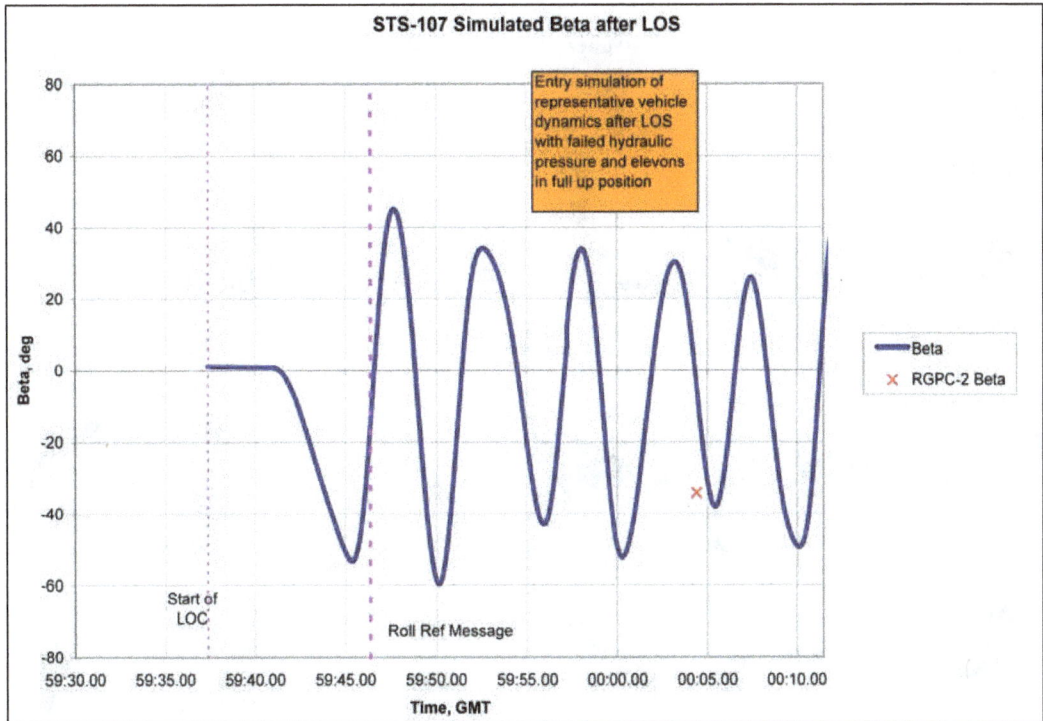

Figure 2.1-12. *STS-107 simulated sideslip angle after loss of signal*. Red data indicate beta condition recorded in the reconstructed general purpose computer-2 data.

Figure 2.1-13 is a second-by-second graphical depiction of the simulated motion of the orbiter during this period. The sequence begins at the start of the LOC period (GMT 13:59:37) and ends at the CE (GMT 14:00:18). The view is from a point in front of the orbiter's direction of travel, looking backward along the velocity vector. The snapshot for GMT 14:00:04 shows the left wing departing intact. In reality, the left wing did not come off all at one time but was shedding debris over a period of time.

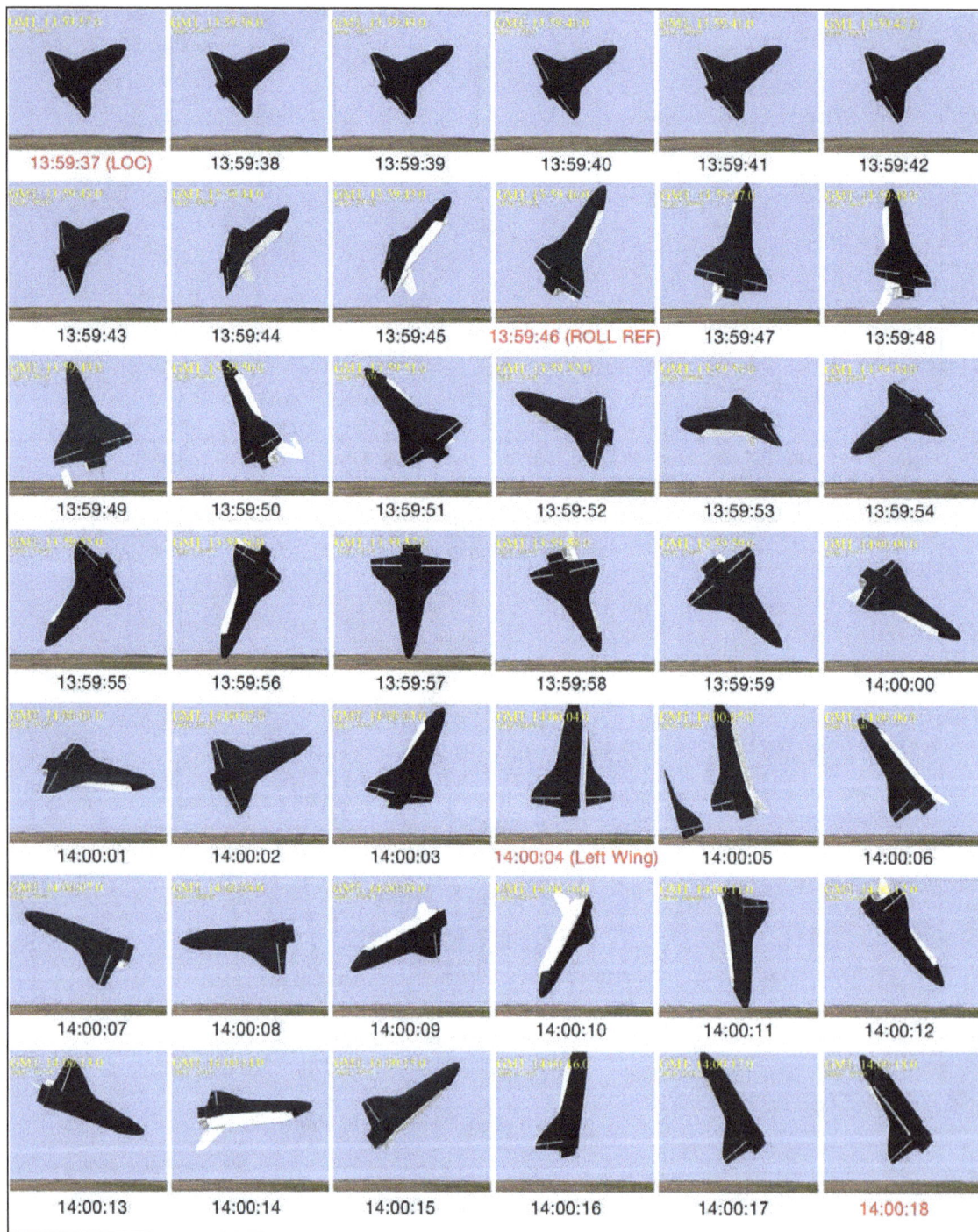

Figure 2.1-13. *Entry simulation snapshot sequence.* Times are in Greenwich Mean Time.

With the sequence viewed from a vantage point looking back up the velocity vector, it is apparent that the predominant orientation of the orbiter remained "belly into the wind" with large excursions in pitch, roll, and yaw. This motion can be characterized[10] as a slow (30 to 40 degrees per second), highly oscillatory spin. As the simulation progresses in time, the rates and attitude excursions increase.

The entry simulation combined with trajectory data yielded the accelerations at the orbiter center of mass, the CM center of mass, and the x-links.[11] The total acceleration consists of the translational acceleration due to drag plus the rotational acceleration due to the rotation of the orbiter. The magnitude of the rotational acceleration is dependent on the distance from the center of rotation, and the direction of the load is always away from the center of rotation. The translational acceleration is always along the velocity vector as the orbiter decelerates due to aerodynamic drag. However, as the orbiter rotated, the direction of this deceleration load rotated with respect to the orbiter body axes. Sometimes the translational and rotational accelerations add together, other times they subtract from one another. The addition or subtraction of the translational and rotational accelerations depends upon the orientation of the orbiter at that particular time. Figures 2.1-14 and 2.1-15 show this concept.

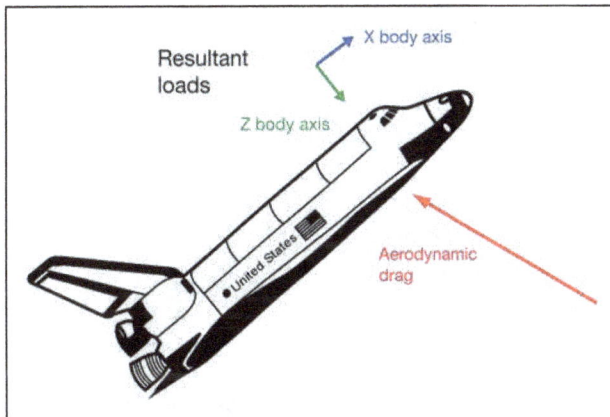

Figure 2.1.-14. Representation of resultant body axis loads imparted from aerodynamic drag in nominal entry attitude.

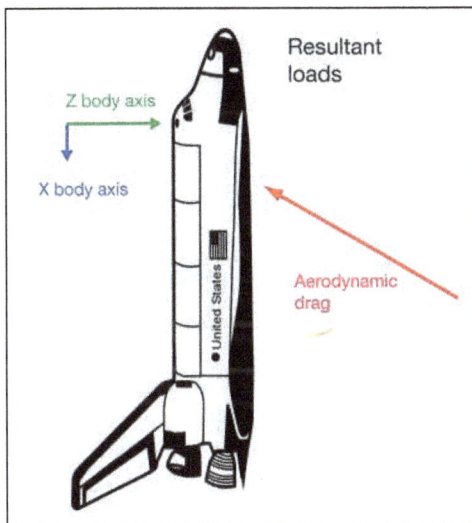

Figure 2.1-15. Representation of change in resultant body axis loads imparted from aerodynamic drag in off-nominal attitude.

---

[10]Flight Test Demonstration Requirements for Departure Resistance and Post-Departure Characteristics of Piloted Airplanes, Air Force MIL-F-83691B, Change 1, 5/31/96.
[11]The x-links connect the crew module to the forward fuselage at the X_o 582 ring frame bulkhead.

The accelerations are shown in figures 2.1-16 through 2.1-18. These figures show the accelerations due to rotational motion of the orbiter plus the translational accelerations due to atmospheric drag. There is one figure for each orbiter axis, and each figure shows the accelerations at the three different vehicle locations – at the orbiter center of mass, at the CM center of mass, and at the x-link location. Included in each figure is a pictorial that helps describe the effective motion of crew members that would result from these accelerations.

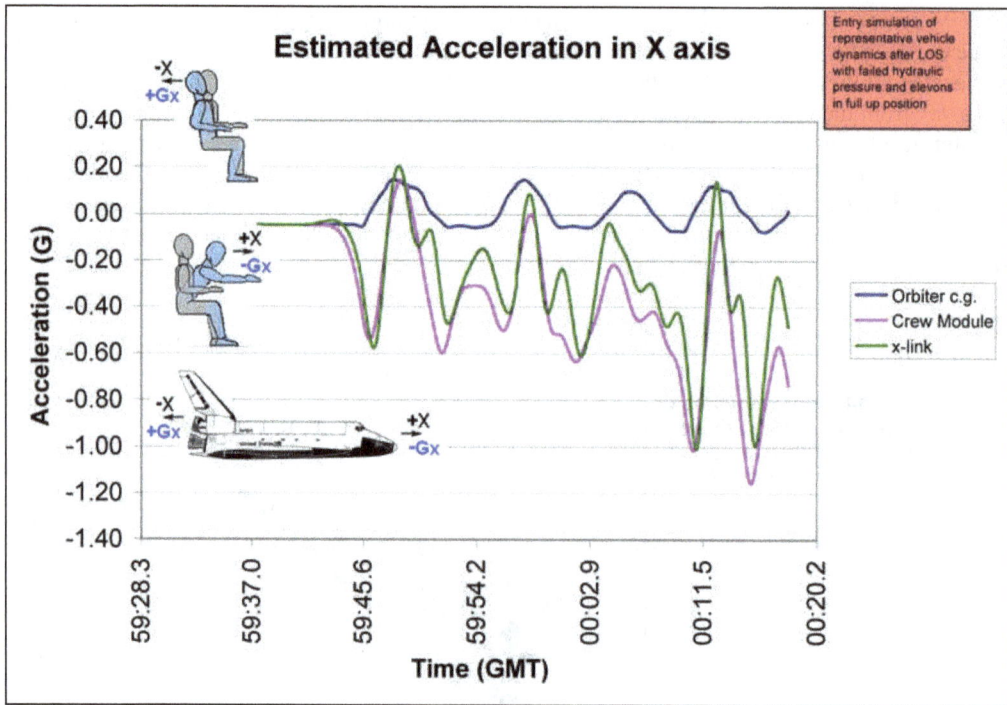

Figure 2.1-16. *Estimated accelerations at the crew module in the X axis from the loss of signal to the Catastrophic Event.*

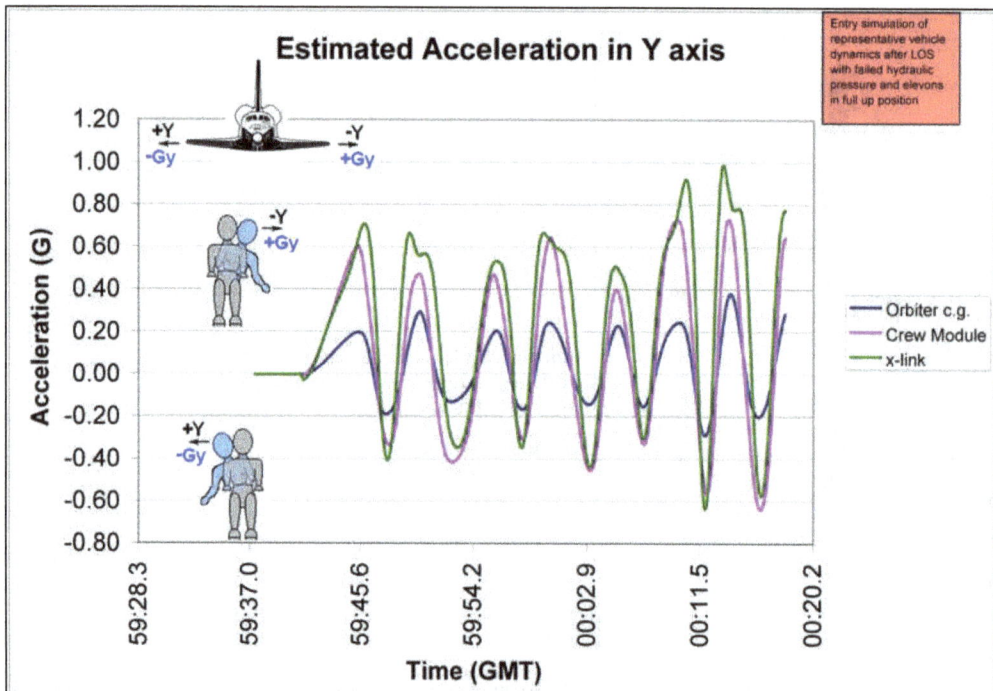

Figure 2.1-17. *Estimated accelerations at the crew module in the Y axis from the loss of signal to the Catastrophic Event.*

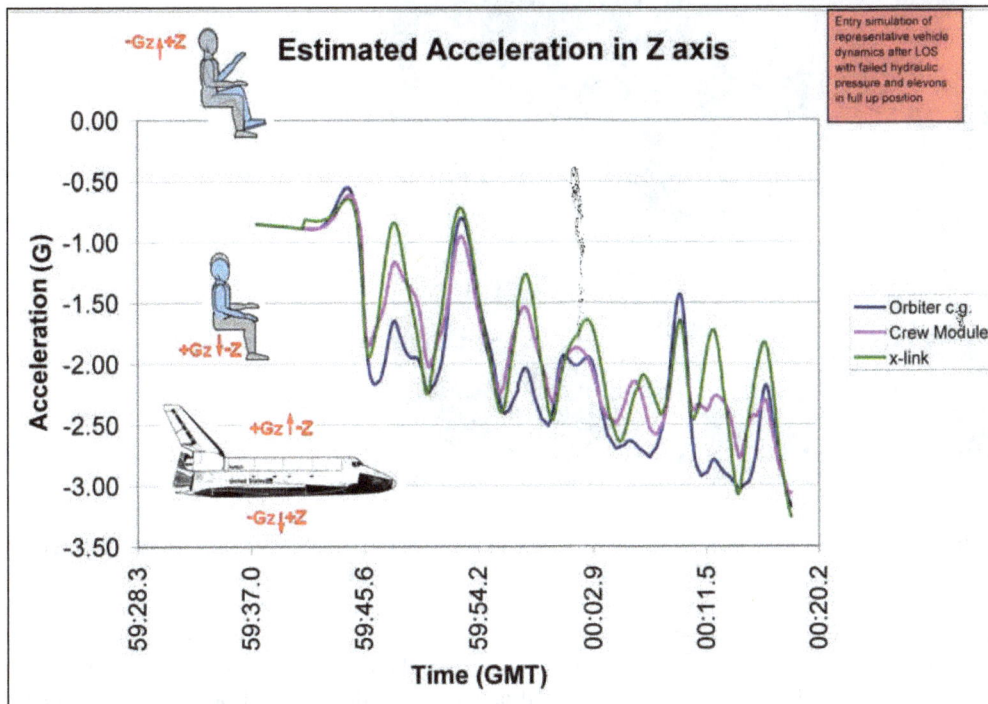

Figure 2.1-18. *Estimated accelerations at the crew module in the orbiter Z axis from the loss of signal to the Catastrophic Event.* [Note: Crew and orbiter Z axes have opposite convention]

As shown in the figures' pictorials, the motion experienced by the crew members inside the vehicle during the intact orbiter LOC time period is best represented by a swaying motion side-to-side, a pull forward in the seat, and a push down into the seat.

2.1.4 Post-Catastrophic Event attitude dynamics of the forebody

The SCSIIT received anecdotal information that the *Challenger* CM was believed to have reached a stable attitude ("trimmed") in a nose-into-the-wind condition. This belief was based on a video that shows a brief clear view of the CM in this orientation (figure 2.1-19), and the fact that the CM appeared to have impacted the water in a slightly nose-down condition.[12] As a result, the assumption upon which crew procedures are based (see Section 3.3) was that in an LOC/breakup scenario, the CM would trim nose-down after reaching terminal velocity. Although many types of data regarding the *Challenger* accident are available, they are almost entirely limited to information about the cause of the accident and are not related to crew survival. Relevant information was generally located by contacting personnel who worked at NASA at the time. No data could be found to support this anecdotal conclusion other than the video. Ground-based video of the forebody of Columbia implied that the forebody was rotating. Consequently, the SCSIIT set out to assess the assumption that the *Columbia* forebody came to an aerodynamically stable attitude.

---

[12]Report of the Presidential Commission on the *Challenger* Accident, Volume III, Appendix O, NASA Search, Recovery and Reconstruction Task Force Team Report.

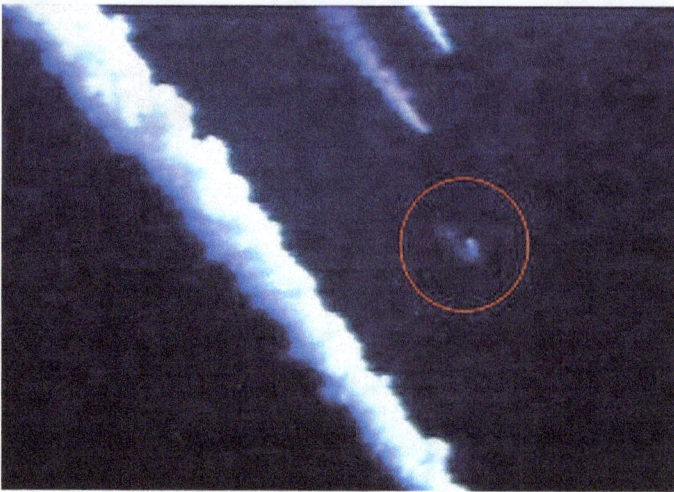

Figure 2.1-19. *Video-capture showing Challenger crew module (circled in red) pointed nose into the direction of travel.*

No telemetry, test data, or previous analysis existed to characterize the behavior of the CM or the forebody after separation from the rest of the orbiter. Therefore, the aerodynamic properties of the forebody had to be estimated and incorporated into an aerodynamic simulation to determine the forebody dynamics. The objective of this study was to examine the stability and possible trim attitudes of the free-flying CM and forebody [CM plus forward fuselage (FF) (figure 2.1-20)] from the CE to the CMCE. These data, when combined with the forebody trajectory aerodynamic drag data, provided the loads applied to the forebody during this phase. For simplicity, the initial conditions for altitude, velocity, and mass properties were held constant through out this 35 second period. The initial condition altitude was 176,790 feet with an initial velocity of Mach 14.4. These initial conditions were extrapolated from the last known vehicle state and projected through the simulation in the preceding section to the CE. The forebody mass properties were estimated using the evaluation of the orbiter breakup (see Section 2.2) and engineering judgment. This motion analysis was based solely on approximations and modeling. Therefore, the results should be considered indicators of general trends, and representative of the approximate accelerations and aerodynamic behavior that likely occurred.

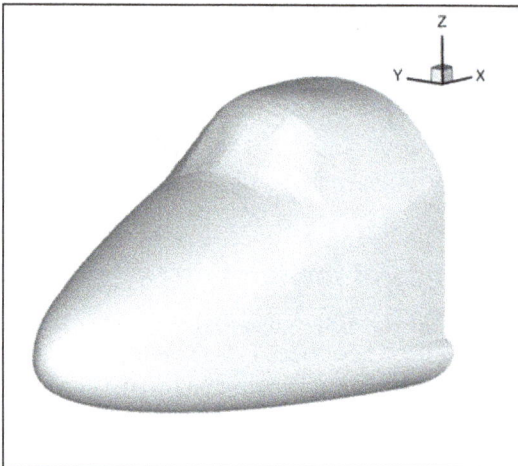

Figure 2.1-20. *Forward fuselage/crew module (forebody) configuration.*

Before any simulations were run, the pitch stability of the CM and the forebody was examined. This was a static, 3-axis uncoupled analysis. Plots of the pitching moment coefficient, C_m, vs. angle of attack showed two possible statically stable attitudes for the forebody at −100° and 125° angles of attack (CM stable attitudes were slightly different). Figure 2.1-21 shows the attitudes with the arrow indicating the direction of the wind.

Figure 2.1-21. *Pitch stable attitudes for the forebody (arrow shows direction of wind).*

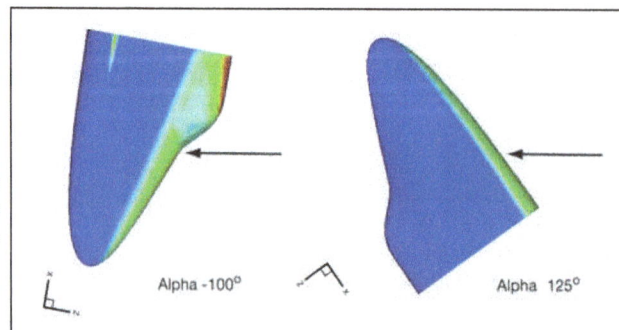

Alpha -100°   Alpha 125°

However, evaluation of the rolling and yawing moment coefficients, C_l and C_n showed that the statically stable attitudes in pitching moments (C_m) are not stable in either yaw or roll. Figure 2.1-22 shows the contour plot of all three moments across an alpha and a beta sweep. Locations where all three moments coincide would be stable in all three axes. Although at alpha approximately +50 degrees the moments appear very close to coinciding, close inspection showed that they do not.

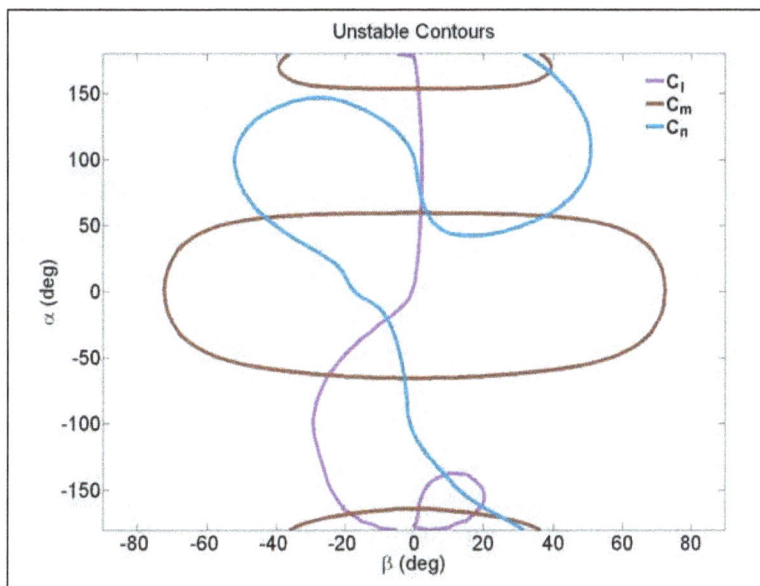

Figure 2.1-22. Contour plot of pitch, roll, and yaw moments for a crew module showing no stability due to slight offset in lateral center of gravity.

To verify this conclusion that the $-100°$ and $125°$ angle-of-attack attitudes were not stable in the other axes, a forebody simulation 6-degree-of-freedom simulation was run. The initial attitude rates were zero, the initial angles of attack were set to the pitch trim points ($-100°$ and $125°$), and the sideslip angle was set to zero. In each case, after a few seconds the forebody began to rotate about the X axis with an increasing rate, and a "wobble" developed about the Y and Z axes. After evaluating the simulations, it was determined that the reason for the instability was a lateral offset of the c.g. The lateral c.g. limit for normal orbiter operations[13] ranges from ±1.0 in. to ±1.5 in. maximum depending on the X_{cg}. The lateral c.g. for *Columbia*'s CM was 0.7 in. from the centerline. This is well within operating limits for the orbiter, but greater than zero.

The simulation was re-run with the lateral c.g. set to zero. Both the $-100°$ and $125°$ angle-of-attack cases were stable when starting with no initial rates. However, when the simulation was initiated with any initial rate in any axis, the forebody became unstable. The forebody never reached a stable attitude when initialized in any other attitude than one of the statically stable angles of attack. Since the orbiter was rotating at the time of breakup, it is expected that the free-flying forebody would also be rotating upon separation from the orbiter. It is concluded that the *Columbia* forebody did not attain a stable attitude.

The team also assessed whether a free-flying CM (with no forward RCS compartment or gear well attached) would come to a stable attitude. This is the *Challenger* case. The analysis was performed using the same tools. The assessment showed that the CM alone would not stabilize either.

Finally, it was speculated that the trailing wires and cables from the CM or forebody may have provided some aerodynamic stability (similar to the tail on a kite). The cables were modeled as 10 stiff poles trailing behind the CM with a diameter of 1.5 in. (based on maximum wire bundle size), a length of 10 feet (figure 2.1-23),

---

[13]Flight and Ground System Specification, NSTS-07700, Volume X, Book I, Section 3.3.1.2.1.5 Orbiter – CG limits, November 10, 1998, p. 3-142.

and a mass of 2.43 lbs. per foot based on a bundle of 715 strands of 22-gauge wire. The resultant simulation showed no significant effect on stability for either the CM or the forebody. This is most likely because of the low mass of the cables compared to the forebody.

Modeling is subject to some inherent uncertainty regarding the actual motion dynamics. However, all analyses consistently showed a lack of stability.

Finding. The forebody and the CM of the orbiter are not aerodynamically stable in attitude with any initial rates or lateral center of gravity (c.g.) other than zero.

To determine the angular velocities and the accelerations at various locations in the forebody for the time period from the CE to the CMCE, two dynamic simulations were performed. Both cases used actual STS-107 c.g. conditions and an initial forebody attitude of $\alpha = 90$ degrees and $\beta = 10$ degrees. This attitude was extrapolated from the final attitude at the end of the LOC-to-CE simulation discussed earlier. Although the attitude of the forebody could be projected from the LOC simulation, the rates imparted to the

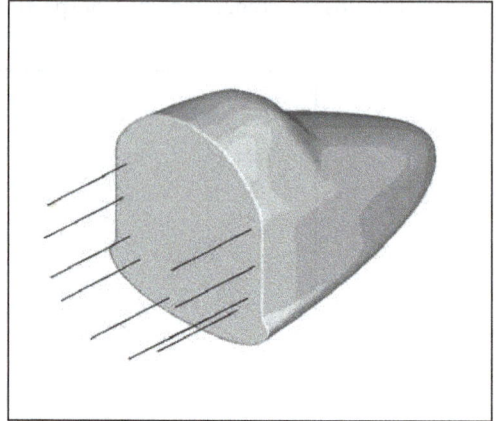

Figure 2.1-23. Modeled trailing wire configuration for forebody.

forebody at the moment of separation cannot be determined. One case was run without any initial rotation rates, and one case was run with initial attitude rates. The initial attitude rates were extrapolated from the final rates from the LOC simulation and are –70 deg/sec in roll, –30 deg/sec in pitch, and 25 deg/sec in yaw. These rates assume that the forebody separated and maintained the same rates experienced by the intact orbiter at the moment of the CE. This is not likely, but the rates were assumed to be representative. Additionally, the shape of the forebody was held constant through the simulation, although it is likely that the forebody's shape and mass properties were changing due to debris shedding. The simulation is believed to address the dominant variables affecting the motion of the forebody. However, because of the many assumptions used, the simulations can only be considered representative of the type of motion that most probably occurred and not an exact determination of a specific attitude or rate at a specific time.

Figures 2.1-24 through 2.1-26 shows the forebody's angular velocity for each case.

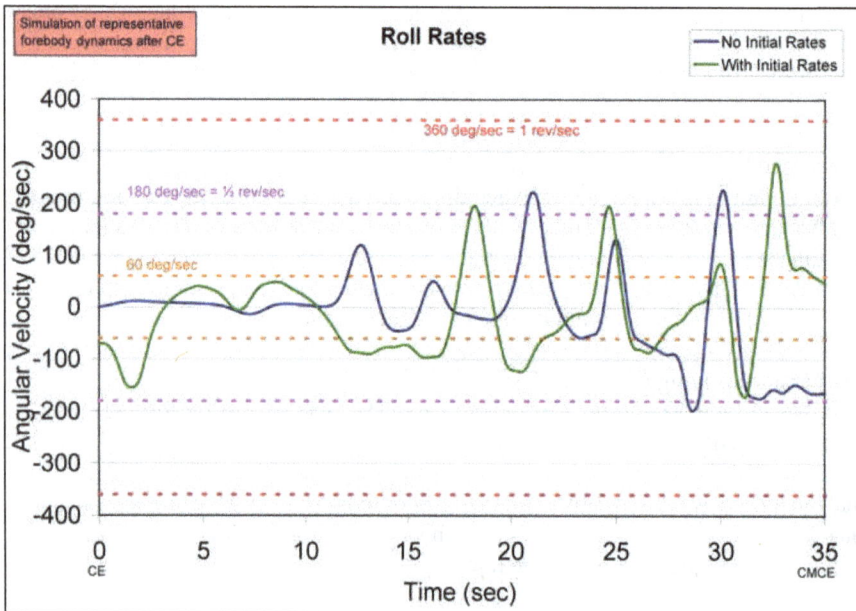

Figure 2.1-24. Free-flying forebody roll rates, with and without initial rates.

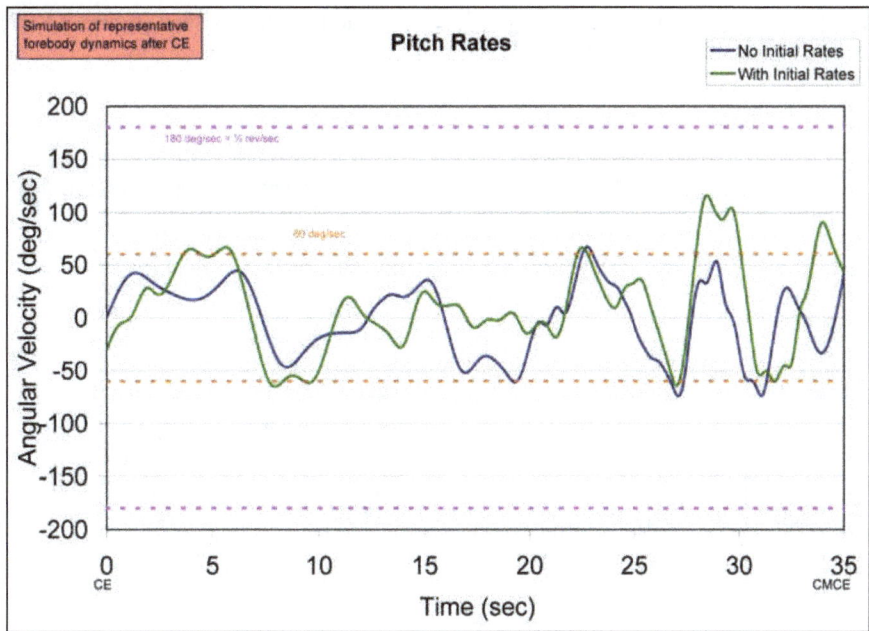

Figure 2.1-25. Free-flying forebody pitch rates, with and without initial rates.

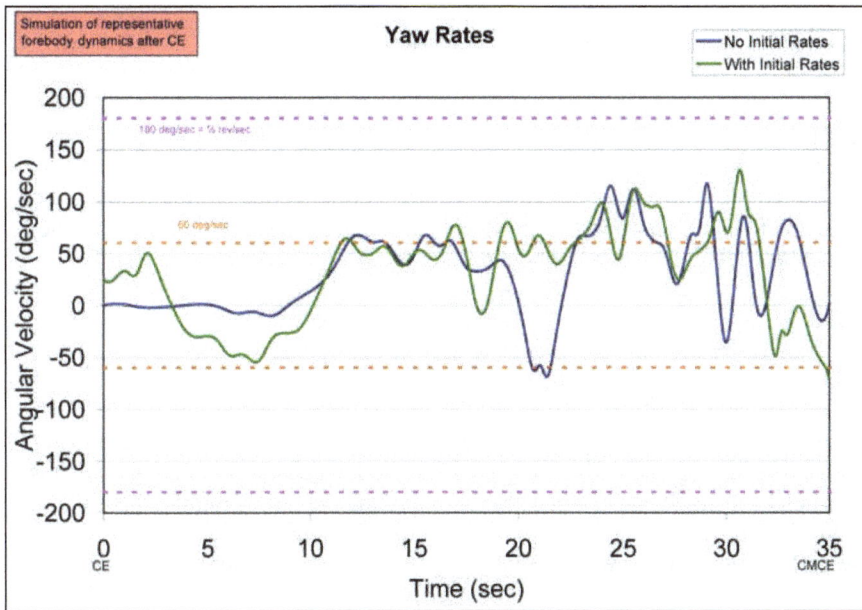

Figure 2.1-26. Free-flying forebody yaw rates, with and without initial rates.

To summarize the data shown above, the analysis shows that for the first 12 to 15 seconds of free-flying motion with no initial rates, the rotation rates remained extremely low. For the case with initial rates, they remained fairly low (<60 deg/sec, or ~0.17 rev/sec) as well. Following that period, with or without initial rates, the rotation rates climbed to as high as 0.5 rev/sec by the CMCE.

This dynamic attitude motion was combined with trajectory data in the same way as previously described for the out-of-control orbiter to determine accelerations at crew seat positions. Each seat was located at a different location from the forebody c.g., so there are slight differences for each location. Figure 2.1-27 shows a chart of the simulation accelerations from the CE to the CMCE for a representative seat position. The accelerations are given in units of G and represent *total* acceleration.

The increasing atmospheric drag experienced by the forebody is seen (figure 2.1-27) in the trend of increasing accelerations. The oscillations are due to the forebody rotation.

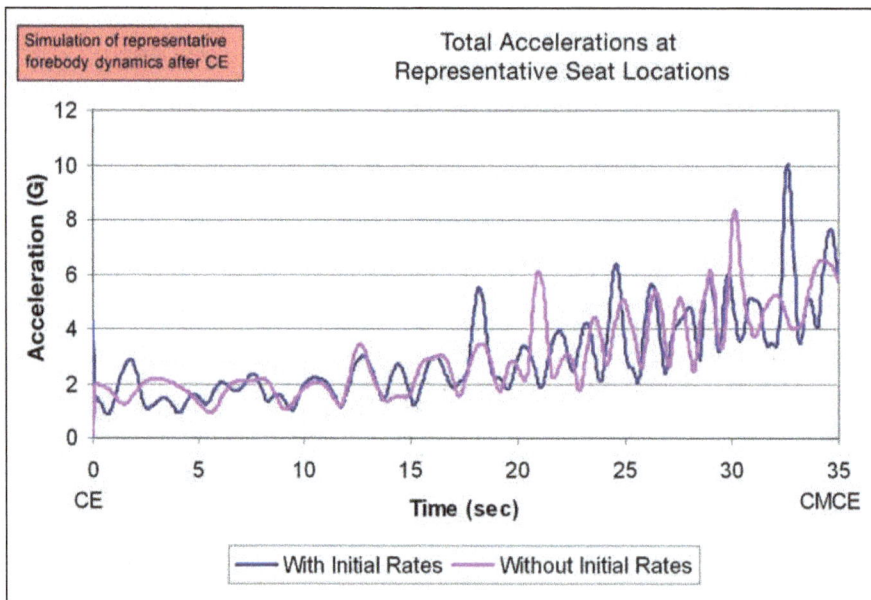

Figure 2.1-27. *Accelerations at a representative seat location.*

As the simulation progressed past 15 seconds, the increased rates and increasing aerodynamic drag resulted in higher accelerations at each crew position.

The conclusion of the forebody motion analysis was that the forebody was not in a stable attitude from the CE to the CMCE, but rotating about all three axes at a range from 0.1 to 0.5 rev/sec, with increasing rates and accelerations.

2.1.5 Relative motion comparison

In ground-based video of the mishap, it was noted that when the free-flying forebody could be identified, the image is periodically fluctuating in brightness. This suggested that the forebody was tumbling, which is consistent with the earlier motion analysis. It was not possible to determine the rate of rotation based on the brightening events because the forebody was irregularly shaped and rotating on more than one axis. Therefore, relative motion was selected as an alternate way of evaluating the motion of the forebody in the available video.

Relative motion analysis compares the rate of change of the movement of objects in the field of view (FOV) of a video. Rate of change can provide an estimation of relative motion experienced by the objects within a single frame of reference. Two relative motion analyses were performed. One applied to deceleration of the forebody during breakup (see Section 2.4). The other was a triangulation of relative motion of the forebody as compared to the main engines from two different ground-based videos between GMT 14:00:27.12 (9 seconds after the CE) to GMT 14:00:52.12 (just prior to the CMCE). This provided insight into the free-flying forebody motion for comparison to the aerodynamic analyses described above.

The basic assumption of the triangulation task was that the selected reference orbiter main engine had a known trajectory that was stable compared to the unknown relative position of the forebody. This is a reasonably good assumption because the engine ballistic number was quite high (> 200 psf) compared to the ballistic number of the forebody (~122 psf), and the engines were assumed to be relatively compact and non-aerodynamic compared to the forebody. Any relative motion between the two trajectories in the video was assumed to be a result of motion of the forebody.

The orbiter engine trajectory was very well understood for several reasons. First, the impact was registered seismically, providing an exact impact time which in turn provided an excellent understanding of its ballistic trajectory. Second, the engines were visible throughout the video through separation and the end of the video, so there was positive identification of the engine. A reference trajectory for the engine was calculated with positions computed at 20-Hz intervals to provide the "known" point in the relative motion.

Two videos with known ground coordinates relative to the trajectory were selected. These videos were known as Hewitt and Mesquite (WFAA4, see Section 2.2, Table 2.2-1) based on the nearest town in Texas where they were recorded (see Section 2.2 and Chapter 4). The FOV was calculated and then trimmed and stabilized from analysis of the imagery, and details about the camera and location were provided by the original photographers. There were a few frames without the engines or the forebody in view; these were not used in the analysis.

An interactive tool was developed to solve for unknown 3-dimensional locations by using 2-dimensional image pairs. The output of this tool led to a reference-trajectory-relative, time-stamped motion path for the forebody.

Figure 2.1-28 provides a snapshot of the tool with known and unknown points selected on still images from the Hewitt and Mesquite videos. The cyan-magenta-yellow axes represent the known engine location specified in the reference trajectory. The red-green-blue axes, with red dot, depict the location of the triangulated CM.

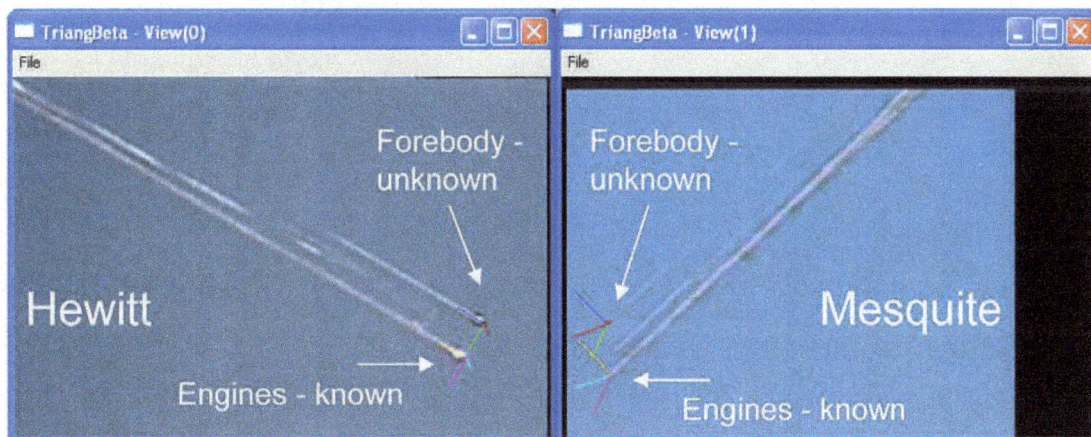

Figure 2.1-28. *Depiction of the triangulation tool.*

With these data, the relative motion in each axis were plotted (figure 2.1-29).

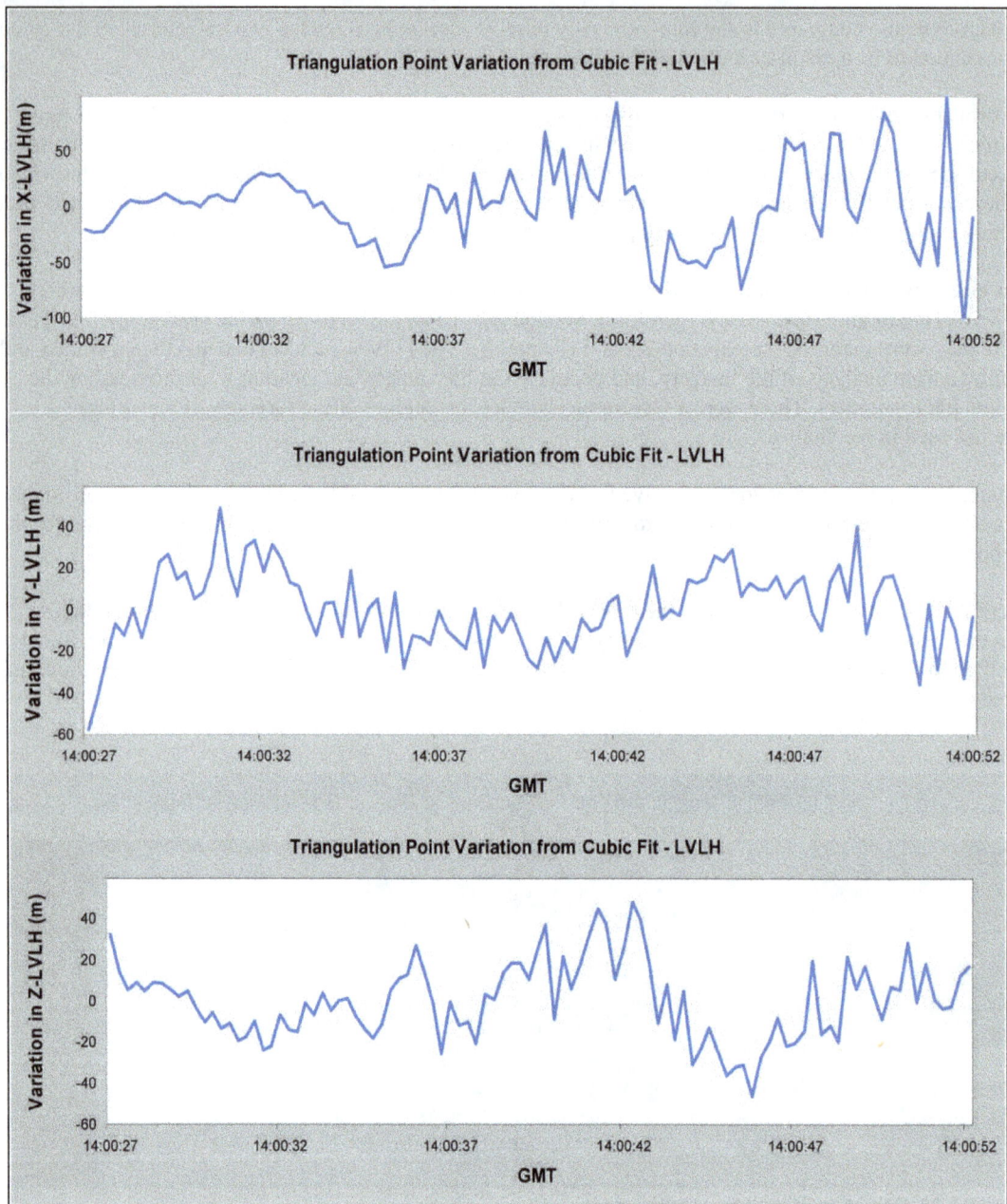

Figure 2.1-29. *Relative motion between the forebody and the main engines.*

These data essentially represent a wobble motion between the two objects. Since the main engines had a much higher ballistic number and a more compact shape, the wobble is assumed to be predominantly due to rotational motion of the forebody changing lift and trajectory properties slightly as the orientation to the drag vector changed.

A Fast-Fourier Transform analysis was performed on the X-, Y-, and Z-axis data to look for frequency of motion. The frequency of the wobble motion was assumed to correlate to the rotation rate in that axis (figure 2.1-30).

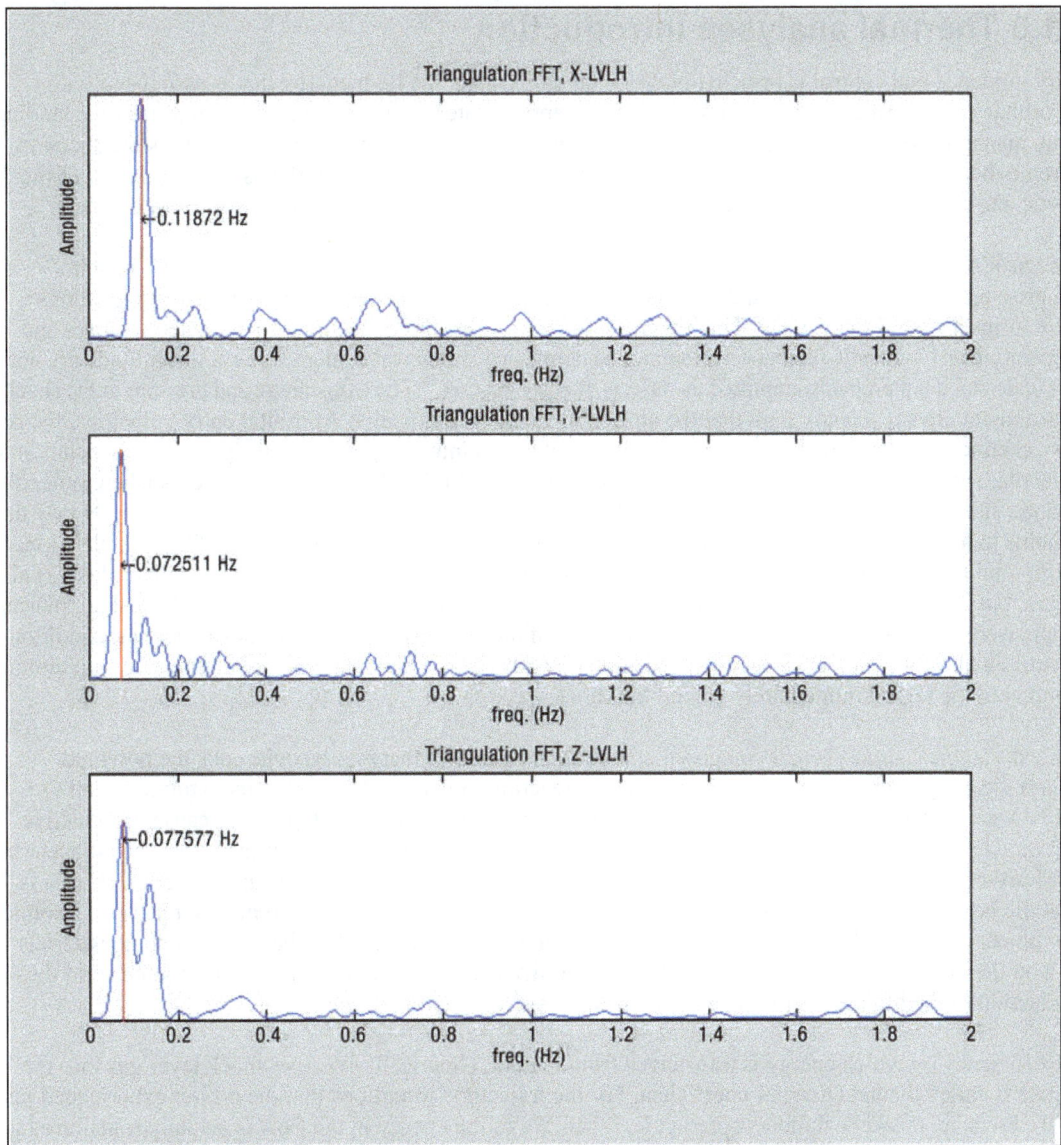

Figure 2.1-30. *Fast-Fourier transform of relative motion in X, Y, and Z axes in hertz.*

The results show a remarkable level of consistency in frequency of wobble motion in all three axes from approximately 0.07 to 0.12 Hz (cycles per second). This wobble cycle is assumed to reflect the rate of rotation in each axis in revolutions per second. The aerodynamic analysis predicted a rotation rate of approximately 0.1 to 0.5 rev/sec in multiple axes. The triangulation analysis thus provided support for the aerodynamic analysis in that the rotation rates are the same order of magnitude as found in the aerodynamic simulation.

Finding. Triangulation results suggest that the free-flying forebody rotated at approximately 0.1 rev/sec in a multi-axis motion.

> **Conclusion L2-1.** Between orbiter breakup and the forebody breakup, the free-flying forebody was rotating about all three axes at approximately 0.1 rev/sec and did not trim into a specific attitude.

2.1.6 Thermal analyses introduction

Entry guidance and control is principally designed to dissipate the high energy that is associated with the orbital velocity of the vehicle, and arrive at the appropriate altitude and velocity conditions for landing, while managing vehicle heating during entry. The atmospheric entry from orbital velocities produces an extreme thermal and chemical environment. At the most fundamental level, this is due to transfer of the kinetic and potential energy of the entry vehicle into thermal and chemical energy in the atmosphere.

The Earth's atmosphere is composed primarily of molecular nitrogen (N_2) and oxygen (O_2), which comprise approximately 78% and 22% of air, respectively. At hypersonic entry speeds, a strong shock wave forms ahead of the vehicle. This shock wave compresses the ambient air to extreme pressures and temperatures. Hypersonic separation dynamics of debris and shock wave interactions at this altitude are not well understood but probably amplified the effects of entry heating.[14] The temperature and pressure in the shock environment are sufficiently high that the air begins to react chemically. At orbital entry velocities, the primary chemical reactions that occur are dissociation of atoms into charged particles (plasma), dissociation of molecular species, and recombination of the resulting atoms. Molecular dissociation occurs when molecules separate into their constituent atomic species ($O_2 \rightarrow O+O$, $N_2 \rightarrow N+N$). Further, monatomic O and N may then combine to form nitric oxide ($N + O \rightarrow NO$). The amount of dissociation is a function of the vehicle's kinetic energy and the strength of the molecular bonds. Dissociation of atoms in charged particles (plasma) also occurs, but is reduced with altitude and Mach number. For the conditions of the *Columbia* mishap, plasma effects were not considered significant. For the case of the space shuttle orbiter at peak heating conditions, essentially all of the molecular O_2 and approximately 50% of the molecular N_2 are dissociated into atomic O and N in the region immediately behind the shock wave.

While the ambient atmosphere is composed of molecular N_2 and O_2 that may be quite cold, the flowfield around the entry vehicle is composed primarily of reacting molecular N_2 and O_2, monatomic N and O, and NO, all at extreme temperatures. At the vehicle surface, the extreme temperature causes conductive heat transfer, a process in which thermal energy is conducted into the vehicle surface from the adjacent hot gas. Further, the presence of dissociated N and O atoms allows for two other heat transfer mechanisms – catalytic heating and oxidation heating. Both of these processes occur when exothermic reactions (atomic recombination and oxidization, respectively) occur at the vehicle surface. The thermal protection materials used on the orbiter are designed specifically to minimize the influence of these two phenomena and the conduction of the high temperatures into the vehicle's structure.

The efficiency by which energy is transferred from the hot, chemically-reactive shock layer gas into the vehicle is called the heat transfer coefficient. For the trajectory conditions that the orbiter experienced just before breakup, theoretical models predict less than 5% of the energy in the flow is actually transferred into the vehicle. The remainder of the energy is left in chemical and thermal modes in the wake of the vehicle. However, at off-design conditions (e.g., vehicle damage or off-nominal flight attitudes), substantially higher heat transfer coefficient values are possible, resulting in much higher energy (or heat) transfer into the vehicle.

In a nominal orbiter entry trajectory, the heating rates vary with time and location on the vehicle. The highest peak heating rate of about 60 British thermal unit (Btu)/ft^2-sec occurs at the wing leading edge. Shock waves forming on different parts of the orbiter can intersect creating shock-shock interactions. These shock-shock interactions influence a vehicle's aerodynamics and increase the heat transfer rate and pressure where the interaction impinges on the vehicle's surface.

A thermal analysis was performed on the CM to understand the maximum survivable altitude for an orbiter breakup. Additionally, thermal analyses were performed to compare predicted entry heating to actual debris condition for several items. The purpose of these assessments was to understand the sequence of events and help verify ballistic release times. Analyses were performed on several specific items of debris, including a

---

[14]NASA TM X-1669, "Flight Experience with Shock Impingement and Interference Heating on the X-15-2 Research Airplane," October 1968, p. 7.

helmet, boot soles, a key piece of middeck floor, a laptop computer, the MADS/OEX recorder, the attach fittings of the CM to the FF, and a payload bay door (PLBD) roller.

Object Reentry Survival Analysis Tool (ORSAT) was used for these thermal analyses. ORSAT incorporates algorithms for trajectory simulation, atmospheres, aerodynamics, aeroheating, and thermal modeling, but its strength is its capability to combine all those algorithms into a time-efficient analysis. For one analysis, the ORSAT outputs were compared to two other models, Boundary Layer Integral Matrix Procedure-Kinetic (BLIMP-K) and Systems Improved Numerical Differencing Analyzer (SINDA).

Simulation and Optimization of Rocket Trajectories (SORT) was used to model the times of release from the reference trajectory for thermal and ballistic analyses. Snewt is a computer program that uses the modified Newtonian method to compute a surface pressure distribution and various aerodynamic coefficients. This was used to predict whether a debris item would stabilize in a given attitude during entry.

For more information on ORSAT, SORT, Snewt, BLIMP-K and SINDA, see Chapter 4.

2.1.6.1 *Thermal analysis – crew module*

The *Columbia* accident prompted a careful look at the question: What is the highest altitude at which an orbiter breakup can occur so that a bare-metal CM can reach the ground without being compromised thermally? The LOC/Breakup Cue Card procedure for crews is based on the estimate that the CM would survive aeroheating if separation took place below 280,000 feet during ascent. This estimate was obtained by a simple thermal analysis, for ascent only, which was performed after the *Challenger* accident.[15] This analysis was based on the assumption that the CM would trim nose into the velocity vector. The thermal element clearly is the most important for entry, but can also play a role on ascent depending on the altitude and speed at the time of vehicle breakup. The team wanted to verify the results of the earlier analysis and to perform an analysis for entry aeroheating.

Obviously, the maximum crew survival altitude is more than just the thermal survival of the CM. Other considerations are the structural capability of the CM to withstand dynamic loads, the crew's ability to escape, and the capability of the existing crew escape equipment (CEE). See the sidebar "Maximum Survival Altitude of the Crew Module" on the following page for a discussion of these issues.

To accomplish the thermal analysis, proper heating rates and sufficient modeling of the CM were required. The heating rates were established by modeling the trajectory of the CM after it had separated from the rest of the vehicle. The initial state vector conditions were dictated by the trajectory profile used. Initially, three trajectory profiles were defined: a typical ascent trajectory; the STS-107 predicted entry from a 39-degree inclination orbit; and a typical International Space Station (ISS) mission entry from a 51.6-deg inclination orbit. Two different inclinations were chosen for entry because entries from higher inclination orbits have higher relative velocities and hotter temperatures. However, analysis showed that there was no detectable difference in the maximum survivable thermal altitudes between the two entry trajectories. Figure 2.1-31 displays the altitude vs. velocity for each of the three trajectory profiles. By design, the ascent trajectory has much lower velocities for a given altitude than the entry trajectories.

---

[15]ES34-87-47M, Crew Escape Thermal Response Study, May 1987.

Maximum Survivable Altitude of the Crew Module

A more detailed discussion of the maximum survivable altitude of the CM must address the structural and equipment capability of the CM as well as the thermal capability. The CM is relatively strong compared to other orbiter structures as it is designed to sustain a significant delta-pressure load against the vacuum conditions in space. Segments of the CM are also designed to withstand crash-landing loads in some axes of up to 20 G. However, the design intent is to have structure absorb enough of the loads so that crew members in their seats are able to egress, not necessarily to have the CM stay structurally intact. The bottom line is that it is not possible to know for certain what load conditions the CM is truly capable of withstanding. According to an estimation of forces on the Challenger *CM immediately after the explosion, the* Challenger *CM likely experienced 16 G to 21 G[16] at orbiter breakup and yet apparently maintained integrity all the way to water impact. Deceleration load spikes in a ballistic trajectory are highly dependent on the initial condition; the lower the velocity at the breakup, the flatter the load spike will be. At higher Mach numbers (generally higher altitudes in the profile) the load spike will be quite high as the change in ballistic number is exacerbated. Since each breakup scenario and resulting trajectory loads case would be different, it can only be said that the higher the altitude, the higher the loads will be, and that based on actual circumstances it may or may not be a factor in loss of CM integrity.*

Rotational forces are yet another constraint to survival. First, they can increase the overall loads that are experienced by the structure and lead to structural breakup. Additionally, they can constrain the crew from moving to the hatch for a bailout. Analysis of loads for the Columbia *mishap showed that rotational loads may be quite high for an unstable, rotating CM (possibly up to 3 G at the seats farthest from the CM c.g.). While it is possible for a restrained crew member to brace and maintain consciousness, this condition would hinder and likely prevent seat egress and bailout. Rotation rate will be dependent both on the initial conditions at the breakup and on the aerodynamic conditions at breakup. In general, at higher altitudes the vehicle will behave more like a spacecraft and less like an airplane. At higher altitudes, momentum and mass properties dominate rotation rate, and rates will take longer to increase. At lower altitudes, the aerodynamic coefficients affecting stability will begin to dominate, and rotation rates can ramp up quickly. In general, the crew should not assume that rotation rates will dampen out and should make every attempt to move to the hatch as quickly as possible for egress.*

The final consideration is the capacity of the suit to support egress (see Section 3.2). The advanced crew escape suit (ACES) is rated to 100,000 feet. This rating is not a performance limit, it is a certification limit. However, no ultimate determination has been made on the maximum temperature and loads at which the suit can survive. Since each bailout situation would result in a different combination of heat and total airloads, it is not possible to pinpoint a specific altitude as a limit.

---

[16]JSC 22175, STS-51L, JSC Visual Data Analysis Sub-Team Report, Appendix D9, June 1986.

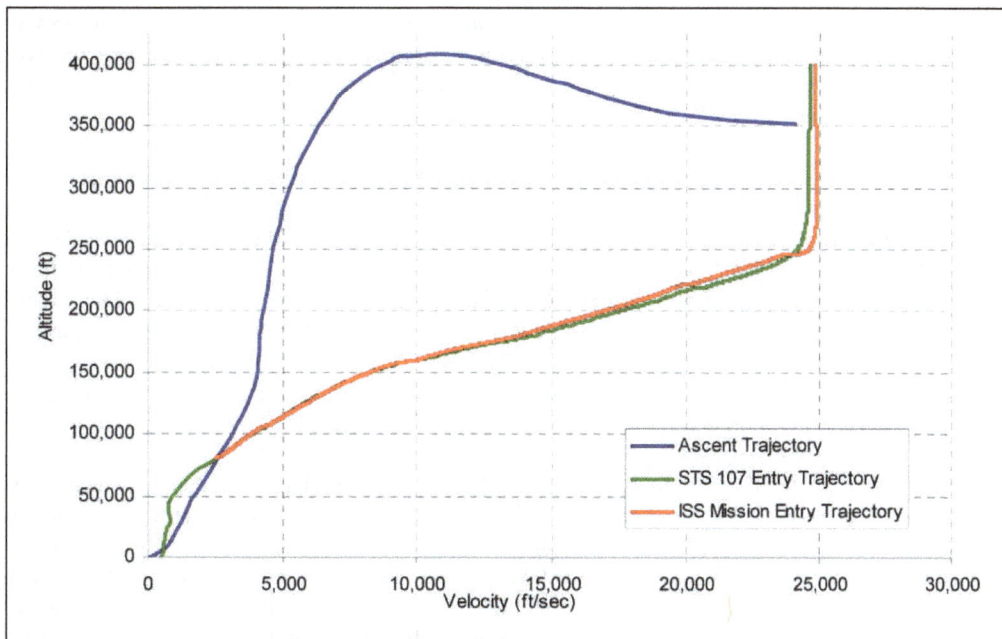

Figure 2.1-31. *Altitude vs. velocity for selected trajectory profiles.*

The analysis was conducted assuming that the CM was intact without any protection from the FF and accompanying Thermal Protection System (TPS). This was a conservative assessment. If the FF and accompanying TPS stays with the CM, it would provide some thermal protection for the CM. This appears to have happened for *Columbia*. However, predicting the exact level of protection provided by the TPS in various scenarios is impossible due to the infinitely variable configurations possible after a chaotic event such as a vehicle breakup. Additionally, the aft bulkhead of the CM is not protected by TPS. Overall, the assumption of no FF/TPS protection results in a conservative (worst-case) estimate of the thermal survival capability of the CM.

The orientation of the CM has a significant impact on the heating model. A stable orientation will result in a more rapid burn-through while a rotating object will distribute heat throughout the object and take longer to burn-through. The preceding motion simulations suggest a slow initial rotation with a gradual increase in rate. However, the *Challenger*-era analysis assumed a stable attitude. *For the purposes of verification and comparison*, both stable and rotating conditions were evaluated. The stable attitudes that were determined in the aerodynamic trim analysis were too complex to model, so the team selected two critical burn-through areas: the aft bulkhead and the forward bulkhead of the CM. These attitudes were defined as the selected bulkhead facing into the velocity vector, the worst-case thermal scenario. Because the post-*Challenger* analysis assumed a stable, nose-down attitude, the analysis assumed the forward bulkhead critical case.

The CM bulkheads have thin sections of skin supported by thicker webs. This complex geometry could not be modeled using the available thermal analysis tool (ORSAT) without substantial modification of the code. Simplifying assumptions were used to bound the problem and provide insight into relative behaviors.

This geometry has the potential for quicker burn-through due to the regions of thin skin, but it also has more mass (due to the thicker webs) that can absorb and distribute the heat away from the thin areas. It is assumed that the actual survivable altitudes would be bounded by analysis on optimistic and conservative cases. Due to these assumptions and constraints, the bulkheads were modeled as flat plates.

For the stable attitude cases, optimistic and conservative bulkhead thicknesses were used to bracket the results. For the optimistic (best-case) modeling, the thickness was established by averaging the entire mass of the bulkhead over the entire area. This optimistic approach produces a large thickness that will distribute heat better and results in higher survivable altitudes because the thicker skin takes longer to burn-through.

It should be noted that aluminum is a good conductor, and it is possible that the bulkhead area can shunt heat to the other areas of the CM, potentially making this "optimistic" estimate conservative. However, heat transport through complex shapes is poorly understood and extremely difficult to model, so it was not included in this analysis.

The conservative (worst-case) thickness for the forward bulkhead critical attitude was approximated by the thinnest portion of the bulkhead. The conservative thickness for the aft bulkhead critical attitude was more complex; it was an average of the thin skin sections of the bulkhead (not using the thickness of the webs). This approach results in lower survivable altitudes because the thinner skin will burn-through sooner. This conservative approach represents failure as a result of hole formation in the thinnest areas. This is conservative because structural failure in a thin individual location is not likely to cause total CM structural failure.

For the case of a spinning, non-stable CM, the CM was modeled as a cylinder. For this case, the skin thickness was averaged based on the mass and dimensions of the entire CM.

The next question was, at what temperature is failure expected to occur? Failure occurs as a result of both thermal and structural loads. There are three types of structural loads: translational loads from aerodynamic drag; rotational loads from vehicle rotation; and loads due to internal cabin pressure. The strength properties of aluminum (the principal structural constituent of the CM) are severely diminished as the temperature increases. Aluminum melts at about 900°F (482°C) to 1,100°F (593°C) depending on the alloy. However, at about 400°F (204°C), the strength of the aluminum begins to seriously degrade and the metal loses its ability to maintain structural integrity. Therefore, when the thermal analysis showed that the selected area on the CM reached 400°F (204°C), the CM was assumed to fail. Figure 2.1-32 shows the estimated strength vs. temperature for aluminum.

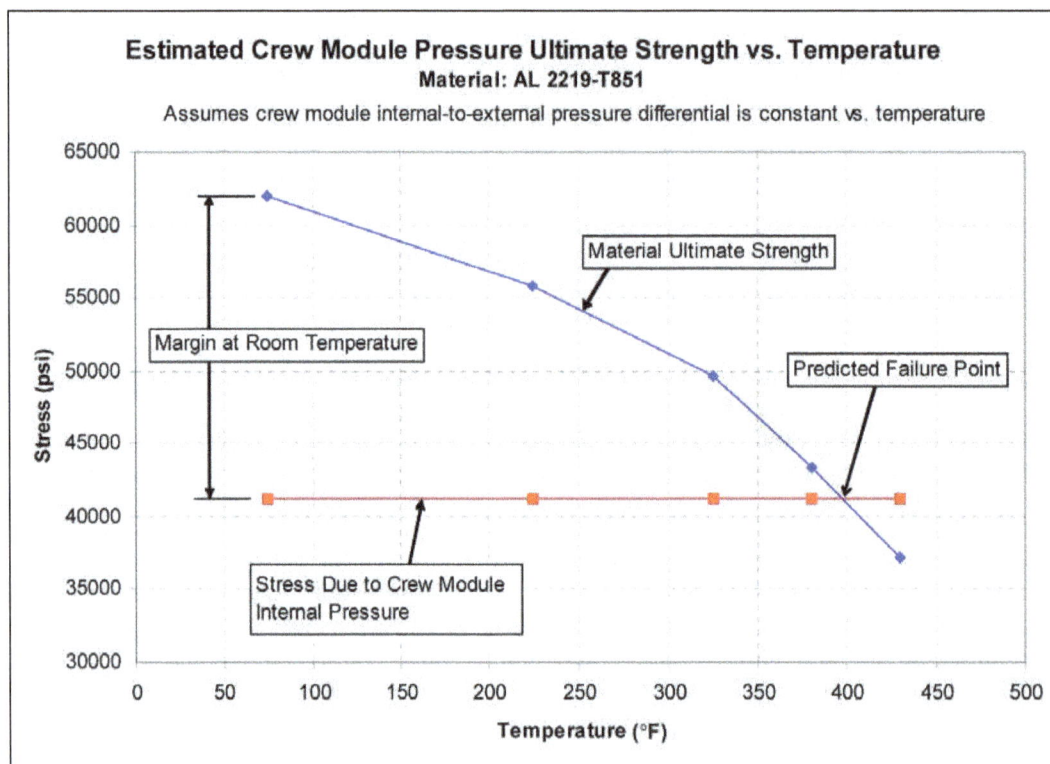

Figure 2.1-32. *Crew module stress vs. temperature.*

Table 2.1-1 shows the results of the analysis. The table is divided into the two different trajectory profiles: ascent and ISS mission entry. Under each trajectory is a listing of all the cases and the values for their respective ballistic number, skin thickness, and resulting maximum survivable CM altitude.

Table 2.1-1. *Maximum Survivable Altitude of Crew Module*

	Ballistic number (psf)	Thickness (in.)	Maximum Altitude (feet)
Ascent			
Aft bulkhead (optimistic)	88	0.34	>300,000
Aft bulkhead (conservative)	88	0.07	250,000
Fwd bulkhead (optimistic)	100	0.25	>300,000*
Fwd bulkhead (conservative)	100	0.03	120,000
Spinning	90	0.15	>300,000
ISS Mission Entry (51.6-deg inc)			
Aft bulkhead (optimistic)	88	0.34	160,000
Aft bulkhead (conservative)	88	0.07	130,000
Fwd bulkhead (optimistic)	100	0.25	140,000
Fwd bulkhead (conservative)	100	0.03	105,000
Spinning	90	0.15	150,000

*Correlates to the post-*Challenger* analysis case.

The analysis showed that the worst case is the forward bulkhead critical stable attitude (nose into the wind, and hole formation resulting in total structural failure). The stable attitude case was done only to compare to the earlier *Challenger* post-accident analysis, which concluded that 280,000 feet was the maximum thermal survivable altitude of the CM on ascent. However, the most relevant and realistic data come from the spinning case since the CM is not believed to have a stable aerodynamic trim condition. This analysis yielded a maximum thermal survivable breakup altitude that was greater than 300,000 feet on ascent and 150,000 feet on entry.

Finding. The estimate for maximum thermal survivability on ascent of 280,000 feet is a reasonable estimate.

Finding. The maximum thermal survivable breakup altitude for the CM on entry is approximately 150,000 feet.

As previously mentioned, this only provides the thermal element for the maximum survivable altitude for a orbiter breakup. See the sidebar, "Maximum Survival Altitude of the Crew Module," for a more detailed discussion.

In summary, the shuttle CM could probably thermally survive a breakup up to 300,000 feet on ascent and below 150,000 feet during entry. Aerodynamic loads may cause rapid structural failure of the CM at lower altitudes. Rotational loads may prevent the crew from translating to the crew hatch. High loads may also result during crew separation if the Mach number is still high (particularly on entry). The ultimate survival limits (altitude, air speed, thermal load, etc.) of the ACES remain unknown.

Conclusion L3-3. The actual maximum survivable altitude for a breakup of the space shuttle is not known.

Recommendation L2-2. Prior to operational deployment of future crewed spacecraft, determine the vehicle dynamics, entry thermal and aerodynamic loads, and crew survival envelopes during a vehicle loss of control so that they may be adequately integrated into training programs.

2.1.6.2 Thermal analysis – helmet

A thermal analysis was performed on a helmet to compare with ballistic estimates of release time. ORSAT predicted that helmets released at GMT 14:01:03 would match the debris results when the thermal decomposition temperatures were used to assume how much of each material would remain (see Section 3.2). This coincides remarkably with the ballistic estimates from SORT, which showed helmet release times within ±10 seconds of this time. This also coincides with the time range determined for the breakup of the forebody.

2.1.6.3 Thermal testing – boot soles

The geometry of the boot soles was too complex to model for ORSAT, especially because of the many alternatives for initial conditions. Boot soles of flight-like boots were thermally tested in an attempt to match the observed thermal damage (see Section 3.2). Test results could not be correlated directly to the debris observations, possibly because the test conditions did not sufficiently approximate the entry environment conditions.

2.1.6.4 Thermal analysis – lithium hydroxide stowage volume door

The lithium hydroxide (LiOH) stowage volume door was recovered with portions of the two seats still attached (figure 2.1-33). The LiOH door is located on the middeck floor with seats 6 and 7 attached to the door (figure 2.1-34). The condition of the door was markedly different from other segments of the middeck floor in that significant thermal erosion of the thinner areas had occurred (see Section 2.4). The hypothesis was formed that additional mass attached to this piece of debris resulted in higher heating. A thermal analysis was performed to determine whether the additional mass provided by the two seats and/or two crew members would result in the observed thermal damage.

Figure 2.1-33. *Recovered lithium hydroxide door with some seat legs still attached.*

Figure 2.1-34. *Example of an intact lithium hydroxide door with seats 6 and 7 attached.* [Crew Compartment Trainer]

Composed of aluminum 7075, the LiOH door is about 41 in. long and 28 in. wide, and has a mass of 41.6 lbs. Figures 2.1-35 and 2.1-38 show the upper surface and lower surface of the LiOH door. The thickness of the door varies and, in some cases, is as low as 0.05 in. The average thickness of the LiOH door is 0.39 in.

Photographs of the recovered LiOH door are shown in figures 2.1-36 and 2.1-38. The thermal failure corresponds to the thin sections of the door. Sections that are thicker show little or no thermal erosion. From observations of the material deposition patterns and the burn-throughs on the recovered door, the major thermal effects were directional, with flow impinging primarily on the bottom side (the side with the stiffeners). This is supported by the highly directional deposition on seats 6 and 7 (see Section 3.1).

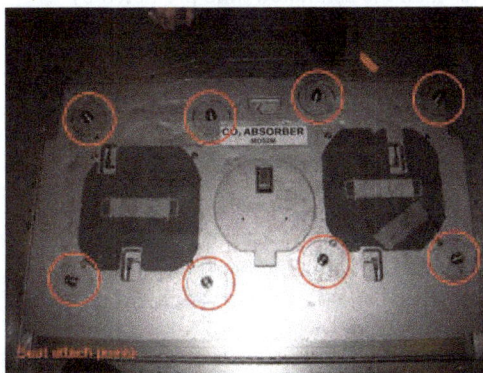

Figure 2.1-35. *Upper surface of the lithium hydroxide door, facing the middeck* (mockup).

Figure 2.1-36. *Top surface of the recovered lithium hydroxide door, facing the middeck.*

Figure 2.1-37. *Lower surface of the lithium hydroxide door, facing the lower equipment bay* (mockup).

Figure 2.1-38. *Lower surface of the recovered lithium hydroxide door, facing the lower equipment bay.*

SORT estimated that the release time occurred at GMT 14:00:59. Four configurations of the LiOH door were analyzed to capture the possible release geometries. Configuration 1 was the LiOH door re-entering by itself. Configuration 2 was the LiOH door with two empty seats attached. Configuration 3 was the LiOH door, the two seats, and a suited occupant seated in one of the seats. Configuration 4 was two suited occupants in the seats attached to the LiOH door. The possible pitch trim attitudes for these configurations were predicted using Snewt, and the results can be seen in figures 2.1-39 through 2.1-43. Untrimmed (tumbling) cases were also evaluated for each configuration to determine the sensitivity of thermal effects to attitude. All these configurations assume that the LiOH door (and the attached seats and crew members) was a single free-flying unit with no other CM structure attached.

The main difference among these configurations is the ballistic number. The ballistic number determines the trajectory, thus the velocity profile. The velocity, in turn, affects the heating rate. For objects with large ballistic numbers, the heating rate is greater than for objects with small ballistic numbers.

Since the minimum thickness of the door is 0.05 in., the door was initially modeled with this thickness to determine whether the aluminum would ablate. Ablation is defined as the amount of energy required for an object to reach its melting temperature *and* overcome its heat of fusion. An object demises when the heat of ablation has been reached; in other words, the objects burns through. This is a lower threshold case than is realistic because the thicker areas of the door would act as a heat sink. If the 0.05-in. plate ablated, the thickness was increased to see at what plate thickness it would not ablate.

The door release time was also adjusted 5 and 10 seconds earlier than the estimated time predicted by SORT to see whether the release time had an impact on the results. The initial temperature in the analysis was 80°F (27°C) for all cases except one, where the initial temperature was increased to 200°F (93°C). This was done to determine the sensitivity to the initial conditions; for example, if the CM interior environment was heated significantly prior to breakup and release of the door. However, there is no evidence to suggest that this was the case.

Configuration 1 (Door alone)
The first configuration was a simple flat plate (representing the door) entering by itself. In this configuration, the door was modeled both for the tumbling case and entering normal to the flow fixed angle of attack of 90 deg (figure 2.1-39). This configuration has the smallest ballistic number, so it will receive less heat than the others. The ballistic number was 2.9 psf for the normal to flow case and 6.1 psf for the tumbling case.

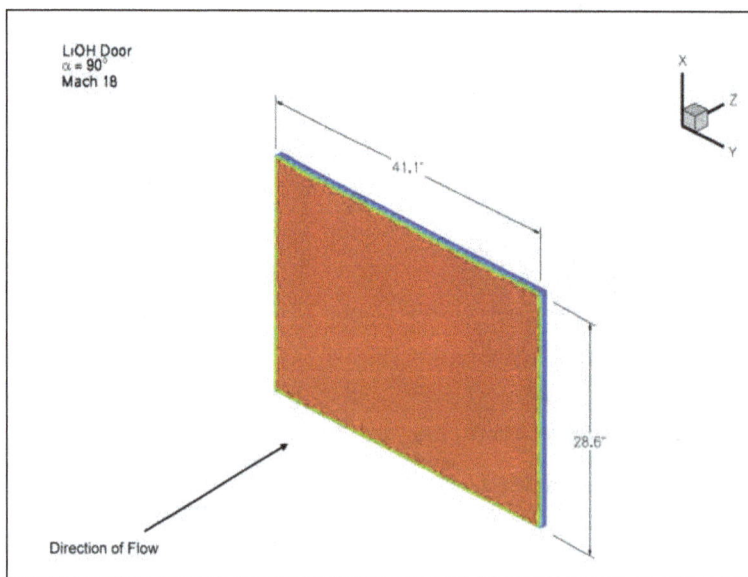

Figure 2.1-39. *Aerodynamic model of the lithium hydroxide door exposed by itself at an angle of attack of 90 degrees.* [Configuration 1]

For this first configuration, neither the tumbling nor the normal-to-flow case predicted that the aluminum melting temperature would be reached.

Configuration 2 (Door with two seats)

The second configuration was the door with empty seats. Since each of the seats has a mass of about 50 lbs., the total aerodynamic weight increased to 141 lbs. This includes the two seats (100 lbs.) plus the door (41 lbs.). In this configuration, the door and the seats were modeled for the tumbling case and also entering at fixed angles of attack, –92 deg or 96 deg (figures 2.1-40 and 2.1-41). This configuration has the second smallest ballistic number. The worst-case (highest) ballistic numbers for the non-tumbling and tumbling cases were 11.5 psf and 20.0 psf, respectively.

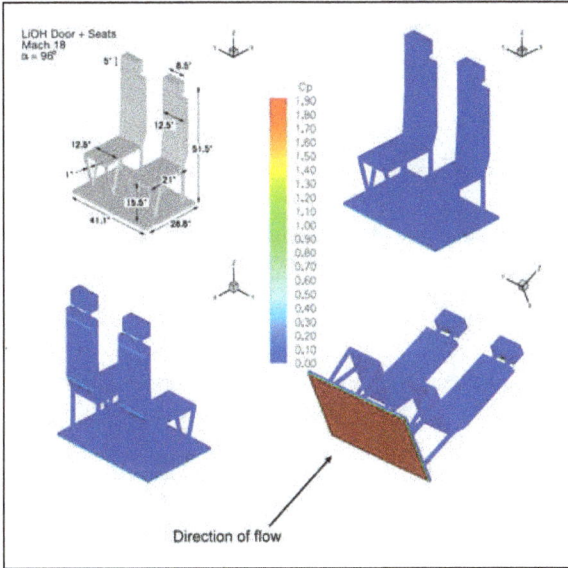

Figure 2.1-40. *Pressure distribution of the lithium hydroxide door and seats at an angle of attack of 96 degrees.* [Configuration 2]

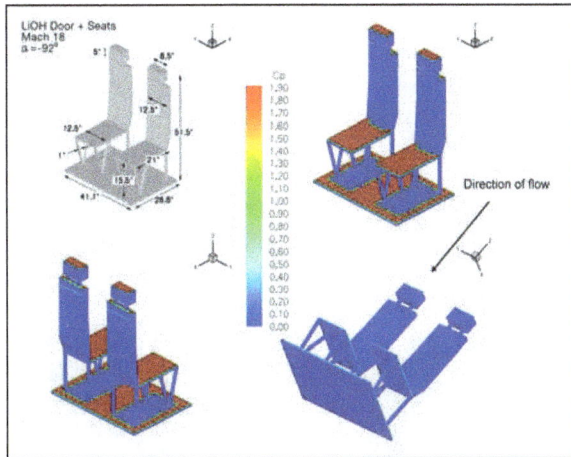

Figure 2.1-41. *Pressure distribution of the lithium hydroxide door and seats at an angle of attack of –92 degrees.* [Configuration 2]

The worst-case tumbling results showed that the melting temperature of the aluminum would be reached, but the estimated heating profile would not overcome the heat of fusion, so no burn-through was predicted.

Configuration 3 (Door with one seated crew member)

In the third configuration, the aero mass was increased to 401 lbs. (41-lb. door with two 50-lb. seats and one suited occupant in seat 7, with a weight of 170 lbs. plus 90 lbs. of launch entry suit). The analysis performed with Snewt did not reveal a stable trim attitude. The ballistic number was 34.0 psf. Figure 2.1-42 shows the geometry.

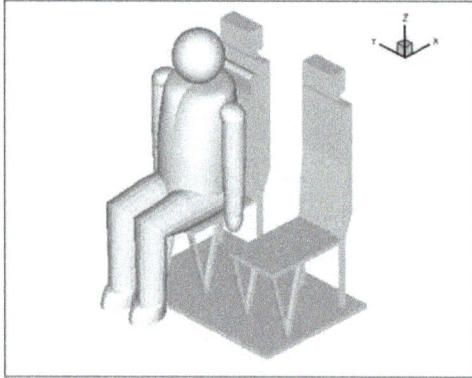

Figure 2.1-42. *Geometry of a single suited occupant on the lithium hydroxide door.* [Configuration 3]

For the given initial conditions, a 0.05-in.-thick plate was predicted to demise. With the thickness increased to 0.07 in., the plate is predicted to survive.

Configuration 4 (Door with two seated crew members)

The fourth configuration was the door with two suited crew members in the seats. Figure 2.1-43 shows this configuration for the trim attitude of 122 deg. In this configuration, the total aerodynamic mass is 690 lbs. This configuration was evaluated for both the tumbling case and a fixed angle of attack in an effort to understand the sensitivity to the attitude condition. This configuration has the largest ballistic number with the worst-case (highest) ballistic number for the non-tumbling case of 42.4 psf and 48.6 psf for the tumbling case.

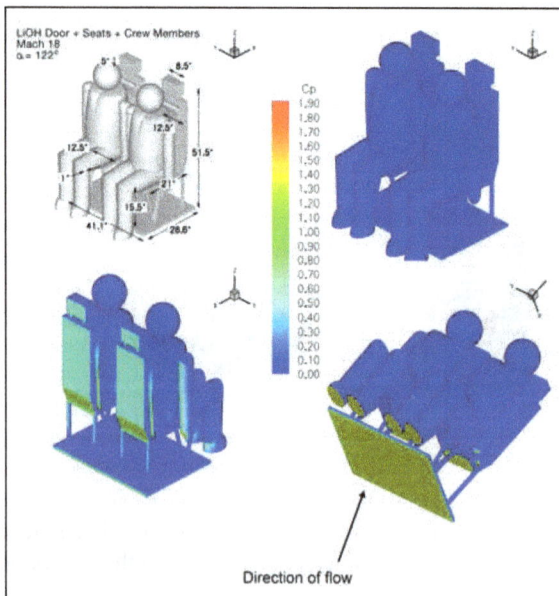

Figure 2.1-43. *Pressure distribution of the lithium hydroxide door and seats with suited crew members at an angle of attack of 122 degrees.* [Configuration 4]

Since the 0.05-in.-thick plate was predicted to demise in the third configuration, it also is predicted to demise in the fourth configuration, which has a larger ballistic number for both the tumbling and the non-tumbling conditions. The 0.05-in.-thick plate is predicted to demise 6 seconds after being exposed. A 0.08-in.-thick plate is predicted to demise 17 seconds after being exposed. However, a 0.09-in.-thick plate is predicted to survive. Increasing the initial temperature to 200°F (93°C) for the 0.05-in.-thick plate case did not change the time predicted for the plate to demise.

The heat rates generated by ORSAT were compared to another model, BLIMP-K. The heat rates predicted by ORSAT were about 12% less than those predicted by BLIMP-K, but were generally comparable. A detailed thermal math model using the SINDA and ORSAT heating rates was constructed for a section of the door and confirmed the results predicted by the ORSAT. In this model, only the fourth configuration was predicted to demise.

For the GMT 14:00:59 estimated release time, the analysis implies the third and fourth configurations (door with one and two crew members, respectively) can produce the thermal damage observed on the recovered LiOH door. Since visual inspection of the recovered door suggests that the object was not tumbling, the stable fourth configuration is the most thermally viable solution. However, both the third and the fourth configurations are improbable because the thermal effects would have melted the seat straps and released the crew members. The surrounding debris was evaluated to determine whether the LiOH inside the compartment caught on fire. Significant portions of LiOH canisters were recovered as well as other items stored in the compartment. Both the canisters and the structures of the compartment were not seriously thermally damaged, which strongly suggests that a fire did not occur.

It is likely that the simplified nature of the assessment could not accurately model this complex object. As a result, the ORSAT analysis on the LiOH door was inconclusive.

2.1.6.5 *Thermal analysis – payload and general support laptop computers*

Some recovered debris items were identified as pieces from the crew's payload and general support laptop computers, none of which were recovered intact. An analysis was performed to determine whether aerothermal heating could cause the destruction of a laptop after it was released from the CM or if it had to be pre-heated inside the CM. If the laptop had to be pre-heated, this analysis could give an indication of what the thermal environment was inside the CM before breakup.

The estimated maximum temperature from aeroheating alone was well above the temperature at which the battery will explode, likely fragmenting the laptop casing. No conclusion could be made about the thermal environment inside the CM before breakup from this analysis.

2.1.6.6 *Thermal analysis – Modular Auxiliary Data System/orbiter experiment recorder*

The OEX recorder, which was part of *Columbia*'s MADS, was found near Hemphill, Texas in near-perfect condition. Figure 2.1-44 shows the recovered OEX recorder. The data on the tape in this recorder were critical to the accident investigation, making the recorder one of the most important recovered items from *Columbia*.

Figure 2.1-44. *Recovered orbiter experiment recorder from STS-107.* [*Columbia* Reconstruction Database, debris item no. 54057]

Recommendation L3-3. Future spacecraft design should incorporate crashworthy, locatable data recorders for accident/incident flight reconstruction.

The rectangular shape of the box made it a good candidate for ballistic analysis providing a release time with high confidence. The box showed no signs of thermal erosion or exposure to high temperature environment. Text labels that have been imprinted on the box still remain. The data that were recovered from the recorder were recorded on magnetic tape that delaminates at 125°F (52°C). An analysis was performed to estimate the thermal damage that a free-flying OEX recorder would receive due to aerothermal heating after release from inside the CM.

The OEX recorder was located in the lower equipment bay (underneath the middeck floor) of the CM. It is loosely covered by a fiberglass shroud that channels cooler air over the recorder. The shroud is bolted over the OEX recorder but is not directly attached.

The OEX recorder case is composed of aluminum 6063. The thermal properties of aluminum 6061 were used in the analysis because the values were already in the ORSAT material database and the differences in thermal properties to aluminum 6063 were negligible. The shell thickness is 0.25 in. and the overall dimensions of the box are 19 in. long, 15.5 in. wide, and 5 in. high. The total weight of the box is 53.8 lbs.

Ballistic analysis produced an estimated release time of GMT 14:01:02. A state vector from SORT was used as a starting point in ORSAT to simulate the trajectory and heating rates of an entry from that time – altitude 131,780 feet, velocity 12,717 feet per second (fps), and flight path angle –6.3 deg. The initial temperature of the OEX recorder used was 80°F (27°C), however this may be a conservative assumption because the internal electronics of the OEX recorder cause the device to run at relatively high temperatures.

Figure 2.1-45 shows the surface temperature profile for the OEX trajectory. The surface temperature was predicted to reach 470°F (243°C). This is well below the melt temperature of the aluminum, 1,100°F (593°C). Aluminum structurally weakens at 400°F (204°C), and some deformation would be expected if the recorder achieved this temperature. However, there is no deformation of the aluminum casing.

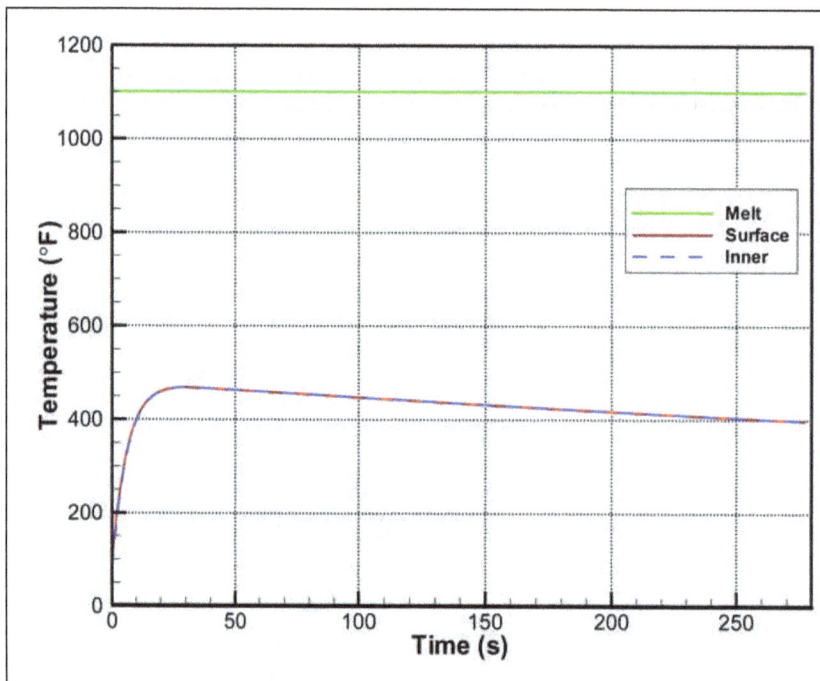

Figure 2.1-45. *Temperature vs. time of the orbiter experiment recorder shell if released at GMT 14:01:02.*

The lack of external thermal damage is quite noteworthy. Anodizing is still mostly present, as are the exterior labels. Exposed wires still have insulation. Finally, since aluminum is a good conductor it is expected that the internal wall temperature would have been close to the surface temperature. Since the recorder's magnetic tape delaminates at 125°F (52°C), this also suggests that the aluminum may not have reached the predicted high temperature of 470°F (243°C).

The conclusion is that the OEX recorder was not released from the CM independently. If the temperatures reached as high as predicted, the magnetic tape would have delaminated and the casing would have likely been deformed. This analysis combined with the relatively pristine condition of the recorder seems to imply that something must have protected the recorder during entry before it was completely exposed to the aero-thermal heating environment. Ultimately, no conclusion can be made as to what ancillary structure or other mechanism protected the OEX, only that it was protected.

2.1.6.7 *Thermal analysis – x-links*

Examination of the recovered CM attachment fittings, known as the x-links, identified intriguing thermal damage. The two x-links attach the CM to the FF and the midbody and carry load in the orbiter X body axis (see Section 2.4 for diagrams of structure). While both x-links showed melting in the same relative locations, the starboard x-link displayed more thermal damage than the port x-link. Because the x-links are composed of Titanium 6Al-4V, which has a melt temperature of approximately 3,000°F (1,649°C), very high heating is required to create the damage that was observed on the recovered x-links. Analyses were performed to determine the thermal mechanism that could have caused this damage. Figure 2.1-46 shows a model of a pristine port and starboard x-link, while figure 2.1-47 shows the two recovered *Columbia* x-links placed side by side.

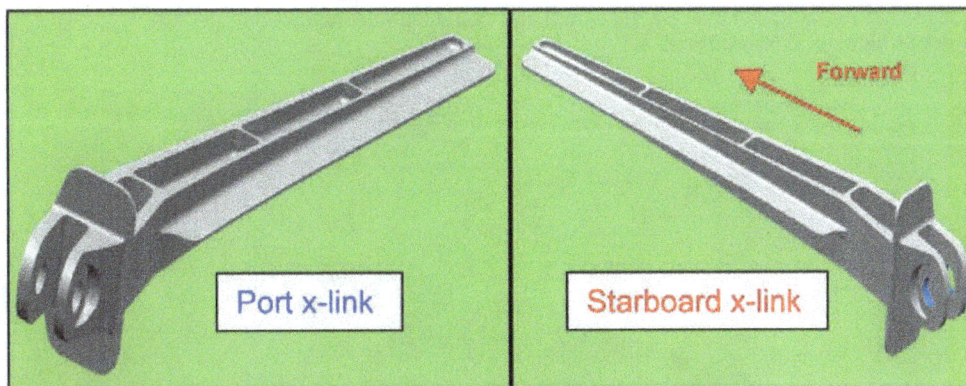

Figure 2.1-46. *Drawings of pristine port and starboard x-links.*

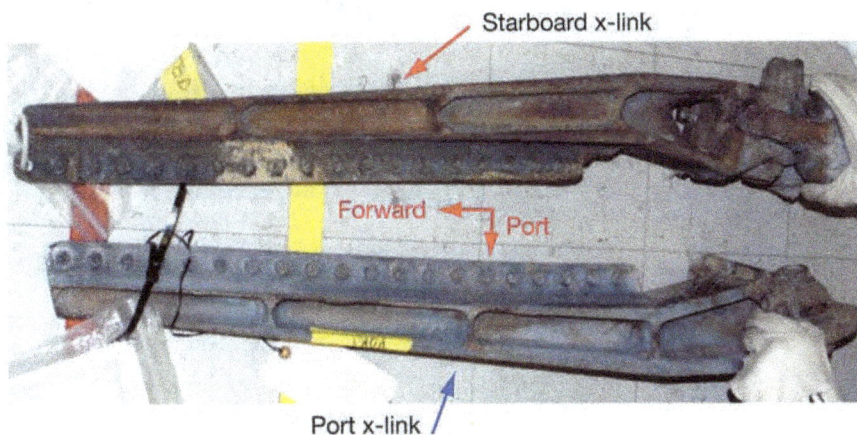

Figure 2.1-47. *Comparison of port and starboard x-links recovered from* **Columbia. [**Columbia **Reconstruction Database, debris item no. 1678, and** Columbia **Reconstruction Database, debris item no. 1765]**

The heating experienced by the starboard x-link was severe enough to melt away the top flange and to burn a hole through the 0.25-in.-thick web. Figure 2.1-48 shows another comparison of the two x-links with a better viewing angle to see the flange damage. Figure 2.1-49 shows a close-up view of the starboard x-link where a hole has been melted through the web.

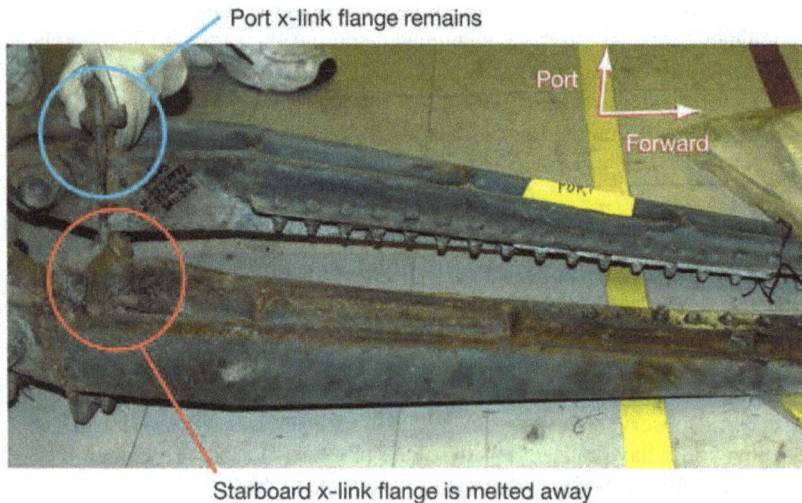

Figure 2.1-48. *Comparison of port and starboard x-link flanges.* [*Columbia* Reconstruction Database, debris item no. 1678 (top) and *Columbia* Reconstruction Database debris item no. 1765 (bottom)]

Figure 2.1-49. *Close-up of the hole in the starboard x-link.* [Columbia *Reconstruction Database, debris item no. 1765*]

At first it was assumed that the x-links received the thermal damage from entry heating as independent free-flying objects. For the ORSAT analysis, an x-link was modeled as a simple box (38 × 3 × 1.2 in.) at a stable attitude with a mass of 22 lbs. It was assumed the x-link broke off the orbiter at 200,000 feet with a relative velocity of 17,145 fps and a flight path angle of -0.68 deg. Based on debris field analysis, the actual separation was likely at a lower altitude and Mach number with lower resultant heat rates, making this assessment conservative.

The ORSAT analysis predicted that entry heating would not be sufficient to cause the damage seen. Also, a free-flying x-link was not predicted to stabilize in one attitude; a tumbling x-link would diffuse the heat better and also not result in the directional heating seen. The similar melting patterns on the two x-links suggests that they were in the same relative orientation at the time the thermal event occurred, presumably still in place attaching the CM to the FF. This led to a more detailed analysis to understand what heat rate would be required to show the damage seen.

The amount of heat to completely melt an object is known as the heat of ablation. The heat of ablation per mass has two components, which are additive: latent heat and sensible heat. Latent heat is the amount of energy in the form of heat released or absorbed by a substance during a phase transition (such as from solid to liquid). The latent heat per mass for titanium is 187 Btu per pound of mass (Btu/lbm). Sensible heat is potential energy in the form of thermal energy or heat for an object. The amount of sensible heat per mass required to raise the temperature of titanium to its melting temperature can be determined by integrating the plot from the initial temperature to the final temperature (figure 2.1-50). Data did not extend beyond 1,600°F (871°C).

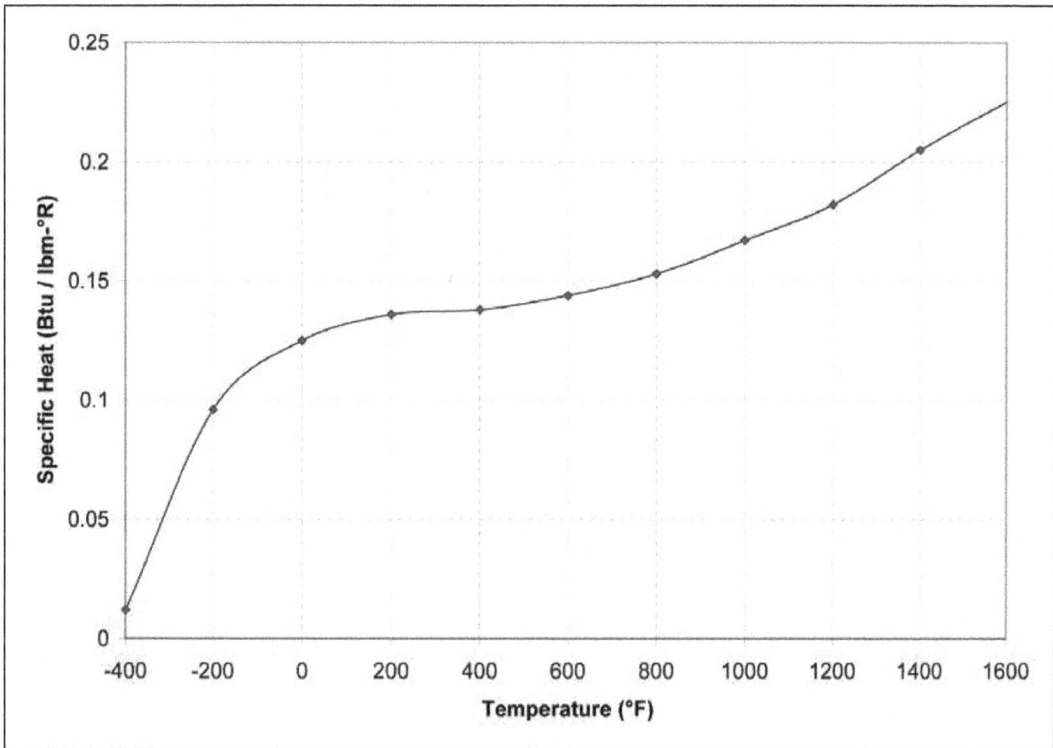

Figure 2.1-50. *Specific heat of titanium vs. temperature.*[17]

Assuming an initial temperature of 80°F (27°C) and that the specific heat at 1,600°F (871°C) does not change until the melting temperature is reached, the amount of sensible heat required to reach the melting temperature of titanium is 564 Btu/lbm, or 750 Btu/lbm to completely ablate the material. It is possible that

---

[17]Thermal Protection Materials, NASA Reference Publication 1289, December 1992.

the x-link was pre-heated by earlier exposure to entry heating as the forebody rotated. The maximum pre-heating considered reasonable was 400°F (204°C), the temperature at which the surrounding aluminum structure would begin to soften and likely release the x-link. If the x-link was pre-heated to 400°F (204°C), the amount of sensible heat per mass required to reach the melting temperature is reduced to 521 Btu/lbm and the heat of ablation per mass is reduced to 710 Btu/lbm.

Knowing the heat of ablation per unit of mass allows a first-order estimate of the required heating that must be applied to cause that burn-through hole in the starboard x-link. The dimensions of the burn-through area are 2 in. by 2 in. and the thickness of the x-link in that area is 0.25 in.

The required heating rates to cause the burn-through hole as a function of time for initial temperatures of 80°F (27°C) and 400°F (204°C) are shown in figure 2.1-51. It can be seen that the heat rates are extremely high for either case, and increase exponentially as the time required to cause the hole decreases. For comparison, figure 2.1-51 also shows the heating rate from entry aeroheating for an x-link exposed at 200,000 feet (dark blue line). This shows that the x-links would have to be exposed to sustained directional entry heating for more than 25 seconds to result in the thermal damage received.

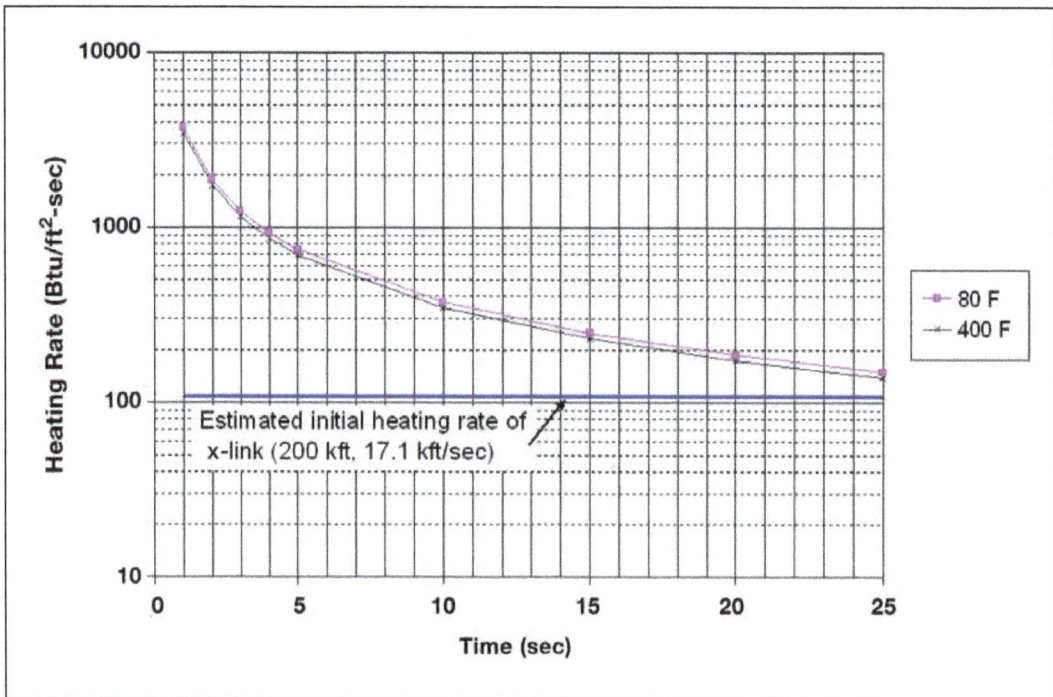

Figure 2.1-51. *Heating rates required to ablate a titanium hole of 0.25 inch in depth.*

The question now became how such severe and similar directional heating of the two titanium x-links occurred in a short time span without melting the surrounding material, which had lower melting temperatures. Because the forebody was not likely to stabilize at a given attitude, there would be very brief durations of exposure as the x-link happened to pass through the velocity vector as the free-flying forebody rotated. The longest reasonable duration for the rotating forebody to experience peak heating was felt to be less than 3 seconds and possibly as short as 1 second. To melt a hole in 3 seconds, the required heat rate is about 1,350 Btu/ft^2-sec. To melt a hole in 1 second, the required heat rate is over 3,900 Btu/ft^2-sec. These values are an order of magnitude higher than the estimated peak heating due to aeroheating of 110 Btu/ft^2-sec.

Finding. Thermal analyses predicted that entry aeroheating alone was insufficient by an order of magnitude to produce the observed thermal damage on the x-links. Therefore, the x-links must have experienced other heating mechanism(s) in addition to normal entry heating.

This was a very surprising result, and led the team to research other heating mechanisms. Two thermal mechanisms, shock-shock interaction and titanium oxidation/combustion, could generate the observed thermal damage either separately or in tandem. Both of these mechanisms are discussed in detail in Section 2.1.7.

2.1.6.8 *Thermal analysis – payload bay door roller*

Another instance of the thermal erosion of high-temperature materials was seen in the PLBD rollers, which contain notable amounts of titanium. Initially, the interest in the PLBD rollers resulted from a search of possible sources for the titanium deposition that was found on the overhead windows (see Section 2.4).[18] A search for forward structures containing titanium showed that the nearest source of titanium material to the windows was the forward PLBD rollers. These rollers are made of aluminum, titanium, and an Inconel sleeve, and the PLBDs rest on them when the doors are closed. The forward rollers are attached to the top of X_o 582 ring frame bulkhead in close proximity to the windows on which the titanium depositions were found (figure 2.1-52). Several recovered rollers from this location showed pronounced erosion of the exposed titanium surfaces, and one roller was recovered with only part of the Inconel sleeve remaining; all titanium and aluminum inner structures were missing (figure 2.1-53). Table 2.1-2 shows the material properties of stainless steel (A286), Inconel 718, aluminum 2000 series, and titanium 6Al-4V. As shown, the Inconel and aluminum alloys used in the rollers both have lower melting temperatures than the titanium alloy.

Figure 2.1-52. *Payload bay door roller showing inner material and Inconel sleeve.*

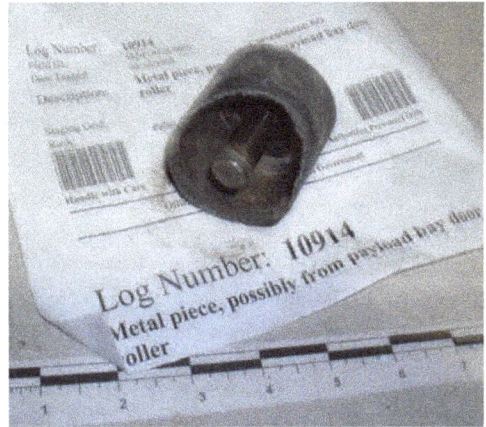

Figure 2.1-53. *Recovered payload cylinder bay door roller with inner material absent and Inconel sleeve remaining. [Columbia Reconstruction Database, debris item no. 10914]*

Table 2.1-2. *Material Properties Important to Thermal Analyses*

Material	Melting Temperature (°F)	Melting Temperature (°C)
A 286 (stainless steel)	2,500	1,371
Inconel 718	2,368	1,298
Aluminum 2024	1,081	583
Titanium-6Al-4V	3,037	1,669

A computational fluid dynamics (CFD) thermal analysis of the rollers was performed to determine the flow field environment and temperature at the face of a roller for an orientation with the front of the roller facing directly into the direction of travel. Figure 2.1-54 shows the predicted heating rate and temperature distribution along the PLBD roller at Mach = 10.5.

---

[18]J. D. Olivas, L. Hulse, B. Mayeaux, S. McDanels, P. Melroy, G. Morgan, Z. Rhaman, L. Schaschl, T. Wallace, and C. Zapata, *Examination of OV-102 Thermal Pane Window Debris – Final Report*, KSC-MSL-2008-0178 (in press).

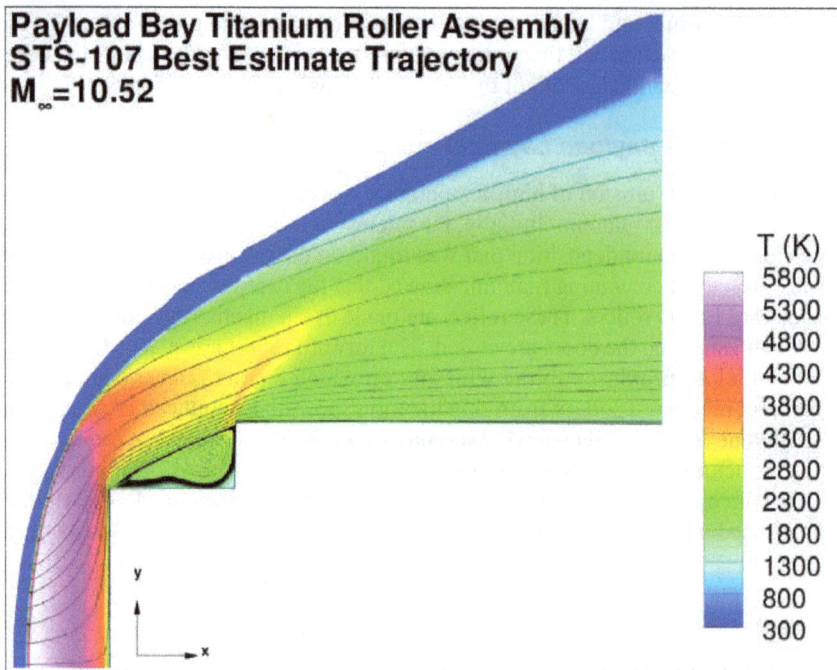

Figure 2.1-54. *Computational fluid dynamics analysis of heating at the tip of the payload bay door roller for orthogonal geometry into the direction of travel at M = 10.5.*

It can be seen from figure 2.1-54 that for Mach numbers greater than 10, the possibility exists that radiative equilibrium surface temperatures in excess of the melting temperature of titanium (~3,000°F) (1,662°C)) can be achieved. Ablation then could occur by pressure forcing away the molten titanium on the surface.

When ORSAT was used to analyze the case of the titanium rollers, the calculated heating rates across the front face compared very well to the results from the CFD case, given the same assumptions regarding geometric orientation and free-stream conditions (figure 2.1-55) for Mach 7.5 to 10.5.

In figure 2.1-55, the heating rate increases along the front face from the center to the edge because the flow is accelerating around the corner of the front face of the roller. The first spike comes from the flow going over the edge of the inner cylinder, and the second spike comes from the flow over the (outer) sleeve cylinder. If a boundary with grid representing the payload bay forward wall had been used in the CFD analysis, it may possibly have shown a shock-shock interaction that might account for the hole found in the surrounding X_o 582 ring frame bulkhead structure on either side of the roller.

The next question was whether the observed amount of material loss was reasonable for the timeframe. As mentioned before, heating only occurs when the object is facing into the direction of travel, yet it is not believed that the forebody stabilized in a specific attitude. Additionally, the nature of the window deposition indicates that the titanium and titanium/aluminum deposition event was continuous and not cyclical. Therefore, the entire titanium/titanium aluminum ablation event had to occur in a relatively short period of time. Given an attitude change of approximately 36 deg/sec (0.1 rev/sec), the maximum duration of heating considered reasonable was about 3 seconds. In any attitude other than directly into the velocity vector, heating is less but is still present. Therefore, the CFD analysis only showed the peak heat was possible. It seems probable that some additional mechanism would be required to deeply erode the rollers.

The analysis of the x-links (see previous section) also suggests that there might be mechanisms other than simple entry aeroheating at work. The first mechanism evaluated was shock-shock interaction.

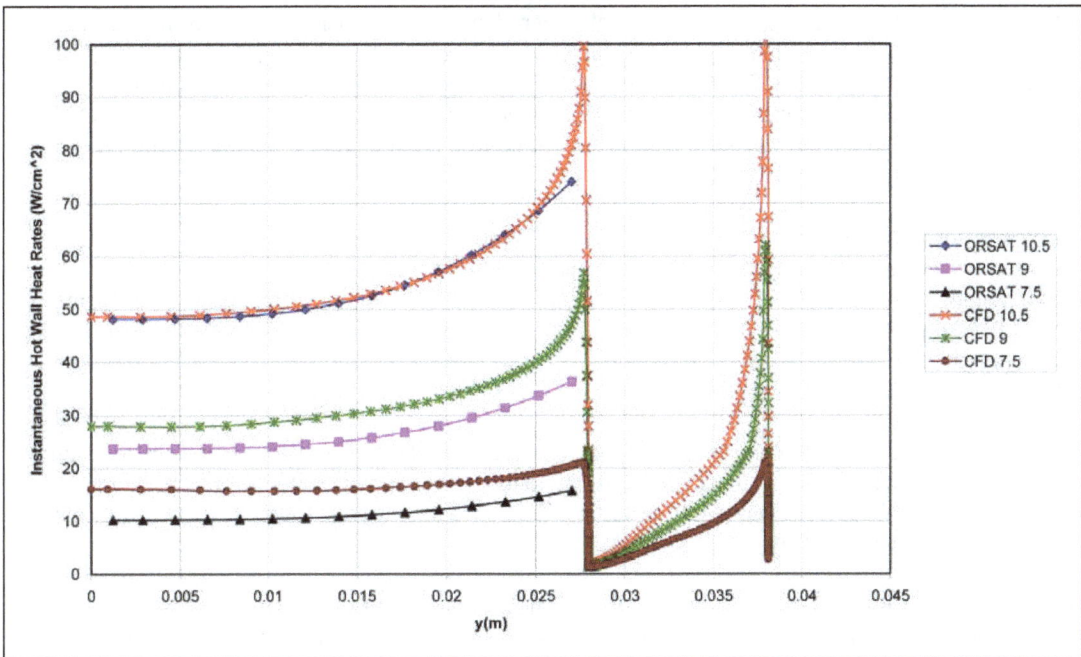

Figure 2.1-55. *Comparison of the object reentry survival analysis tool and computational fluid dynamics heating rates for the front face of the payload bay roller for various Mach numbers.* Y axis is radial distance from the center of the roller.

2.1.7 Shock-shock interaction and combustion

Shock-shock interaction occurs when two shock waves intersect. Figure 2.1-56 shows an example of one interaction on the orbiter leading edge.

Figure 2.1-56. *Schlieren photograph of the orbiter showing a shock-shock interaction region for nominal entry.*[19]

---

[19]This profile was taken from orbiter aerothermodynamic heating tests at the Calspan-University of Buffalo Research Center, shock tunnel facility at Buffalo, New York at a Mach number of 10.

These shock-shock interactions influence a vehicle's aerodynamics and increase the heat transfer rate and pressure where the interaction impinges on the vehicle's surface. There are six types of shock-shock interactions.[20] These types are defined by where the incident and bow shock waves intersect, the strength of the incident shock, and the angle on the blunt leading edge surface with respect to the impinging shear layer. Shock-shock interaction was examined to determine whether it could explain the thermal damage observed on the x-links and PLBD rollers.

The best candidate for the extreme heat rates is called a Type IV shock-shock interaction. Type IV shock wave interference can cause stagnation rates to be amplified by a factor of 38 for a Mach number of 16, which is slightly higher than the Mach number at orbiter breakup. This amplification factor could increase the baseline heating rate to almost 4,000 W/cm^2, so it would only take about 1 second to burn-through the titanium. These Type IV shock wave interference scenarios cause a jet that is very small and will result in highly localized heating.

Type IV shock waves are highly dependent on geometry relative to the airflow. Due to the chaotic nature of the vehicle breakup, it was not considered reasonable to attempt to determine the precise geometry for the PLBD rollers or the x-links that would result in a Type IV shock wave. However, experts did determine that there was no reason to rule out shock-shock interactions based on geometry.[21]

Therefore, theoretically a shock-shock interaction can help explain the type of damage seen on the rollers and starboard x-link. However, two questions still remained. The first was the original mass loss rate question from the PLBD rollers. The second question was why surrounding materials with lower melt temperatures, such as aluminum and the Inconel roller sleeves, remained relatively intact in such an intense environment. The aluminum condition could be explained by the fact that it has much great thermal conductivity than titanium. With sufficient mass, it is capable of withstanding higher heating rates because it can create a more effective heat sink. Since most of the CM is made of aluminum, the aluminum portions of the structure may have been very effective at shunting heat away from locally exposed regions during a period of high localized heat exposure. However, the thermal conductivity of Inconel is much lower, closer to that of titanium, and the relative lack of damage to the Inconel cannot be explained by heat shunting. This introduced the possibility that the titanium may have reacted different chemically to the environment than the other materials around it.

Crystallographic assessment by the JSC Materials and Processing Office, the JSC Astromaterials Research Office, the Kennedy Space Center (KSC) Failure Analysis and Materials Evaluation Branch, and the White Sands Test Facility on the nature of the oxide formation on the window indicated that deposition occurred at a temperature that was well above the melt temperature of the titanium alloy. The crystallographic and microscopic work further indicated that the oxide species were fluid once deposited on the windows and did not experience solidification during transportation from the source to the deposition site. These additional clues further implied some significant secondary mechanism.[22]

One important difference between titanium and the other materials is that titanium is highly reactive with O_2. In fact, titanium is often used as an igniter or a promoter, much like magnesium. Oxidation is an exothermic reaction, and the chemical reactions that occur can introduce heat into the material through bulk material conversion from a molten metallic state to an oxidized state. This could explain why the Inconel (which is not particularly reactive in terrestrial O_2 environments) did not thermally erode, although the melt temperature is more than 600°F (316°C) lower than that of titanium (Table 2.1-2), and why the titanium alloyed material reacted and deposited on the windows earlier than other materials.

It is very difficult to predict the combustion process of metals with the analytical tools used for this investigation. The ORSAT code used for much of the aerothermal analysis in this study does have a

---

[20]Barry Edney, "Anomalous Heat Transfer and Pressure Distributions on Blunt Bodies at Hypersonic Speeds in the Presence of an Impinging Shock," The Aeronautical Research Institute of Sweden, 1968.

[21]Allan Wieting, "Shock-Shock Wave Interference Heating Possibilities," March 2007.

[22]J. D. Olivas, M. C. Wright, R. Christoffersen, D. M. Cone, and S. J. McDanels, *Crystallographic oxide phase identification of char deposits obtained from space shuttle* Columbia *window debris*, Acta Materialia, 2008 (in press).

primitive function for determining oxidation, so the phenomenon is not unforeseen. However, there are many limitations to the model. The ORSAT function only accounts for the mass loss of the material as it is consumed by oxidation, not by a combustion reaction. ORSAT accounts for this by creating an additional component of heat that is formed by the reaction of the metal and molecular O_2. During vehicle deceleration from high Mach numbers, O_2 is dissociated into its monatomic form (O) which is much more chemically reactive than molecular O_2. A literature search and survey of major testing facilities proved that investigation into the effects of high enthalpy-low pressure environments on bare titanium has been very limited in the past.[23]

The shock-shock interaction discussed previously was not easy to test. However, simple experiments could be conducted, using an arc-jet facility, to understand whether titanium combustion could have occurred in the *Columbia* entry environment.

A selection of enthalpy-pressure test points was chosen based on the predicted trajectory and ballistic number of *Columbia* and the free-flying forebody. A series of tests was conducted at the Boeing St. Louis Large Core Arc Tunnel (LCAT) plasma arc facility.[24] This testing showed that in some higher enthalpy conditions the titanium test article eroded within a few seconds and exhibited behavior that is characterized as combustion. Although mass loss rates are not known, this type of oxidation can become a sustained reaction and does not necessarily require the continued exposure to heating to continue.

An analysis was done using ORSAT to see whether the model could account for the thermal effects seen in the arc-jet test without combustion. In lower enthalpy cases, the model matched arc-jet test results very well. However, at the higher enthalpy cases, there were incongruities that could not be accounted for by entry heating alone.[25] The conclusion was that titanium combustion was possible for the *Columbia* entry environment. This appears to be the first documented characterization of this material physical property.

Additionally, the arc-jet test team noted that geometry also appeared to play a possible role in the initiation of oxidation. This provides the intriguing possibility that momentary shock-shock interactions, which might not erode titanium significantly on their own, may have acted as a trigger for oxidation, combustion, and heavy erosion.

The fact that combustion was only seen at higher enthalpies and pressures suggests that if oxidation occurred, it was most likely when the forebody was still intact as the smaller individual objects are not likely to generate the heat required. The fact that the rollers deposited material on the windows combined with the similar directional heating on both x-links implies structural integrity of the forebody, which is consistent with this conclusion.

In conclusion, some of the thermal mechanisms experienced during the *Columbia* entry were different from those of a nominal entry, implying the possibility of shock-shock impingement, titanium oxidation, or a combination of both effects resulting in rapid and selective melting of titanium prior to the aluminum around it.

> *Conclusion A13-1.* Titanium may oxidize and combust in entry heating conditions dependent on enthalpy, pressure, and geometry.

> *Conclusion A13-2.* The heating from a Type IV shock-shock impingement and titanium combustion (in some combination) likely resulted in the damage seen by the forward payload bay door rollers and the x-links.

---

[23]William Rochelle, "Survey of Titanium Testing at NASA, DOE, and DOD Test Facilities," ESCG-4380-06-AFD-MEMO-0011, April 20, 2006.

[24]J. D. Olivas, B. Mayeaux, P. Melroy, and D. Cone, Study of Ti Alloy Combustion Susceptibility in Simulated Entry Environments, AIAA, 2008 (in press).

[25]W. Rochelle, J. Marichalar, M. Larin, A. Dobrinsky, J. Dobarco-Otero, and Ries Smith, "Comparison of LCAT Arc-Jet Titanium Plate Test Data with Aerothermal Predictions," ESCG-4380-07-AFD-MEMO-0012, April 20, 2007.

Recommendation A13. Studies should be performed to further characterize the material behavior of titanium in entry environments to better understand optimal space applications of this material.

2.1.8 Synopsis of motion and thermal analyses

In summary, the motion and thermal analyses of the vehicle and forebody produced significant results and findings. The loads experienced by the intact orbiter and the free-flying forebody were estimated based on a reference trajectory and attitude analyses. The forebody and CM of the orbiter are not expected to trim into a single aerodynamic attitude upon breakup of the vehicle. Thermal analyses results were multifaceted. In some cases, thermal models showed good agreement with debris condition, such as for the crew helmets. In other cases, geometry and protection from other structure proved too complex to model accurately. Further research is warranted to investigate other thermal mechanisms (shock interaction, combustion) to provide greater understanding of aerothermal effects on entry.

2.2 Orbiter Breakup Sequence

This section discusses the breakup sequence of the orbiter following the LOC. The CAIB concluded that the orbiter was shedding debris throughout entry.[1] LOS occurred at GMT 13:59:32. RGPC on-board data indicate that the Freon coolant loops in the PLBDs were still intact at GMT 14:00:04.8, indicating that the radiators and PLBDs were still intact. After GMT 14:00:04.8, no GPC data could be recovered. The on-board MADS/OEX recorder was powered until GMT 14:00:18, the time known as the CE. Based on video analysis, at GMT 14:00:18 the orbiter was no longer intact.[2] At this point in the entry, the orbiter was estimated to be at an altitude of 181,000 feet and traveling at Mach 15.

The SCSIIT relied upon analysis of ground-based video, analysis of debris recovery locations, ballistic analysis, and structural analysis to reconstruct the sequence of events experienced by *Columbia*. This section provides the framework for the sequence of structural failure and covers large-scale events that are related to orbiter structure and the separation of the forebody from the intact orbiter. Section 2.4 discusses events specifically related to the forebody breakup. These sections overlap in certain areas. Detailed description of the analysis techniques are contained in this section and are not repeated in Section 2.4.

The orbiter breakup sequence discussion is presented by type of analysis. The first is the *video analysis*. Analysis of ground-based videos consisted of reviewing all ground-based video of *Columbia*'s entry. Ground-based video analysis allowed a time-tag to be assigned to specific visual events that were seen in ground-based video. Considerable elements of the analysis were performed for the CAIB[3] although the SCSIIT performed additional analysis. On-board video is discussed in Sections 3.2 and 3.4.

The next analysis presented is the *ballistic analysis*, which used ground location, size, shape, and mass of recovered debris items. As an individual piece of debris was shed, it took on a unique ballistic trajectory based on its ballistic number (see Appendix). Heavier items with larger ballistic numbers travel farther downrange (in this case, east), while lighter items with smaller ballistic numbers decelerate quickly and achieve terminal velocity, traveling essentially straight down to the ground. Therefore, comparing the impact location alone of any one piece of debris relative to another will not provide information about the time or sequence of release; a detailed ballistic analysis comparing their trajectories is required. The modeled debris was iteratively connected to the reference trajectory (see Section 2.1) until the calculated ground impact longitude matched the actual recovered longitude of the item. This is a lengthy process and is heavily dependent on understanding the aerodynamic characteristics of the object in question. It is easiest to do this analysis on simple shapes, such as spheres and boxes. As a result, only select items have ballistic assessments. Conducting ballistics analysis on several debris items that came from the same general zone on the orbiter (such as the right wing, or aftbody) could provide a general release time for that specific structure. These times can be correlated with video events to produce a time-based breakup sequence.

The team elected to perform ballistics only on key items of debris and *cluster analysis* on the remaining objects. Cluster analysis assumes that when the recovery locations of a large number of debris items from the same structural element (e.g. tail, wings, payload bay, CM) are considered, the debris will have a similar range of ballistic numbers. Evaluating the centroid of clusters of structural elements *relative* to each other can provide a relative sequencing of key events. Specifically, because the centroid of the cluster for one

---

[1]*Columbia* Accident Investigation Board Report, Volume I, August 2003, p 12.
[2]*Columbia* Accident Investigation Board Report, Volume II, Appendix D.7, Working Scenario, October 2003, p. 209.
[3]*Columbia* Accident Investigation Board Report, Volume III, Appendix E.2, STS-107 Image Analysis Team Final Report, October 2003.

structural element X is west of the centroid of recovered debris from structural element Y, X was released before Y. Cluster analysis of the debris field consisted of plotting the latitude and longitude of recovered debris items that originated from a specific location (such as one of the wings) on the orbiter, identifying the centroid of the debris cluster, and comparing that to the cluster centroid of items from a different source (such as the other wing). A cluster identifies a ground recovery zone or "footprint" of debris from a specific source.

The conclusion was that comparing the locations of these clusters does, in most cases, produce accurate results of *relative* sequencing. It does not provide a time-based sequence. The relative sequence may not exactly coincide with the ballistically determined time-based sequence, which is considered more accurate. The inherent uncertainty contained in the ballistic time sequence due to the reference trajectory assumptions is an error range of ±5 seconds. Other errors in the ballistic release times include the effects of cascading failures and the effects of aerodynamic lift, both of which bias release times later than the actual release time.

The *Columbia* Reconstruction Database (CRD) was critical to both the ballistic and the cluster analyses because it contains the records for ground recovery locations. When items were recovered in the field, typically their Global Positioning System (GPS) latitude and longitude were recorded. All recovered debris (approximately 84,000 pieces, or 39% of the orbiter[4]) were entered into the CRD, which is located at KSC. For details regarding the debris collection and processing, reference the *CAIB* Report, Volume II, Appendixes D.10 and D.11, and the Reconstruction Report.

Finally, *structural analysis* was based upon detailed engineering knowledge of the orbiter and forebody structures compared to the condition of the recovered debris. Given the knowledge gained from the video analysis, ballistic analysis, and subsequent cluster analyses, the overall structural analysis provided the most likely scenarios for failure modes of the orbiter.

The following is a summary of findings, conclusions, and recommendations from this section:

Finding. Ground-based video was a vital resource of data for understanding the accident, especially after telemetry was no longer available.

> **Recommendation A11.** All video segments within a compilation should be categorized and summarized. All videos should be re-reviewed once the investigation has progressed to the point that a timeline has been established to verify that all relevant video data are being used.

> **Conclusion L3-2.** The breakup of both *Challenger* and *Columbia* resulted in most of the X_o 582 ring frame bulkhead remaining with the crew module or forebody.

> **Recommendation L3-1.** Future vehicles should incorporate a design analysis for breakup to help guide design toward the most graceful degradation of the integrated vehicle systems and structure to maximize crew survival.

2.2.1 Ground-based video analysis

There was no NASA ground-based video imagery of the entry and breakup of *Columbia*. Shortly after the *Columbia* accident, NASA issued a request to the public to submit any photographic stills or videos taken of the vehicle's launch and entry that might aid in the mishap investigation. More than 170 videos and 1,500 stills as well as verbal accounts of the entry were submitted by the public to NASA. Most submitted video data had a variety of limitations for analysis, e.g., changing zoom factors and poor tracking. However, these videos were invaluable and contributed significantly to an understanding of the events of the last few minutes of *Columbia* and the crew.

---

[4]STS-107 *Columbia* Reconstruction Report, NSTS-60501, June 2003, p. 143.

The ISAG generated a master timeline that was comprised of two timelines; a "western" and an "eastern" timeline.[5] The western timeline spans the orbiter's trajectory from off the coast of California to New Mexico. A 2-minute-and-12-second gap in ground-based imagery coverage occurred as *Columbia* traveled from south of Santa Fe, New Mexico to Palo Pinto (just west of Dallas, Texas), GMT 13:57:31 to GMT 13:59:43. Video from the eastern timeline ends in Louisiana, where the engine powerheads impacted the ground.

Late in the SCSIIT investigation, the team discovered an additional video that provided useful information that had not been included in previous analyses because it was not classified as important initially. After the mishap timeline had been established, it was more evident that this video provided unique data for an important timeframe. This video, which is referred to as NBC (National Broadcasting Corporation) (EOC2-4-0076-B), adds several seconds of good-quality imagery to the eastern timeline and begins at GMT 13:59:32.5 (± 1 second), just after LOS at GMT 13:59:32. The gap in video coverage was reduced by 11 seconds. More importantly, this video provided insight into the LOC. The vantage point of the NBC video is directly under the flight path of the orbiter. Unfortunately, when the forebody separates after the CE, its trajectory is above other debris pieces, so this video adds little additional information after the CE.

Video frames shown in this section are taken from various videos that capture several events. Each frame represents 1/29.95 of a second. The frames shown come from compressed video; the frames are not representative of the higher-quality frames that were used for evaluation. The "times" shown in the frames reflect the time code synchronization protocol of the Society of Motion Picture and Television Engineers (SMPTE), not GMT times. Even in the few videos where the GMT code is displayed, the time is only as accurate as the time to which the camera was set. The GMT times reported below the video frames shown in this report have been time-synchronized and should be accurate to within approximately 1 second. The one exception is the Apache video. The source of Apache was a military helicopter and the time was GPS-based and, therefore, highly accurate (<0.3 second).

Errors associated with video event times are impacted by the magnification of the FOV, resolution, and viewing geometry. These factors can introduce some error into the timing. A combination of error sources can lead to an accuracy of ±1 second for defined events within a single video or between videos, although the actual error may be better or worse by up to an additional 1 second. This is more accurate than the ballistic trajectory times, which have a minimum error of ±5 seconds. As a result, the video analysis was used to determine the timing of key events.

Finding. Ground-based video was a vital resource of data for understanding the accident, especially after telemetry was no longer available.

> **Recommendation A11.** All video segments within a compilation should be categorized and summarized. All videos should be re-reviewed once the investigation has progressed to the point that a timeline has been established to verify that all relevant video data are being used.

2.2.1.1 *Significant events seen in the video*

Eastern timeline video coverage starts at GMT 13:39:32.5, immediately after LOS. There are four significant events for which there is video coverage: LOC, CE, CMCE, and Total Dispersal (TD). The CMCE and TD are discussed in Section 2.4. Significant debris shedding events are visible throughout the timeline.

Although SMPTE times that are seen in the frames are video-specific relative tape times, they can be correlated to GMT times if a visual event can be synchronized to a known GMT. The bright "dot" seen in the videos is the envelope of hot gases surrounding the orbiter during entry. Both the "dot" and the trail left by the orbiter are typical and expected visual signatures during entry. The apparent "star" around the orbiter is a lens flare, a common image effect due to the orbiter's brightness and the lens system of a camera. This "star" is seen in most videos.

---

[5] *Columbia* Accident Investigation Board Report, Volume III, Appendix E.2, STS-107 Image Analysis Team Final Report, October 2003.

Loss of control

Figure 2.2-1 is the first frame of video for the eastern timeline. Figure 2.2-2 shows GMT 13:59:37, 135 frames (4.5 seconds) after figure 2.2-1. Figure 2.2-3, which shows 13 frames (approximately 0.5 second) after figure 2.2-2, correlates to LOC at GMT 13:59:37. The width of the trail increases at this time, which likely indicates a change in the flight condition. In addition to the change in the width of the trail, the trail appears to pulse or "corkscrew" over a period of less than 1 second (figure 2.2-4). This is consistent with the motion analysis, which suggests that the orbiter went into a flat spin.

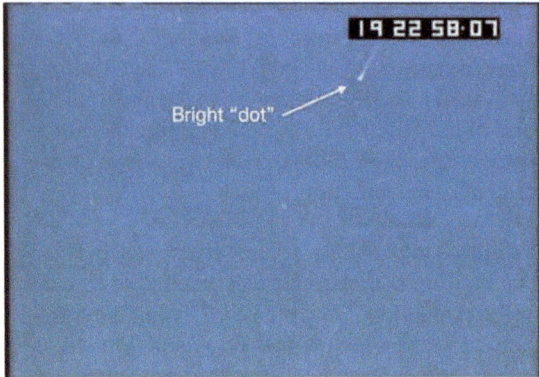

Figure 2.2-1. *First frame of video of the Eastern timeline, GMT 13:59:32.5.*

Figure 2.2-2. *GMT 13:59:37.00.*

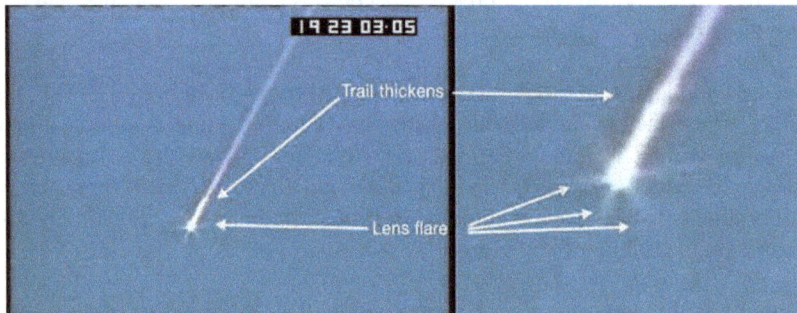

Figure 2.2-3. *GMT 13:59:37.43, 13 frames after figure 2.2-2. Right image is a zoomed view of the left image.*

Figure 2.2-4. *GMT 13:59:37.80.*

Catastrophic Event

The second significant event that is seen in video is the CE. Figure 2.2-5 shows the CE; there is no change in magnification between frames. The video of the CE is a distinct visual event in which the orbiter envelope brightens significantly and the trail width doubles. Additionally, some color change to orange is evident just prior to the CE. The color change is not readily visible in the still image frames that are taken from the video. In videos of the *Challenger* accident, an orange color was also seen in the videos and was assumed to be hydrazine, the material that is contained in the Orbital Maneuvering System (OMS) pods and RCS tanks. It is unknown whether the orange color seen in *Columbia*'s trail correlates to OMS or RCS tanks, or to some other cause.

Shortly after the brightening event and color change, a split trail is seen (figure 2.2-6). This strongly supports that the orbiter broke into multiple pieces at the time of the CE. The NBC video shows that the orbiter has an intermittent split trail prior to the CE. Prior to the CE, the split trails are tied to the separation of pieces of debris, and a singular main body trail continues after the separation of the debris. The split trail seen after the CE is constant; it does not dissipate and does not resolve back to a singular trail (figure 2.2-7).

GMT 14:00:18.23   GMT 14:00:18.26   GMT 14:00:18.30

Figure 2.2-5. *The Catastrophic Event is depicted in these three frames of video that cover 0.1 second. There is no change in the magnification/zoom factor. The third frame represents GMT 14:00:18.3.*

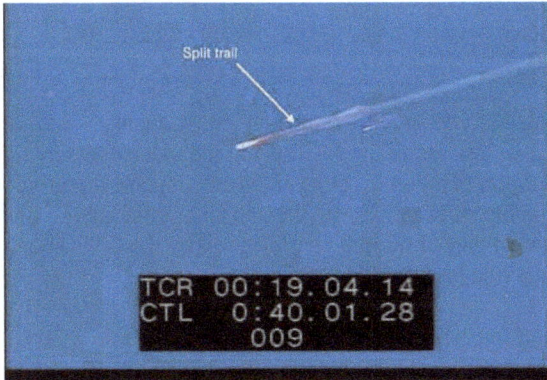

Figure 2.2-6. *Split trail after the Catastrophic Event.* [GMT 14:00:18.6]

Figure 2.2-7. *Persistence of the split trail 23 seconds after the Catastrophic Event.* [GMT 14:00:41.1]

Due to camera effects of saturation of pixels, magnification settings, distance to the orbiter, and angle of view of the orbiter, visual separation of the major pieces of the orbiter is not identifiable until 8 seconds after the CE. The visual separation of the orbiter into separate identifiable pieces is first made evident by the appearance of a second "star" beside the first "star." Figure 2.2-8 uses reversed color ("inverted") images of the orbiter to more readily show the "star" (lens flares). The dark lines on the images are shown in an effort to illustrate the lens flares.

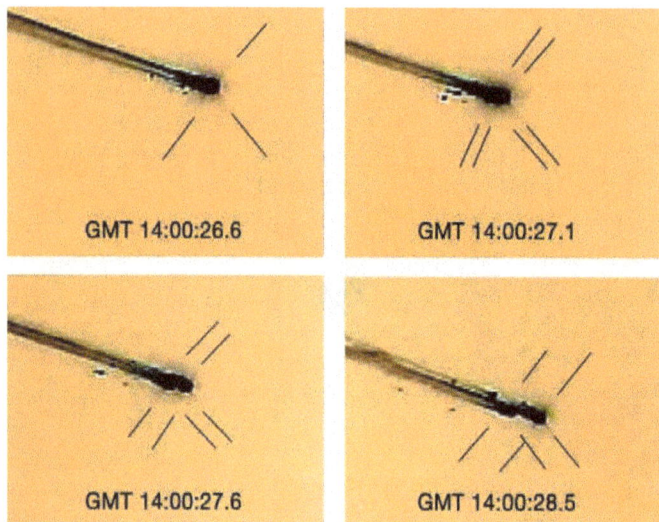

GMT 14:00:26.6

GMT 14:00:27.1

GMT 14:00:27.6

GMT 14:00:28.5

Figure 2.2-8. *Color-inverted video images of the start of the double star event (GMT 14:00:26.6).* Black lines have been added to more clearly identify the two separate objects.

The forebody breakup was identified at GMT 14:00:53 (see Section 2.4).

Other Major Structure Identification

Once the key events had been identified, an effort was made to tie specific debris shedding events with specific recovered debris. In particular, the team wanted to identify the forebody in the video. Heavier objects have a higher ballistic number and slower deceleration than lighter objects. Heavy objects, therefore, would be visible longer than other objects and could possibly be identified by their distinct trajectory in the video compared to other objects. The aftbody is the heaviest part of the orbiter, followed by the CM. The next heaviest item was SPACEHAB, which was located in the payload bay.

Six entry videos were found to offer the most information for debris shedding identification (Table 2.2-1). Videos are referred to by the name listed in the table. Figure 2.2-9 shows the debris tree created from these videos. The end time for a debris piece generally indicates when the object left the FOV of the video rather than the object breaking into pieces too small to be seen by the camera.

Table 2.2-1. *Videos Used to Create the Debris Tree in Figure 2.2-9*

Reference EOC No.	City in Texas	Name	Latitude	Longitude
EOC2-4-0024	Arlington	Arlington	32.7	−97.1
EOC2-4-0209-B	Hewitt	Hewitt	31.4	−97.2
EOC2-4-0221-4	Mesquite	WFAA4/Mesquite	32.8	−96.6
EOC2-4-0221-3	Fairpark	WFAA3	32.8	−96.7
MIT-DVCAM	Fort Hood	Apache	31.2	−97.6
EOC2-4-0077	Burleson	NBC	32.5	−97.3

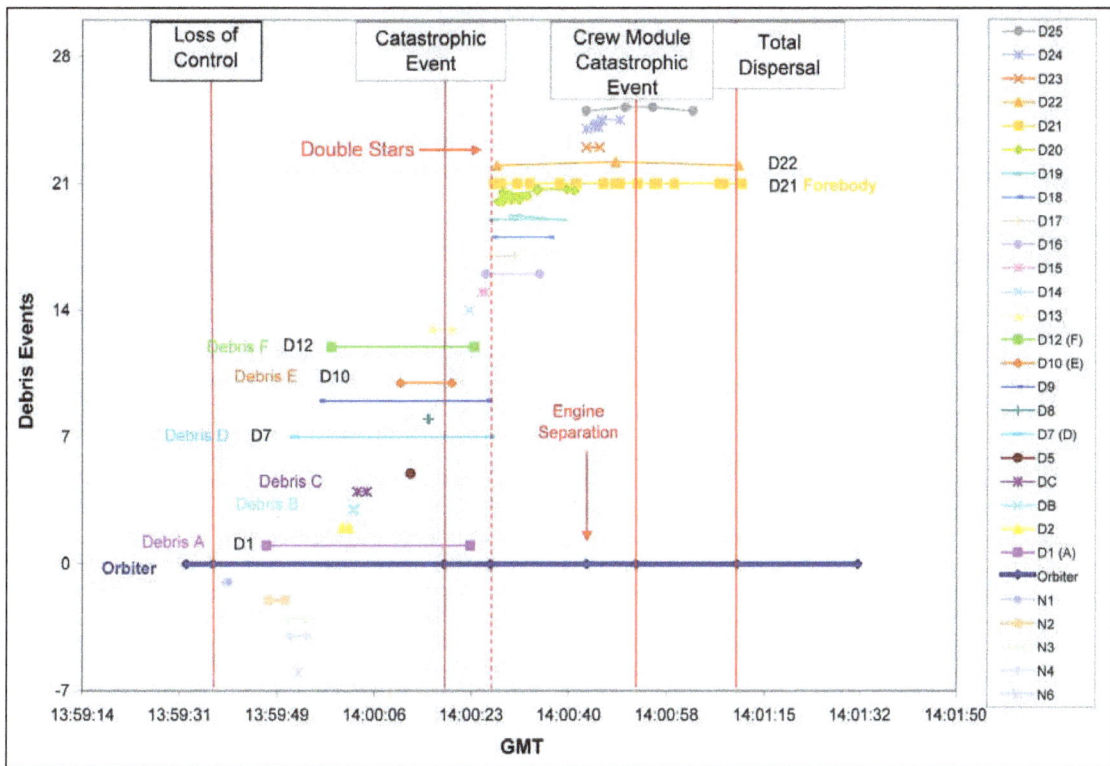

Figure 2.2-9. *Debris tree.*

The points that are labeled Debris A, B, etc. are the names that correlate to the master timeline provided to the CAIB.[6] These were the starting point for the SCSIIT video analysis. When other videos were reviewed, they received a different numbering system. For example, debris shedding events that were identified in Hewitt were labeled D1, D2, D3, etc. In the NBC video, debris events were labeled N1, N2, N3, etc. When an object could be positively correlated with the originally identified debris events (A, B, C, etc), they were merged into that line, showing the full range across all videos for how long a specific debris object was visible in any video. If an object could not be clearly correlated with another video, it was kept separate in the debris tree and labeled accordingly.

The points that are seen on the various horizontal lines typically represent a significant piece separating from that debris piece. In some cases, these significant pieces are graphed with the parent piece. All the major events are marked on the timeline. Debris D through F and Debris 8 through 12 might have some overlap as objects but, due to different angles, were difficult to correlate.

Two pieces of debris (Debris A and, shortly afterward, Debris D) are seen in the NBC video (and some other videos) well before the CE. Debris A "flashes," brightening and completely disappearing repeatedly, which is suggestive of tumbling. At times, the track and movement of the piece possibly suggests that it generates lift. At one point, the single flashing debris disappears and multiple smaller flashing debris appear, suggesting that Debris A broke into pieces. Debris D, however, paces the orbiter, suggesting that it is an object with significant mass and a high ballistic number. Given that the left wing and OMS pod were known to be structurally degrading in this timeframe, Debris A may possibly be the left OMS pod cover or a piece of the left wing, and Debris D may possibly be the left OMS pod, which is fairly heavy and substantial.

---

[6]*Columbia* Accident Investigation Board Report, Volume III, Appendix E.2, STS-107 Image Analysis Team Final Report, October 2003.

The aftbody is a part of the line labeled "orbiter" in the figure, since it was traced back to the intact orbiter and was the most persistent piece of the orbiter after the CE. It was easily identifiable because the engines separate from it late in the video.

Debris item 21 (D21) is identified as the forebody. This identification is based upon the mass of the forebody and trajectory in the video. Ballistic and debris field evidence later confirmed this assessment.

The points seen on the debris item 21 line represent all events that could be correlated to more than one video. Because of various camera viewing angles, correlation of debris pieces was difficult. Cross-correlation with three other videos showed that brightening events could be seen in many videos while few specific debris events could be cross-correlated.

The last time at which the main engines visually appeared to be a single unit is GMT 14:00:50.6; this is defined as the starting point for engine separation. Figure 2.2-10 shows the engines at GMT 14:00:53, the three pieces below the forebody which are circled in green. Their separation is already well under way. Considering that visual separation of the orbiter pieces after the CE took about 10 seconds, the actual time of separation of the engine components from one another may have been

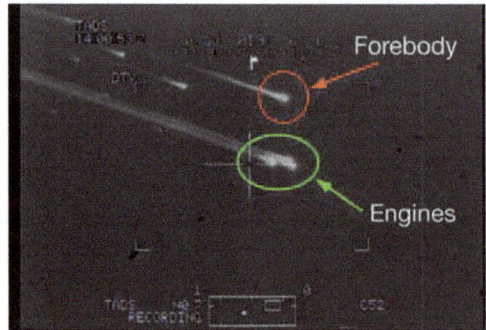

Figure 2.2-10. *Clear visual separation of the three main engines.*

much earlier than GMT 14:00:50.6. Available data do not allow for a more accurate time of separation.

In summary, specific debris shedding events were very difficult to tie to specific recovered debris with a few exceptions. The forebody was positively identified, as was the aft section/engines. Debris A is possibly the left OMS pod cover. Debris D is possibly the left OMS pod. Debris D22 is suspected to be the SPACEHAB. The sequence, as derived from the video analysis, is presented in Table 2.2-2. It should be noted that the OMS pod time represents a time at which the debris piece is already clearly separated from the orbiter. The actual separation time could not be identified. Both forebody and (potentially) SPACEHAB are identified as separating from the rest of the orbiter at the CE. The times shown below indicate the video-based time of structural breakup.

Table 2.2-2. *Orbiter Breakup Event Sequence and Times Determined by Video*

Event	GMT
LOC	13:59:37
OMS pod cover	13:59:46
OMS pod	13:59:51
Orbiter breakup (CE)	14:00:18
SPACEHAB breakup (D22)	14:00:48
Main engines separate	14:00:50
Forebody breakup (D21)	14:00:53

2.2.2 Ballistic analysis

Ballistic analyses were conducted on selected recovered debris objects to help define the orbiter breakup sequence. The ballistic analysis determined an object's approximate time of release from the vehicle given certain initial conditions. It should be noted that this type of analysis does not account for serial (cascading) debris events where a large object is released on its own trajectory and the object then breaks up into multiple smaller objects. Therefore, assessing an individual object has inherent uncertainty. For objects that left the vehicle in almost the same configuration as they were recovered (i.e., an individual item of crew equipment), the ballistic numbers and times of release have a higher confidence level. However, conducting ballistics analysis on several debris items from a specific zone of origin on the orbiter can provide a general release time for that specific structure.

The process is called "ballistic analysis" because all the objects are assumed to become ballistic upon release. The trajectory of a ballistic object is unique to the properties of that object and is not controlled by power or directional steering. A good example would be a ball fired from a cannon. The cannonball's trajectory is determined by the forces acting on it: its momentum (mass × velocity), the drag from the atmosphere, Earth's gravity, and winds (figure 2.2-11).

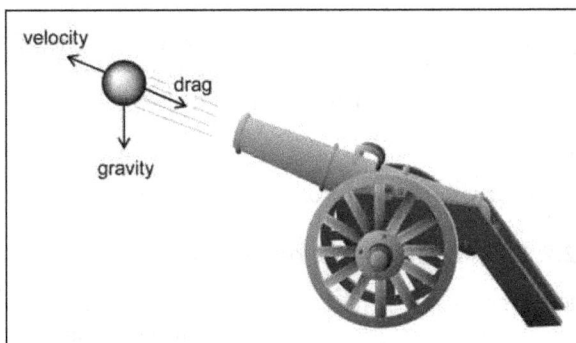

Figure 2.2-11. *A cannonball is an example of a ballistic object.*

The modeling tools used in the ballistic analysis were Snewt and SORT (see Chapter 4).

Objects were selected for ballistic analysis based upon weight, shape, original location on the orbiter, and possible relevance to the breakup sequence. For each debris object, a reference area, reference length, and c.g. location were specified. The c.g. location of some analyzed debris items was already known. For others, the c.g. location was measured from the debris recovered, or the c.g. location was estimated. The aerodynamic coefficients were generated at different orientations through an alpha (angle-of-attack) and beta (sideslip angle) sweep or at a significant number of random orientations. The ballistic number can be determined using aerodynamic coefficients. It is possible to average the ballistic number of the various orientations to come up with an average ballistic number. The more "regularly" shaped the object is (sphere, cube, rectangular prism, flat plate, etc.), the more precisely the average ballistic number can be computed. For most of the objects analyzed, the average ballistic number was used because it was assumed that the objects were tumbling and, therefore, were constantly changing their orientation. Very few objects had directional thermal erosion, indicating that they had aerodynamically trimmed in a specific orientation. For these objects, the ballistic number for that orientation was used. The average ballistic number assumption can introduce errors in both directions for release time because the actual ballistic number may have been either larger or smaller than the assumed average.

After the aerodynamic coefficients were calculated, the object's approximate release time could be determined. As previously stated, there is an error bar of ±5 seconds in release times due to the uncertainty in the reference trajectory used as a baseline for the item to be released from.

Release times for large structures were determined using the estimated release time of their subcomponents. However, these estimated times do not take into consideration cascading failures of these large objects. The recovered debris objects were separated into categories of large orbiter structure, such as left wing, OMS pod, forward bulkhead, aft bulkhead, etc. Each category of objects recovered on the ground made a cluster along the ground track. It was first thought that objects in the westernmost part of a given cluster would have the lower ballistic numbers, and objects to the east would have the larger ballistic numbers. However, it was found from examining the database of recovered items that each debris cluster included many low-ballistic-numbered objects, meaning that cascading failure was common. Items may in fact be part of a larger object that separated earlier, for which a trajectory is not known. Only the trajectory of the individual item can be computed. Cascading events generally result in calculated release times for individual items that are later than the actual release times.

Another key assumption in the ballistic analysis was that the analyzed objects did not have lift. To determine whether an object could have generated lift, the shape and especially the orientation of the object during flight has to be known. Since that knowledge was not available, drag-only or ballistic flight was assumed. In general, an object with lift will travel a greater downrange distance than that same object on a purely ballistic trajectory, causing an error in release time that is biased to a later time than actual release.

Since cascading failures and lift would send an object farther downrange (east) than expected from a simple ballistic trajectory, a decision was made to use items from the westernmost (earliest) edge of each cluster and to use easily modeled items (such as flat plates) when possible to estimate the release time of the large orbiter structure. A conscious effort was made to choose items such as structural skin rather than tiles or reinforced carbon-carbon (RCC), which were continually shed during the entire entry. If three or more debris items were found that gave the same estimated release time, confidence was increased that the large orbiter structure came off at that time. The estimated time of release of large structures is referred to as the major structural release time.

For further information on ballistics, see Appendix A.

2.2.2.1 *Major structural release times*

The estimated release time is the calculated time at which an object was released from the reference trajectory, given the previous assumptions and conditions. Table 2.2-3 shows the estimated release times for large orbiter structures. After the CE, the vehicle was no longer intact and was in at least three major pieces (forebody, midbody/right wing, and aftbody). Some of the subsequent release times reflect separation from those major pieces, not the intact orbiter.

Table 2.2-3. *Release Times of Large Structures*

Major Structural Release Time (GMT)	Vehicle Structure	Time of Release (GMT) and Debris Object Number[7]
13:59:48	Left OMS Pod (skin, structure, honeycomb)	13:59:48 (78899) 13:59:57 (85446) 14:00:16 (84132)
14:00:04	Left Wing (RCC, upper and lower wing skin)	14:00:02 (70391) 14:00:05 (81331) 14:00:05 (11525)
CE		
14:00:26	Midbody Fuselage	14:00:20 (82427) 14:00:33 (82172) 14:00:35 (38315)
14:00:27	Right Wing (RCC, skin, structure)	14:00:24 (68702) 14:00:26 (24508) 14:00:28 (8172) 14:00:30 (49833)
14:00:31	SPACEHAB	14:00:29 (65045) 14:00:33 (7641) 14:00:57 (22900)

---

[7]The debris object numbers are the numbers that are assigned to each recovered object when its recovered location, description, weight, dimensions, etc. was entered into the *Columbia* Reconstruction Database for tracking and identification.

Table 2.2-3. *Release Times of Large Structures* (Continued)

Major Structural Release Time (GMT)	Vehicle Structure	Time of Release (GMT) and Debris Object Number[8]
14:00:34	Vertical Tail (structure, drag chute panel)	14:00:26 (26078) 14:00:27 (77800) 14:00:38 (85279) 14:00:39 (1633) 14:00:40 (45837) 14:00:48 (52092)
14:00:34	Aft Fuselage (Carrier Panel, structure)	14:00:32 (43091) 14:00:33 (31308) 14:00:36 (79178)
14:00:36	Internal Airlock soft stowage items	14:00:36 (65900) 14:00:36 (31297)
14:00:43	Tunnel Adapter structure	14:00:42 (64966) 14:00:44 (69606)
14:00:54	Aftbody forward bulkhead (X_o 1307)	14:00:52 (12877) 14:00:52 (83678) 14:00:53 (14559) 14:00:57 (64156)

Large structural pieces, such as the left OMS pod or left wing, were not released from the orbiter at a single time as is implied by the major structural release time. The objects in this section should not be considered as being released from a single complete orbiter, but from (at minimum) three separate objects. Table 2.2-4 compares the sequence suggested by video against the sequence suggested by ballistic assessment.

Table 2.2-4. *Comparison of Video and Ballistic Sequence*

Video Sequence (relative)	Ballistic Sequence (time-based)
OMS Pod	OMS Pod
	Left Wing
Forebody/Midbody/Aftbody (CE)	Midbody
	Right Wing
SPACEHAB	SPACEHAB
	Vertical Tail
	Aftbody
Main Engines	Main Engines
Forebody	Forebody

The differences between the two sequences can be explained by cascading events where major portions of the vehicle, such as the aftbody and the forebody, remained intact for some time. One conclusion from this comparison is that the relatively lighter components making up the midbody, including the wings and payload bay, disintegrated fairly rapidly and as individual objects while the aftbody, SPACEHAB, and the forebody maintained some integrity.

---

[8]The debris object numbers are the numbers that are assigned to each recovered object when its recovered location, description, weight, dimensions, etc. was entered into the *Columbia* Reconstruction Database for tracking and identification.

2.2.3 Cluster analysis

This section addresses the cluster analysis that shows how the pattern of recovered debris aided in understanding the sequence of the orbiter breakup.

An evaluation of the debris field was conducted to match debris sources (orbiter structural zone) with debris recovery locations. In all cases for the debris maps that follow, the orbiter travels from approximately the upper left corner (northwest) to approximately the lower right corner (southeast). Although debris was seen falling away from *Columbia* in ground-based video as far west in its trajectory as off the coast of California, the farthest west confirmed *Columbia* debris was recovered in Texas.

The location of debris on the ground is influenced by when the debris separated from the orbiter and how far the object traveled downrange. As a general rule, heavier objects travel farther downrange (to the east, in this case) than lighter items of similar shape and size. In addition, wind effects and, possibly, lateral forces exerted during orbiter breakup had some effect as evidenced by the lateral dispersion of objects.

The center of a large debris group was considered reasonably accurate for relative sequencing, with ballistics allowing for refinement and correction, if needed. Cascading debris failures and lift generation will result in biasing the debris cluster farther east (artificially elongated at the trailing end) and greater lateral spread for the same reasons as the effects on the ballistics analysis. Wind, which at the time of the mishap was prevailing from the southwest, can also affect the debris field cluster shape and centroid.

However, cross-referencing ballistic release times across a range of clusters showed that as few as five objects in a debris group can allow some conclusions to be drawn *if the objects have approximately the same range of ballistic numbers.* When five or more items were available to determine a cluster, there was fairly high confidence in comparing that cluster to another.

Knowledge of the topographic conditions was important in some instances, where water hazards or thick brush interfered with debris recovery efforts. Some graphs show that fewer debris items were recovered in certain areas, suggesting a decrease of debris when, in fact, the debris field thinning is related to field conditions and obstacles that impeded the ability to locate the debris items.

2.2.3.1 *Debris clusters*

The orbiter is made up of multiple structural elements (figure 2.2-12). The recovery location data of the debris were divided into many of these structural elements to determine the apparent sequence in which the orbiter failed.

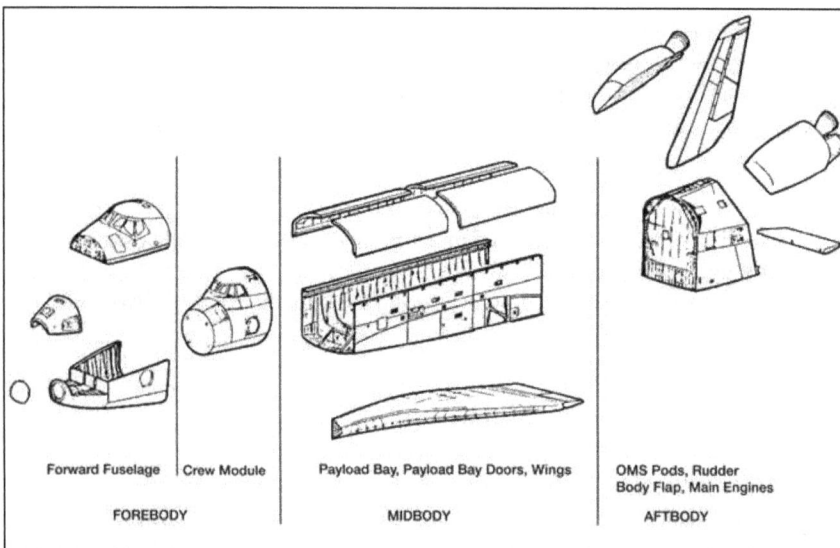

Forward Fuselage	Crew Module	Payload Bay, Payload Bay Doors, Wings	OMS Pods, Rudder Body Flap, Main Engines
FOREBODY		MIDBODY	AFTBODY

Figure 2.2-12. *Depiction of the orbiter forebody, midbody, and aftbody elements.*

First, recovered debris longitude data were sorted by respective structural sources. Then, each item became a data point placed into longitude "bins," each having a range of 0.5 degree. Each bin was then graphed by what percent of the recovered debris fell into what longitude bin. The numbers of debris items in a group ranged from fewer than 100 into the thousands. One bias introduced is that the data were not corrected for these differences in total number when comparing datasets; only the percent of the total recovered from that zone is compared. It is unknown how this difference affects the data overall.

Details of each debris field cluster are presented in the following sections.

Wings

Figure 2.2-12 identifies where the wing structure is in relationship to the other orbiter structural components. TPS tile is included in these figures, as some tile adhered to structure and it was not easily possible to separate out individual tiles. Figure 2.2-13 shows a ground debris map that is similar to that produced in the CAIB Report, Volume I, p. 75; however, figure 2.2-13 shows all the debris for the wings whereas the CAIB map shows only the RCC panels. As the CAIB reported, the left wing failed before the right wing. This is implied in figure 2.2-14 as well. Overlay mapping of debris at their correct latitudes and longitudes prevents easy recognition of their actual ground footprints. The bars shown on the map and in later ground debris maps represent the main cluster of the debris. If coverage extends beyond the map shown, an arrow indicates that the footprint continues off the map. The bar shown above each histogram indicates the longitude range covered by the overlay mapping.

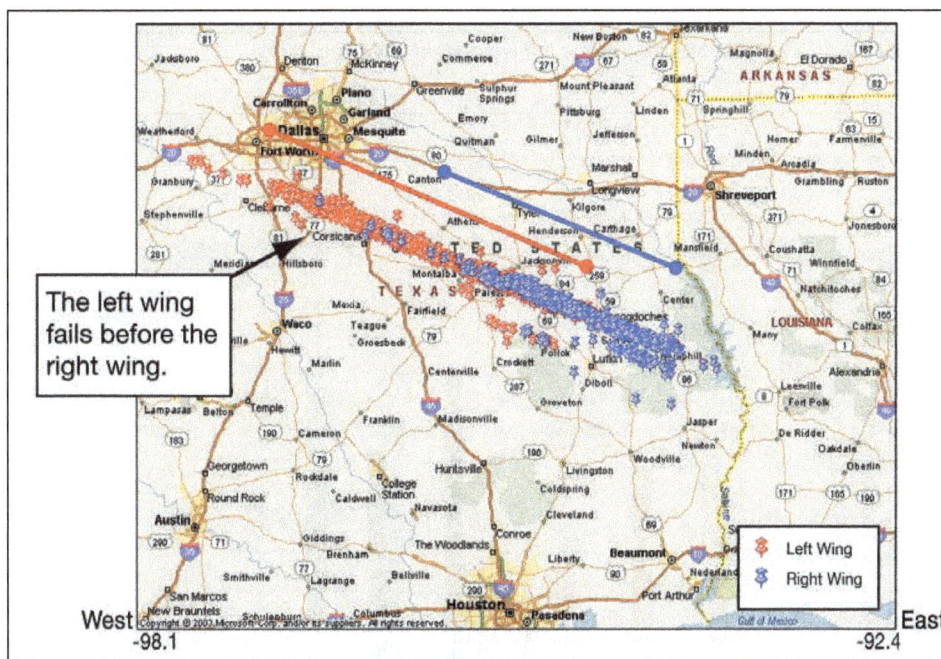

Figure 2.2-13. *Debris field of the left and right wings.*

Tail (vertical stabilizer)

Figure 2.2-14 illustrates the debris of the tail section in relation to the wing debris. The bulk of the tail debris impacted west of the bulk of the right wing debris. Figure 2.2-15, which is a histogram of longitude vs. the number of debris items, shows this as well. The lines above the histogram indicate the approximate area of the histogram that the debris map covers. Figure 2.2-15 suggests that the tail possibly began to depart between the left wing and the right wing.

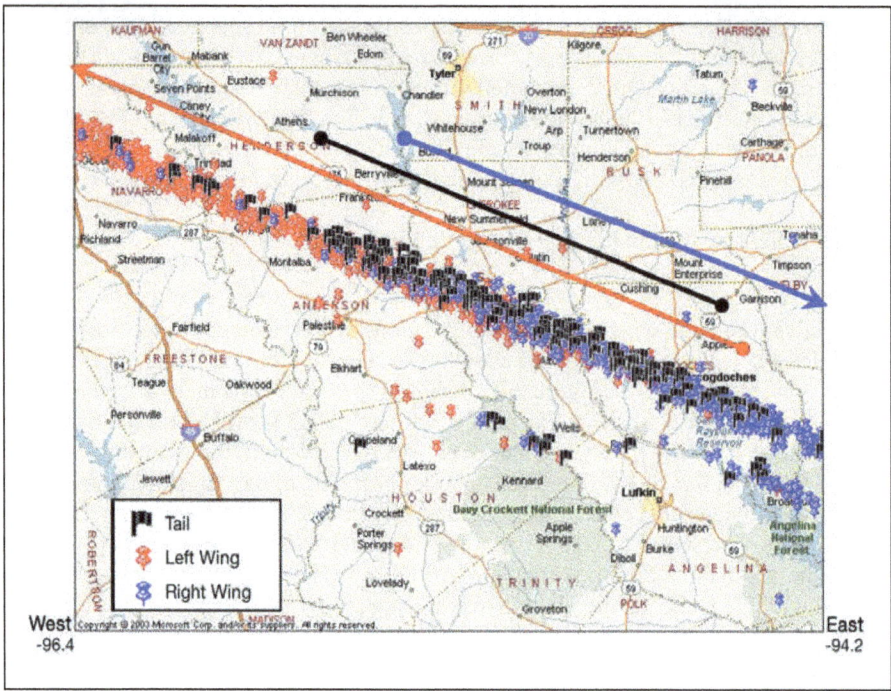

Figure 2.2-14. *Partial debris field of the left wing, right wing, and tail.*

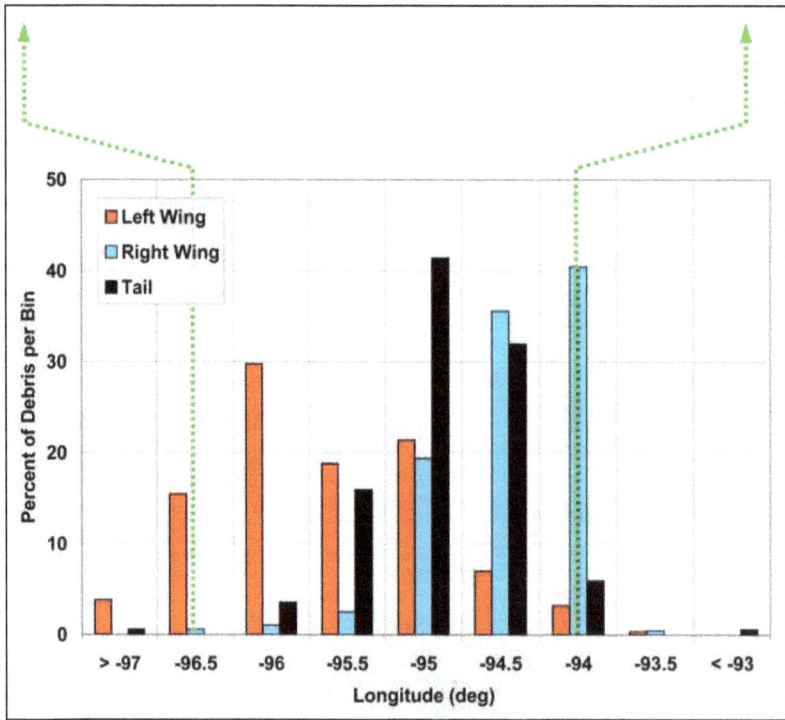

Figure 2.2-15. *Histogram of the left wing, right wing, and tail debris.* The green dotted lines are the projection onto the histogram of the longitude range shown in the map.

Midbody

Figure 2.2-12 identifies where the midbody is related to the other orbiter structural components. The midbody consisted of the mid fuselage, payload bay, and PLBDs (SPACEHAB is not included in this chart). The midbody began losing structure and shedding objects after the left wing but prior to the tail (figures 2.2-16 and 2.2-17). The midbody debris shown is comprised predominantly of PLBD structure, although pieces of the midbody are scattered throughout the entire debris field.

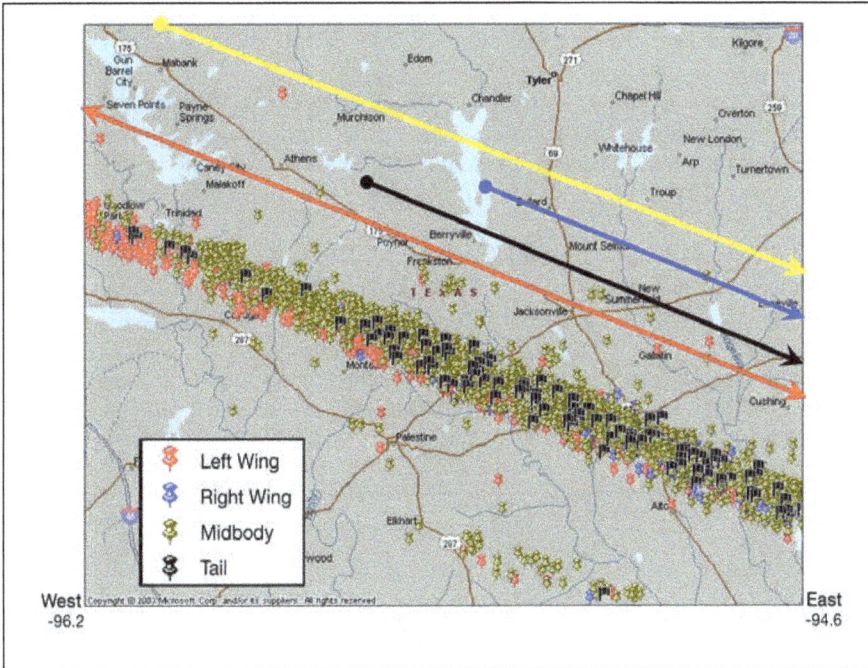

Figure 2.2-16. *Partial debris field of the left wing, right wing, midbody, and tail.*

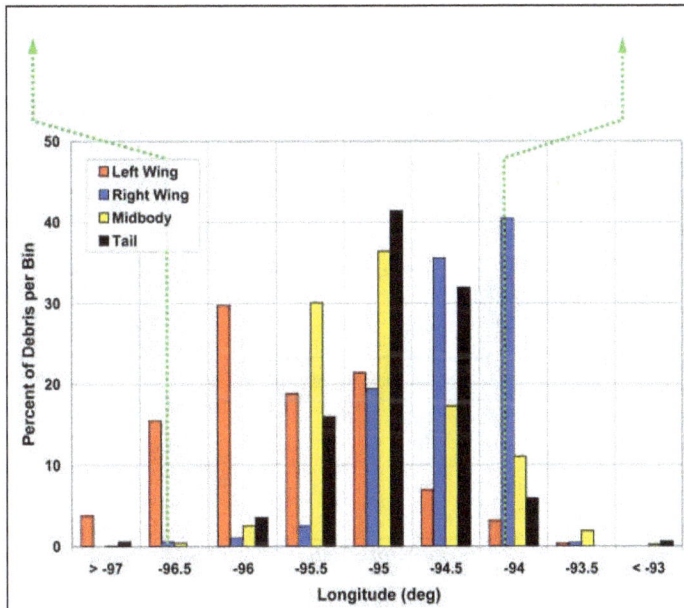

Figure 2.2-17. *Histogram of the percentage of debris found vs. longitude for the left wing, right wing, midbody, and tail.* The green dotted lines are the projection onto the histogram of the longitude range shown in the map.

Figure 2.2-18 shows the on-ground sequence of the midbody structural debris with right (starboard) and left (port) midbody origins marked where known. Beside the chart is a schematic of the midbody with some positions marked in X_o coordinates for reference. The X_o terminology refers to the X position in inches in the orbiter coordinate frame, where the X axis runs the length of the orbiter from fore to aft in the orbiter X-axis measurements. The nose of the orbiter is at X_o 236″ as the coordinate system extends beyond the nose. X_o 582″ is the demarcation between the forebody and the midbody. The port lower wing chine and midbody floor skin stretches from X_o 582″ to X_o 807″ (just forward of the SPACEHAB location). The wing attachment region spans X_o 807″ to X_o 1365″. The error bars shown on some pieces indicate the range of X_o positions that a piece of debris covered or could originate from, if the range was noted in the database.

The westernmost piece of ground debris from the midbody structure with a known X_o location was a port location just aft of SPACEHAB. However, this appears to be an outlier data point. Figure 2.2-18 indicates that based upon recovered midbody structure, the midbody first suffered a major failure ranging from about X_o 582″ to X_o 919″.

The next grouping of debris predominantly originates from X_o 776″ (towards the aft of SPACEHAB) to aft of the Fast Reaction Experiments Enabling Science, Technology, Applications, and Research (FREESTAR) payload near X_o 1124″, although there are a few pieces near the forebody bulkhead (X_o 582″).

The final cluster is from the aft region of the midbody. As debris originating from near the midbody aft bulkhead (X_o 1307″) appears farthest east in the debris field, this suggests that some midbody payload bay structure remained attached to the aftbody.

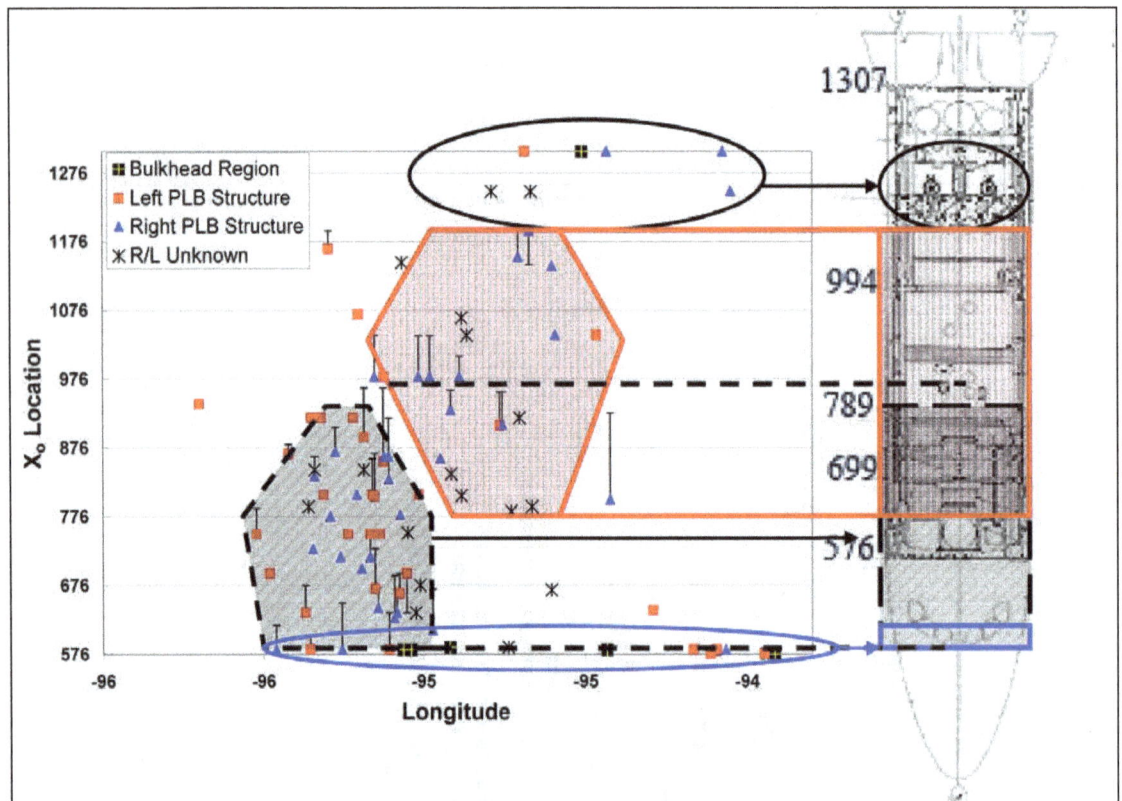

Figure 2.2-18. Longitude vs. the X_o origin of the structural debris of the mid fuselage. To the right is a drawing of the mid fuselage with key X_o positions marked.

The data in figure 2.2-18 cannot be used to determine a port or starboard initiation of breakup of the midbody. Size of individual items was also examined with the assumption that ballistic numbers of objects of roughly the same size and shape will land in the order in which they separate from the orbiter. The number of objects is small, so this may be an invalid assumption. For smaller-sized midbody debris ($\leq$500 in^2), port debris appears west of starboard debris. For larger-sized debris ($\geq$1000 in^2), starboard debris appears west of port debris. Overall, it is indeterminate whether the port or starboard side failed first based upon ground debris.

SPACEHAB debris was also evaluated (not shown here). SPACEHAB structural debris is concentrated near the forebody debris field, suggesting that the element maintained structural integrity for some period of time as did the forebody.

The midbody *structural* debris (vs. objects contained within the midbody such as SPACEHAB) and the right wing debris fields increase in density at the same time. As shown above, the port and starboard midbody structural debris originating from the same X_0 locations begin appearing in the same longitude range (95.5 to 95W). However, in the next longitude range sector (95 to 94.5W), predominantly starboard pieces were recovered. This is the same longitude range sector where the largest number of right wing debris items was recovered. This suggests that a portion of midbody structure and right wing may possibly have initially remained together as a unit when *Columbia* experienced the CE.

Aftbody

The midbody ground debris analysis indicates that at least some of the midbody may have remained with the aftbody. As observed on video, the main engines (the heaviest individual components of the orbiter, with a correspondingly high ballistic number) remained a cohesive unit until GMT 14:00:50.6. Aftbody shedding did occur throughout the period following the CE. Main engine pieces, which were among the easternmost debris objects recovered in the debris field, were found in Fort Polk, Louisiana. Figures 2.2-19 and 2.2-20 show the distribution of the debris field of the aftbody and the left and right wings relative to each other.

Forebody

The forebody breakup sequence is discussed in Section 2.4. The forebody is composed of the FF, CM, nose cap, nose landing gear, and forward RCS.

The debris field shows that the CM and FF failed at nearly the same time, with their debris footprints overlaid on one another. This supports that the forebody remained an integral unit until structural failure, which began at the CMCE. The debris field showed that about 87% of the forebody structural debris appears suddenly; degradation may have occurred, but it was minor until the start of total forebody structural failure. Video corroborates this, showing only minor intermittent debris loss from the forebody until its failure.

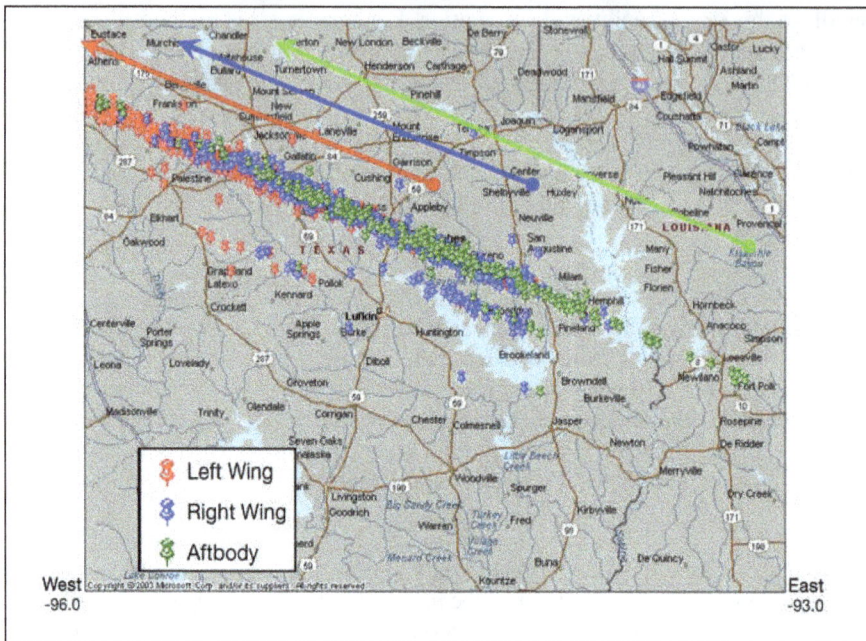

Figure 2.2-19. *Partial debris field of the left wing, right wing, and aftbody.*

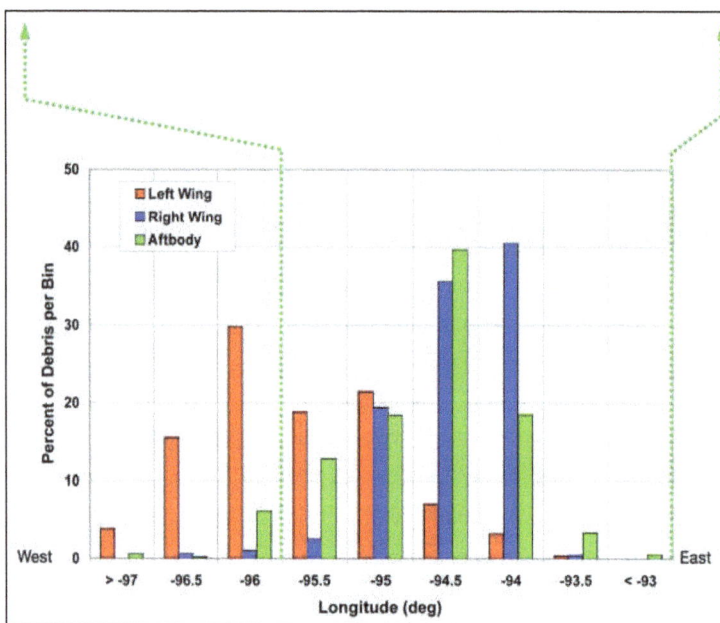

Figure 2.2-20. *Histogram of debris of the wings and aftbody (including orbiter main engines).* **The green dotted lines are the projection onto the histogram of the longitude range shown in the map.**

In summary, figure 2.2-21 shows an overall assessment of debris clusters. The westernmost longitude bin of each group that exceeded 30% of the recovered debris group is referenced with an arrow; the label above the bin identifies the debris group. It was observed that when the number of recovered debris items from one orbiter structural zone reached approximately 30%, the next group began to appear in rapidly increasing numbers. The significance of this observation is not known. It may be an artifact of data processing, or it may be characteristic of a cascade effect of structural breakup. This is a topic that may warrant further research and investigation.

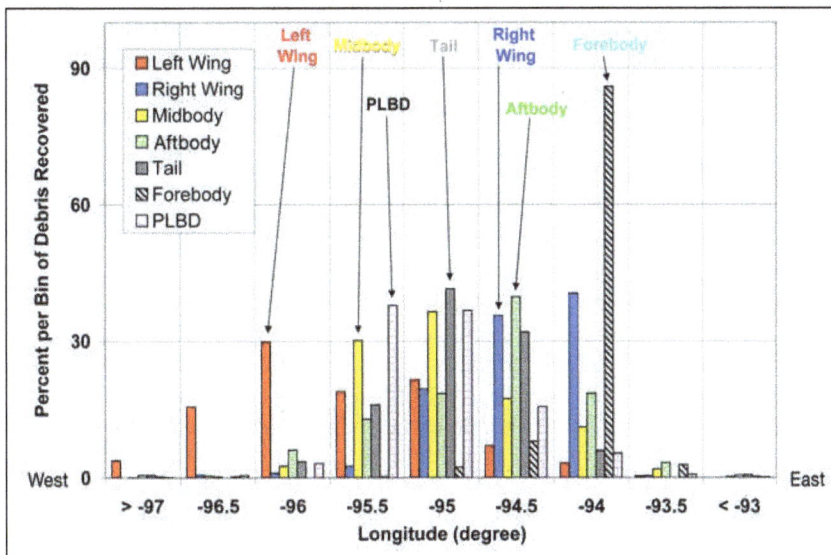

Figure 2.2-21. *Histogram of the relative order of the major debris groups recovered.*

Cluster analyses show that following the left wing, midbody items (predominantly PLBD pieces) departed the vehicle. The tail, the right wing, and the aftbody shed a significant number of pieces, followed by SPACEHAB and the forebody. Table 2.2-5 shows a comparison of sequencing from cluster analysis compared to video and ballistic analyses.

Table 2.2-5. *Comparison of Orbiter Breakup Sequence from Video, Ballistic, and Cluster Analysis*

Video Sequence (relative)	Ballistic Sequence (time-based)	Cluster Sequence (relative)
OMS Pod or Left Wing	OMS Pod	Left Wing
	Left Wing	
Forebody/Midbody/Aftbody separation (CE)	Midbody/PLBDs	Midbody
	Right Wing	Vertical Tail
SPACEHAB	SPACEHAB	
	Vertical Tail	Right Wing
	Aftbody breakup	Aftbody breakup
Main Engines separation	Main Engines separate	
Forebody breakup	Forebody breakup	SPACEHAB/Forebody breakup

This relative sequence of cluster-based groups does not match the ballistics reported time-based sequence (Table 2.2-3) of a few debris groups. Specifically, ballistics shows the tail failing before the right wing. This is probably the result of a cascading failure. Based on video, ballistic, and cluster analysis, the conclusion was that the left OMS pod and left wing departed first. It is possible that the PLBDs departed next. At the CE, the orbiter appeared to separate into three main components: forebody, midbody including the right wing, and aftbody. The midbody/right wing and aftbody failed in cascade following the primary breakup event. The forebody failed last.

2.2.4 Structural analysis

2.2.4.1 *Background*

This section provides a discussion of an analysis and scenario for the orbiter breakup. In addition to the data presented in previous sections, this analysis also draws on the orbiter's design strengths and weaknesses.

This structural assessment predominantly focuses on events leading up to the CE and the forebody release at the CE, since that was the area of most interest to the team. Little to no assessment was done by the SCSIIT structures team on the wings or aftbody since they were not considered relevant enough to this study to warrant the resources.

There are difficulties in understanding and supporting the various theories concerning the orbiter vehicle breakup. Except for the left wing, the vehicle was essentially structurally intact with most systems functioning normally at the time of the last telemetry received (GMT 14:00:05). No recorded on-board data were recovered after GMT 14:00:19, 1 second after the orbiter breakup began. Ground-based video footage does not provide adequate detail to determine the exact sequence of events other than as previously outlined.

Some background regarding the orbiter structure is necessary for this discussion. Orbiter structure is mainly constructed of aluminum components such as riveted skins and stringers, integrally machined plates, honeycomb sandwich panels, frames, bulkheads, and trusses. The exceptions are the OMS pod skins and PLBDs, which are made from graphite epoxy honeycomb sandwich panels, and the aft fuselage thrust structure, which is diffusion-bonded titanium that is reinforced with a boron-epoxy laminate. The payload bay is bounded by a forward bulkhead, the X_o 582 ring frame bulkhead (adjacent to the aft bulkhead of the CM), and an aft bulkhead (located at X_o 1307).

As detailed in figure 2.2-22, the FF is a semi-monocoque structure consisting of aluminum (2024) skins, stringers, longerons, bulkheads, and frames. Its structural purpose is to withstand the loads from the nose landing gear, CM, aerodynamic loads, and CM venting pressure. It also supports the associated portion of the TPS. The forward RCS is considered an integral part of the FF. The FF skin panels were designed primarily based on stiffness requirements to minimize local skin deflection to avoid the cracking or loss of the thermal protection tiles.

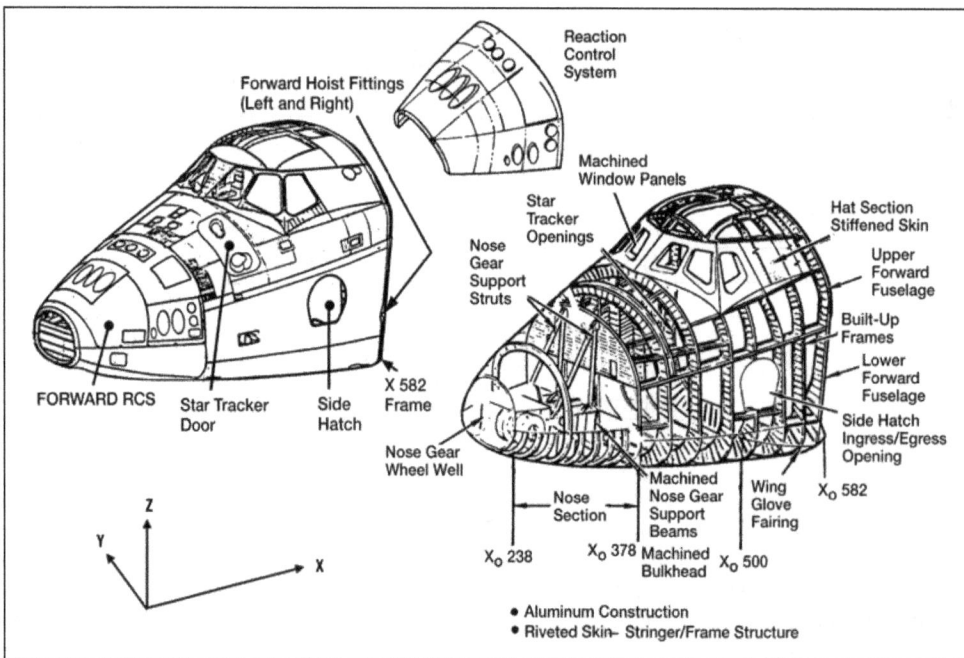

Figure 2.2-22. Forward fuselage structure.

The CM is an airtight pressurized compartment that is constructed from welded aluminum panels (2219 aluminum alloys) with integral stringers, frames, and longerons. It is enclosed and protected by the FF skin panels and the TPS. It is supported and suspended inside the FF by a combination of fittings and linkages.

The CM is supported within the FF at four main attach points made of titanium (figures 2.2-23 and 2.2-24). Side links provide secondary stabilization but do not provide significant structural support. Two major attach points are at the aft end of the flight deck floor level. These links support loading in the X (longitudinal) direction. They also carry 52% of the Z (vertical) direction loading and 34% of the Y (lateral) direction loading. Because these two links carry loads in the X, Y, and Z directions, they are called xyz links, but are generally referred to simply as "x-links." The x-links bridge the boundary of the X_o 576 bulkhead (aft bulkhead of the CM) and the X_o 582 ring frame bulkhead (the forward bulkhead of the midbody (figure 2.2-25)). The x-links physically connect the CM, the FF, and the midbody structure.

The other links are the z-link and y-links. The z-link (vertical load reaction) is on the centerline of the CM forward bulkhead. The y-links (lateral load reaction) are on the lower portion of the X_o 576 bulkhead; they attach to the X_o 582 ring frame bulkhead. These attach fittings are designed to crash requirements to 9 G in –X (within a 30-deg cone), and ±3 G in the Y and Z axes.

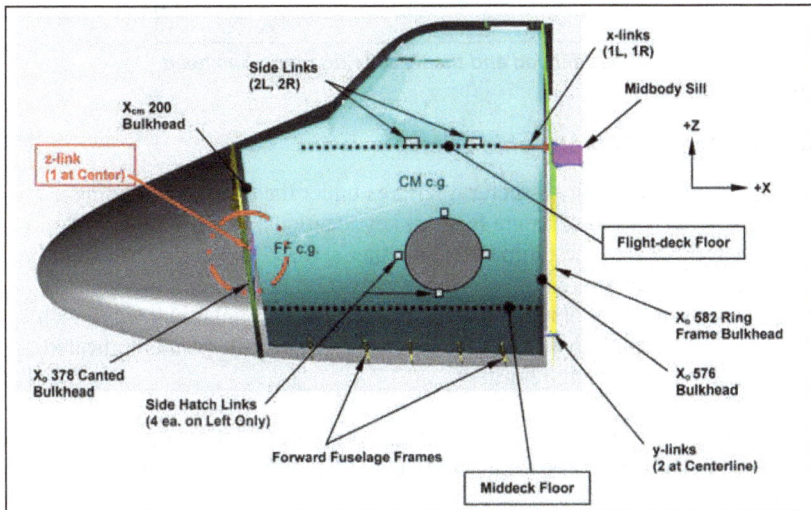

Figure 2.2-23. *Thirteen links to the crew module inside the forward fuselage.*

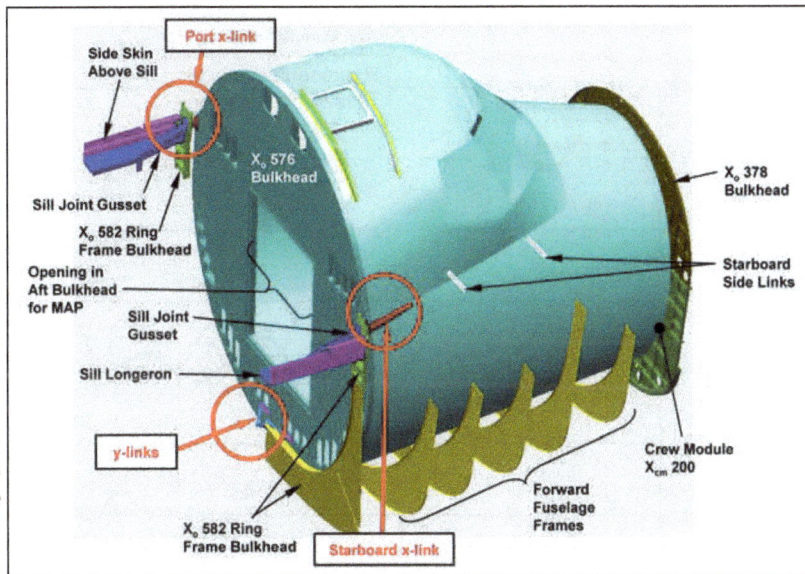

Figure 2.2-24. *Crew module support links, starboard-side view.*

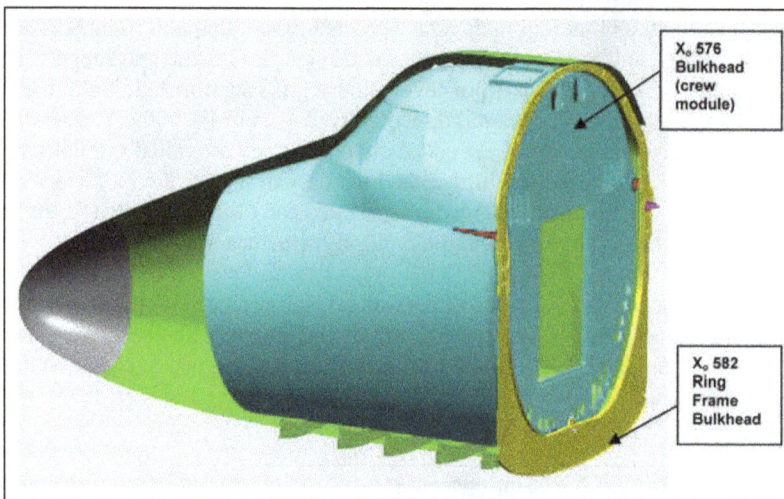

Figure 2.2-25. *Picture of the X_o 576 bulkhead and the X_o 582 ring frame bulkhead relative to each other.*

2.2.4.2 *Forebody structure prior to the Catastrophic Event*

It is unlikely that any part of the FF experienced major structural failures under the estimated loading before the CE. A motion analysis was performed to compute loads at key locations after the LOC of the orbiter (see Section 2.1). Figure 2.2-26 shows that the estimated G loading at the CM increased to slightly more than 3 G just prior to the CE. Acceleration along the vehicle's Z axis (Gz) is the major component. This analysis assumed intact PLBDs. Loss of the doors may have increased loads in unknown ways such that the motion may have been affected. Assuming that the PLBDs were still intact, the results indicated that inertial loads were within the orbiter vehicle design limits.

Stress analysis also shows that it is unlikely that any of the major CM support links failed before the CE. Even if the temperature at the titanium links were elevated to about 600°F (316°C), the links still would have a high margin of safety before failure.

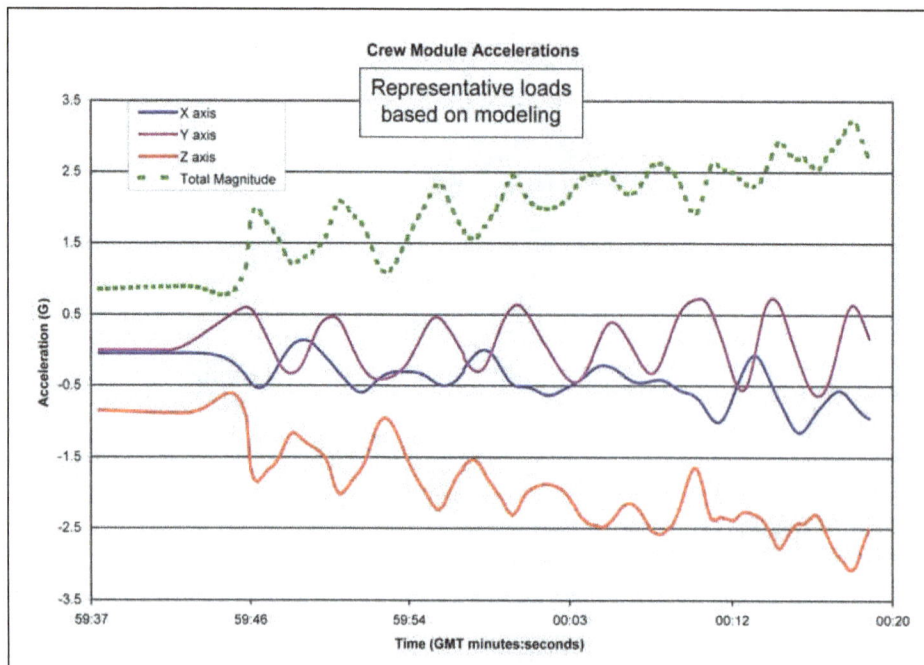

Figure 2.2-26. *Resultant acceleration (G) vs. time prior to the Catastrophic Event in the orbiter coordinate frame.*

There was no evidence from the ballistic or debris cluster analysis that the forebody suffered any major breach before it separated from the midbody. Figure 2.2-21 (in the debris cluster assessment section) shows that forebody debris does not begin to appear in the debris field until other midbody debris is present in large quantities.

Most of the FF debris components that were recovered west of the main forebody debris field were TPS fragments and external items such as star tracker doors that came from the forward section of the FF, probably due to warping of the skin and aerodynamic loads as the forebody rotated (figure 2.2-27). Approximately 87% of the remaining FF structural debris has the same footprint as the CM debris footprint. This implies that the significant FF structural breakup occurred nearly coincidental with the CM shell breakup.

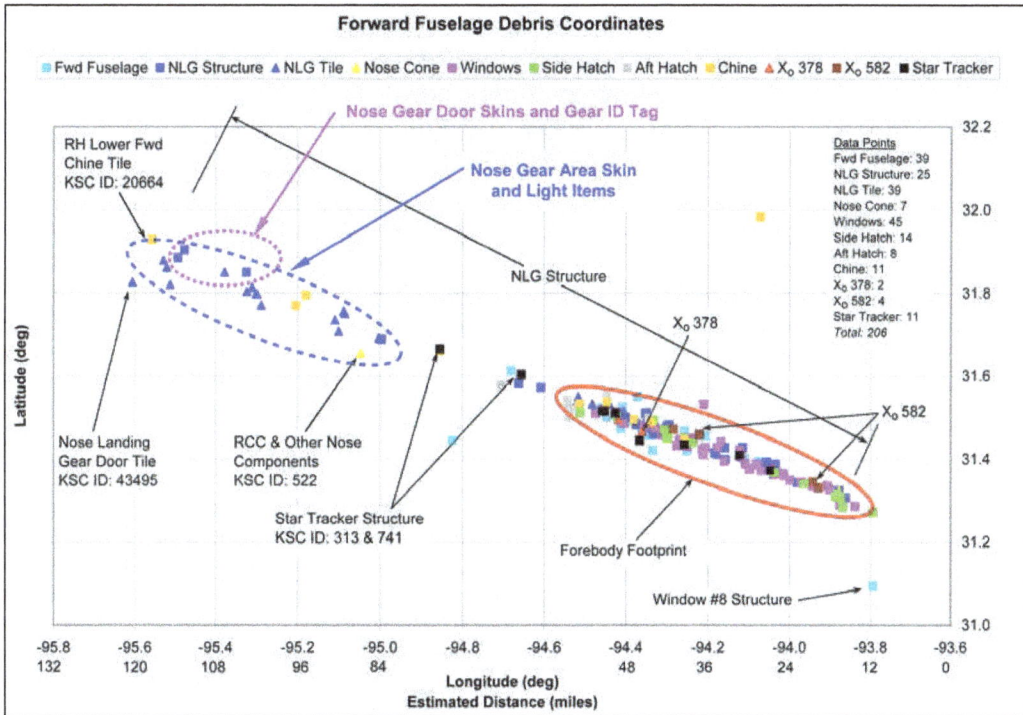

Figure 2.2-27. Forward fuselage debris field.

Two exceptions are the nose landing gear skin items and one landing gear nametag foil that was recovered farther west. These are two very thin pieces of nose landing gear skin and one thin foil piece with very low ballistic numbers, similar to that of the TPS debris.

2.2.4.3 Payload bay doors

During normal entry, the PLBDs and the sidewalls of the midbody are not exposed to high-temperature air flows. After the vehicle lost control, the PLBDs were exposed to abnormal midbody flexing. Also, as the orbiter changed attitude, parts of the vehicle that were not designed for high heating were periodically exposed to the velocity vector and the full force of entry heating. From a structural standpoint, the PLBDs are the most likely place for the first orbiter structural failure that is unassociated with the left wing degradation. RGPC data indicate that the Freon loops in the doors were intact at GMT 14:00:05. Since the Freon radiator loops are an integral part of the doors, this time bounds the earliest time at which the doors could have failed.

Figures 2.2-28 and 2.2-29 show the debris field coordinates for the recovered PLBD debris and a comparison of the PLBD vs. payload bay structures. Ballistics analysis showed very late release times (well past the CMCE) that were considered suspect and assumed to contain errors due to lift generation and/or cascading events. Evaluation of the PLBD debris vs. payload bay structures shows that the debris field was extremely long, suggesting a very complex breakup and cascading event.

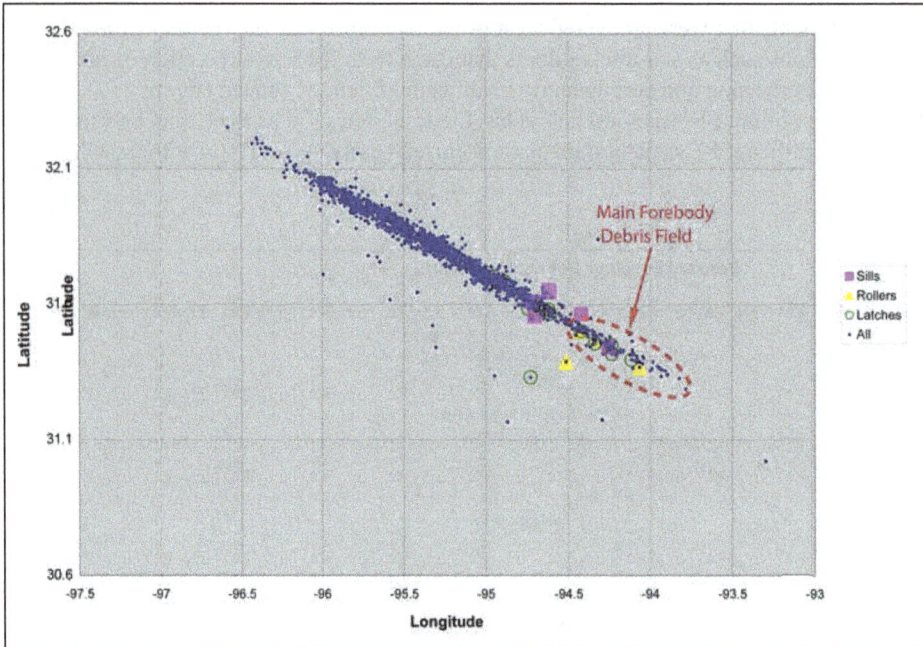

Figure 2.2-28. *Payload bay door debris footprint relative to the main crew module debris field.*

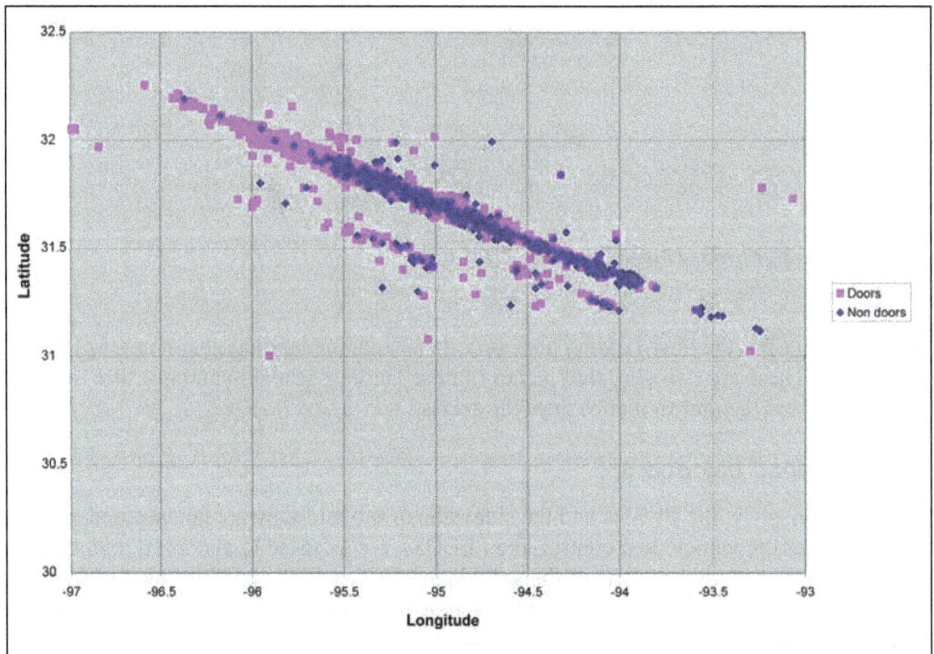

Figure 2.2-29. *Payload bay door debris footprint relative to the payload bay structure.*

The PLBDs could have begun failing as early as immediately after the last available RGPC data, at GMT 14:00:06, or about 12 seconds before the CE. If this were the case, the debris cluster evidence could be interpreted to indicate that the sills and structural elements of the doors remained intact as a part of a larger section of structure after the departure of the doors.

2.2.4.4 *Forebody separation event*

Comparison to *Challenger*

Very little *Challenger* data regarding the CM structure were found. Most data were gathered by locating individuals who had performed some part of the assessments. In many cases, anecdotal reports were all that were available. It was anecdotally reported to the team that in the *Challenger* mishap, the CM separated from the orbiter between the X_o 576 and X_o 582 ring frame bulkheads. Figure 2.2-25 shows a picture of the two bulkheads relative to each other. This anecdotal information was subsequently found to be incorrect.

During the CAIB investigation, all *Columbia* debris were evaluated, but only items deemed significant to the cause of the accident were placed on the reconstruction grid and easily accessible. Significant efforts were carried out to identify CM structural debris, but not much was done on the midbody structure. As a result, the initial CSWG debris review resulted in the belief that no material had been recovered from the X_o 582 ring frame bulkhead, although significant elements had been found of the X_o 576 bulkhead. This belief was incorrect. X_o 582 ring frame bulkhead items were recovered; they were simply not readily available for inspection. This oversight led to the assumption that the failure mode was the same for *Columbia*, and this assumption was reported to the CAIB.[9]

However, the SCSIIT review of *Challenger* debris photographs clearly showed that part of the X_o 582 ring frame bulkhead was recovered with the *Challenger* X_o 576 bulkhead debris (figure 2.2-30).

Figure 2.2-30. *Recovered X_o 576 bulkhead of the* Challenger *crew module showing a portion of the X_o 582 ring frame bulkhead (circled in red).*

This discovery led to a detailed search through *Columbia* debris for any recovered portions of the X_o 582 ring frame bulkhead. It should be noted that the X_o 582 ring frame bulkhead contains less mass than the X_o 576 bulkhead (a ring shape rather than a flat plate (figure 2.2-25)) and so significantly less debris would have been expected. As a result of this extensive search, conducted with the assistance of the *Columbia* Research and Preservation Team, four items of X_o 582 ring frame bulkhead debris were found. Plotting the ground coordinates on the debris field (figure 2.2-28) showed that these recovered elements of the X_o 582 ring frame bulkhead impacted the ground well within the recovered forebody debris cluster. This strongly

---

[9]*Columbia* Accident Investigation Board Report, Volume I, August 2003, p. 77.

suggests that most of the X_o 582 ring frame bulkhead stayed with the forebody and only broke up when the forebody broke up. This is consistent with what appears in the *Challenger* photograph.

Load path assessment

With or without the PLBDs, stress concentrations and the highest structural loading would have occurred in the areas adjacent to the X_o 582 ring frame bulkhead (attachment of the forebody to the midbody) and X_o 1307 bulkhead (attachment of the aftbody to the midbody). At the X_o 1307 bulkhead, the wings and wing carry-through structures provide the area with some additional strength. Structural assessment shows that the weaker link is at the X_o 582 ring frame bulkhead area because this area must react to a more concentrated loading from the CM through the two x-links attaching to the midbody sill longerons.

By design, the forward splice of the X_o 582 ring frame bulkhead is much stronger than the aft splice because this bulkhead was built integrally with the FF shell and attached to the FF by multiple longitudinal frames or longerons, while the aft of the bulkhead is spliced to the midbody by only two sill longeron joints and the lower/side skin splice.

Without the PLBDs, the weak zone between the forebody and the midbody is the sill joint gusset on the aft side of the X_o 582 ring frame bulkhead (figure 2.2-31). This gusset is an offset structural connection that transfers the major X-axis loads from the CM to the midbody sill longeron via the x-links.

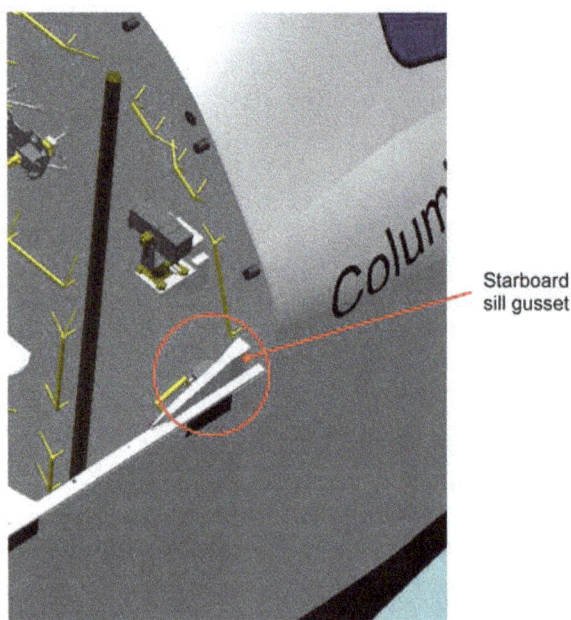

Figure 2.2-31. *Close-up of the starboard sill gusset area.*

This assessment agrees with the debris field assessment finding that the X_o 582 ring frame bulkhead stayed with the forebody at the CE.

Midbody and X_o 582 ring frame bulkhead structure comparison

The midbody and forebody skin debris that originated from near the X_o 582 ring frame bulkhead was evaluated for confirming evidence. Specifically, the two forward hoist fittings that were used to lift the orbiter during ground processing and were aligned with the X_o 582 ring frame bulkhead, and the lower midbody skin from Bay 1, the forward-most section of the payload bay that is adjacent to the X_o 582 ring frame bulkhead, were evaluated.

The recovered starboard forward hoist fitting (figure 2.2-32) is still attached to FF skin. This is evidence that the midbody skin separated from the FF structure aft of the X_o 582 ring frame bulkhead. It was recovered from the westernmost portion of the forebody debris field. The green-colored Koropon primer shown in the photograph is similar to the condition of the side skin a few feet above it (figure 2.2-33), which was recovered near the center of the forebody debris field. These elements most likely stayed with the forebody after the CE and only departed at the forebody breakup event. Temperatures above 400°F (204°C) degrade the primer's appearance, and high temperatures (>900°F) (486°C) will completely ablate it. The presence of Koropon indicates that the breakup of these elements was caused by mechanical overload rather than thermal effects. The presence of intact primer indicates that the FF TPS protected this area from entry heating until the forebody breakup at the CMCE.

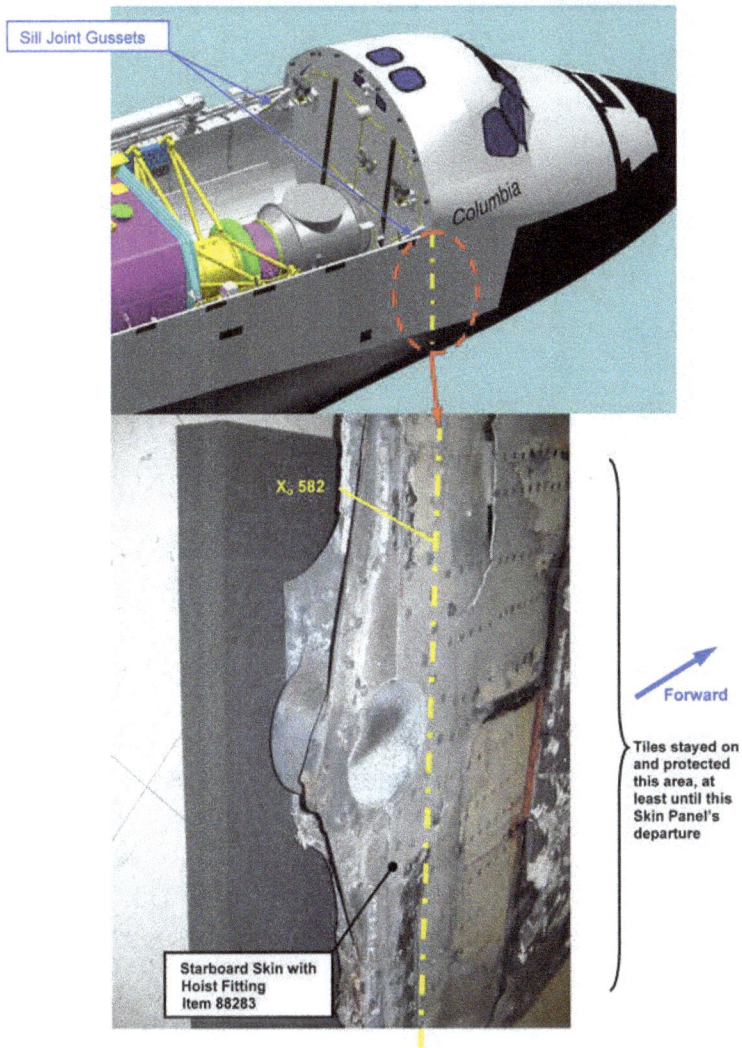

Sill Joint Gussets

X₀ 582

Forward

Tiles stayed on and protected this area, at least until this Skin Panel's departure

Starboard Skin with Hoist Fitting Item 88283

Figure 2.2-32. *Forward fuselage skin still attached to the starboard hoist fitting.*

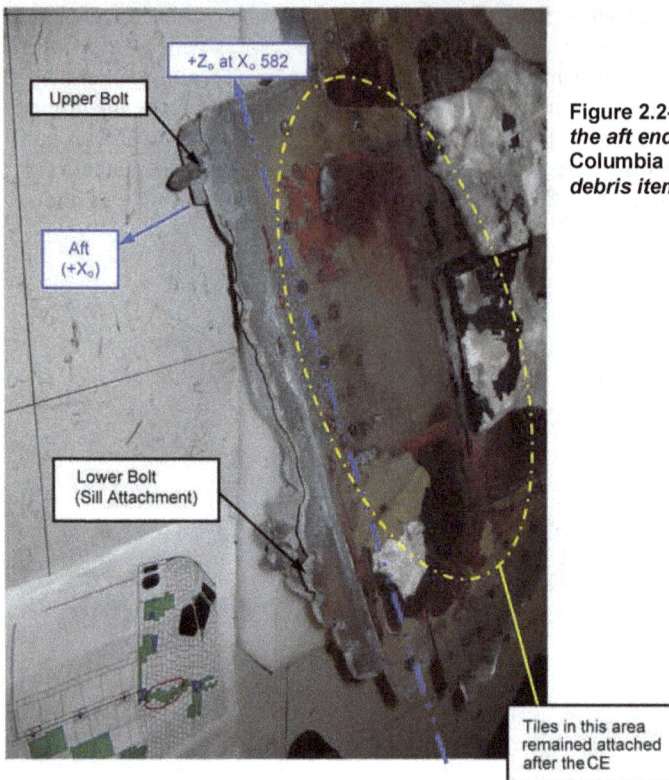

Figure 2.2-33. *Thermal erosion on the aft end of the starboard-side skin,* Columbia *Reconstruction Database debris item no. 2436.*

The condition of the recovered forward portside hoist fitting (figure 2.2-34) indicates that the Xo 582 ring frame bulkhead likely broke away from the midbody skin via mechanical failure, prior to any thermal effects. The Bay 1 side skin pulled away because of mechanical failure. The midbody skin splice severed the fasteners from the titanium hoist fitting by mechanical force, some Hilok bolts were cut off, and some flush bolt heads pulled through the skin splice. This damage, which may have occurred at the CE, suggests that the forebody yawed left and pitched down relative to the midbody. This motion may also have broken the port x-link lug (discussed in the next section). This kind of mechanical failure could only happen when the hoist fitting was still attached to the FF, when sufficient mass existed to exert this level of force.

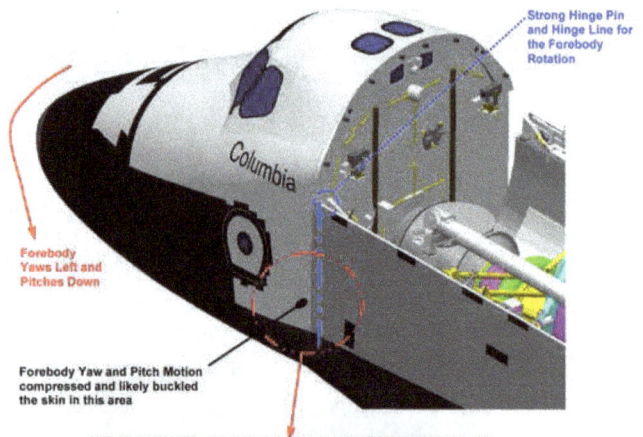

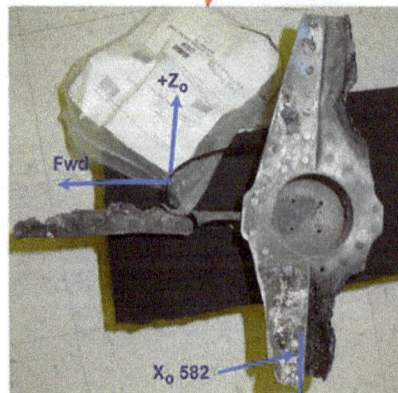

Figure 2.2-34. *Forward port hoist fitting,* Columbia *Reconstruction Database debris item no. 32038.*

Unlike the starboard hoist fitting, the skin panel forward of the port hoist fitting was not recovered. One possible explanation is that the FF skin on the forward side of this item was deformed or wrinkled by compression loading at the CE when the forebody yawed left and pitched down. This may be supported by the fact that the port forward hoist fitting was recovered in the middle of the forebody debris field while the starboard hoist fitting was found in the western end of the debris field. Wrinkled skin most likely debonded TPS tiles, so this area would be subjected to thermal flow damage on at least the skin-side surface starting at the CE, then on both sides after it departed from the forebody after the CMCE. Alternatively, the port area may have simply stayed intact longer and received more entry heating following the CMCE.

The recovered midbody bottom skin pieces from Bay 1 (the most forward bay in the midbody, immediately aft of the X_o 582 ring frame bulkhead (figures 2.2-35, 2.2-36, and 2.2-37)) show little or no heat damage to the skin at the splice aft of the X_o 582 ring frame bulkhead. The damage is mainly mechanical breakup of the frame to skin splice.

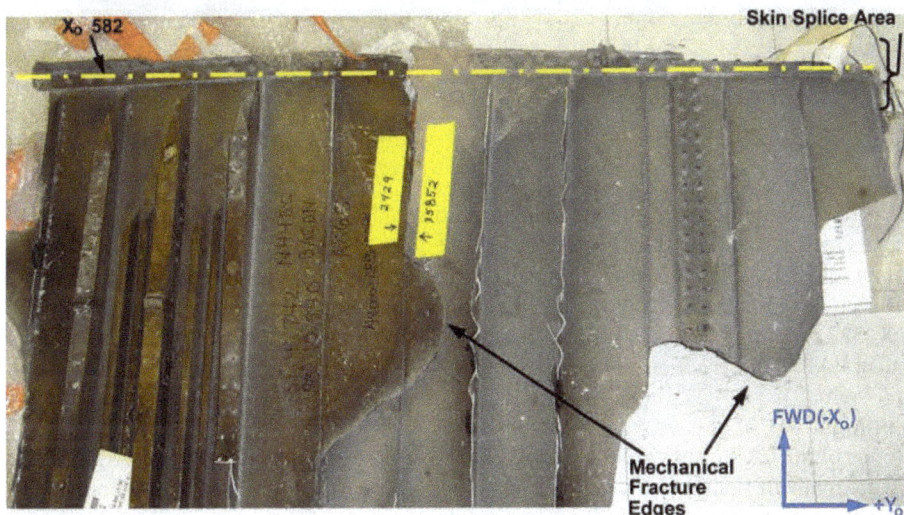

Figure 2.2-35. *Bay 1 midbody bottom skin,* Columbia *Reconstruction Database debris item nos. 2429 and 33852.*

Figure 2.2-36. *Bay 1 midbody skin,* Columbia *Reconstruction Database debris item no. 88290.*

Figure 2.2-37. *Bay 1 midbody skin,* Columbia *Reconstruction Database debris item no. 14908.*

The *Columbia* Reconstruction Team noted that thermal damage to the midbody was not generally severe.[10] The thin skin and stringers of Bay 1 experienced some heating but not sufficient to melt the material. One scenario that may explain this is that this bay corresponded to the X location of the structural failure and disintegrated quickly during or immediately after the CE. Even if the bay skin remained with the midbody at the CE, without the X_o 582 ring frame bulkhead frame aerodynamic loads would peel off the Bay 1 skin quickly. This is consistent with the interpretation that the forebody and midbody separated in this area.

Crew module attach fittings

The primary attach fittings of the CM to the forebody are the x-links, the y-links, and the z-link (figures 2.2-23 and 2.2-24). All are made of titanium. The x-links also connect the FF and CM to the midbody and will be addressed here. The y-links and z-link will be discussed in Section 2.4.

The x-links are normally protected by the PLBDs, which also provide stiffness for the orbiter. By design, the x-link attachment to the CM is stronger than its attachment to the FF and midbody because it is attached to the extension webs of the CM flight deck floor and X_o 576 bulkhead.

Comparison of the recovered port and starboard x-links (figure 2.2-38) shows two important differences. First, the starboard x-link was recovered with a portion of the X_o 582 ring frame bulkhead and starboard sill attached, while the port x-link failed forward of the X_o 582 ring frame bulkhead. Second, the starboard x-link experienced much more thermal erosion than the port x-link. It is clear that the directionality of the thermal damage (aft to forward flows above the x-links, and forward to aft flows along the bolt heads) is the same on both links. This orientation was shown to be unlikely due to a common trim attitude when free-flying (see Section 2.1). This strongly suggests that they were in the same relative orientation (still attached to forebody structure) when exposed to heating. The common thermal pattern on the x-link body is discussed in Section 2.4.

---

[10]STS-107 *Columbia* Reconstruction Report, NSTS-60501, June 2003, p. 55.

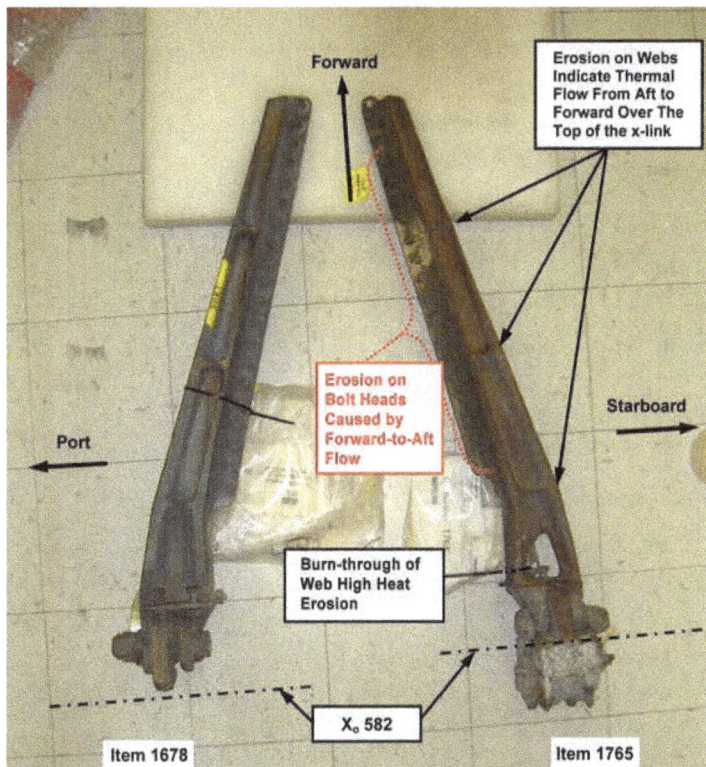

Figure 2.2-38. *Port and starboard x-links*.

Starboard x-link assessment

Figures 2.2-39 and 2.2-40 show the heavy thermal erosion on the sill joint gusset that is attached to the starboard x-link. The portion of side skin splice from just outboard of the starboard x-link was also recovered. Heavy heat erosion was noted on the midbody side (aft side of X_o 582 ring frame bulkhead), but was not significant elsewhere on the debris. This skin panel was recovered near the center of the forebody debris field. Together, this confirms that the CM stayed inside the FF shell until the forebody breakup, and suggests that the trailing edge (midbody side) was exposed to heating.

The structure aft of the X_o 582 ring frame bulkhead on both the starboard covering side skin and the starboard x-link was heavily damaged by heat. The heat erosion pattern on bolt heads on the aft side shows that the higher (+Z) and outboard (+Y) bolts experienced more heat erosion than the lower/inboard bolts. The resulting shape of eroded bolt heads indicates

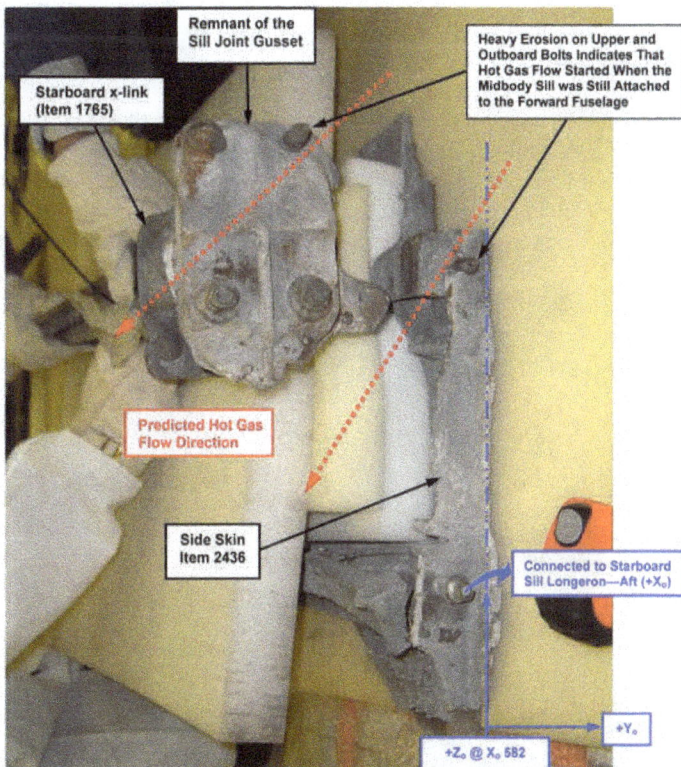

Figure 2.2-39. *View looking forward at the starboard x-link and side skin*.

that hot gas flowed down and inboard (figures 2.2-39 and 2.2-40). This suggests that the aft side of the starboard x-link area was exposed to hot gas while the starboard midbody sidewall/sill was still attached to the X_o 582 ring frame bulkhead and FF.

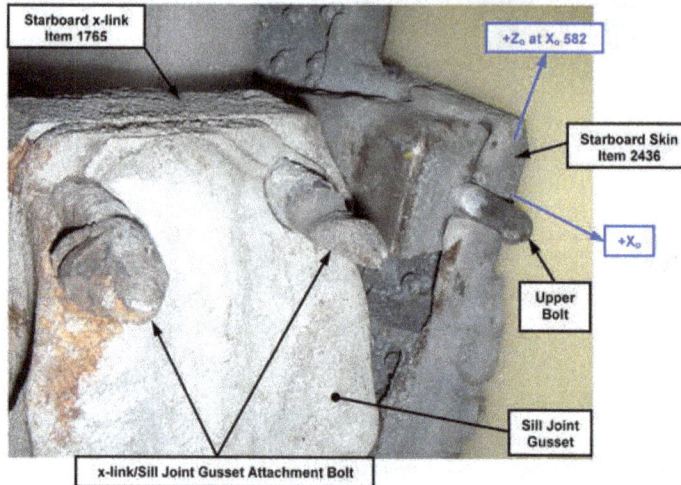

Figure 2.2-40. *Detail of thermal erosion on the upper bolts at the starboard x-link and skin.*

Hot gases flowing over the starboard sill and X_o 582 ring frame bulkhead could elevate the temperature of the starboard structural joint between the sill and the FF behind the starboard x-link while the sidewall shielded the lower portions of these structures. It appears that the sill and the sidewall were still attached to the FF when this thermal erosion took place. The condition of the remaining bolt heads indicates that these bolt heads and shanks did not fail by mechanical loading.

To summarize the assessment, the point of the orbiter that was weakest under dynamic rotational loading was likely to be the sill gusset immediately aft of the x-link and the X_o 582 ring frame bulkhead. Debris shows that for the starboard sill, the failure of the midbody sill joint occurred at a location aft of the X_o 582 ring frame bulkhead, perhaps near the middle of the gusset (figures 2.2-41 and 2.2-42). Since orbiter load limits were apparently not exceeded, this implies that thermal degradation was required.

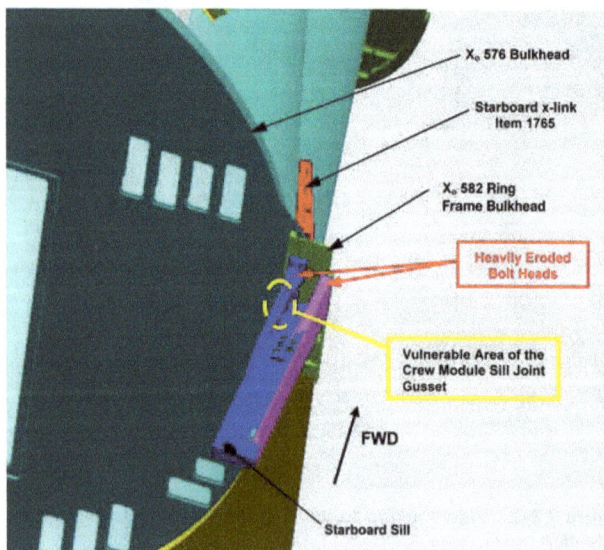

Figure 2.2-41. *Weak links – sill joint gusset, starboard side.*

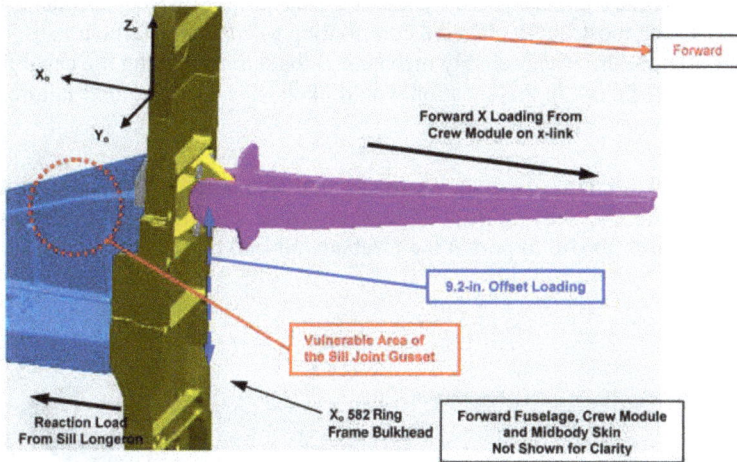

Figure 2.2-42. *Structural loading from starboard x-link to sill longeron*.

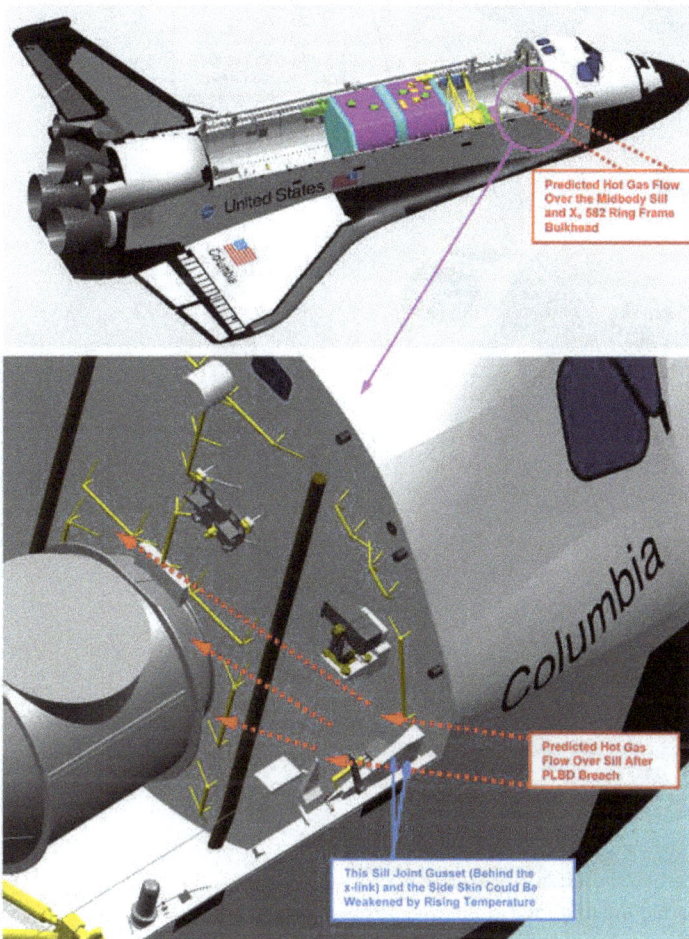

This thermal degradation could have occurred prior to the CE if the PLBDs departed or were compromised in that area (figure 2.2-43). Debris evidence shows that the starboard sill joint experienced more thermal damage than the port sill joint. This suggests the possibility that rising temperature on the starboard sill area weakened the sill joint until it failed mechanically. Although orbiter motion generally resulted in "belly-into-the-wind" orientation (see Section 2.1), with the large pitch and roll oscillations hot gas could flow beyond the belly of the orbiter, up and over the sill. However, hot gas exposure to the sill area could only be short and intermittent (<3 seconds at a time). Debris field and ballistic assessment of PLBD debris is inconclusive. However, a local failure or compromise on the starboard PLBD near the sill could have resulted in this scenario without evidence being apparent in these assessments.

Figure 2.2-43. *Hot gas flow heats up starboard sill and the adjacent x-link*.

It is unlikely that the temperature increase on the starboard-side sill joint was the primary cause in the orbiter breakup. High heating most likely played a contributing role for the starboard sill joint failure. The main cause for the sill joint failure was probably mechanical loading exceeding the capability of the thermally weakened structure. A failure in this location would likely cascade structural failures along load paths throughout the vehicle.

As illustrated in figure 2.2-44, failure of the starboard sill joint area would trigger a separation of the forebody away from the midbody. Starting from the starboard side, the midbody skin splice would fail progressively, opening from starboard to port at the forebody aft bulkheads.

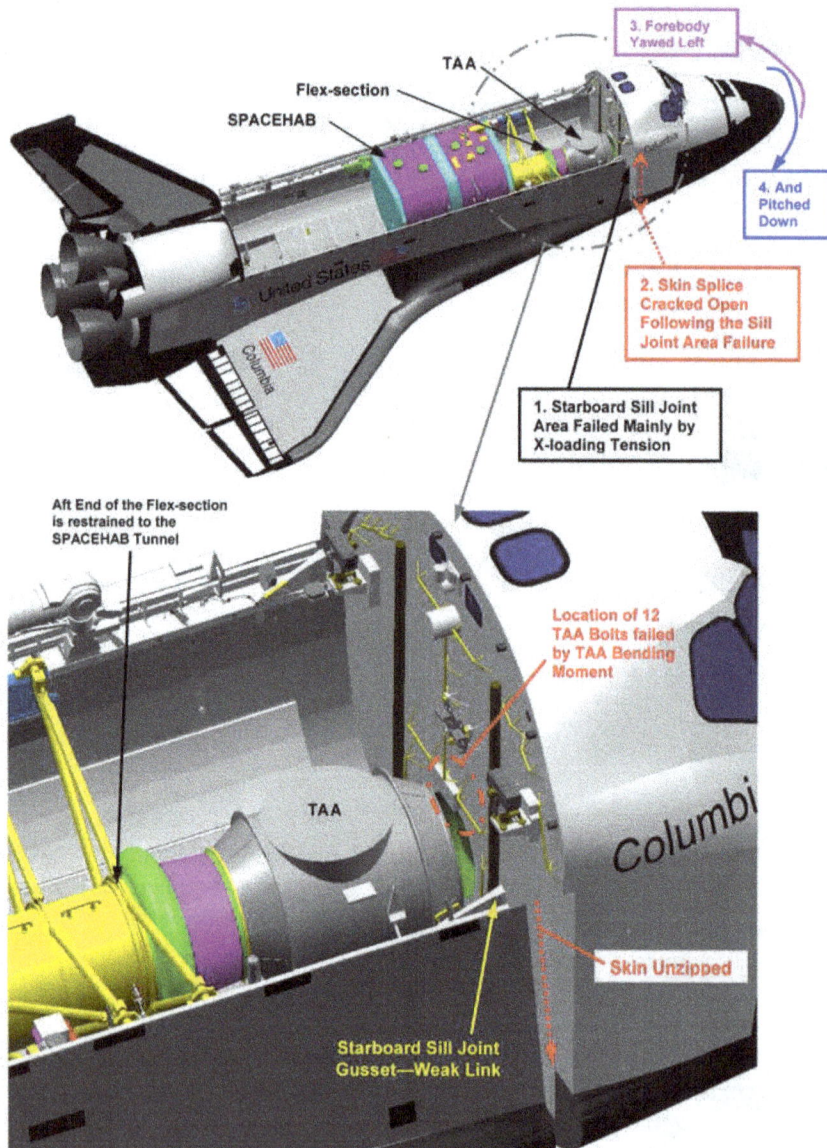

Figure 2.2-44. *Forebody departed the midbody at the Catastrophic Event.*

The other possibility is that a portion of the starboard midbody remained with the forebody after the CE and failed during subsequent motion of the forebody. However, it is unlikely that the forebody separated the midbody farther aft of the starboard sill joint area unless the farther aft area had been damaged earlier because of other thermal or mechanical causes, for which there is no clear evidence.

Port x-link assessment

A portion of the X_o 582 ring frame bulkhead and portside sill fitting, which has a lug that supports the port x-link, was recovered and inspected. The lug broke near its base mainly because of a bending load caused by the x-link and the forebody yawing to the left (figure 2.2-45). This yawing motion was identified by the yielding on the outboard flange of the clevis on the aft end of the port x-link (figure 2.2-46).

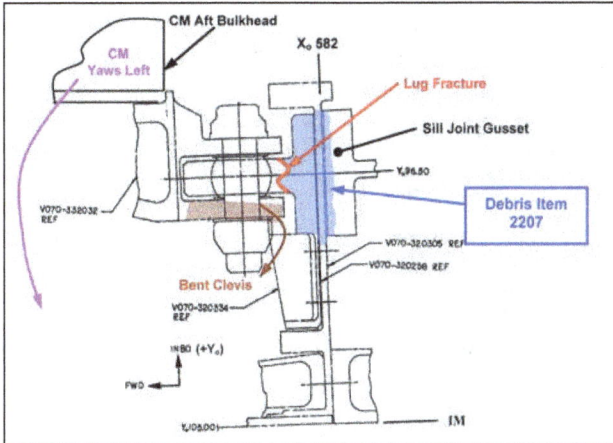

Figure 2.2-45. *How the port x-link failed, cross-section view looking down.*

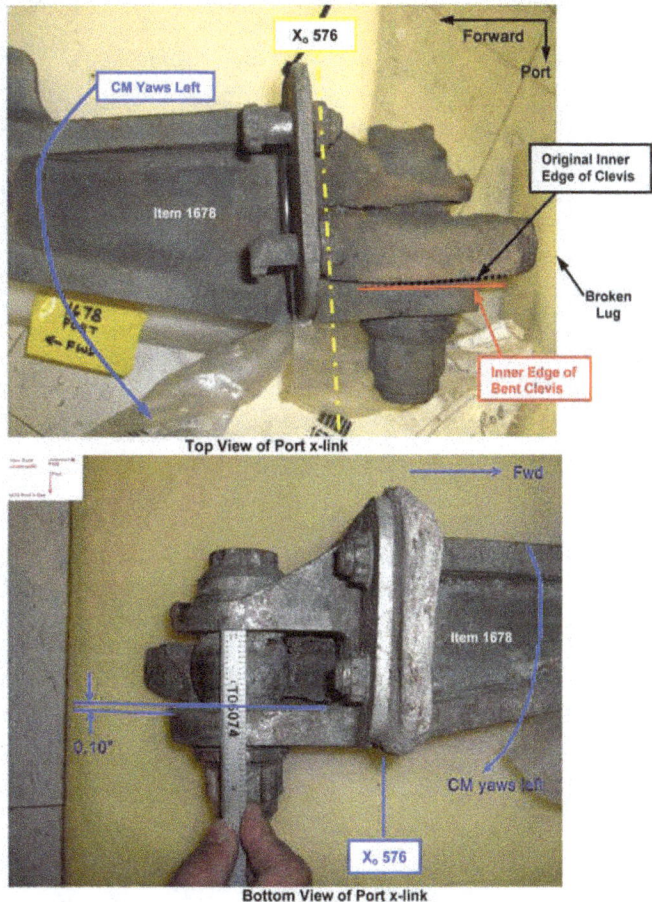

Figure 2.2-46. *Views of the port x-link with about a 0.10-inch yield on the outboard side of the clevis.*

As shown in figure 2.2-47, the outboard upper bolt head was gouged and deformed by an impact from the broken lug, which remained in the port x-link clevis. This indicates that the forebody not only yawed to the left but also shifted to the left when it moved away from the midbody.

Gouged by impact from the port x-link after separation

Figure 2.2-47. Columbia *Reconstruction Database debris item no. 2207, which includes the remainder of the lug fitting that supports the aft end of the port x-link.*

This implies that when this lug failed, the port x-link was still attached to the CM and the lug still had the structural support of the midbody sill or a large portion of the portside midbody/sill attached. On the portside, the FF and x-link separated from the midbody sill on the *forward* side of the X_o 582 ring frame bulkhead. The remaining part of the portside X_o 582 ring frame bulkhead and the sill joint gusset experienced little heat erosion (figures 2.2-48 and 2.2-49). These two items were found at longitude 94.0W and 94.3W, which are in the central portion of the midbody debris field (figure 2.2-16). It is possible that they stayed with the portside midbody sill, which is one of the heaviest parts of the midbody shell. No ballistic assessment was performed to confirm or deny this theory.

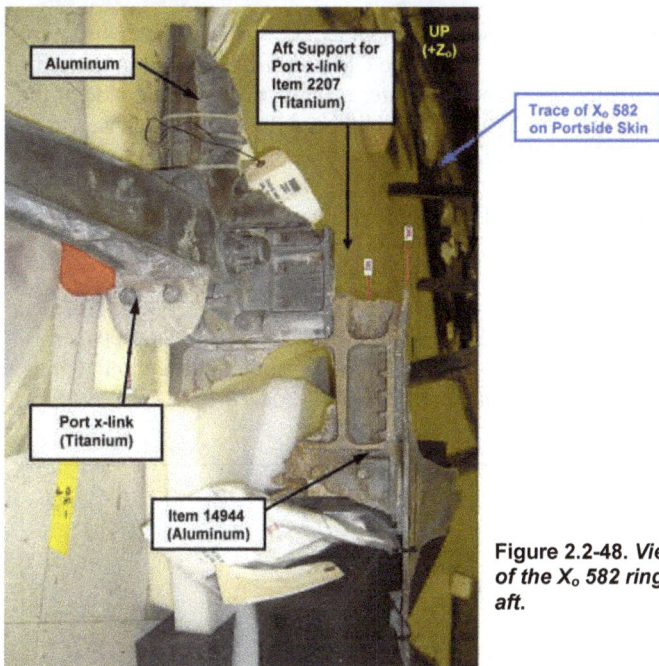

Figure 2.2-48. *View of the port x-link and a portion of the X_o 582 ring frame bulkhead debris items, looking aft.*

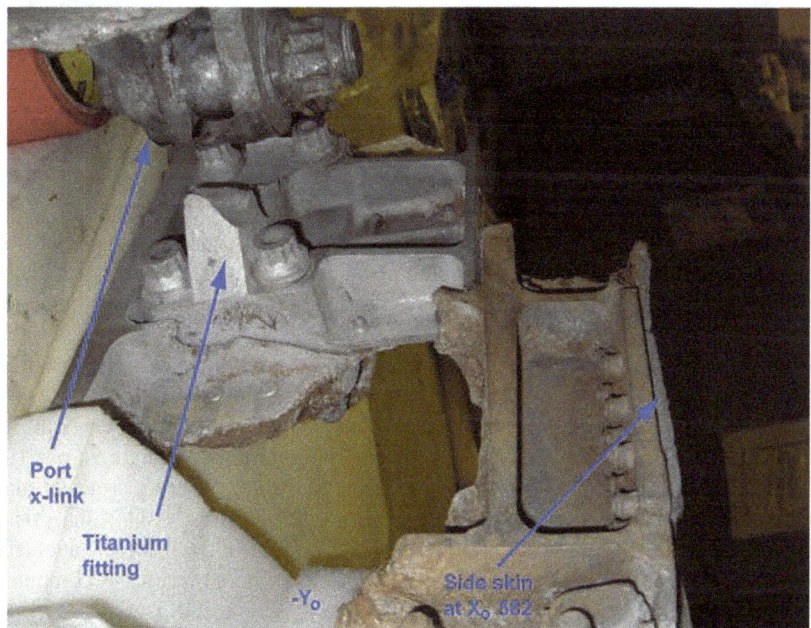

Labels in figure: Aluminum; Aft Support for Port x-link Item 2207 (Titanium); UP (+Z_o); Trace of X_o 582 on Portside Skin; Port x-link (Titanium); Item 14944 (Aluminum)

Figure 2.2-49. *View looking up and aft at the port x-link.*

Labels in figure: Port x-link; Titanium fitting; -Y_o; Side skin at X_o 582

These pieces provided a more subtle understanding of the separation event between the forebody and the midbody.

The conclusion was that the forebody separation started *aft* of the X_o 582 ring frame bulkhead on the starboard-side sill area, and then spread across the fuselage along the aft portion of the X_o 582 lower ring frame bulkhead. The breakup progressed toward the portside sill while the forebody yawed to the left and pitched down, causing the portside lug fitting to fail by bending to the portside. This resulted in failure *at* the X_o 582 ring frame bulkhead on the portside of the bulkhead, leaving a portion of the bulkhead with the midbody.

Immediately after the CE, the starboard CM x-link pulled forward through the X_o 582 ring frame bulkhead as soon as the CM began to move forward inside the FF shell. The CM port x-link was unsupported by midbody structure after the CE and was only attached to the CM. Interestingly, the starboard side of the *Challenger* X_o 582 ring frame bulkhead also remained with the X_o 576 bulkhead. However, because the breakup events had different causes, this is assumed to be a coincidence.

> **Conclusion L3-2.** The breakup of both *Challenger* and *Columbia* resulted in most of the X_o 582 ring frame bulkhead remaining with the crew module or forebody.

Tunnel adapter

The tunnel adaptor assembly (TAA) provided the internal pressurized path for the crew to move between the middeck to the SPACEHAB in the payload bay (figure 2.2-50). The tunnel adaptor attached to the middeck access panel (MAP), which is a large removable panel in the X_o 576 bulkhead.

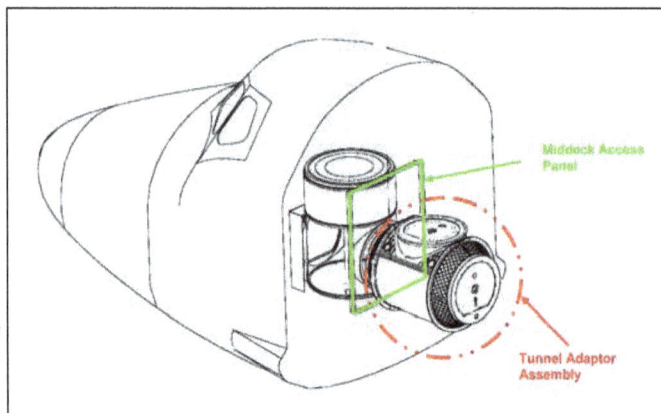

Figure 2.2-50. *Tunnel adaptor assembly and middeck access panel.*

One note of importance is that a breach in the TAA would cause the immediate rapid depressurization of the internal airlock, since there was no hatch between it and the TAA volume. The airlock inner hatch to the middeck was closed, so this would not result in the depressurization of the main CM, only the airlock.

Some items that were stowed in the airlock were recovered in the western portion of the debris field, prior to the main body of forebody debris. These items were either soft goods that were stowed in bags taped or strapped to the wall or floor, or bonded items (no airlock structure), which indicates that although the airlock rapidly depressurized, the structure remained intact. A headrest pad (used for ascent only) was stored in the internal airlock for entry according to the crew's entry stowage plan. Ballistic analysis was performed on this one item because of its easily described shape. This analysis determined that this item separated from the airlock/TAA compartment at GMT 14:00:36. Ballistic analysis was also performed on two small pieces of TAA structure. The release time for both pieces was computed to be GMT 14:00:43.

The debris field analysis shows that most of the TAA was recovered west of the main forebody debris field, suggesting that the TAA did not remain with the forebody but departed before the CMCE.

As shown in figures 2.2-51 through 2.2-54, the 12 high-strength bolts (180,000 pounds per square inch (psi)) that attached the forward ring of the TAA to the MAP failed with shanks still remaining in the inserts. This suggests that this area was subjected to high tension and bending while the remaining bolts were not subjected to the same loading. It implies that this TAA joint reacted to a bending moment (M_{TAA}) plus aft tension loading (figure 2.2-54). When the aft end of the TAA was restrained by the SPACEHAB tunnel to the midbody, the forebody yawing left and pitching down relative to the midbody could create the same bending effects. This suggests that the TAA broke away from the MAP when the forebody separated from the midbody at the CE or shortly afterwards.

Figure 2.2-51. *Tunnel adapter assembly and the manufacturing access panel debris, view looking forward at "B" hatch opening.*

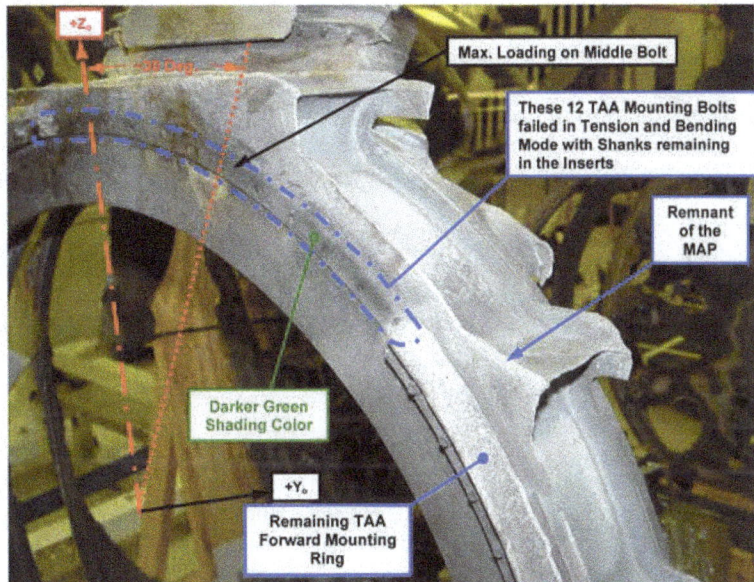

Figure 2.2-52. *Tunnel adapter assembly debris showing that the bolt shanks failed mainly by tension and bending.*

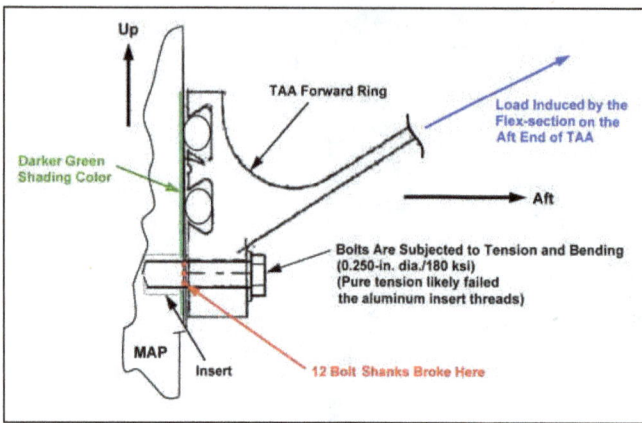

Figure 2.2-53. *Cross section through tunnel adapter assembly mounting bolt.*

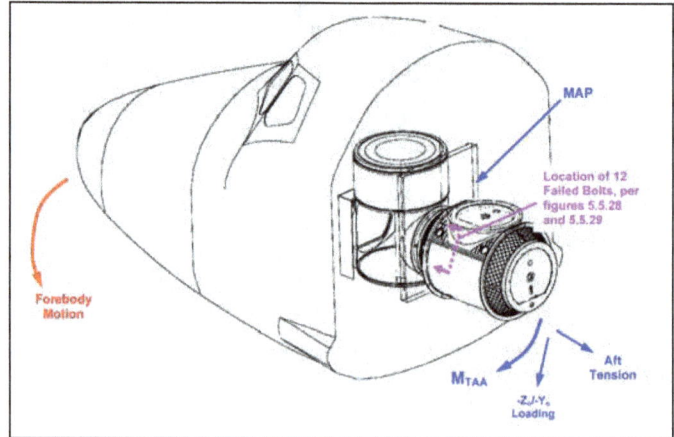

Figure 2.2-54. *Predicted motion based on debris evidence.*

Darker shading color on the surface of the MAP, as was noted on figures 2.2-51 and 2.2-52, indicates that this area was shielded by the missing portion of the TAA ring and suggests that the MAP surface outside of the TAA ring was subjected to thermal flow before the TAA departure. If the TAA departed at the CE, the MAP or the aft bulkhead of the CM was exposed to the thermal flow for some period of time prior to the CE (figure 2.2-44). This would agree with the theory that the PLBDs were compromised prior to the CE.

The failure of these TAA mounting bolts also indicates that it is unlikely that the X_o 576 CM aft bulkhead suffered any major structural failure at the CE, although it was subjected to CM internal pressure and apparently some thermal exposure. By design, the X_o 576 bulkhead can handle 24 psi of internal pressure. The TAA aft loading required to fail the TAA mounting bolts would impose less load on the X_o 576 bulkhead than the maximum cabin internal pressure load case. The large mass of the bulkhead would also effectively absorb and diffuse heat.

As shown in figure 2.2-50, the MAP is a significant element of the X_o 576 bulkhead. The bulkhead is linked to the flight deck floor, the middeck floor, and the two partitions on both sides of the airlock. It is not likely that the MAP would fail without causing a massive failure of the whole aft section including the X_o 576 bulkhead and the airlock, affecting the floors and partitions (figure 2.2-55). This supports the conclusion that no significant damage occurred to the bulkhead at the CE.

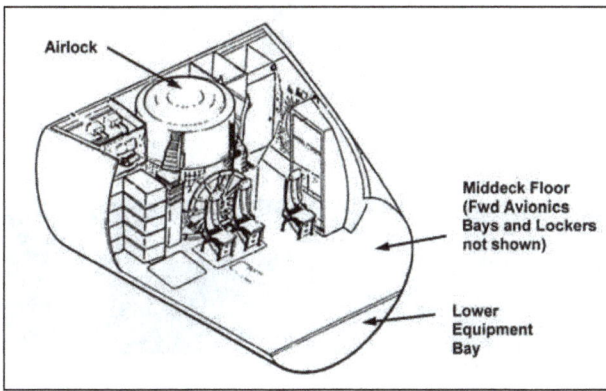

Figure 2.2-55. *Airlock in the middeck.*

Extreme thermal erosion at the lower edge opening on the MAP where the forward end of the TAA is attached indicates that after the TAA broke away from the MAP, hot gas flowed in over the lower edge of the "B" opening in the MAP. This thermal flow also caused heat erosion to the two aft corners of the stowage platform, which was mounted to the airlock floor (figure 2.2-56). This is consistent with the rotational motion of the forebody post-CE, when such an exposure could have occurred.

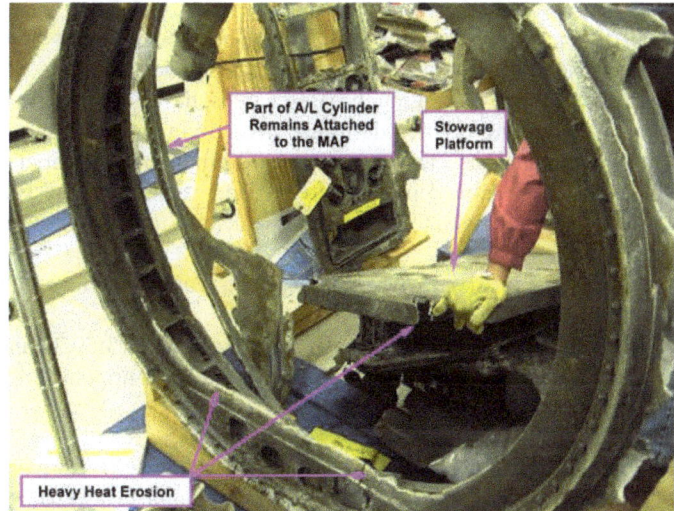

Figure 2.2-56. *View looking forward at the "B" hatch opening on the X_o 576 bulkhead.*

2.2.5 Synopsis of orbiter breakup sequence

As reported by the CAIB, the left wing was gradually failing until orbiter breakup. At GMT 13:59:37, the orbiter lost control. Around GMT 13:59:49, the left OMS pod probably departed. It is possible that the PLBDs departed next or at least degraded enough to allow hot gas to flow over the starboard sill as the orbiter rotated. The orbiter then broke up at GMT 14:00:18 (CE) into aftbody, forebody, and midbody/right wing components.

It is likely that the initial failure was at the weakest area of the overall structure, just aft of the X_o 582 ring frame bulkhead. This failure progressed from starboard to port by unzipping the skin splice at the X_o 582 ring frame bulkhead. The forebody yawed left and pitched down, hinging at the portside x-link. This motion broke the lug fitting that supports the aft end of the port x-link, leaving the base of the lug and the local X_o 582 ring frame bulkhead portside with the midbody. The aft and right wing attachments successively failed as abnormal load paths were propagated through the orbiter from the forebody separation. The next failure was the TAA, depressurizing the airlock. The forebody remained intact at orbiter breakup (see Section 2.4).

> **Recommendation L3-1.** Future vehicles should incorporate a design analysis for breakup to help guide design toward the most graceful degradation of the integrated vehicle systems and structure to maximize crew survival.

2.3 Crew Cabin Pressure Environment Analysis

An analysis of the crew cabin pressure environment in *Columbia* was critical to formulating an understanding of what happened to the crew. This analysis was particularly important to acquiring insight into how the crew cabin environment affected the crew's ability to make decisions, at what point during the depressurization the crew's ability would have been permanently compromised, and when the crew members would have lost consciousness. To formulate this understanding, various aspects of the accident were analyzed. These included telemetry, ground-based videos, the debris, medical evidence, and structural analysis. This section draws on information described in detail in other areas of this report (structural analysis, medical findings, video analysis, etc.) to determine the timeline of the cabin depressurization. The timeline includes times for when the cabin depressurization began, when the depressurization was complete, and the rate of depressurization. Related to this analysis is determining the location(s) of the CM breach(es).

The conclusions relative to the cabin depressurization timeline are provided below:

> *Conclusion L1-1.* After loss of control at GMT 13:59:37 and prior to orbiter breakup at GMT 14:00:18, the *Columbia* cabin pressure was nominal and the crew was capable of conscious actions.

> *Conclusion L1-2.* The depressurization was due to relatively small cabin breaches above and below the middeck floor and was not a result of a major loss of cabin structural integrity.

> *Conclusion L1-3.* The crew was exposed to a pressure altitude above 63,500 feet, indicating that the cabin depressurization event occurred above this altitude.

> *Conclusion L1-5.* The depressurization incapacitated the crew members so rapidly that they were not able to lower their helmet visors.

2.3.1 Depressurization timeline boundaries

The depressurization timeline boundaries were identified with reconstructed telemetry and ground-based video. RGPC-2 data indicate that *Columbia*'s cabin pressure was normal (~14.7 psi) until GMT 14:00:04.826. Therefore, cabin depressurization started no earlier than (NET) GMT 14:00:05. No visual events in ground-based videos were identified positively as evidence of cabin depressurization. However, the videos show that at GMT 14:01:10, the CM image vanishes while it was still clearly in the camera's FOV. The image loss was due to the CM being broken into subcomponents that were too small and dispersed to be visible on the video. This event was defined as TD. After this time, the CM no longer had any structural integrity. Thus, cabin depressurization was complete no later than (NLT) GMT 14:01:10. This analysis establishes the absolute NET and NLT times for the start of the cabin depressurization and the completion of the depressurization. The effort to narrow these boundaries to the maximum extent possible is discussed in subsequent sections.

2.3.2 Start of depressurization

Medical evidence and debris field and ballistic analyses were used to determine the NET and NLT times for the beginning of the cabin depressurization.

Medical evidence suggests that the cabin pressure condition at CE was within the bounds of human survival (see Section 3.4). Therefore, the cabin depressurization started NET the CE at GMT 14:00:18.

> **Conclusion L1-1.** After loss of control at GMT 13:59:37 and prior to orbiter breakup at GMT 14:00:18, the *Columbia* cabin pressure was nominal and the crew was capable of conscious actions.

Recovered objects originating from inside the CM are a positive indication of a breach in the CM. The 20 westernmost debris items originating from within the CM were evaluated.[1] All items were small (< 8 in.) in size and none were structural elements of the CM. The debris item located farthest west in the debris field was a piece of reflective tape from a crew helmet. Because this item was not a good candidate for ballistic analysis, the next farthest west item, a mission patch, was used for ballistic analysis. All crew members wear a mission patch attached by VELCRO® to the ACESs; virtually all other patches are stored in plastic-wrapped packages of 25 patches, which are stowed in a stowage compartment below the middeck floor called Volume E. Of the 20 debris items, nine were patches that probably originated from Volume E (Table 2.3-1).

Table 2.3-1. *Westernmost Crew Module Debris*

Item description	CM location
Launch/entry helmet reflective tape	Crew helmet
STS-107 mission patch fragment	Volume E (below middeck floor)
STS-107 mission patch fragment	Volume E
Harness or parachute webbing piece	Crew
STS-107 mission patch fragment	Volume E
STS-107 mission patch fragment	Volume E
Payload patch	Volume E
STS-107 mission patch fragment	Volume E
Payload patch	Volume E
Life raft spray shield fragment	Crew parachute pack
Panel illuminator fragments	Flight deck panel
Middeck ceiling luminous panel	Middeck ceiling
STS-107 mission patch fragment	Volume E
Air duct fragment	Middeck port
Payload checklist fragment	Middeck locker MF43K
STS-107 mission patch fragment	Volume E
Life raft reflective tape	Crew parachute pack
In-flight maintenance tool, no. 0 screw driver	Middeck locker MF43C
Sleep station closeout material	Middeck starboard
Sleep station light cover	Middeck starboard

Ballistic analysis produced a separation time of GMT 14:00:35 ± 5 seconds for the westernmost patch. Therefore, the depressurization of the CM started NLT GMT 14:00:35 ± 5 seconds.

---

[1]*Columbia* was traveling west-to-east, so debris that was recovered in the western portion of the debris field generally are items that departed from the vehicle earlier than items that were recovered farther east.

2.3.3 Depressurization rate

Debris items and medical evidence were analyzed in the hope that they would aid in determining the cabin depressurization *rate* and thereby aid in determining when the depressurization was complete.

Some intact packages (drink pouches and personal hygiene bottles) were recovered as was the CM cabin altimeter. Rapid depressurization tests were performed on new packages to determine the depressurization rate required to rupture these types of packages. Determining the depressurization rate sufficient to cause rupture also identifies the maximum rate that will *not* cause rupture, yielding an upper bound for the cabin depressurization rate. However, even at the maximum rate that the test chamber provided (almost 32 psi/sec[2]), the packages did not rupture, so an upper bound for the cabin depressurization rate could not be determined from these tests.

The recovered middeck cabin altimeter was disassembled and compared to a new cabin altimeter to determine whether the recovered altimeter contained any evidence of the cabin pressure environment. No pressure-related differences were noted between the recovered altimeter and the new altimeter. Neither the packages nor the altimeter analyses could provide any conclusions regarding depressurization rates.

Medical forensic evidence was studied to determine the rate of the cabin depressurization. Information on the effects of a rapid depressurization to vacuum is limited to postmortem analysis of isolated accidental occurrences and animal studies. A literature search revealed a case of apparent lung trauma occurring at slow depressurization rates.[3] Additionally, information on the fatal depressurization accident of *Soyuz 11* in 1971 revealed that although the *Soyuz 11* cabin depressurization was relatively slow (reportedly taking more than 3.5 minutes to depressurize to 0 psi), it was reported that the depressurization was fatal to the *Soyuz* crew in roughly 30 seconds.[4] Further research indicates that the specific circumstances (depressurization rates, the magnitude of the pressure differentials, absolute pressures, etc.) that result in the type of depressurization-related tissue damage seen in the *Soyuz 11* and *Columbia* crews have not been fully characterized. Because the exact scenario cannot be positively identified, no conclusions with respect to cabin depressurization rates or timing can be made from the medical findings.

The 51-L *Challenger* accident investigation showed that the *Challenger* CM remained intact and the crew was able to take some immediate actions after vehicle breakup, although the loads experienced were much higher as a result of the aerodynamic loads (estimated at 16 G to 21 G).[5] The *Challenger* crew became incapacitated quickly and could not complete activation of all breathing air systems, leading to the conclusion that an incapacitating cabin depressurization occurred.[6] By comparison, the *Columbia* crew experienced lower loads (~3.5 G) at the CE. The fact that none of the crew members lowered their visors[7] strongly suggests that the crew was incapacitated after the CE by a rapid depressurization.

Although no quantitative conclusion can be made regarding the cabin depressurization rate, it is probable that the cabin depressurization rate was high enough to incapacitate the crew in a matter of seconds.

> **Conclusion L1-5.** The depressurization incapacitated the crew members so rapidly that they were not able to lower their helmet visors.

---

[2]This rate would result in a shuttle cabin depressurization in less than half a second – a scenario that is contradicted by debris and video evidence. The shuttle cabin depressurization rate was probably an order of magnitude less than 32 psi/sec.

[3]"Survival Following Accidental Decompression to an Altitude Greater than 74,000 Feet (22,555m)."

[4]http://history.nasa.gov/SP-4209/ch8-2.htm.

[5] JSC 22175, STS-51L, JSC Visual Data Analysis Sub-Team Report, Appendix D9, June 1986.

[6]Report from Dr. Joe Kerwin to Rear Adm Truly, http://history.nasa.gov/kerwin.html, July 28, 1986.

[7]See Section 3.2, Crew Worn Equipment.

2.3.4 Depressurization complete

Because the depressurization rate was concluded to be high enough to incapacitate the crew within seconds, the depressurization complete NET time is some number of seconds after the earliest initiation of the depressurization, (i.e., several seconds after GMT 14:00:18). However, no direct debris evidence or analysis provided conclusive results that could refine the NET time for when the depressurization could have been completed.

Debris field analysis, ballistics analysis, and video evidence were used to provide the NLT time for when the depressurization completed. The presence of significant CM structural debris items in the debris field indicated a loss of structural integrity of the CM and, therefore, the inability to maintain cabin pressure. Ballistic analysis on a middeck floor panel indicates a release time of GMT 14:01:02 ± 5 seconds. More than 200 pieces of CM structure were recovered west of this item, strongly suggesting that the CM lost structural integrity prior to this time. Video evidence indicated that major changes in the appearance of the CM and significant debris shedding occurred from GMT 14:00:58 to GMT 14:00:59. It is probable that the CM lost structural integrity and was fully depressurized NLT GMT 14:00:59. The CM was estimated to be over 135,000 feet altitude at GMT 14:00:59, so the crew was exposed to a high-altitude environment. Figure 2.3-1 shows the depressurization timeline with the start of depressurization NET and NLT times and the NLT time for when the cabin depressurization was complete.

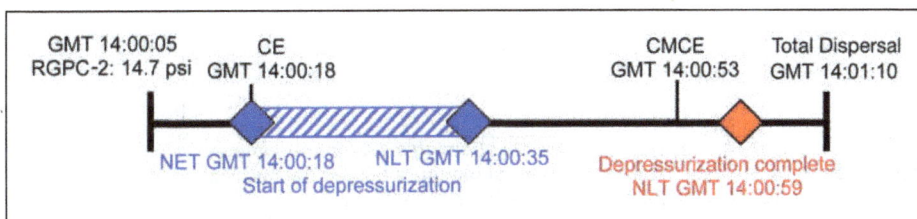

Figure 2.3-1. *Cabin depressurization timeline.*

Conclusion L1-3. The crew was exposed to a pressure altitude above 63,500 feet, indicating that the cabin depressurization event occurred above this altitude.

2.3.5 Location(s) of the cabin breach(es)

Structural analysis was performed to evaluate the overall loads for the CM to determine whether a structural breach occurred in the CM due to forces during the vehicle LOC. The CM was a strong pressure vessel that was designed to withstand the loads of a crash landing. Aerodynamic modeling of the vehicle provided loads for the period from LOS (GMT 13:59:32) to the CE. Deceleration loads were increasing and peaked at 3.5 G at the CE. None of the loads exceeded the CM crash loads structural limits, but CM structural deformation was possible. Impacts between the CM and the FF during and shortly after the CE likely caused damage to the CM, resulting in a depressurization.[8] Internal damage from the sudden load change also accounts for objects breaking free from Volume E and other stowage areas.

Depending on the size and location of breaches in the CM, a depressurization can result in differential pressures across the flight deck and middeck floors. The middeck floor had very few openings; the vent path between the middeck volume and the lower equipment bay consisted primarily of narrow gaps around the panels and access doors. Totaling all the gaps, this venting area was approximately 50 in². The middeck floor structure could withstand a differential pressure of 0.32 psi without suffering deformation.[9] A cabin depressurization computer model was used to determine the maximum hole sizes that would exceed the

---

[8]See Section 2.4, Forebody Breakup Sequence.
[9]Rockwell internal letter SAS/AERO/88-469.

capability of the floor.[10] This model predicted that a hole larger than 12.4 in. diameter (or several smaller holes equivalent to 12.4 in.) *above* the middeck floor, or a hole larger than 4.8 in. in diameter *below* the middeck floor would result in a differential pressure across the floor greater than 0.32 psi (figures 2.3-2 and 2.3-3).

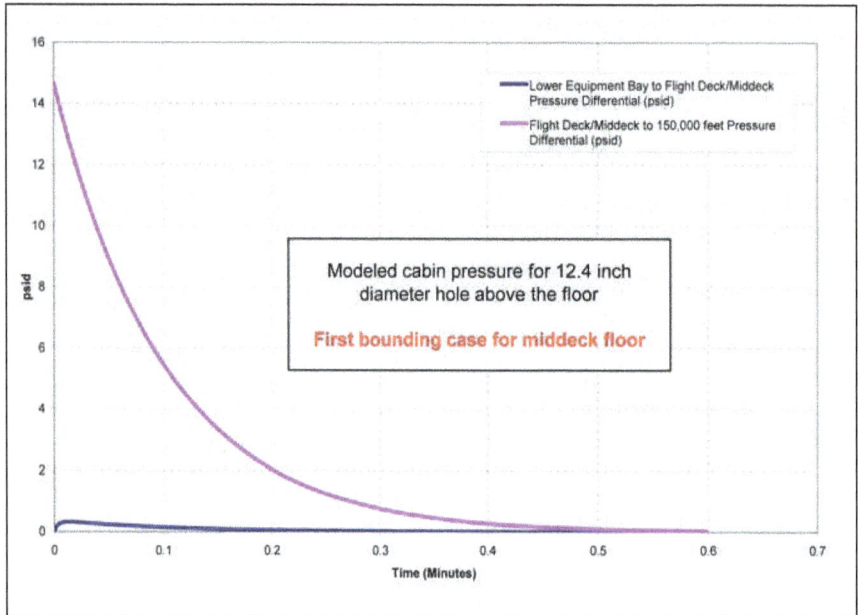

Figure 2.3-2. *Plot from a cabin depressurization model showing a 12.4-inch-diameter hole above the middeck floor.* The maximum differential pressure across the middeck floor is 0.317 pounds per square inch (within middeck floor capability).

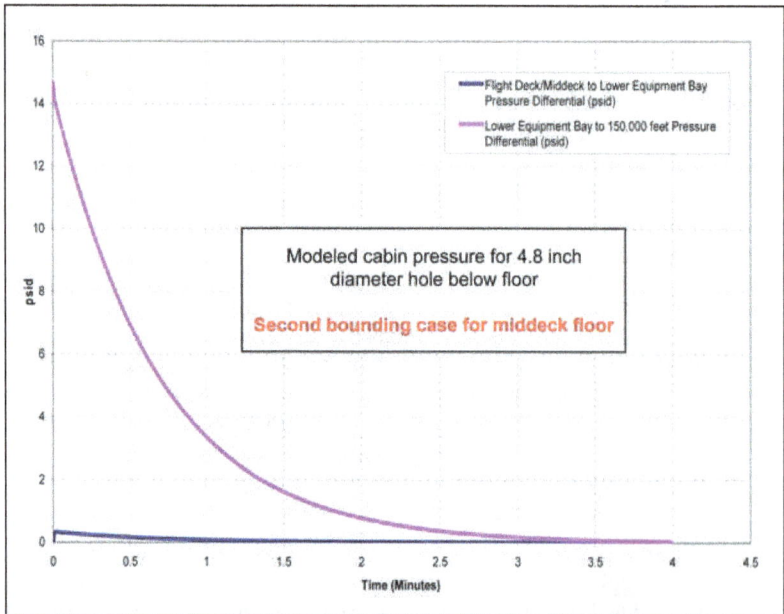

Figure 2.3-3. *Plot from a cabin depressurization model showing a 4.8-inch-diameter hole below the middeck floor.* The maximum differential pressure across the middeck floor is 0.319 pounds per square inch (within middeck floor capability).

[10]The shuttle life support system includes emergency regulators that can maintain the cabin pressure at 8 psi in the event of a cabin breach. The combined maximum flow rate for the system is less than 200 lbs./hour. Cabin breaches discussed in this report were much larger (> 50,000 lbs./hour) and would have overwhelmed the system. Additionally, because the O_2 and N_2 supply tanks were separated from the forebody at the CE, the cabin depressurization analysis presented here does not include gas introduction from the emergency system.

More than 65% of the middeck floor was recovered, and there was no evidence of buckling due to differential pressure. This indicates that the CM depressurization rate did not exceed the structural capability of the middeck floor.[11] Therefore, the cabin depressurization was not caused by an instantaneous hole above the middeck floor larger than 12.4 in. diameter (or several holes, all above the middeck floor, with a combined effective hole size of 12.4 in. diameter). Additionally, the depressurization was not due to an instantaneous hole (or several holes combined) below the middeck floor larger than 4.8 in. diameter. However, holes on both sides of the floor could cause cabin depressurization to occur more quickly and still not cause differential pressures sufficient to damage the floor. The analyses assumed that the hole sizes remained constant; if the holes were to enlarge gradually, a depressurization could occur faster without exceeding the middeck floor's 0.32 pounds per square inch differential (psid) capability.

Based on the fact that the westernmost CM debris field contained items originating from stowage volumes below the middeck floor and from areas above the middeck floor, and the lack of evidence of floor deformation, it is probable that the cabin breach involved holes above and below the middeck floor. The small size of the items and the distinct lack of CM shell structure or elements of heavy internal structure suggest that the individual breaches were not large.

The CE was the most probable time at which a structural failure would occur that would result in structural warping of the CM and/or CM/FF impacts, resulting in one or more breach locations. This is consistent with the previous conclusion that the crew was conscious at the time of the CE (GMT 14:00:18).

> **Conclusion L1-2.** The depressurization was due to relatively small cabin breaches above and below the middeck floor and was not a result of a major loss of cabin structural integrity.

2.3.6 Synopsis of crew cabin pressure environment analysis

Prior to the CE (GMT 14:00:18), the *Columbia* cabin pressure was nominal and the crew was capable of conscious actions. The CM depressurization began NET GMT 14:00:18 and NLT GMT 14:00:35, and was due to cabin breaches above and below the middeck floor. The depressurization rate was high enough to incapacitate the crew members within seconds such that they were unable to perform actions. Although the CM lost structural integrity and was fully depressurized no later than GMT 14:00:59, it is highly probable that the depressurization was complete earlier.

---

[11]The flight deck floor can be damaged by a differential pressure greater than 0.81 psi. Due to the much larger venting area between the flight deck and the middeck, holes much larger than 12.4 in. diameter would be required to deform the flight deck floor. Very little flight deck floor debris was recovered, so no conclusions regarding the deformation of the flight deck floor could be made.

2.4   Forebody Breakup Sequence

This section discusses the breakup of the forebody of *Columbia*. The findings in this section are based on ground-based video analysis, ballistic calculations, cluster analysis of the debris field, and structural analysis.

The format of this section follows the format of Section 2.2, Orbiter Breakup Sequence. *It is strongly recommended that the reader review Section 2.2 before reading this section.* Complete descriptions of the types of analysis can be found in the introduction of Section 2.2 and are not repeated here.

Based on video, the forebody broke away from the intact orbiter at the CE, GMT 14:00:18. Because the forebody's ballistic number was significantly higher than that of the intact orbiter (more dense, with less drag), at the moment of separation the deceleration due to drag decreased suddenly. The loads experienced by the forebody dropped from the peak load of approximately 3.5 G just before the CE to approximately 1 G after the CE. As the forebody began its own unique ballistic trajectory, deceleration forces began to build again. Additionally, the forebody was rotating in all axes at approximately 0.1 rev/sec with an increasing rate (see Section 2.1).

The forebody breakup was initiated at the CMCE at GMT 14:00:53. This was the beginning of a sequence of events resulting in the dispersal of the forebody into multiple smaller components. NLT GMT 14:01:10, the CM had lost all structural integrity and had been broken into subcomponents. This time was defined as the TD.

The following findings and recommendation are in this section:

Finding. The *Columbia* windows remained largely intact up until the CMCE and were not a cause of cabin depressurization.

Finding. Windows 7 and 8 experienced a titanium deposition event that occurred prior to window breakup.

Finding. The most probable source for the titanium deposition on Windows 7 and 8 was PLBD rollers. These rollers were not exposed to heat flow until after the PLBDs were compromised.

Finding. All the windows had an aluminum-rich deposition, which was consistent with a turbulent process.

> **Recommendation L3-1.** Future vehicles should incorporate a design analysis for breakup to help guide design toward the most graceful degradation of the integrated vehicle systems and structure to maximize crew survival.

2.4.1 Ground-based video analysis

This section addresses the detailed analysis of ground-based video related to the forebody after separation from the orbiter at the CE. This section only covers major events seen in the video that occur relative to the forebody breakup, including the CMCE and the TD.

A video triangulation analysis on the motion of the free-flying forebody was discussed in Section 2.1. A different relative motion analysis related to the TD is discussed here.

Catastrophic Event to Crew Module Catastrophic Event

It was initially anticipated that the CM depressurization could be identified in videos as a halo or other visible effect. However, detailed review of the video showed a visually complex event at the CE. The forebody was not visually distinct from the rest of the orbiter's pieces until about 8 seconds after it separated from the intact orbiter. It is unknown whether a depressurization event would be visible; but if it was and it occurred during this time, any indication of the event would be lost in the merged signals of the orbiter's pieces. None of the changes in appearance at the CE or afterward could be positively identified as a depressurization event (see Section 2.3).

Crew Module Catastrophic Event

Video analysis established the precise time for the CMCE (the initiation of the breakup of the forebody) as GMT 14:00:53. Thermal and ballistics analyses of forebody debris items were consistent with this, supporting the video-based time.

At the CMCE, the video shows that the forebody began to visibly brighten and the envelope of gases around the forebody appear to increase in size. This brightening event was followed quickly by a significant debris shedding event that was likely to be the FF and the CM separating from one another.

Figure 2.4-1 shows a set of five paired images covering less than 2 seconds of time. On the left is the original image and on the right is a close-up of the forebody and CM. The first frame shows when the forebody begins to brighten. The second frame, which was taken one-third of a second later, shows the slight increase in apparent size of the envelope of gases of the orbiter. The third image, almost 1 second later than the previous image, shows the first definitive indication (although it can be recognized a few frames earlier) that the forebody is beginning to fail. The magnified image has been inverted to emphasize the split in the trail of the forebody. The next frame, which occurs about one-third of a second later, shows what are believed to be two significant portions of the FF separating from the CM. The last image, taken one-fifth of a second later, shows how quickly the FF is breaking into pieces that are too small to be seen by the camera. The CM breaks up over the next 16 seconds.

Total Dispersal

As objects separated from each other during the breakup, each took on its own trajectory based on its unique ballistic number. The high initial speed immediately resulted in a wide dispersion of trajectories as lighter and smaller items decelerated very rapidly, while heavier and larger items decelerated less and, hence, traveled farther. Evaluation of the debris appearance confirmed that very little debris-to-debris interaction (impacts) occurred. As subcomponents decelerated, the entry heating began to decrease quickly, resulting in a loss of visual signal in the video. The CM image vanishes while it was still clearly in the camera's FOV. The image loss was not due to a major deceleration taking the intact CM out of the frame, but was due to the CM being broken into subcomponents that were too small and dispersed to be visible on the video. This event was described as the TD. After this time, the CM no longer had any structural integrity.

It should be noted that cascading structural failures were still occurring following the TD, as well as frictional heating on individual objects with high ballistic numbers that decelerated more slowly. These ongoing failures could not be seen on video.

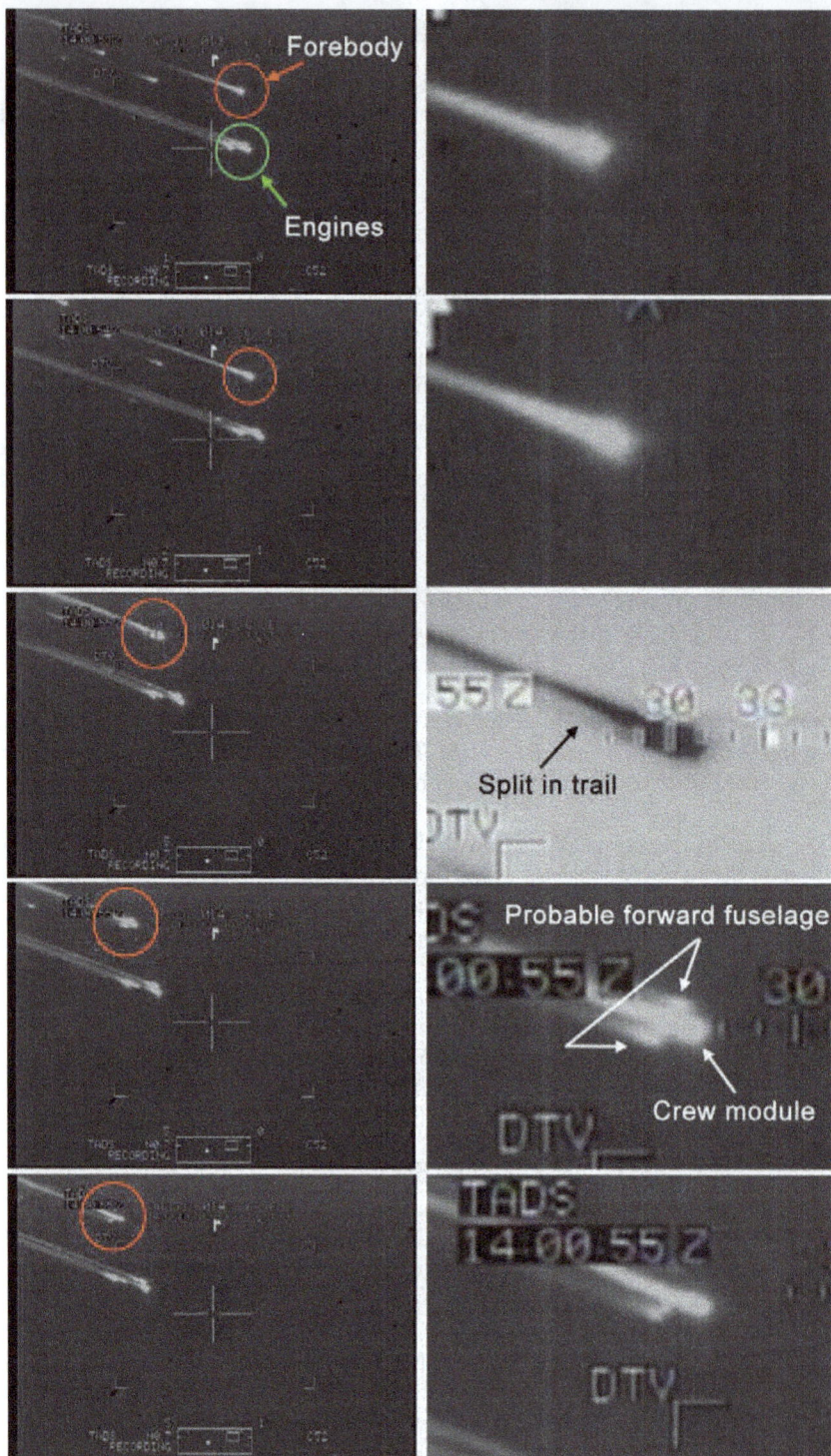

Figure 2.4-1. *Five time-paired images covering under 2 seconds of time showing the Crew Module Catastrophic Event.* Image on the left is the full frame, image on the right is an enlarged view.

Figure 2.4-2 shows the last few seconds of the CM as seen in the Apache video, ranging from GMT 14:01:06 through GMT 14:01:09. The images on the right side are a magnified view of the original images on the left. The CM has been circled in red.

As the images illustrate, when structural integrity of the CM completely failed, it did so in a fraction of a second. The times for the images (figure 2.4-2), from top to bottom, are GMT 14:01:06.73, GMT 14:01:06.87, GMT 14:01:06.97, GMT 14.01.08.3, and GMT 14:01:09.67. In the next frame, the CM is no longer visible.

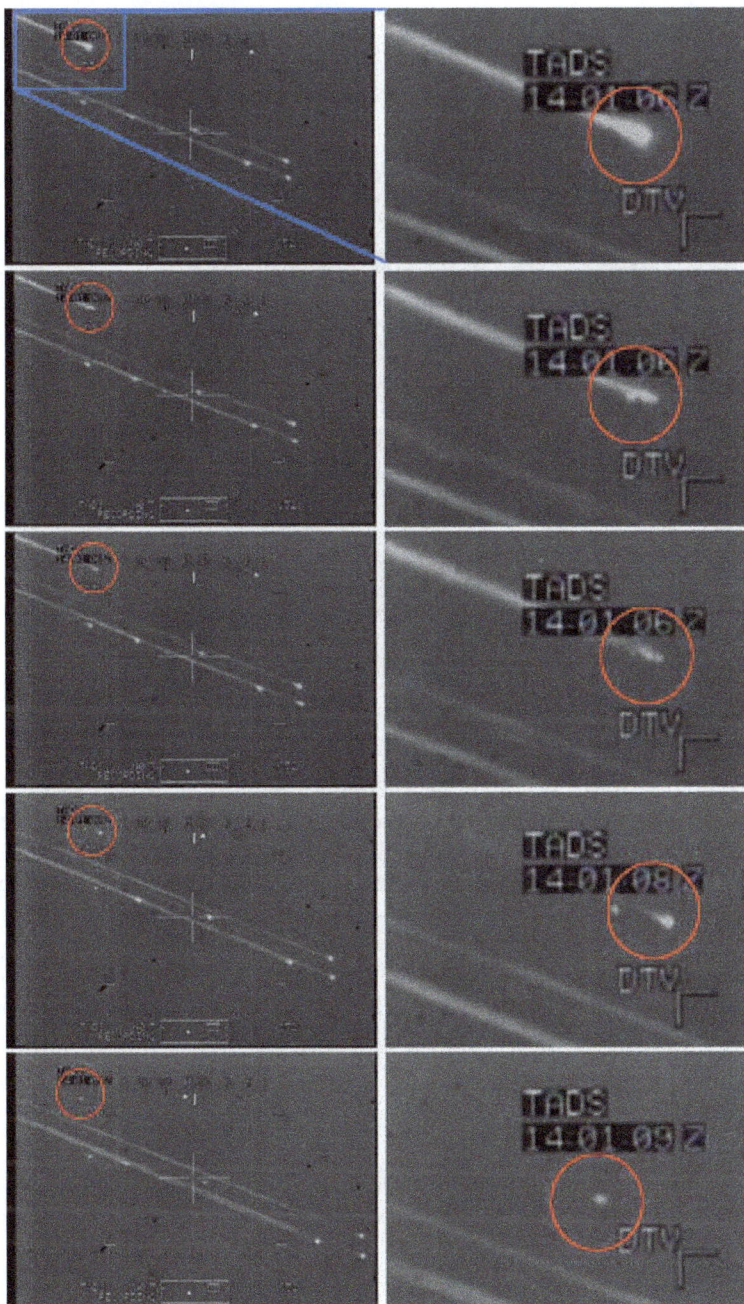

Figure 2.4-2. *Apache video of the Total Dispersal, spanning from GMT 14:01:06.73 to GMT 14:01:09.67. The forebody/crew module is circled in red.*

Relative motion analysis

Relative motion analysis compares the rate of change of the movement of objects in the FOV of a video. Rate of change can provide an estimation of the G-load that is experienced by the bodies within a relative frame of reference; e.g., a relative difference might be that one object experiences a deceleration of 3 G relative to another object. While the analysis indicates that the first object experiences three more Gs than the other object, it does not define how many total Gs either object is actually experiencing.

Two relative motion analyses were performed. The first regarded a triangulation of the motion of the forebody relative to the engines between the CE and the CMCE (see Section 2.1).

The second relative motion analysis was performed on the CM (identified as D21) and the aft engines post-CMCE in a stabilized Apache video (figure 2.4-3).

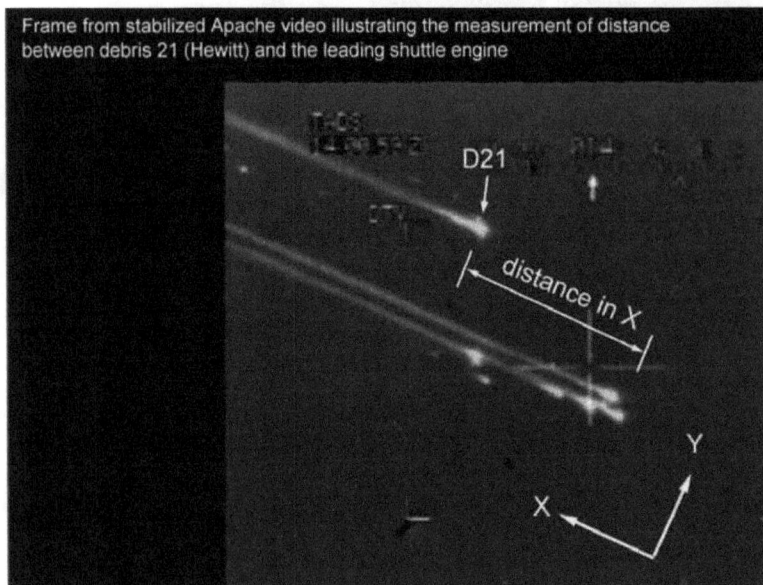

Figure 2.4-3. *An anayzed frame from the Apache video.*

The relative motion of the CM to the engines appears to suggest a high deceleration event occurred during the breakup of the forebody (between the CMCE and the TD). Later understanding of the breakup sequence revealed that this deceleration was related to the disintegration of the CM and the resulting cloud of debris. Rather than a specific high-G event that was experienced by an intact CM, it represents the cloud of individual items separating and rapidly decelerating. Heating was sufficient to keep the individual objects visible for a short period of time, and these items were close enough together that they could not be distinguished as separate items. This also explains why the CM's visible disappearance occurred in a fraction of a second, as the deceleration passed the threshold of sufficient heat generation for visibility in the video.

2.4.2 Ballistic analysis

Refer to Section 2.2.2 for a full description of the techniques, assumptions, and limitations for ballistic analyses.

Forebody structure
Table 2.4-1 shows the debris ballistic timeline for some selected forebody structures. Ballistic analysis could not be done on all recovered debris because ballistic numbers are hard to estimate for irregularly shaped objects and the analysis is time-intensive. For some structures, there is a major structural release time. For other objects, the major structural release time is the same as the debris object release time since there is only one object.

Table 2.4-1. *Ballistic Timeline for Post-Crew Module Catastrophic Event Forebody Events*

Major Structural Release Time (GMT)	Vehicle Structure	Debris Object Number and Time of Release (GMT)
14:00:54	FF	11119 (14:00:51) 81231 (14:00:55) 26099 (14:00:55)
14:00:57	Forward RCS strut	2170
14:00:59	Forward RCS thruster	2167
14:01:05	Thermal pane glass (outer pane)	14:00:57 (65012) 14:01:02 (68534) 14:01:04 (1574) 14:01:05 (73192) 14:01:16 (77517)
14:01:07	FF port ejection panel[1]	51987
14:01:12	Nose cap	257
14:01:13.5	Forward RCS helium tank	14:01:13 (1481) 14:01:14 (1209)
14:01:35	FF starboard ejection panel	71801

These individual release times alone did not provide significant insight into the breakup event, but will be compared to other analyses in this section. However, the times generally span the video determination for the time between the CMCE and the TD. Although a few items appear to have been released after the TD, cascading failures were expected to occur even after the TD, and this accounts for those times being later (see Section 2.2.2).

Crew module interior items

Objects that are discussed in this section originated from inside the CM but are not directly associated with a seat or a crew member. All of the items listed were selected because their shapes were easily modeled for ballistic analysis, and all are believed to have been stowed or located in the middeck. No flight deck structural debris items were good candidates for ballistic analysis because of their irregular shapes. For a detailed discussion of individual crew seat and crew equipment recovery, see Sections 3.1 and 3.2. The results of that ballistic analysis concluded that the middeck seats and equipment were released prior to the flight deck.

The release times and descriptions of these interior items are in Table 2.4-2. The mission patch (debris item no. 31539) in this table should not be confused with the separate mission patch that is identified in the cabin depressurization analysis, which was one of the most westerly objects recovered.

---

[1]*Columbia* was the only orbiter in the fleet with ejection hatches for the Commander and Pilot stations. The ejection seat systems were disabled and the ejection hatches were deactivated when the orbiter program was deemed operational. Eventually, the ejection seats were removed from *Columbia*, but the hatches remained as an integral part of the structure of the FF.

Table 2.4-2. *Individual Debris Objects for the Crew Module Post-Crew Module Catastrophic Event*

Release Time (GMT)	Debris Number	Description	Shape
14:00:56	55952	Toilet handle ball	sphere
14:01:01	31539	STS-107 mission patch	triangular
14:01:02	1170	Middeck floor section	rectangular
14:01:07	1155	Middeck accommodation rack (MAR)	rectangular
14:01:07	2499	Middeck floor section	rectangular
14:01:08	44199	Volume E access door	rectangular
14:01:10	23262	Locker door	rectangular
14:01:10	7717	Photo TV floodlight	rectangular
14:01:13	7662	Light emitting diode (LED) indicator panel	square
14:01:15	8820	Window shade bag	rectangular

Again, no major conclusions could be drawn from these data alone. However, like the structures ballistic analysis, these times are consistent with the times for the CMCE and the TD as determined by video.

2.4.3 Cluster analysis

For a complete discussion of the techniques, assumptions, and limitations of cluster analysis of the debris fields, see Section 2.2.3.

Catastrophic Event to Crew Module Catastrophic Event
The most significant finding from debris field analysis showed that the FF and CM appeared to remain relatively intact from the CE to the CMCE, a period of 35 seconds (see Section 2.2.3). Small amounts of forebody structure, such as some nose landing gear door tile and small structure, were released earlier. However, 87% of the FF structure and all of the CM pressure vessel structure debris clusters overlap completely (figure 2.4-4), strongly suggesting a relationship between their structural failure. The western end of the main portion of the forebody structure debris field is at a longitude of 94.5W.

The fact that the forebody maintained structural integrity immediately after the orbiter breakup may be explained by the reduction in deceleration loads occurring at the CE due to the change in ballistic number. However, thermal effects would begin to increase, and deceleration loads once again began to build up to a peak of 3.5 G at the forebody c.g. at the CMCE. Note that loads experienced at the farthest outboard edges of the forebody could be as high as 10 G to 12 G at the CMCE due to the moment arm from the c.g. as the forebody rotated.

Cabin pressurization
Cabin depressurization was an important event to identify. The initiation of the depressurization of the cabin could not be observed in the video. However, the debris field provided the opportunity to evaluate when depressurization may have started, because items that were originally stowed inside the CM and recovered west of the main forebody debris field could indicate a CM breach.

Items that were stored on board the orbiter were carefully packed, stowed, and documented prior to launch. Some items on STS-107 were stowed in the middeck and the SPACEHAB (laptop computers and LiOH canisters). Most items also had a designated entry stowage location. Common use items (such as pens and

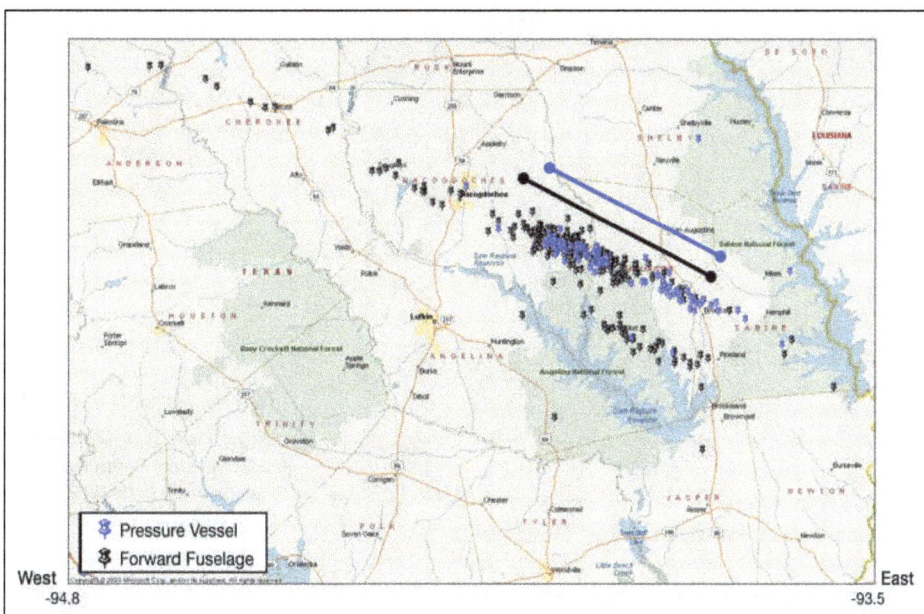

Figure 2.4-4. *Debris field of the forward fuselage and crew module pressure vessel.*

pencils) are not carefully tracked and not necessarily re-stowed to launch configuration for entry. In particular, trash was stored in multiple areas such as the SPACEHAB, the airlock, and the middeck volumes, some of which extend into the lower equipment bay below the middeck floor. The SCSIIT chose to investigate only nonstructural items with accurately known storage locations. No CM pressure vessel structure or significant interior structure was recovered west of the main forebody debris field. Nonstructural items that were recovered west of the main forebody debris field were assumed to have been evacuated from the CM due to decreasing pressure and a small breach in the CM rather than major structural degradation. Study of these objects allowed the team to better understand when pressure inside the CM was lost, and to determine the areas where suspected breaches might have occurred.

Many "crew equipment" items that are listed in the database west of longitude 94.5W may not have originated from *Columbia*. Because they represent common use personal items, they cannot positively be identified as from the orbiter and may be discarded items that were present in the debris field. The items that can confidently be concluded came from the orbiter include empty food packets, LiOH canisters, clothing, and laptop computer debris. The bulk of these items were stored in the middeck lockers, the lower equipment bay volumes, the SPACEHAB, the airlock, and the Waste Collection System (WCS).

Seventy-one items were recovered west of longitude 94.6W and were positively identified from the middeck, flight deck, or CEE. The specific locations for these items were Volume E, middeck lockers MF43K and MF43C, port middeck and stowage volumes from the lower equipment bay, flight deck illuminator panels (i.e., acrylic sheeting), and equipment from the flight deck. These items were all small (< 8 in. and, in most cases, much smaller).

In general, small and light items did not travel downrange significant distances due to the ballistic properties of the object. Some paper/lightweight items were offset slightly northeast of the main debris footprints, probably due to the prevailing winds at the time of the accident. Figure 2.4-5 shows the debris field for nonstructural internal CM debris.

Many of the westernmost debris items came from internal stowage volumes. This implies that internal structural damage occurred at the CE. A full description of cabin depressurization is contained in Section 2.3. The integrated assessment concluded that depressurization occurred NET GMT 14:00:18 (CE) and NLT GMT 14:00:35. It ended NLT GMT 14:00:59, and most likely earlier.

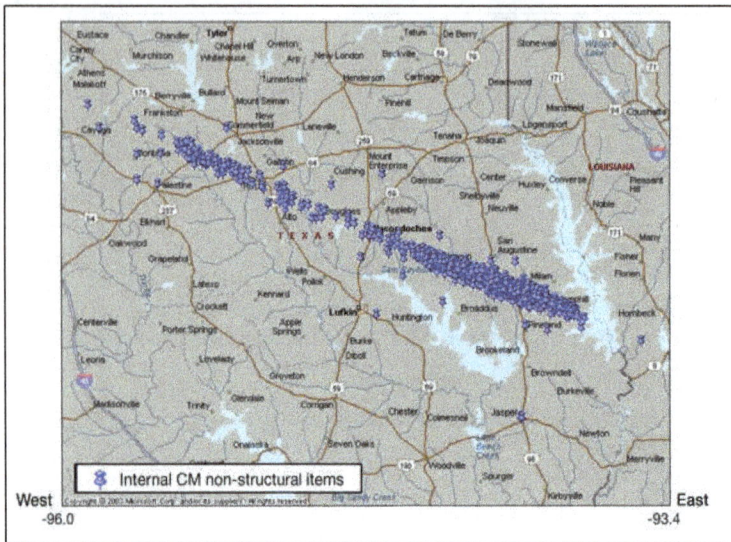

Figure 2.4-5. *Debris field of the nonstructural items from inside the crew module pressure vessel.*

West -96.0

East -93.4

Internal CM non-structural items

Window panes

A perceived vulnerable area of the forebody was the windows. Most of the windows in the forebody are triple-paned. Only the two aft windows, which are protected by the PLBDs during entry, are double-paned. For the three-paned windows, the exterior panes are "thermal" panes that provide thermal protection. The inner panes are "pressure" panes that provide structural support for the pressure inside the CM. The middle pane is a "redundant pane" as it is intended to be redundant for both the thermal and the pressure pane. Figure 2.4-6 shows the schematic of the three-paned window of an orbiter.

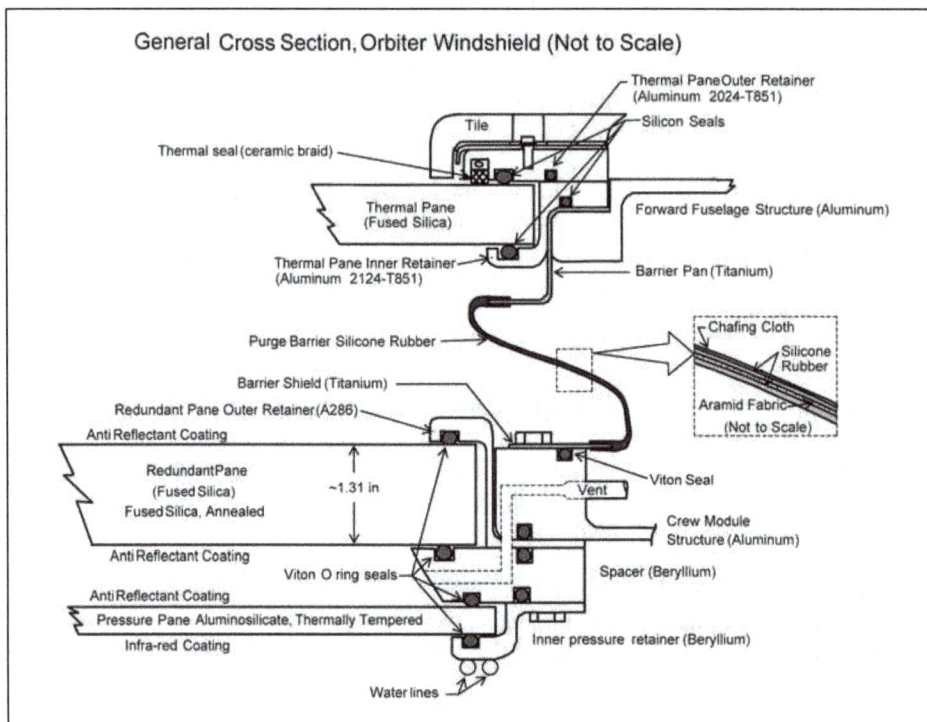

Figure 2.4-6. *Schematic of the orbiter three-paned window.*

Figure 2.4-7 shows the numbering system for the windows; the forward windows are numbered 1 through 6, the overhead windows are 7 and 8, and the aft-facing windows are 9 and 10. Window 11 (not shown) is the side hatch window. The thermal, redundant, and pressure panes vary in thickness depending on location on the vehicle. Table 2.4-3 shows the specified thicknesses for the various panes from the 11 different windows.

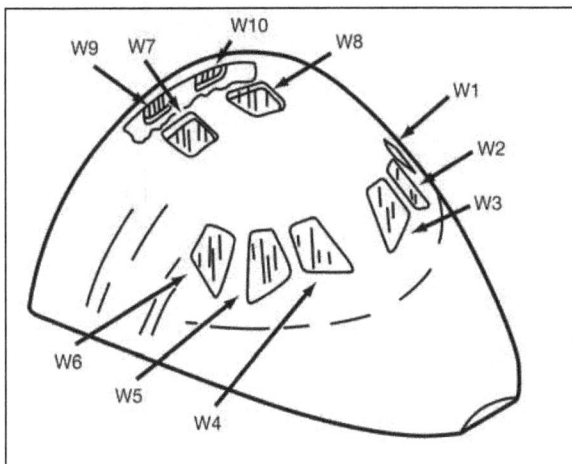

Figure 2.4-7. *Window numbering system.*

Table 2.4-3. *Specified Orbiter Window Design Thicknesses*

Size (in.)	Window pane
0.29	Side hatch thermal pane
0.33	Windows 9, 10 redundant pane (tempered)
	Windows 9, 10 pressure pane (tempered)
0.47	Windows 7, 8 redundant pane (tempered)
	Windows 7, 8 pressure pane (tempered)
0.49	Windows 7, 8 thermal pane
	Side hatch redundant pane
0.56	Windows 1, 6 thermal pane
0.61	Windows 2, 5 thermal pane
0.62	Side hatch pressure pane
0.63	Windows 3, 4 pressure pane (tempered)
0.65	Windows 1, 2, 5, 6 pressure pane (tempered)
0.69	Windows 3, 4 thermal pane
1.30	Windows 3, 4 redundant pane
1.32	Windows 1, 2, 5, 6 redundant pane

The orbiter windows are made of a compositionally unique fused silica that is highly thermally stable and not expected to thermally degrade under entry heating. Material analysis was not a primary means of assessment due to limited resources. Recovered glass was identified by measuring the thickness and comparing it to the data shown in Table 2.4-3. The highest confidence was in the approximately 1.3-in.-thick redundant panes of glass since this is not a common commercially available thickness. Due to the shattered nature of the tempered-glass fragments, further identification was difficult for loose fragments (those not retrieved from within the frames).

Initially, the windows were considered as a potential site of cabin breach due to thermal failure. The thermal panes, while designed to withstand thermal conditions, are not normally exposed to the highest entry heating conditions because of the geometry of the vehicle and its nominal attitude during the heating phase on entry. However, a preliminary thermal analysis showed that the structure around the windows would fail thermally before the window panes.

A different concern was whether the flexing of the structure under the varying loads at the LOC and the CE would cause the glass to shatter. The debris field analysis refuted that theory as well. Figure 2.4-8 shows the debris field coordinates for recovered glass.

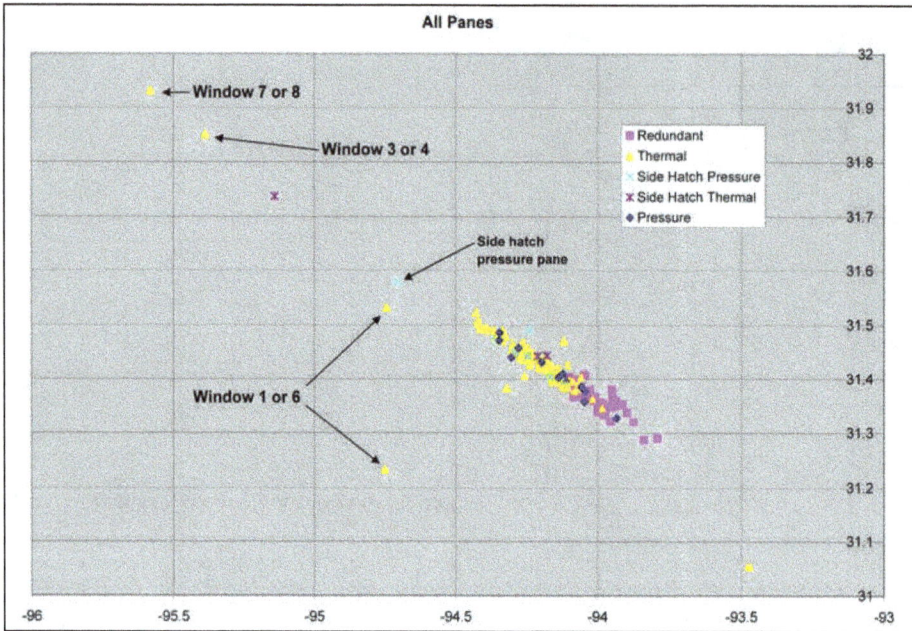

Figure 2.4-8. *Recovered debris field for window glass.*

Of the 201 recovered glass fragments, only six pieces of loose glass that are confidently believed to be from the orbiter were found west of the main forebody debris field. All were thermal (outer) panes with the exception of one piece of probable side hatch pressure pane. However, a piece of pressure pane that was recovered in the western end of the debris field does not alone indicate a source of depressurization. Pressure panes may be damaged due to objects moving around inside the CM as it rotated, impact-ing the windows. If the cabin is depressurized from a different source, these pieces can evacuate through other breaches.

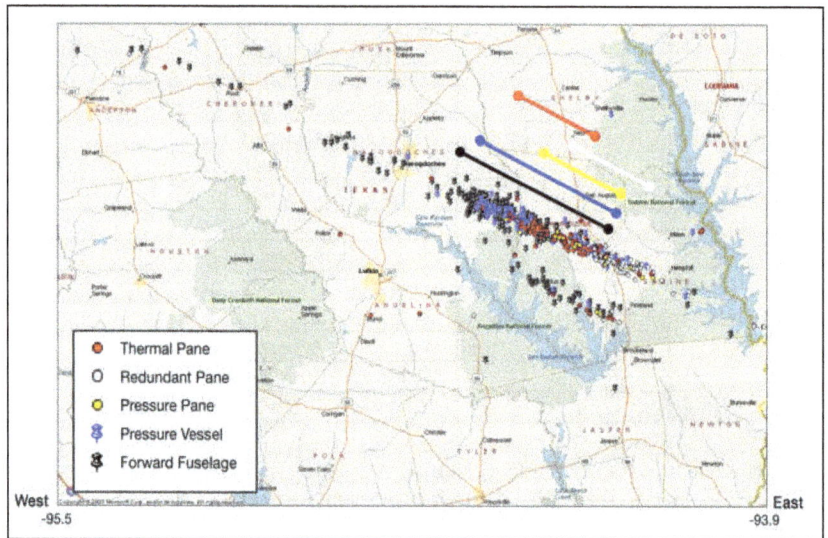

Figure 2.4-9. *Debris field of the window panes, forward fuselage, and crew module pressure vessel.*

If the source of the cabin depressurization had been a window, both the pressure pane and the redundant pane would have to have failed. However, redundant pane is not seen until the far eastern end in the fore-body debris field. Also, in such a case, the pressure from inside the CM would likely blow the entire pane out and result in significant amounts of glass that were not seen west of the main forebody debris field.

Figure 2.4-9 shows the window debris field overlaid on the FF and pressure vessel debris fields.

Figure 2.4-9 shows that thermal pane debris was recovered east of the beginning of the main forebody debris, suggesting that the windows broke at or after the CMCE. Pressure panes were recovered west of the bulk of the redundant panes, possibly because the redundant panes were extraordinarily thick and heavy and may have traveled farther east because their ballistic number was higher. Determining the failure sequence of the windows based on ground plots alone was somewhat suspect because some of the window frames remained attached to other window frames as a unit, and the shape of the windows can potentially generate lift. Regardless, because the debris field does not contain significant pane debris west of the main CM debris field, it is concluded that the windows were not the source of a CM breach or a part of the initiating event of the forebody breakup.

Finding. The *Columbia* windows remained largely intact up until the CMCE and were not a cause of cabin depressurization.

Post-Crew Module Catastrophic Event forebody breakup

This section discusses the debris field clusters related to the CMCE and the breakup of the forebody. Figure 2.4-10 shows the major elements of the forebody including the CM pressure vessel, FF, forward RCS, nose, and nose landing gear debris. Figure 2.4-11 shows greater detail regarding specific items and highlights the scarcity of items recovered prior to the main FF/CM debris field.

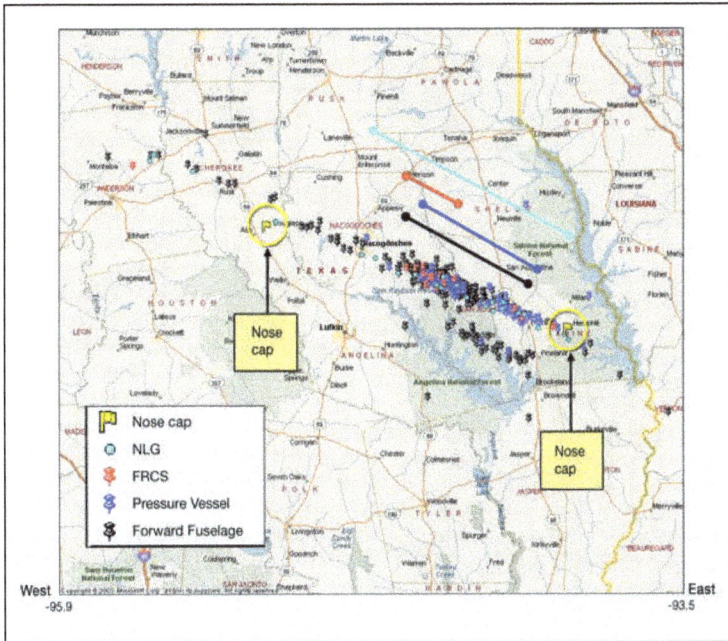

Figure 2.4-10. *Debris field of the forebody components.*

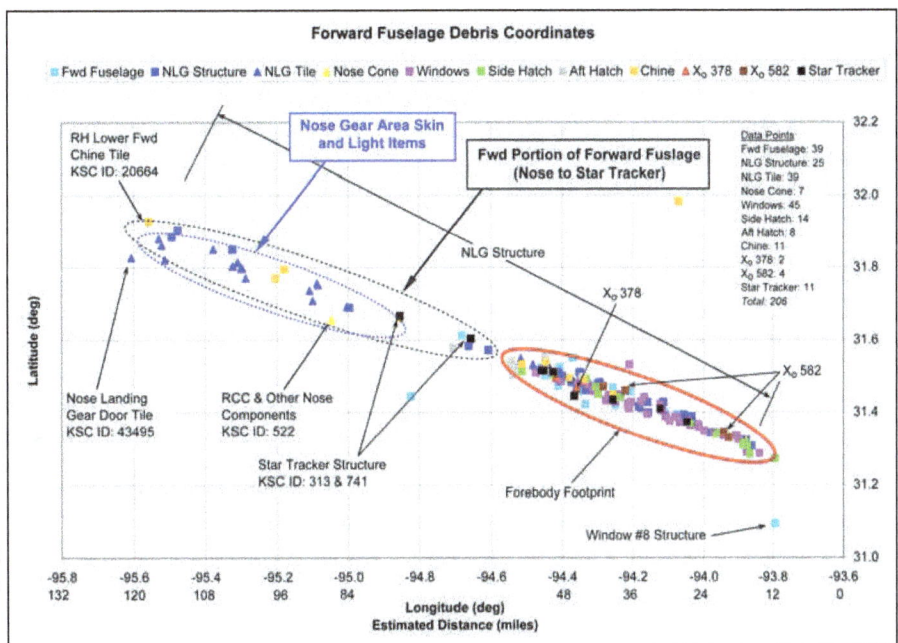

Figure 2.4-11. *Forward fuselage debris field.*

The debris field analysis suggests that the order of separation was as follows: nose landing gear structure and nose cap, FF, and forward RCS, followed by the CM pressure vessel. This is consistent with the video that appeared to show FF separating, followed very quickly by failure of the CM.

Crew module interior structure
The CM interior structure discussion includes the flight deck, middeck, and airlock. *Columbia* was the only orbiter with an internal airlock, meaning that the airlock structure was located inside the middeck of the CM rather than in the payload bay like the other orbiters.

Middeck and flight deck
Figure 2.4-12 shows the distribution of middeck and flight deck structure. The middeck structure appears farther west in the debris field, while the flight deck debris cluster is concentrated at the eastern end. This suggests that the flight deck remained intact longer and traveled farther downrange than the middeck.

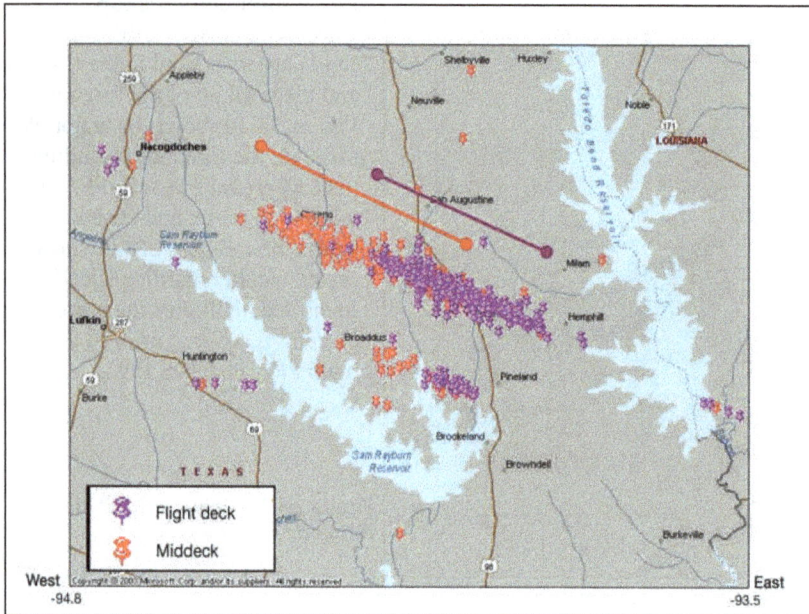

Figure 2.4-12. *Debris field of the middeck and the flight deck.*

This conclusion is supported by the fact that almost all of the middeck floor debris was in fair to very good condition. However, very little of the structurally stronger flight deck floor was recovered. Furthermore, the flight deck items that were recovered were heavily thermally eroded. This would be the case if the flight deck stayed together and received more thermal damage as a result of having a higher ballistic number than the individual middeck components.

Ground debris footprints for the crew worn equipment and seats are not addressed in this section. The analyses and conclusion from the seat and suit assessments (see Sections 3.1 and 3.2) also concluded that the middeck elements separated before the flight deck elements. This supports the conclusions from the structural debris fields.

The debris field was also evaluated based upon whether debris items came from the port or starboard side of the orbiter. Figure 2.4-13 shows the middeck structure debris field. Figure 2.4-14 shows the flight deck panel debris field.

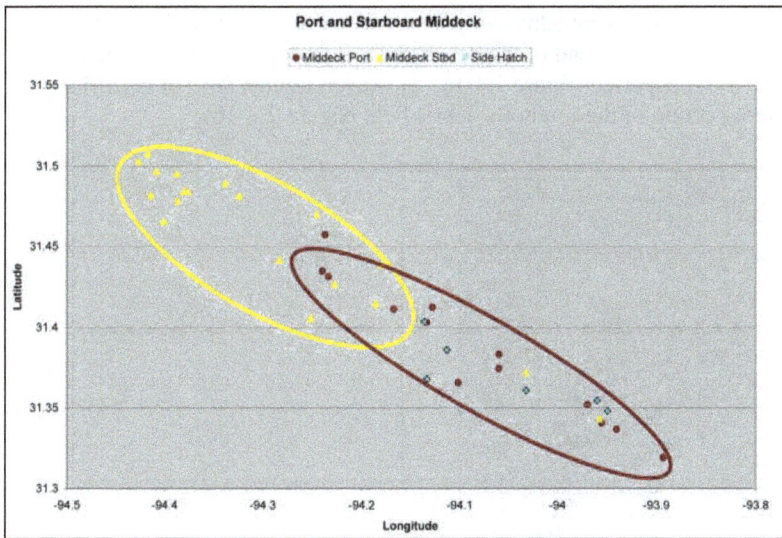

Figure 2.4-13. *Middeck structural debris port vs. starboard.*

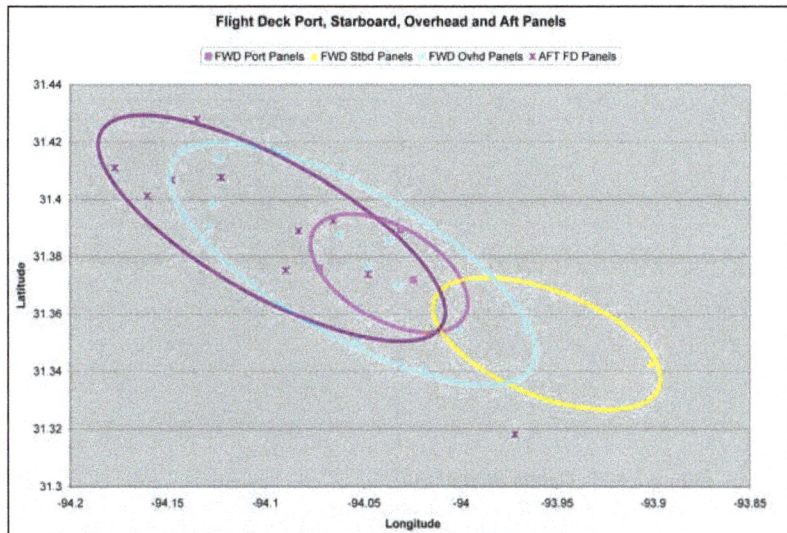

Figure 2.4-14. *Flight deck panels by overhead, port, and starboard.*

There was a substantial amount of debris for the middeck analysis. Middeck starboard debris was recovered west of the middeck port debris, suggesting that the starboard side failed before the portside. Many of the items from the starboard side were lightweight items from the sleep station with low ballistic numbers, which may also affect this debris field. The debris field is consistent with a yaw to the portside, exposing the starboard side to greater thermal and aerodynamic loads. However, the flight deck panels appear inconclusive relative to a specific failure sequence of forward to aft or port vs. starboard. Not many panels were recovered from port or starboard, which makes interpretations risky.

Airlock

The airlock was located on the middeck. The aft opening of the airlock had no hatch and was open to the tunnel to the SPACEHAB in the payload bay. The forward hatch, leading to the CM, was closed and locked. Therefore, the airlock could lose pressure without impacting the internal conditions of the CM. See Section 2.2.4 regarding the failure of the TAA and subsequent depressurization of the airlock at or closely after the CE.

A few "loose" items, as well as a few items that were adhesively bonded to structure and a few pieces of secondary structure, were recovered in the western end of the debris field, indicating that these items were evacuated from the airlock when the TAA departed (figure 2.4-15). However, almost 80% of the airlock structure was recovered in the eastern portion of the forebody debris field (figure 2.4-16).

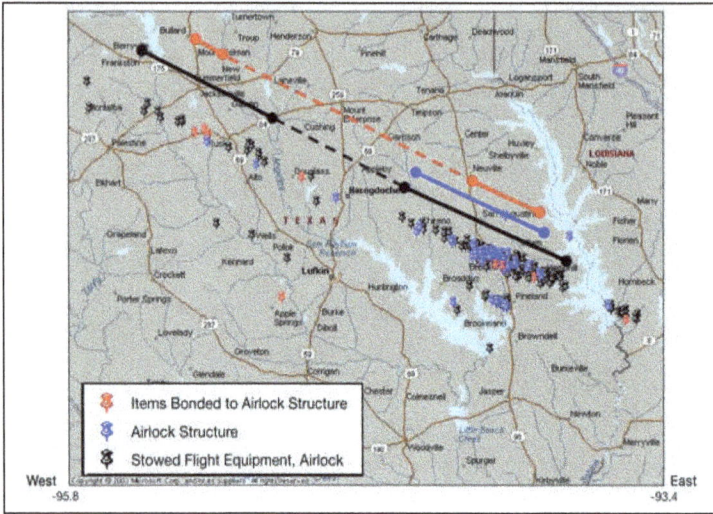

Figure 2.4-15. *Debris field of the airlock structure, flight crew equipment stowed in the airlock, and items bonded to airlock structure.*

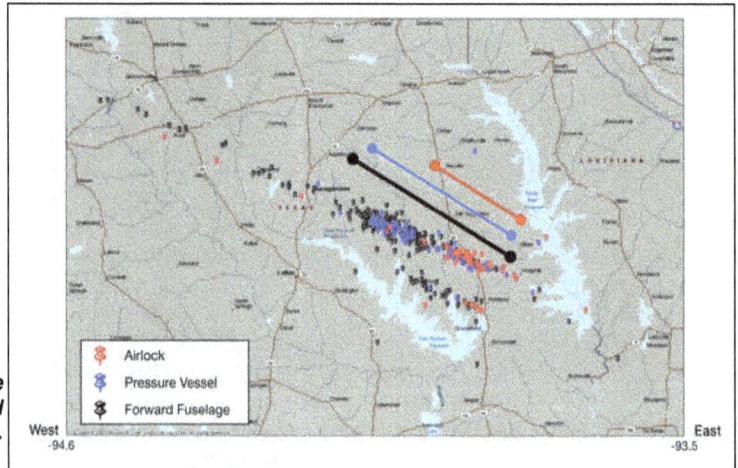

Figure 2.4-16. *Debris field of the forward fuselage, pressure vessel, and airlock structure.*

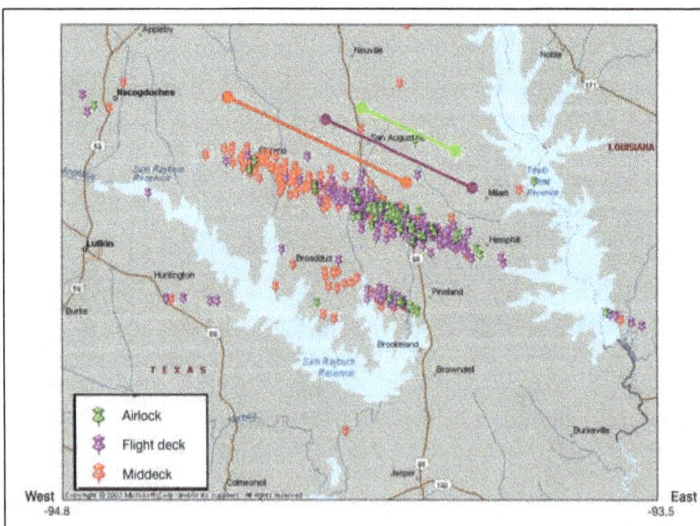

The internal airlock is structurally attached to the CM aft of the X_0 576 bulkhead. It likely stayed with the aft bulkhead after the middeck departed. Figure 2.4-17 indicates that the airlock structural debris is also squarely in the middle of the flight deck structural debris. This suggests that the flight deck and airlock remained connected to each other by the aft bulkhead.

Figure 2.4-17. *Debris field of the middeck, flight deck, and airlock.*

Forebody structural bulkheads

The forebody contained four major bulkheads. The bulkhead that was aft of the forward RCS compartment and immediately in front of the CM forward bulkhead was the X_o 378 bulkhead. The forward bulkhead of the CM was the X_{cm} 200 bulkhead.[2] The aft bulkhead of the CM was the X_o 576 bulkhead, and the bulkhead immediately aft of it was the X_o 582 ring frame bulkhead (figure 2.4-18).

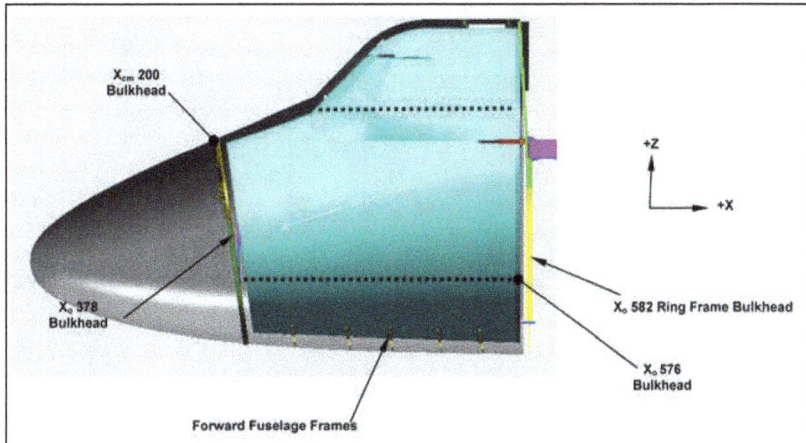

Figure 2.4-18. *Forebody bulkheads.*

Figure 2.4-19 shows the bulkhead debris field clusters relative to each other. Figure 2.4-20 shows the CM (X_{cm} 200 and X_o 576) bulkhead debris fields relative to the middeck and flight deck debris fields.

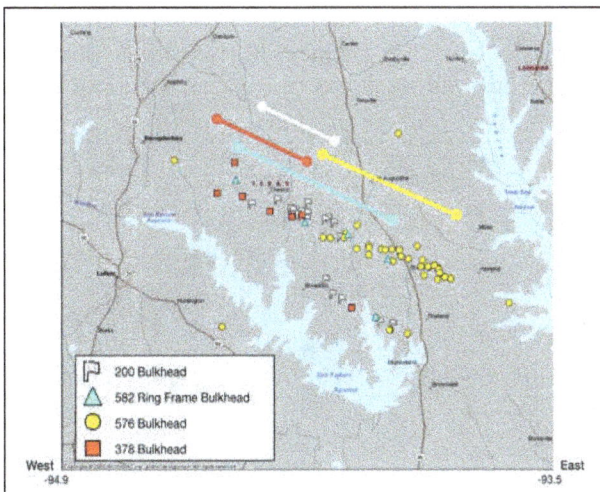

Figure 2.4-19. *Debris field of the key forebody bulkheads.*

---

[2]The X_{cm} reference frame is the CM reference frame. The X_{cm} reference frame axes are aligned with the X_o reference frame, but the X-axis origin is different.

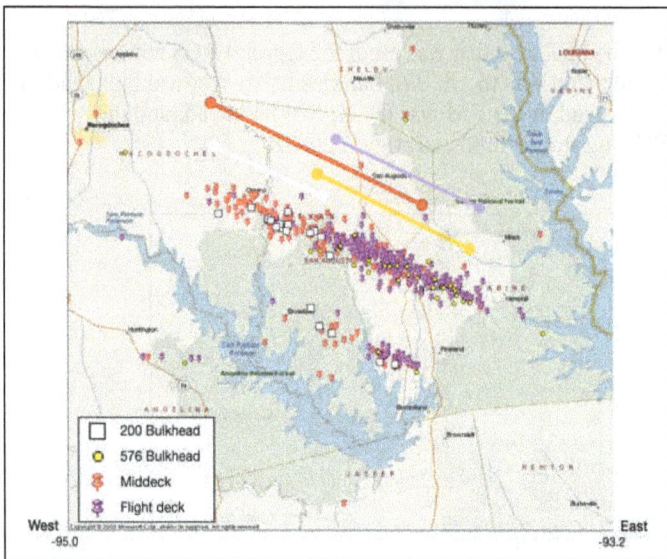

Figure 2.4-20. *Debris field of the crew module bulkheads, middeck, and flight deck.*

The debris field clusters suggest that the two forward bulkheads appeared to fail earlier than the aft bulkheads. Also, the outer (fuselage) bulkheads failed prior to their corresponding inner (CM pressure vessel) bulkheads. The debris footprint for the CM forward bulkhead (X_{cm} 200) coincides with the center of the middeck debris field, suggesting that the X_{cm} 200 departed with the middeck. The CM aft bulkhead (X_o 576) appears to have remained with the flight deck until it disintegrated, likely also connected to the airlock since that debris field coincides as well.

This agrees with the general conclusion that the FF failed followed by the CM pressure vessel failure.

Avionics bays

The CM avionics bays debris plots were also evaluated to see whether they might provide insight into the CM failure sequence. These bays are in immediate proximity to the forward and aft bulkheads of the CM.

Debris plots represent debris from four avionics bays that are located on the middeck. Avionics Bays 1 and 2 are in the forward-most portion of the CM, immediately aft of the forward X_{cm} 200 bulkhead. Avionics Bays 3A and 3B are in the aft of the middeck on starboard and port sides of the airlock, respectively. These bays are immediately forward of the aft X_o 576 bulkhead. Although Bays 3A and 3B are smaller than Bays 1 and 2, a greater number of items were recovered from Bays 3A and 3B than from the forward bays. The recovered debris (mostly avionics boxes) were remarkably consistent in shape and subsequently estimated ballistic number, so the clusters were assumed to be adequate for a relative assessment (figure 2.4-21).

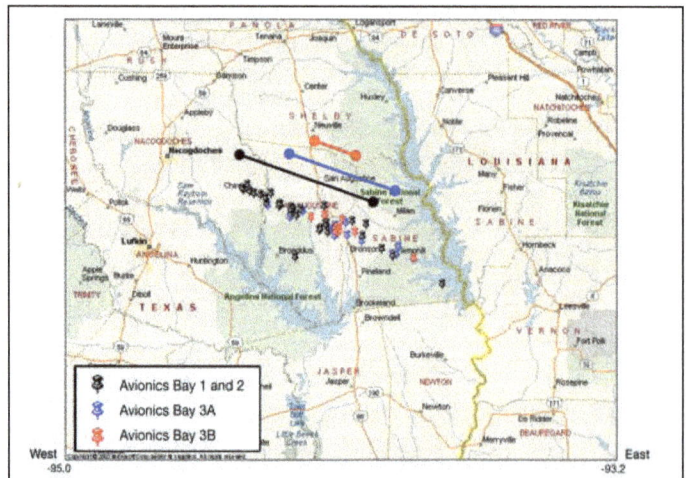

Figure 2.4-21. *Debris field of the avionics bays.*

Debris from Bays 1 and 2 was recovered west of Bays 3A and 3B. This order matches the CM bulkhead order, supporting that the front (X_{cm} 200) bulkhead departed before the aft (X_o 576) bulkhead.

Sequencing based on cluster, video, and ballistic analysis

Figure 2.4-22 shows the relationships of some of the addressed forebody structures. The CM bulkheads are included in CM structure, and the X_0 378 bulkhead is included in the FF structure. The percentages listed in the figure refer to percentage of recovered debris, not percentage of the original intact area.

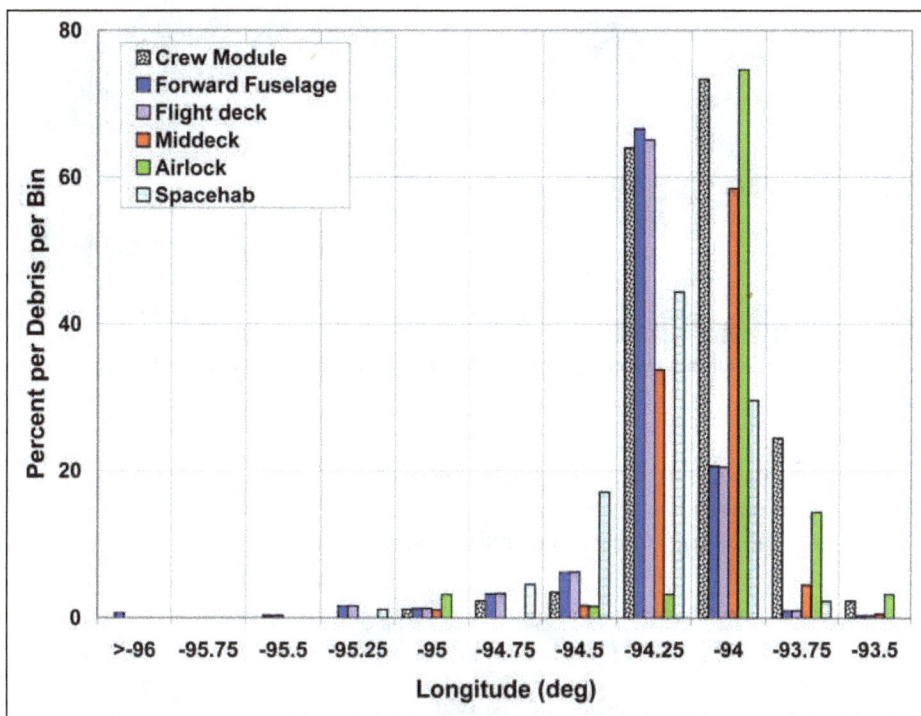

Figure 2.4-22. *Histogram of the forebody structural components, plus SPACEHAB.*

To summarize the video and debris field findings, at the CE the CM was contained within the FF structure. The CM began to depressurize through a series of small breaches, with small amounts of FF debris being shed. Eventually, the FF structure failed, and the CM itself failed within a few seconds afterwards. The middeck and forward bulkhead (X_{cm} 200) of the CM departed, while the airlock, flight deck, and aft bulkhead (X_0 576) remained together for a short period longer until all elements separated and the TD occurred.

2.4.4 Structural analysis

This section discusses the breakup of the forebody based on structural analysis, using supporting evidence from the video, ballistic, and debris cluster analyses. See Section 2.2 for a more detailed discussion of the separation of the forebody from the orbiter. This analysis specifically discusses events that are related to the forebody at the CE and ending at the TD.

2.4.4.1 *Events at the Catastrophic Event*

General condition of forebody recovered debris

The recovered FF components are predominantly skin/stringer segments. Most components exhibited mechanical overload as the primary failure mechanism. Roughly 40% of the FF was recovered with no difference in damage levels comparing left to right or upper to lower. Two recovered RCC components, the nose cap and the chin panel (figure 2.4-23), show evidence of mechanical breakup with low thermal damage. It should be noted that the nose cap, based on field reports, apparently hit a tree before hitting the ground. The presence of intact Koropon primer on many FF components indicates that significant heating did not occur. Temperatures above 400°F (204°C) degrade the primer's appearance, and high temperatures

(>900°F) (486°C) will completely ablate it. The presence of Koropon indicates that the breakup of these elements was caused by mechanical overload rather than thermal effects.

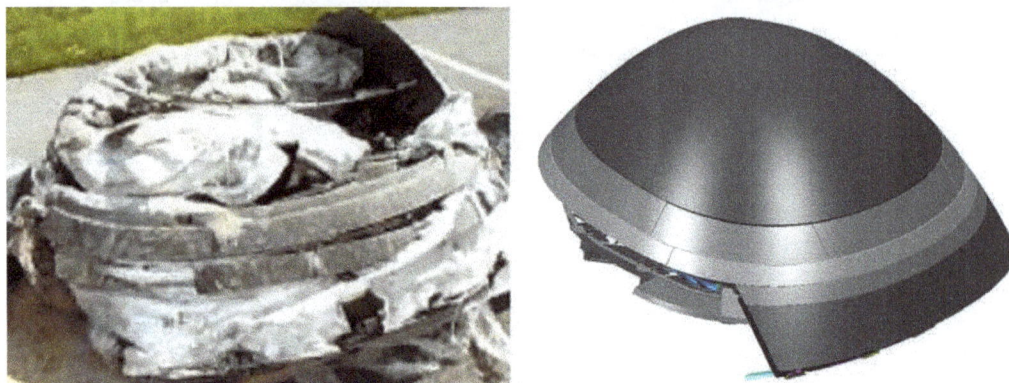

Figure 2.4-23. *Nose cap* Columbia *Reconstruction Database debris item no. 1114 vs. the original nose cap with chin panel.*

Some large pieces of the forward RCS were recovered (figure 2.4-24). These exhibit evidence of mechanical overload as the primary failure mechanism. Heating did not appear to play a significant role in the component degradation and appears to have occurred during or subsequent to the mechanical breakup.

Figure 2.4-24. *Two large forward Reaction Control System debris pieces,* Columbia *Reconstruction Database debris item nos. 792* (left) *and 82061* (right).

Very few CM skin pieces were recovered. Most skin components were identified as portions of the thicker sections of the aft and forward bulkheads; the pieces all exhibit heavy thermal erosion. Several skin pieces that were identified as part of the CM center/bottom strip ("keel") were recovered. These pieces exhibit some mechanical breakup along with heat erosion within this thicker strip.

More than 65% of the middeck floor panels were recovered with paint and Koropon primer still intact, indicating that they were exposed to low thermal erosion.

Small numbers of the flight deck floor structural pieces were recovered; all recovered pieces exhibit heavy thermal erosion.

Crew module attach fittings (x-links, y-links, z-link)

All four main CM/FF attach links were recovered. In Section 2.2, it was concluded that the attach fittings, which are known as the x-links, y-links, and z-link, stayed with the forebody. These fittings attach the CM to the FF. The x-links also bridge to midbody structure. For a more detailed discussion of these attach fittings, refer to Section 2.2.4.

X-LINKS

Both the port and the starboard x-links (figure 2.4-25) were recovered nearly intact with evidence of high heating. The titanium fittings on both links experienced significant thermal exposure/melting, predominantly on the upper surfaces. Additionally, the starboard side fitting experienced significantly greater heating and erosion than the portside.

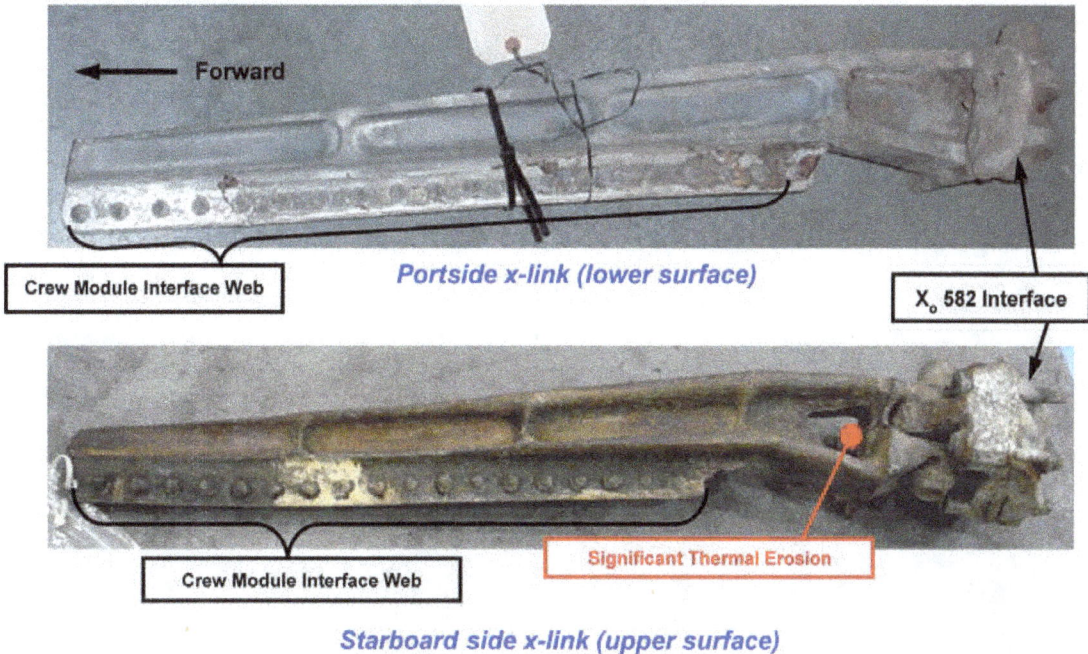

Figure 2.4-25. *Crew module x-links,* Columbia *Reconstruction Database debris item nos. 1678* (top) *and 1765* (bottom).

The titanium x-link components did not fail. The attach area on the CM side of the x-link is stronger than the FF side since it is reinforced by the flight deck floor and the X_o 576 bulkhead. The failures occurred at the weaker X_o 582 ring frame bulkhead connection points. The port x-link experienced a mechanical lug failure at the X_o 582 ring frame bulkhead interface, while the starboard side fittings pulled through the X_o 582 ring frame bulkhead at the CE. The starboard x-link retained some of the sill and X_o 582 ring frame bulkhead while the portside did not. Furthermore, it was concluded that at the CE, the forebody rotated left and pitched down, separating from the midbody (see Section 2.2.4).

Heat damage patterns on the webs of both x-links indicates that, at some point, there was hot gas flow from aft to forward above both x-links. The starboard x-link has more damage than the port x-link. Thermal analysis (see Section 2.1.6.7) of the x-links shows that entry heating alone was not capable of causing such heavy erosion. Additionally, because the surrounding structure was aluminum with a much lower melting temperature, entry heating alone would have resulted in the weakening of the surrounding structure first and release of the x-links. Section 2.1.7 discusses the other thermal mechanisms that were likely involved, shock-shock interactions and combustion. Both of these mechanisms are possible with the forebody geometry as understood from the debris.

Debris item no. 2436 is a piece of FF skin panel outside of the starboard x-link (figure 2.4-26). It was subjected to heat erosion mainly at the aft edge (along the X_o 582 ring frame bulkhead) (figure 2.4-27). The pattern of deposited molten aluminum on the inboard side indicates that it was subjected to thermal flow aft to forward over the x-link (see arrows, figure 2.4-27).

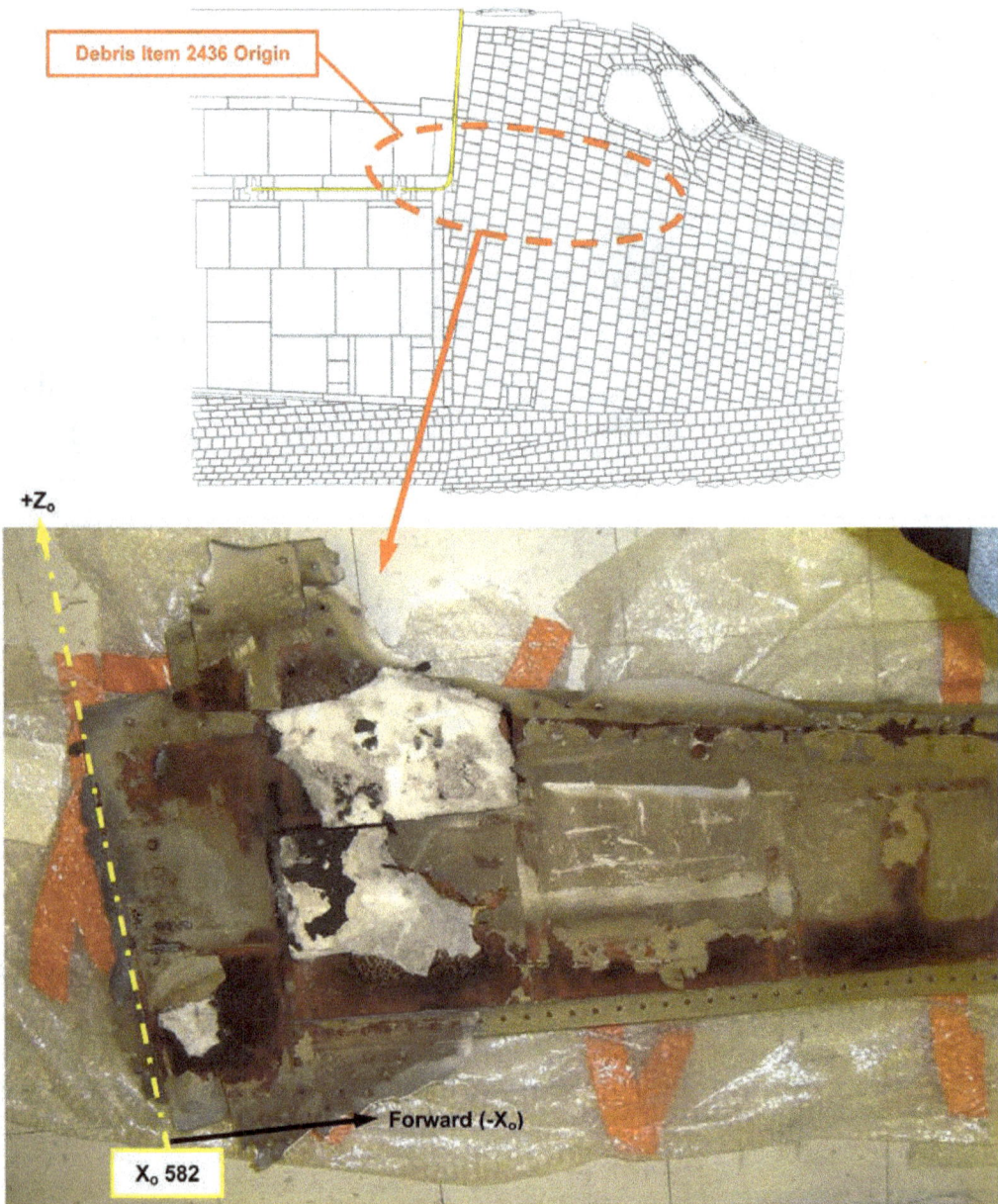

Figure 2.4-26. *Starboard-side skin, view looking inboard,* Columbia *Reconstruction Database debris item no. 2436.*

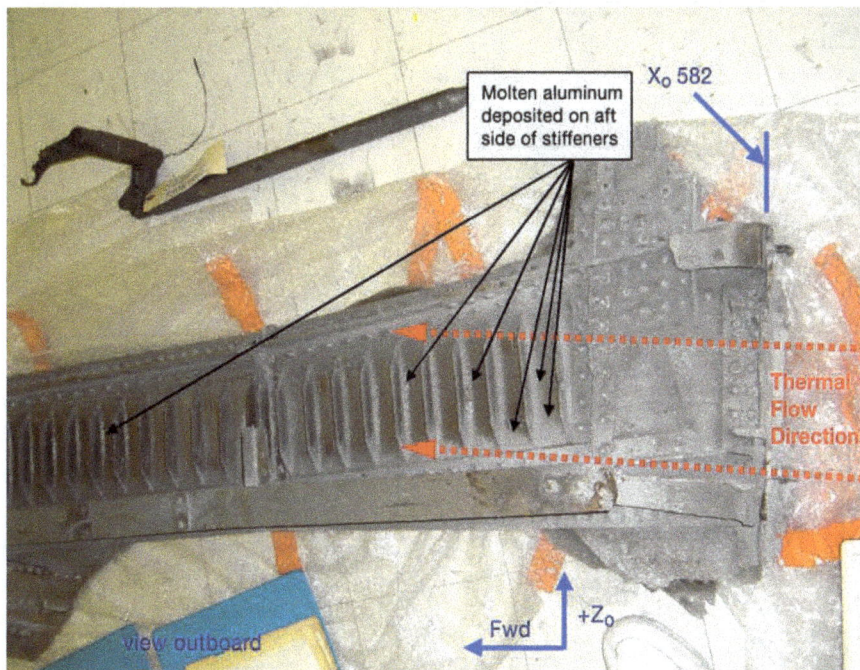

Figure 2.4-27. *Inboard side of the starboard-side skin,* Columbia *Reconstruction Database debris item no. 2436.*

This debris shows high heat erosion at the aft end while the remaining edges failed mechanically. The pattern of molten aluminum deposition indicates that the hot gas flowed mainly from aft to forward. If the thermal event happened after this panel broke away from the FF shell, all edges should experience a similar level of thermal damage, and the deposition of molten aluminum on the inside surface of the skin should have a random pattern instead of a directional pattern (figure 2.4-27). This suggests that this piece of FF skin was in place when the thermal event occurred. It is not known when this thermal event occurred. Since the forebody was rotating, the aft portion of the forebody may have periodically been presented to thermal flow for brief periods between the CE and the CMCE. The thermal event may also have happened around the CMCE, when the FF pulled away and exposed the area to thermal flow.

Y-LINKS

The y-links attach the aft bulkhead of the CM (X_o 576) to the X_o 582 ring frame bulkhead at the lower central portion of the bulkhead. It is possible that the two y-links remained intact at the CE, since most of the X_o 582 ring frame bulkhead is believed to have stayed with the FF. After the CE, any Y-direction movement of the CM relative to the X_o 582 ring frame bulkhead would exert high tension or compression loads on these links. Failure by tension is evident on the recovered debris of these links. The portside y-link shows that it also was softened by thermal exposure along with the tension failure. This portion protrudes beyond the X_o 582 ring frame bulkhead, which would have been exposed periodically to hot gas flow following the CE as the forebody rotated (figures 2.4-28 and 2.4-29).

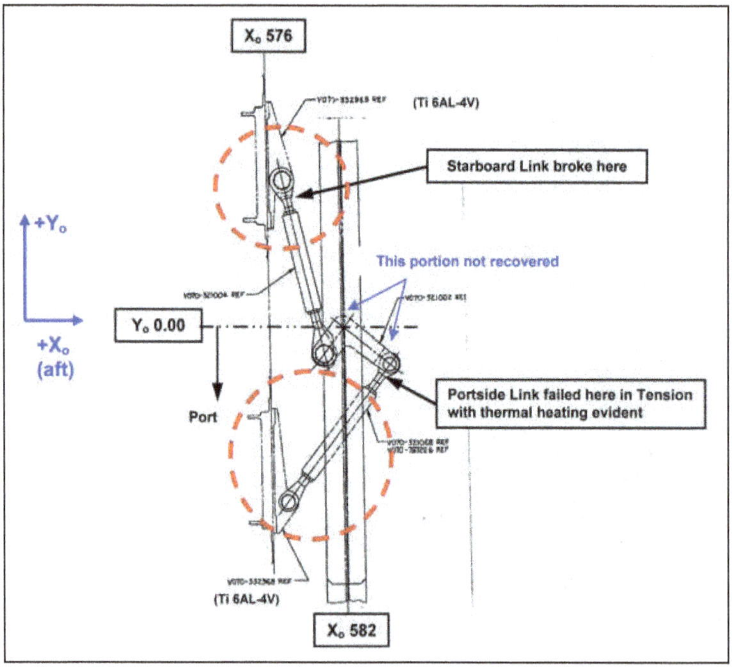

Figure 2.4-28. *View looking from the top at the y-links.*

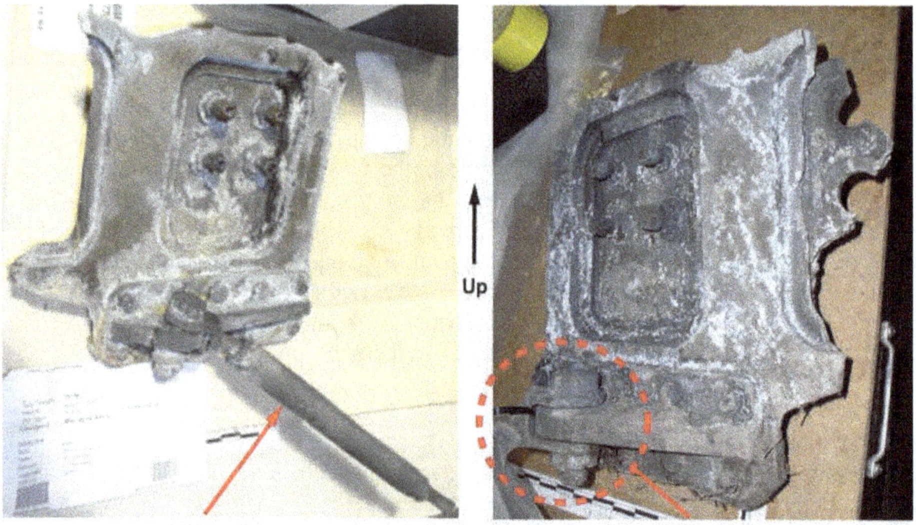

Portside y-link   **Starboard side y-link**

Figure 2.4-29. *The y-link debris, still attached to the mounting on the X$_o$ 576 bulkhead,* Columbia *Reconstruction Database debris item nos. 72770 (left) and 9280 (right).*

Z-LINK

The z-link failed at the attach point to the X$_{cm}$ 200 bulkhead (figure 2.4-30). The joint failed by a combination of fastener tensile failure and fastener insert pullout. Insert pullout requires thermal weakening of parent material because, by design, the insert is stronger than the fastener (under normal temperatures, the fastener should fail before the insert pulls out). Heating may have been occurring as the forebody rotated, allowing hot gas into the plenum between the CM and the FF. Failure resulted at the CMCE.

Alternately, after the FF separated, the X$_o$ 378 bulkhead may have remained in place attached to the X$_{cm}$ 200 bulkhead, connected by the z-link, and exposed to entry heating flow. The two bulkhead debris fields support this conjecture because the X$_o$ 378 bulkhead and the X$_{cm}$ 200 bulkhead debris fields overlap, with the X$_o$ 378 debris field being slightly west of the X$_{cm}$ 200 debris field. No ballistic analysis was done on either bulkhead elements to support or refute this conjecture.

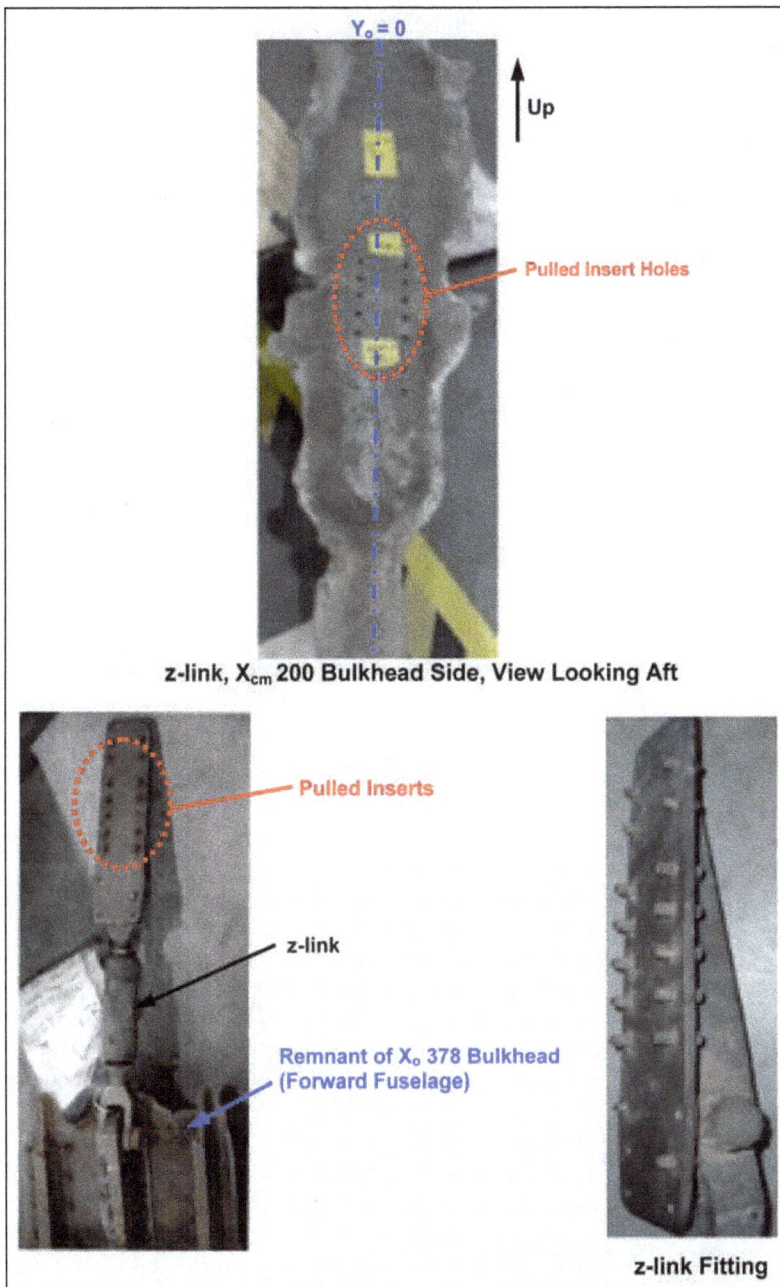

Figure 2.4-30. *The z-link debris photographs,* Columbia *Reconstruction Database debris item no. 53828* (top).

z-link, X_{cm} 200 Bulkhead Side, View Looking Aft

z-link Fitting

Forward fuselage and crew module interaction at the catastrophic event

The cabin depressurization analysis (see Section 2.3) led to the conclusion that the cabin began to depressurize NET the CE (GMT 14:00:18) and NLT GMT 14:00:35; and probably depressurized through several small breaches. There is limited debris evidence to pinpoint locations for these breaches.

Items that originated from inside the CM and were recovered in the western portion of the debris field were reviewed to determine a potential breach area. These items included multiple STS-107 mission patches. Most of the mission patches were stowed in a middeck sub-floor stowage volume called Volume E. In-flight access to Volume E was not possible in the *Columbia* cabin configuration, so the crew could not have accessed the patches and stowed them elsewhere. Review of the stowage configuration documents for this compartment revealed that the patches were stowed near the bottom of Volume E. Therefore, the bottom portion of the Volume E structure had to be compromised to release the patches.

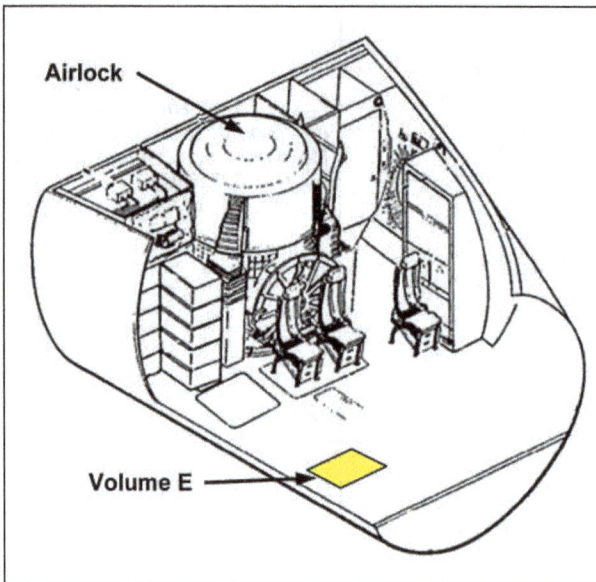

Figure 2.4-31. *View of middeck floor and Volume E, looking down and aft.*

Volume E is a container with an opening on the top. The top part of this volume is a machined aluminum upper frame with a door, a hinge, and latches to cover the opening. Four sides and bottom panels are made of aluminum honeycomb core and aluminum facesheets. The upper edges of the side panels are attached to the upper frame vertical flanges by rivets. As illustrated by figure 2.4-31, the door of Volume E is at the level of the middeck floor. The upper frame of the box is attached to the floor beams by 18 Milson bolts and receptacles. The bottom of Volume E is tapered to match the curvature of the lower portion of the CM pressure shell.

Detailed inspection of Volume E debris indicates that it was subjected to mechanical damage prior to thermal damage. The outer perimeter "picture frame" of the bottom panel was recovered and shows evidence of impact from below. Impact from the CM skin below Volume E would push the aft/outboard corner of the volume upward and then split the edge of the upper frame, as seen in the debris (figures 2.4-32 and 2.4-33).

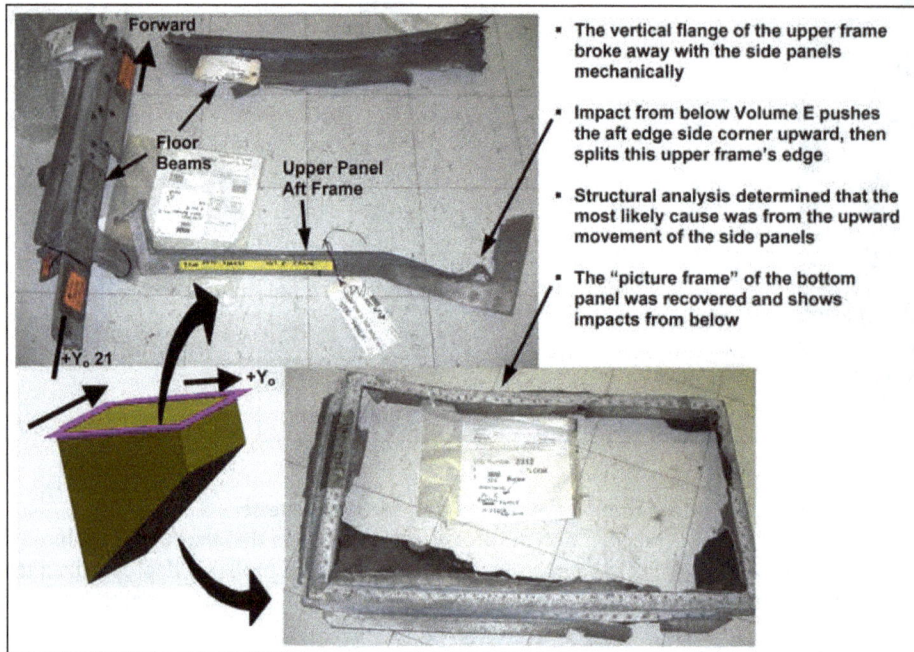

- The vertical flange of the upper frame broke away with the side panels mechanically

- Impact from below Volume E pushes the aft edge side corner upward, then splits this upper frame's edge

- Structural analysis determined that the most likely cause was from the upward movement of the side panels

- The "picture frame" of the bottom panel was recovered and shows impacts from below

Figure 2.4-32. *Volume E debris damage.*

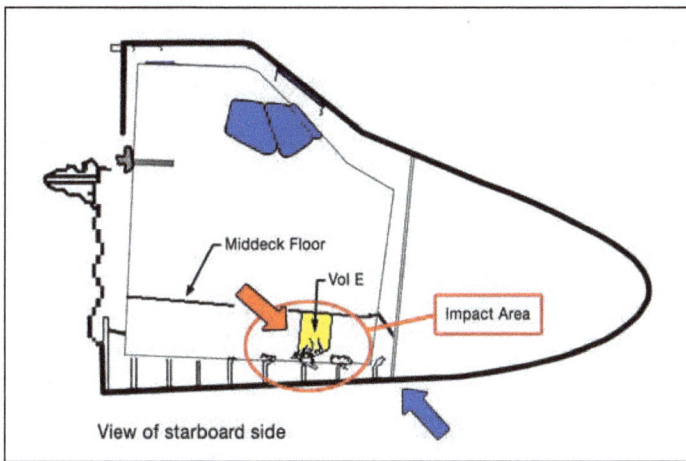

Figure 2.4-33. *Scenario showing how the crew module pressure vessel could impact the forward fuselage, and the middeck Volume E could impact the crew module pressure vessel, with resultant damage.*

Sudden changes to the rotation of the forebody complex probably occurred at the CE because of the asymmetric release from the mid-fuselage. Due to c.g. locations of the FF and CM, centrifugal forces would tend to pull one away from the other as the forebody rotated. The large portion of the X_o 582 ring frame bulkhead was still attached to the FF. Thus, it and the other remaining attached linkages may have prevented the CM extraction from the FF.

It is likely that at or shortly after the CE, the FF structure impacted the CM skin below Volume E and, in turn, caused an impact between the CM skin and Volume E. The impact was severe enough to crack open the Volume E box, possibly near the lower portion, spilling some of its contents, including the patches. The patches then escaped the CM during CM depressurization.

The impacts under Volume E could possibly have breached the CM skin, providing a local vent path for the stowage volume contents such as the crew patch.

More impacts likely occurred as the forebody continued to rotate. From the CE to the CMCE, it is likely that the CM swung back and forth inside the FF, with both the CM and the FF experiencing impacts at multiple locations. The FF structures would have received damage, possibly creating paths for thermal inflow, as the forebody rotated. Impacts from the relative motion between the CM and the FF would damage the forward bulkheads (X_o 378 and X_{cm} 200), the fuselage frames, and the thinner areas of CM aluminum skin, star tracker well area, etc., depending on the attitude and the rotational movement. Recovered crew equipment debris also came from the forward lockers on the portside, not far from the star tracker wells, lending credence to this. Frame and bulkhead damage would destabilize the FF structures, perhaps popping open numerous items, including the star tracker panels, the TPS around the forward windows, the forward RCS module, the nose gear doors, and the side hatch outer layer. Review of forebody items found in the debris field prior to the main debris field support this conclusion, as portions of the nose gear, star tracker, and multiple tiles and portions of TPS made up this early released debris (figure 2.4-34).

Between the CE and the CMCE, it is not clear when the remaining links (z-link, four side links, and four side hatch links) failed. Obviously, sufficient integrity was maintained to hold the CM inside the FF.

Mechanical and thermal degradation of the crew module pressure vessel between the Catastrophic Event and the Crew Module Catastrophic Event

The airlock depressurized rapidly following the CE with the departure of the TAA (see Section 2.2.4). It might be speculated that rapid depressurization of the airlock with the interior of the CM still pressurized might implode the airlock structure. However, the airlock is intended for depressurization during spacewalk activities and its interfaces are reinforced. Empirical review of airlock structural capacity shows that implosion of the airlock is an unlikely scenario.

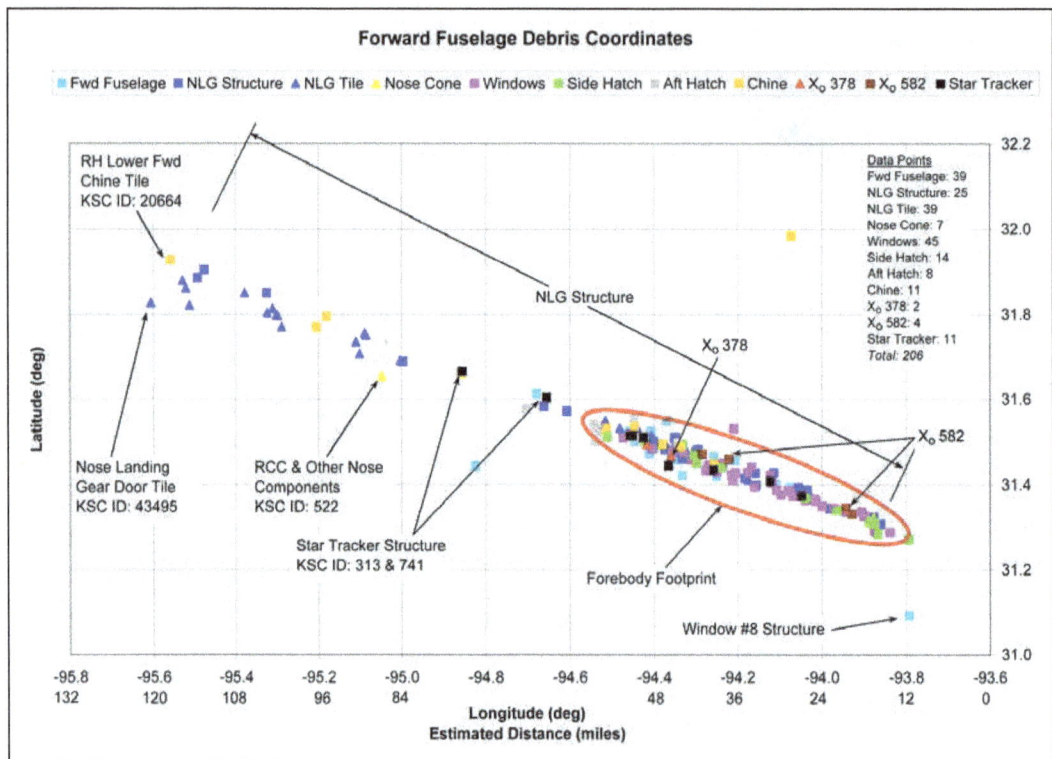

Figure 2.4-34. *Forward fuselage debris field showing tile, nose landing gear door structure, and star tracker door structure west of the main forebody debris field.*

Other warping as a result of the stresses of the CE cannot be ruled out. However, impacts as a result of the CM moving inside the FF appear to be the best candidate for mechanical breaches because they are the best supported by the debris field.

Thermal exposure of the CM pressure vessel may have weakened certain areas resulting in mechanical failure. This would not have happened immediately at the CE but would have taken time to develop as the forebody rotated, periodically exposing unprotected segments of the forebody.

The CM aft bulkhead was a clear immediate candidate for thermal breach since, unlike the rest of the CM, it was not protected by the FF and accompanying TPS. However, the bulkhead was extremely heavy, because it was made of aluminum waffle with additional reinforcing beams, and would act as a good heat sink. Debris field evidence does not support significant thermal erosion of the CM aft bulkhead prior to CM breakup.

There are gaps between the X_o 576 bulkhead and the X_o 582 ring frame bulkhead and directly over the x-links that would have allowed hot gas entry into the plenum between the forebody and the CM when rotation resulted in the aft forebody being presented to the directional thermal flow. Mechanical breaches resulting from the impacts of the CM and the FF could also have allowed thermal flow into the plenum if the breach were presented to the velocity vector.

The CM pressure vessel skin was designed with a main cone shape that has low thickness and minimized integral stringers to save weight (figures 2.4-35 and 2.4-36). Many large areas of the cone shape have uniform low thickness—as thin as 0.039 in. Skin thickness was controlled tightly by a chemical milling process during manufacturing. Because of low and uniform thickness, a large area of skin can be heated up quickly and uniformly. This can result in rapid thermal failure of the skin panel, and heat erosion can propagate rapidly.

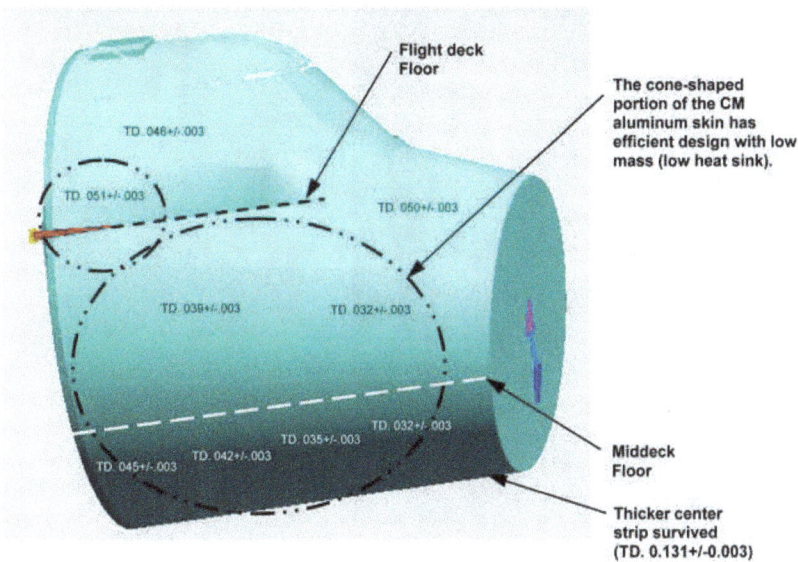

Figure 2.4-35. *Crew module pressure vessel skin lightweight design.* The thickness dimension (TD) is represented in inches.

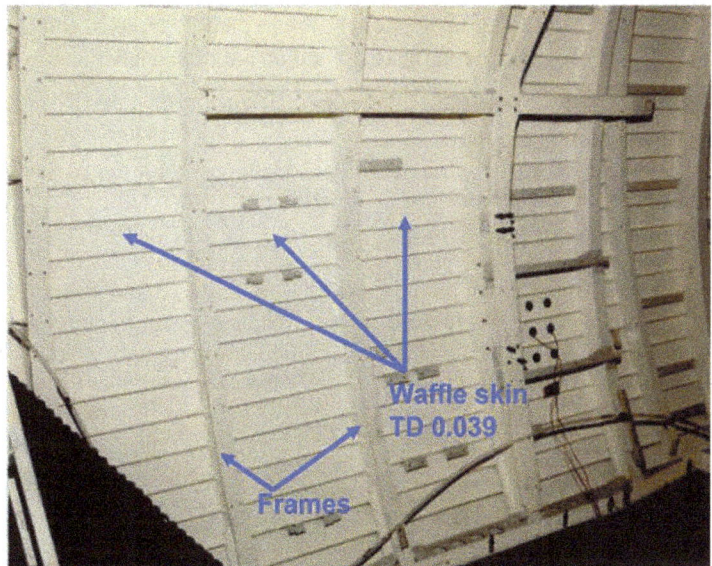

Figure 2.4-36. *Typical crew module pressure vessel skin in middeck, looking outboard.*

Very little CM skin structure survived. It is very thin, and probably melted quickly after breakup when exposed to entry heating. Only a relatively thicker region of skin strip (0.131-in. thick) along the bottom at the centerline was recovered in multiple pieces (figure 2.4-37).

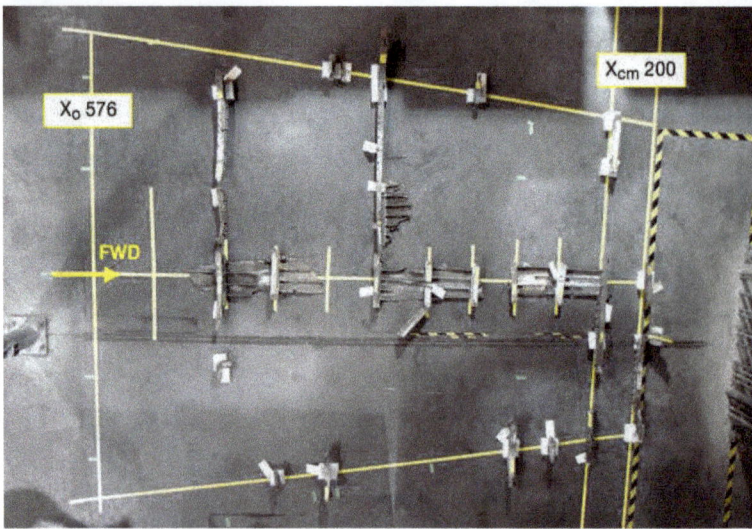

Figure 2.4-37. *Recovered crew module pressure vessel base skin.*

It is possible that areas of the CM skin were weakened or softened uniformly by thermal flow into the plenum between the CE and the CMCE. However, many recovered FF skin areas exhibited intact primer on the inner surfaces. This implies that most of the forebody TPS still remained to provide thermal protection on the outboard side of the FF panels before the FF breakup, and that the loss of CM pressure vessel skin was a result of exposure during the breakup after the CMCE.

2.4.4.2 *Crew Module Catastrophic Event/forebody breakup*

Forward fuselage failure

It is unclear where the initial failure of the FF occurred. Loads due to deceleration of the forebody were increasing. Although much of the forebody was protected by the TPS of the FF, periodic exposure to heating was increasing as well.

The "arrowhead" skin/TPS panel between Windows 3 and 4 (debris item no. 65049) (figures 2.4-38 and 2.4-39) on the front of the forebody was recovered near the middle of the forebody debris field. It appears to have been torn on the portside and pulled through on the starboard side by mechanical loading, probably initiated by a relative movement between the port Windows 1 through 3 and the starboard Windows 4 through 6.

Figure 2.4-38. *Forward fuselage arrowhead panel and the forward two windows, facing in.*

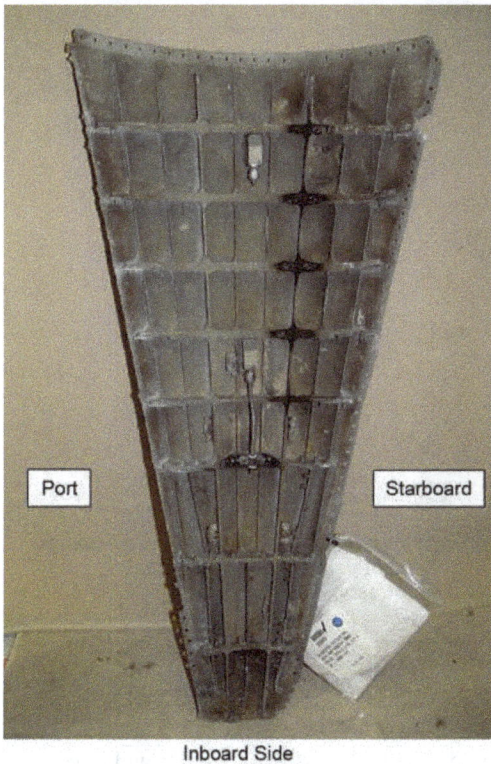

Figure 2.4-39. *Low-heat damage on the forward fuselage arrowhead panel, facing out,* Columbia *Reconstruction Database debris item no. 65049.*

Port

Starboard

Inboard Side

The departure of the FF Windows 1 through 6 and the FF arrowhead panel would likely trigger the departure of the rest of the FF canopy by aerodynamic forces. This may have been the initiating event of the CMCE, or it might have been simply one event during the breakup. Without the upper FF half, the lower half would immediately peel away. This is supported by the video image, which shows what appear to be two large and symmetrically sized objects separating from the forebody at the CMCE.

Figures 2.4-40 and 2.4-41 are of the FF skin debris items that lie below the CM. Many panels have little molten metal deposition on the inside surfaces. Broken edges show little thermal damage, and some still have intact primer. This indicates that the plenum behind these areas was not exposed to long durations of thermal flow. Some skin pieces even had stringers that were removed mechanically. Most of the recovered skin debris appears to have mechanically fractured edges. Many forward RCS structural items were recovered; it appears that the forward RCS was pulled away mechanically and broke up into pieces with low thermal effects. The conclusion was that it is likely that the FF skin panel peeled away and broke up quickly.

Figure 2.4-40. *Lower forward fuselage skin and frame debris.*

Figure 2.4-41. *Lower forward fuselage skin panel near external tank attachment.*

Crew module breakup

Unlike the FF, very little CM pressure vessel skin was recovered. The lack of thermal effects and interior deposition on the FF debris indicates that it was no longer with the CM when the pressure vessel skin melted. Once the FF separated, the CM would become highly susceptible to thermal heating, especially in the areas of low skin thickness. Multiple additional thermal breaches probably appeared on the CM skin within a few seconds because of the low heat sink and uniform thickness of the CM pressure vessel skin along the CM sides and bottom.

Failure modes that were assessed on the CM pressure vessel and secondary structural components suggest that fractures occurred subsequent to elevated temperature exposure (corresponding to a significant reduction of material properties). This clearly suggests that following FF separation, breakup of the CM structure occurred as a consequence of combined aerothermal heating and aerodynamic loading.

Middeck and forward bulkhead

Figure 2.4-42 shows recovered CM forward bulkhead (X_{cm} 200) debris on a grid that is the size of the original bulkhead. As described in the debris field cluster analysis, the CM forward bulkhead debris was recovered west of the aft bulkhead debris, and the middeck starboard debris was recovered west of the port debris. This suggests that the CM middeck starboard/forward was shed first and that the starboard and forward CM areas were exposed to higher heating first and/or faced the velocity vector at the time of the CM major breakup. The main CM breakup involved the departure of the middeck area, including the middeck floor and everything that was attached to that floor. Anything not firmly attached to the flight deck floor also came out with the middeck. Most equipment that was above and below the middeck floor was supported by the middeck floor; avionics bays are also supported by the side skins and bulkheads.

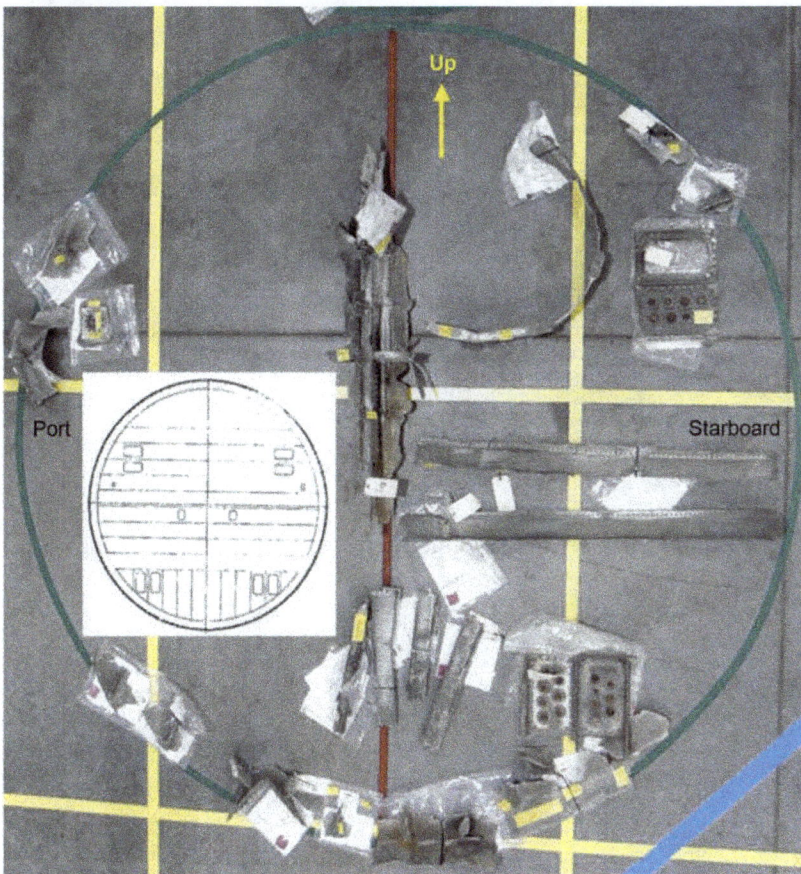

Figure 2.4-42. *View looking forward of the recovered crew module X_cm 200 bulkhead debris.*

Structurally, the flight deck section is connected to the middeck section and the lower equipment bay is mainly connected by the CM side skin, avionics bay partitions, and the aft and forward bulkheads. The side skin and the forward bulkhead could be easily damaged by aerodynamic heating after the CM lost the protection from the FF shell. Therefore, as soon as the middeck side skin and forward bulkhead were compromised structurally, the lower half of the CM could have swung away from the flight deck portion by the effects of aerodynamic drag. It is likely that the whole middeck floor assembly came out together with the lower equipment bay. The hinge line appears to be at or near where the middeck floor attaches to the aft bulkhead. This motion would expand the middeck compartment (similar to opening a clam shell) and release all items attached to the middeck floor as well as other items stowed beneath that floor. Because the middeck floor disintegrated into many smaller parts without being significantly heated, it appears that the middeck floor failed as a result of structural loading. This may help to explain why more than 65% of the middeck floor was recovered without significant thermal erosion (figures 2.4-43 and 2.4-44).

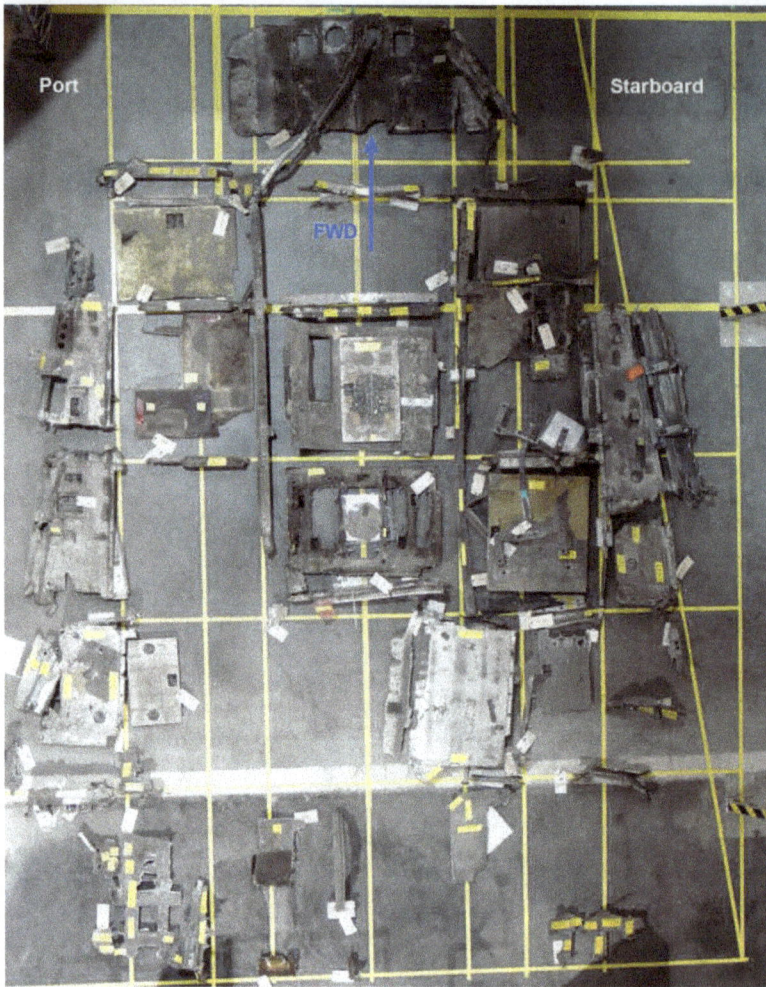

Figure 2.4-43. *Top view of the crew module middeck floor debris.*

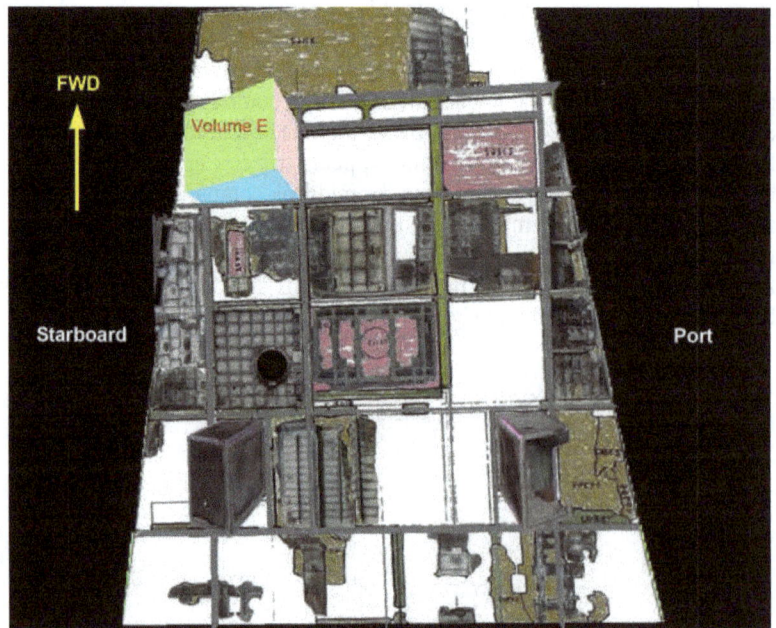

Figure 2.4-44. *Bottom view of the recovered middeck floor items from virtual reconstruction.*

The middeck floor and the lower equipment bay quickly disintegrated, with middeck floor panels, crew escape pole, MAR, and other crew equipment items (window shade bag, middeck lockers, sub-floor components, MADS/OEX recorder, etc.) departing quickly from the CM.

Flight deck and aft X_o 576 bulkhead

The CM debris field suggests that after the forward bulkhead and the middeck floor departed, it was followed by the airlock, the flight deck, and the aft bulkhead. The flight deck floor debris exhibited much more thermal damage than the middeck floor debris.

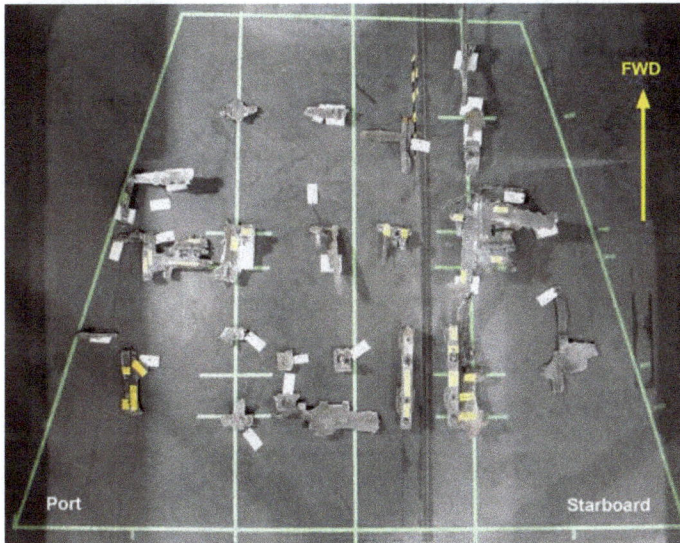

Figure 2.4-45. *Top view of the crew module flight deck debris.*

From the thermally eroded state of the flight deck debris, it appears that the flight deck stayed nearly intact for a period of time following the departure of the middeck area. This explains why so little of the flight deck floor was re-covered (figure 2.4-45). Based on debris field evidence, the internal airlock, which was supported by the MAP (which is part of the X_o 576 bulkhead), likely stayed together with the bulkhead and the flight deck.

Significant mechanical damage was noted on the starboard aft panels of the flight deck. These panels were recov-ered in much smaller segments than other flight deck panels. This suggests that the starboard side of the flight deck near the aft bulkhead (near the starboard x-link) experienced a more dynamic failure than the portside. This may have resulted from structural degradation that occurred when the forebody separated from the midbody at the CE in this location. It is also consistent with a starboard-to-port failure for the middeck.

In general, heavy portions of the X_o 576 bulkhead (including the MAP) survived the entry heating (figure 2.4-46). This is possibly due to the high heat absorption property of the aluminum bulkhead.

Some "T"-section stiffeners survived with little heat damage, possibly because of the early departure of these elements before the high thermal event that consumed the bulkhead.

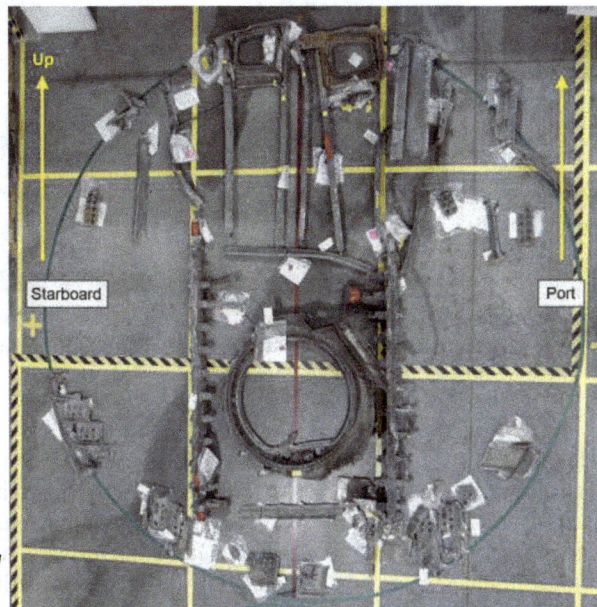

Figure 2.4-46. *View looking aft of the recovered X_o 576 bulkhead debris.*

Crew module crew equipment

The CM contains many items that were installed to facilitate the space shuttle crew's on-orbit operations. This equipment is generally termed "crew equipment" and includes the crew seats, middeck stowage lockers, the MAR, sleep stations, the galley, the WCS, the ergometer, the Crew Escape System (CES) pole, crew worn equipment, and loose equipment that was stowed in various locations. Attention was focused on the MAR and the CES pole because they have substantial attachments to the CM structure. Generally, the debris from the sleep stations, galley, WCS, ergometer, lockers, and loose equipment was highly fragmented and did not provide significant insight into the CM breakup. Analysis on those items was limited to identification and ballistics analysis on a few select items. Because the SPACEHAB module contained loose equipment that was stowed in middeck lockers, the SPACEHAB and the CM debris footprints both contained loose equipment and locker structure debris. Therefore, loose equipment and locker structure items were excluded from analysis on the SPACEHAB and CM debris footprints.

Recovery locations of suit and seat components indicate that the middeck crew members separated from the CM before the flight deck crew members. Additionally, flight deck seats experienced higher heating than middeck seats. These findings support the debris field cluster analysis conclusion that the middeck broke up before the flight deck broke up (see Sections 3.1 and 3.2).

Middeck accommodations rack

The MAR is pinned to the port wall and the middeck floor forward of the side hatch and aft of the galley (figure 2.4-47) on the middeck. The MAR spans from middeck floor to ceiling.

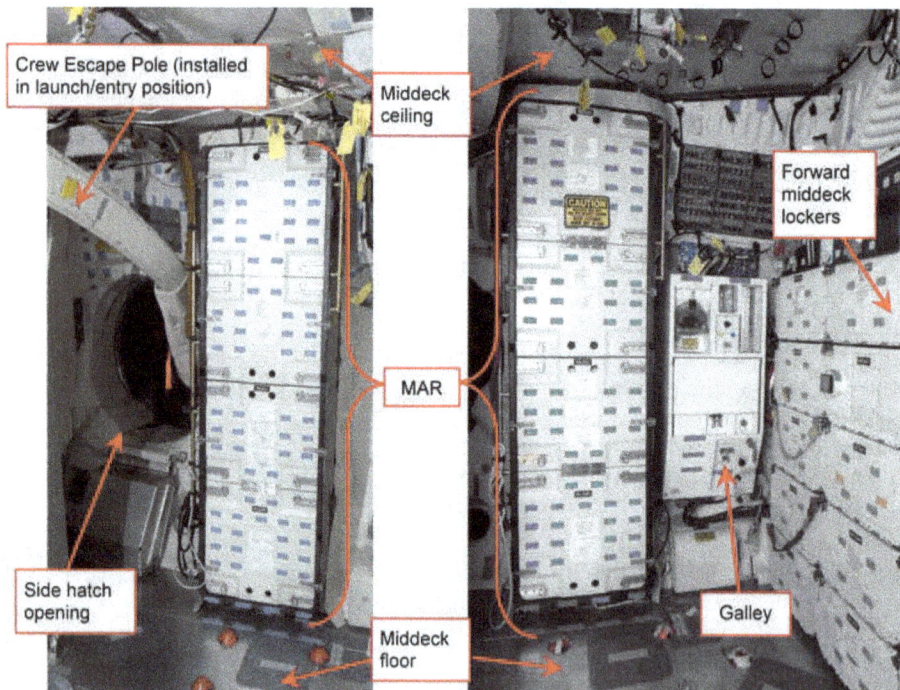

Figure 2.4-47. *Middeck accommodations rack.* [Picture from a shuttle training mockup in the JSC Space Vehicle Mockup Facility]

The MAR structure was made of a carbon fiber/epoxy composite with an aluminum honeycomb core. Two doors, which were hinged in the middle, face inboard into the CM, and were each held closed by eight locking spring latches. The MAR had two aluminum handrails that were attached to facilitate crew restraint and mobility. One handrail was mounted on the forward edge of the MAR and spanned from the top to the middle of the MAR. The other handrail was mounted on the aft edge of the MAR and spanned approximately three-quarters of the height of the MAR.

The MAR weighed 105 lbs. empty and could accommodate 12 ft^3 and over 200 lbs. of stowed items. Shelves could be bolted inside the compartment to subdivide the compartment. Stowed items were packed either in cargo transfer bags (CTBs) or foam cutouts to protect against damage.

For STS-107, the MAR contained payload general support computers (PGSCs), cables, a printer, the vacuum cleaner and attachments, medical kits, shuttle urine pretreat assembly (SUPA) hoses, and one can of LiOH. The total weight of the MAR (structure and return contents) was approximately 200 lbs.

Approximately 75% of the MAR structure was recovered, mostly intact. The blue areas in figure 2.4-48 represent the MAR structure that was recovered. The portions that were not recovered included the upper one-fourth of the MAR structure (including the upper attachment bracket), the top half of the upper door, a portion of the bottom half of the lower door, and the bottom surface of the MAR structure (including floor attachment brackets).

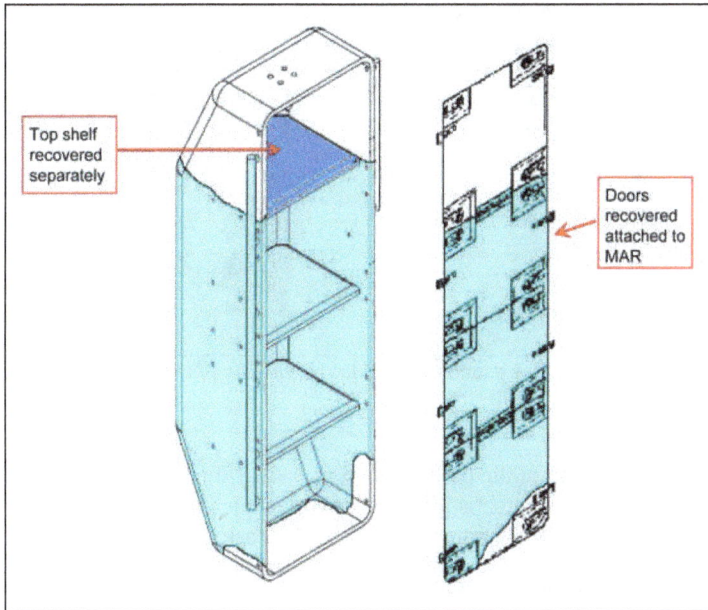

Figure 2.4-48. *Recovered middeck accommodations rack structure* (shown in blue).

The MAR was recovered with the middle and bottom shelves still attached. The top shelf was recovered separately. The lower door and the bottom half of the upper door were recovered with the MAR structure, as were the contents of the compartment between the middle and bottom shelves (CTBs with the printer and the vacuum cleaner (figure 2.4-49)). The contents of the compartment between the middle and top shelves and the contents of the compartment below the bottom shelf were recovered separately from the MAR structure and were highly fragmented.

Figure 2.4-49. *Main middeck accommodations rack structure, as found.*

The lack of impact witness marks on the MAR aft handle and the aft wall upper surface indicates that the MAR did not impact the CES pole, which is mounted just inches aft of the MAR. The lack of impact witness marks on the MAR starboard surfaces indicates that the MAR did not impact the seat 5 structure, which is mounted inches starboard of the MAR. Ballistic analysis indicates that the MAR separated from the CM shortly after the CMCE. Based on the failure of the fasteners securing the MAR attachment bracket to the middeck floor, it is concluded that the MAR separated from the middeck floor mostly to fully intact. These conclusions support the general findings that the middeck breakup was rapid and expansive, with very little interaction between major structures, and occurred at the onset of the CMCE.

Crew escape system pole

The purpose of the crew escape pole is to guide the crew member under the left wing when bailing out of the orbiter during controlled, gliding flight. For launch and landing, the escape pole is located on the middeck and is mounted to the starboard (right) ceiling and port (left) wall just forward of the side hatch (figure 2.4-50). During on-orbit operations, the escape pole is removed and stowed against the middeck ceiling.

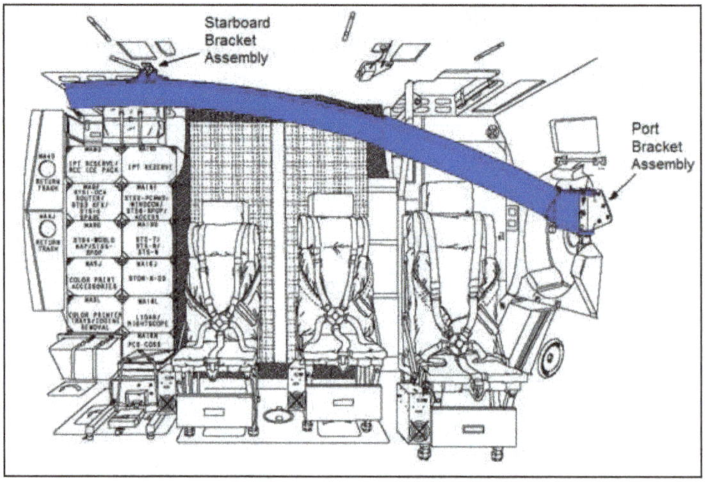

The escape pole consisted primarily of a curved, spring-loaded telescoping aluminum cylinder and steel spring. It was contained within an aluminum housing. The complete assembly weighed 267 lbs. A magazine con-

Figure 2.4-50. *Crew escape pole in launch/landing position, middeck, looking aft.*

taining eight crew lanyards was attached to the port end of the pole housing near the side hatch tunnel (figure 2.4-51). In the event of a bailout, the orbiter side hatch is jettisoned pyrotechnically, the pole is deployed, and each crew member extends a D-ring and bridle from his/her parachute pack and attaches it to a

snap hook on the outermost lanyard. As the crew member egresses the orbiter, the pole directs him/her beneath and away from the vehicle. Upon bailout, the forces on the lanyard and bridle initiate automatic parachute deployment.

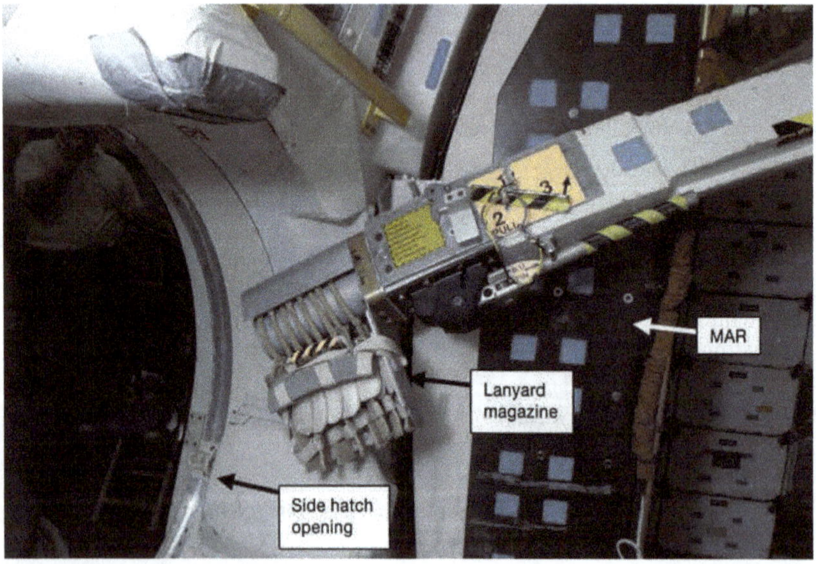

The SCSIIT database contains detailed information regarding the analysis performed on the CES pole and the conclusions made from the analysis. The conclusions and their relevance to the CM breakup are presented here.

Figure 2.4-51. *Port end of the crew escape pole, showing lanyard magazine, middeck, looking forward.* [Picture from a shuttle training mockup in the JSC Space Vehicle Mockup Facility]

The debris indicates that the pole was installed in the launch/landing position. There is no evidence to suggest that the pole had been deployed (or that the side hatch had been jettisoned[3]). The pole housing, the main pole, and the extension pole (figure 2.4-52) were recovered separately. The starboard end of the housing, including the middeck ceiling mounting bracket, shows evidence of thermal damage. The main pole and extension pole (figure 2.4-53) were recovered in relatively good condition, showing little mechanical or thermal damage. The main pole deployment spring was recovered separately from the housing, with the starboard end cap of the pole housing attached to the spring. Additionally, the portion of the CM port wall to which the pole attaches was recovered and analyzed. Several lanyards were recovered separately, but the lanyard magazine was not recovered.

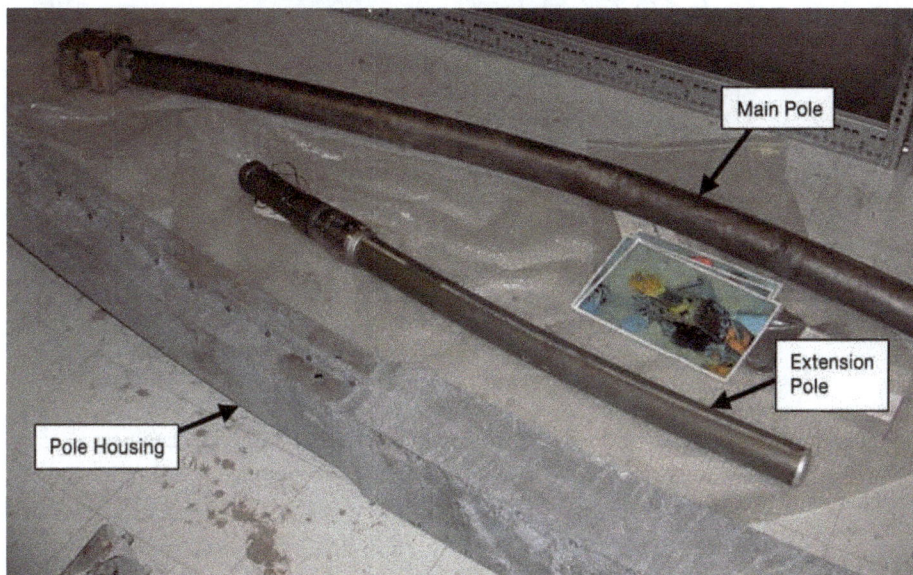

Figure 2.4-52. *Pole housing, main and extension poles.*

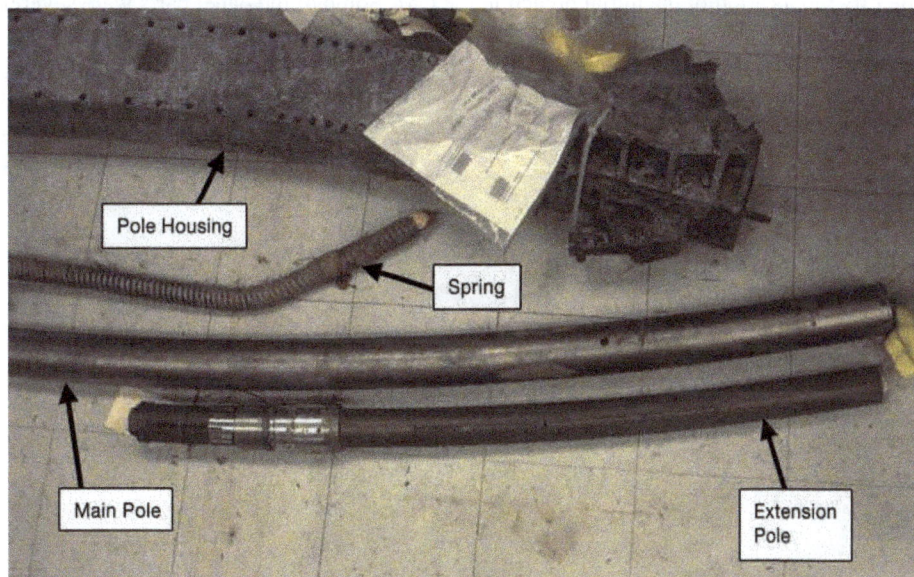

Figure 2.4-53. *Port ends of housing, deployment spring, and main and extension poles.*

---

[3]Several pyrotechnic components were recovered. All indicated that the hatch jettison system had not been activated.

Deformation of the upper portion of the knuckle that attaches to the CM indicates that it experienced an upward cantilever load (figure 2.4-54). This could be caused by the starboard end of the pole moving upward. This suggests that the flight deck separated from the middeck roughly at the flight deck floor/middeck ceiling level.

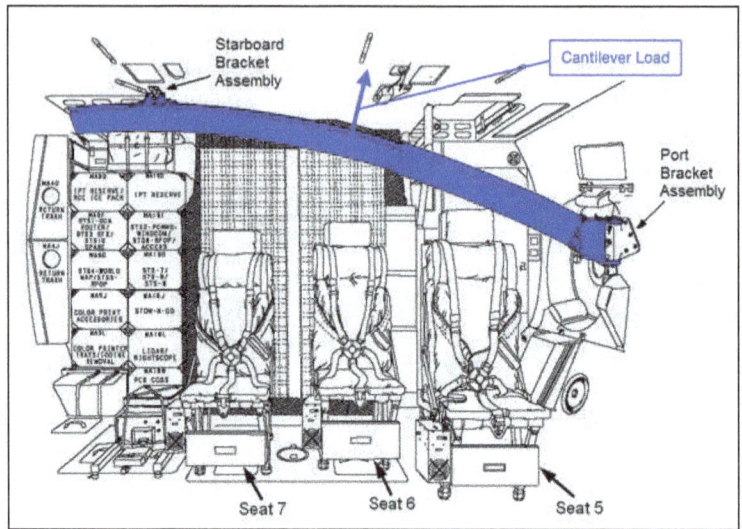

Figure 2.4-54. *Crew escape pole in launch/landing position, middeck, looking aft, with direction of cantilever load noted from debris.*

The relative lack of significant witness marks on the pole housing indicates that it experienced very few impacts. These conclusions support the earlier conclusions that the middeck breakup was rapid and expansive, with very little interaction between major structures.

Flight deck instrument panels

Many of the flight deck panels were recovered and identified. Mechanical and thermal damage to the recovered panels was evaluated to assist in understanding the sequence of the breakup in an attempt to identify the location of the initial breach in the CM. Some of the panel parts were severely torn and deformed, yet some were mostly intact with less damage. Most of the recovered panels were photographed as orthogonally as possible under consistent lighting conditions, and the images were imported into a 3-dimensional computer model of the orbiter flight deck to create a virtual reconstruction of the flight deck (figures 2.4-55 and 2.4-56) (see discussion of virtual reconstruction in Chapter 4).

Figure 2.4-55. *Intact orbiter flight deck from the Shuttle Mission Simulator.*

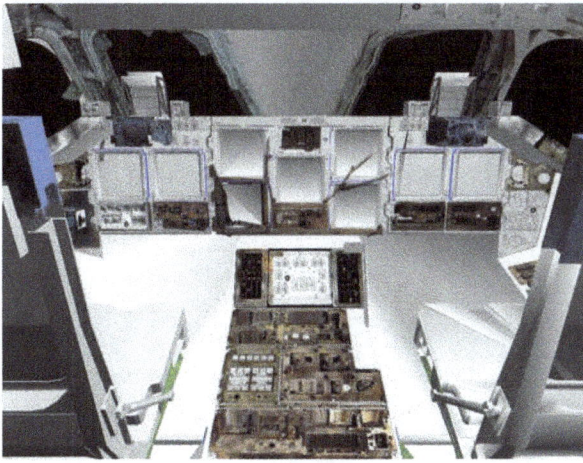

Figure 2.4-56. *Virtual reconstruction of the recovered* Columbia *flight deck panels.*

This virtual reconstruction was studied to investigate whether there were clear indications of a thermal or structural breach. However, adjacent panels were seen to have received significantly different amounts of thermal damage. This indicated that some panels broke off earlier than other panels, or were temporarily shielded from the thermal flow during the breakup. The damage varied greatly from one panel to the next, indicating a chaotic breakup sequence. No clear evidence of the initial breach location could be determined.

2.4.4.3 *Orbiter window analysis*

At least one window frame (CM or FF) was recovered from every window. Two large debris assemblies were recovered (figures 2.4-57 and 2.4-58). Each large piece is a complete assembly of three CM forward window frames (Windows 1, 2, and 3 and Windows 4, 5, and 6) with some broken glass pieces still captured in their retainers. Three CM window frames of Windows 7, 9, and 10 were recovered in separate pieces since they were not connected to each other by heavy CM skin structures (figure 2.4-59). Five thermal frames (FF) were also recovered.

Figure 2.4-57. *Port view looking aft, recovered* Columbia *crew module Windows 1, 2, and 3.*

Figure 2.4-58. *Starboard view looking aft, recovered* Columbia *crew module Windows 4, 5, and 6.*

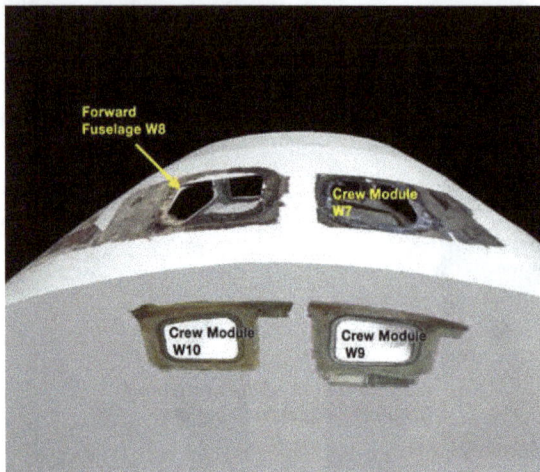

Figure 2.4-59. *View looking forward, recovered* Columbia *crew module Windows 7, 9, and 10 and forward fuselage Window 8.*

All CM window frames are heavily reinforced to limit the window retainer deflection when the CM is exposed to internal pressure load and flight loads. The redundant panes are also made from the same material as the outer thermal pane (fused silica) so that if the outer thermal pane failed, the middle redundant pane still could be able to take some limited thermal flow for a short time.

Debris evidence suggests that most of the CM skin structures surrounding the CM flight deck windows had been melted away during the event. One exception is the upper edge of CM aft Windows 9 and 10, which shows mechanical fracture edges.

To summarize previous conclusions, there is no evidence to suggest that failure of the CM window panes was the cause of cabin depressurization. The debris field shows that no glass from the redundant panes, which would have had to have failed to lose pressure, was recovered west of the main forebody debris field. Most of the loose glass from all panes was recovered from the main forebody debris field, suggesting that the windows shattered during the CMCE. The departure of the FF Windows 1 through 6 and the FF arrow-head panel may have triggered the departure of the rest of the FF canopy by aerodynamic forces.

The recovered glass showed a marked discoloration. Discussions with the window subsystem manager confirmed that the window appearance did not match the appearance of thermally discolored glass. Close inspection showed that the discoloration appeared to be deposited material. Analysis of the deposition on the glass was conducted to determine the sequence of events experienced by the windows.

Evaluation was restricted to only thermal (outer) pane glass from which the location was positively identified. Based on thickness, one piece of glass was identified as either Window 3 or Window 4, but for purposes of this assessment this was considered sufficiently specific. When possible, glass samples were obtained from what remained in the various window frames.

For purposes of this report, the deposition that coated the glass was referred to generically as the "char layer," regardless on which window it formed. Because the char layer was suspected to have formed at relatively high altitudes and temperatures, it was presumed that it remained intact from formation through eventual ground impact. Therefore, it was assumed that the char layer was sufficiently adhered to the glass surfaces such that all lightly attached particles were either deposited later in the breakup sequence or were field contamination. The harvested samples were cleaned in a laboratory setting using standard preparation techniques. Figure 2.4-60 is an example of an extracted thermal pane in its cleaned state, prior to sectioning. Black lines on the images denote approximate sectioning planes.

Figure 2.4-60. *Window fragment removed from the thermal pane frame from Window 8.* The black dashed line denotes the sectioning plane. The red dot was used to indicate the outward face of the pane.

Various aspects of char layer characterization were performed by the JSC Materials and Processing Office, the JSC Astromaterials Research Office, the KSC Failure Analysis and Materials Evaluation Branch, and the White Sands Test Facility. Electron and light microscopy, X-ray diffraction, powder diffraction, layer metrology, and phase characterization were all performed.[4,5]

The char layer coverage observed on Windows 3, 4, and 5, appeared relatively translucent when placed in front of a light source. The coloration of the char layer from these samples varied from a brownish tan to a dark brown hue. The translucent characteristics of these samples implied a relatively thin deposit thickness. By comparison, samples from Windows 7 and 8 were notably opaque and had more of a blackened appearance. The surface texture of the window samples appeared relatively rough, consistent with re-solidified molten deposition. Visual examination alone was not able to assess the relative thickness of the deposits on Windows 7 and 8.

Cross sectioning of the various thermal pane window remnants was performed using standard cross-sectioning metallographic techniques. The char layer deposits on all thermal panes examined were evaluated and characterized based on mean thickness and constitution of the deposits (voids, inclusions, etc.). In general, the thermal panes for Windows 3, 4, and 5, were covered by a char layer deposit that ranged from a few microns (μm) to nearly 100 μm. The deposit in the char layer for these forward-facing windows was not continuous; the areas without deposition retained their translucence. In the regions of continuous deposit, the thickness profiles were irregular, indicating sporadic deposition of the material. For regions where deposits were thick, void entrainment (porosity) was evident. In contrast to the exterior surfaces of the forward-facing windows, the exterior surface of the thermal panes for Windows 7 and 8 samples were covered by a char layer deposit that ranged from 30 μm to nearly 500 μm and appeared continuous. Although the char layers on these windows varied in thickness, the mean thickness was on the order of 50 to 100μm. The measurements taken on the interior and fracture surfaces of Windows 7 and 8 samples were consistent with those of the forward-facing windows.

In-depth materials analysis was performed on the char layers for the forward windows to compare to the char layer on the overhead windows.

While almost every window sample contained multiple metallic species, aluminum was the predominant component of the char layer with other metal species existing in either discrete features or within a very narrow region of the layer. Interspersed in the layers were globules of silicon throughout the thickness. The spectroscopic signature indicated that the majority of this layer was a heavily oxidized aluminum amalgam consistent with a 2000-series aluminum alloy. This series aluminum alloy is used in the FF and CM structure.

A porous layer of aluminum was deposited on all window pane (forward and overhead thermal) samples, including both external and fracture surfaces. The porous nature of this feature in the char layer was considered to be a result of a dynamic process of deposition when the fragments of the glass, and the structure that contained them, possessed a high relative and turbulent motion to the deposition source. The deposition source, likely a 2000-series aluminum alloy and probably from the aluminum 2024 FF structure, was dispersed in the form of molten/semi-molten particles that partially cooled and/or oxidized prior to

[4] J. D. Olivas, L. Hulse, B. Mayeaux, S. McDanels, P. Melroy, G. Morgan, Z. Rhaman, L. Schaschl, T. Wallace, and C. Zapata, *Examination of OV-102 Thermal Pane Window Debris – Final Report*, KSC-MSL-2008-0178 (in press).

[5] J. D. Olivas, M. C. Wright, R. Christoffersen, D. M. Cone, and S. J. McDanels, *Crystallographic oxide phase identification of char deposits obtained from space shuttle* Columbia *window debris*, Acta Materialia, 2008 (in press).

impacting the glass substrate. Additionally, this feature of the char layer was irregular in thickness and also had distinct particles of other oxidized metal systems.

The formation of this porous char layer can be explained by several scenarios. The deposition source could have been nearby, but the area had highly turbulent relative motion. Or, the deposition source could have been a significant distance from the window, creating high dispersion in the molten material flow. This is less likely because liquid droplets would cool quickly and be less likely to adhere. Finally, the deposition could have resulted from the glass passing through a rapidly solidifying vapor surrounding the forebody as a result of thermal erosion of materials. It is conceivable that all three processes were occurring, either simultaneously or discretely. Given the debris field of recovered glass and the presence of the deposition on the fracture surfaces and inner panes, it is concluded that this deposition event occurred between the CMCE and the TD.

In addition to this porous layer, the two overhead thermal panes showed unique layers not seen on the forward windows. On these two panes, two additional layers, which contained titanium in appreciable quantities, were identified below the porous aluminum-rich layer. The overhead thermal pane window char was loosely characterized into three layers: a titanium-rich layer closest to the glass, a titanium-aluminum-rich layer, and the porous aluminum-rich layer that was described above. Crystallographic investigation of the titanium-rich region adjacent to the glass indicates that the nodules are consistent with TiO_2, a titanium oxide.

The aluminum-rich, thinner outer char layer appeared to be deposited via a different mechanism than the lower titanium-rich layers. Based on the lack of porosity, the environment in which the lower titanium-rich layers were deposited was likely substantially less dynamic than the environment during the deposition of the porous aluminum-rich top layer. This continuity of the titanium-rich layers suggested that the source was likely in close proximity to the windows; longer distances from the source would result in a more dispersed and turbulent flow, and probably result in material cooling/solidifying prior to impact on the window, which was not seen. The transition from titanium to a mixture of aluminum/titanium seems to imply that some time after the titanium began to deposit on the windows, aluminum from a nearby source also began to deposit on the windows. Analysis of the deposition on the carrier panel tile surrounding the overhead window panes showed similar deposition layer patterns. This deposition occurred prior to the breakup of the windows and structure, which was presumed to have begun at the CMCE. The existence of this distinctive three-zone char layer only on the exterior surfaces of the thermal panes from Windows 7 and 8 also supports the previous debris field finding that the windows were largely intact through the CMCE.

The previous findings were considered interesting enough to lead to a search for the source of the titanium and aluminum. The intention was to discover whether there was any information that was suggestive of the structural condition and the orientation of the forebody prior to the CMCE. A search for forward structures containing titanium showed that the nearest source of titanium material to the windows was the forward PLBD rollers. These rollers are made of titanium and aluminum with an Inconel sleeve. The PLBDs of the orbiter, when closed, rest on these roller mechanisms, which are attached to the X_o 582 ring frame bulkhead upper arch in the forward part of the payload bay. These rollers are in close proximity to Windows 7 and 8. The structure that supported the roller components was primarily composed of 2024 aluminum alloy and also displayed evidence of significant thermal erosion. However, regions farther away from the rollers showed minimal signs of thermal erosions; green Koropon primer was still present on portions of these remote regions (figures 2.4-61, 2.4-62, and 2.4-63).

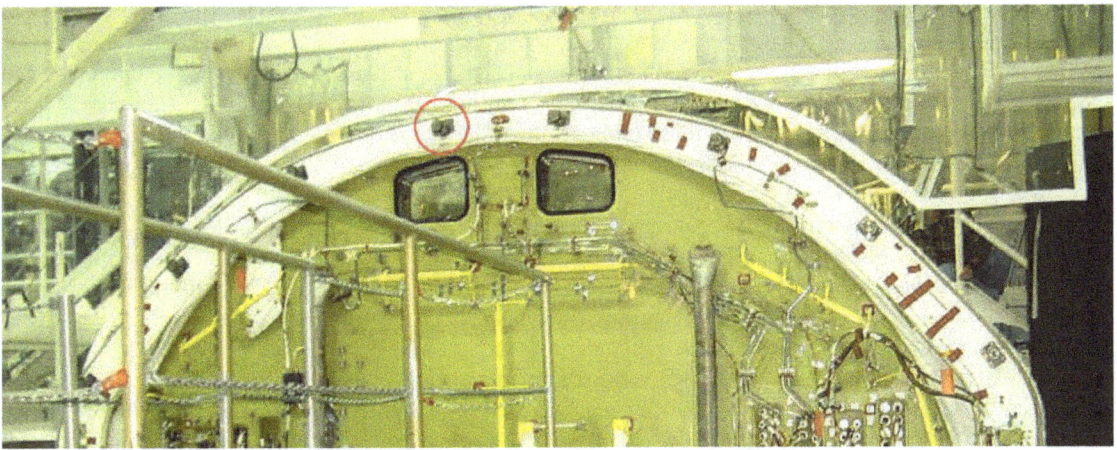

Figure 2.4-61. Endeavour, *OV-105, X$_o$ 582 ring frame bulkhead arch with rollers*. Circled in red is one of the eight rollers.

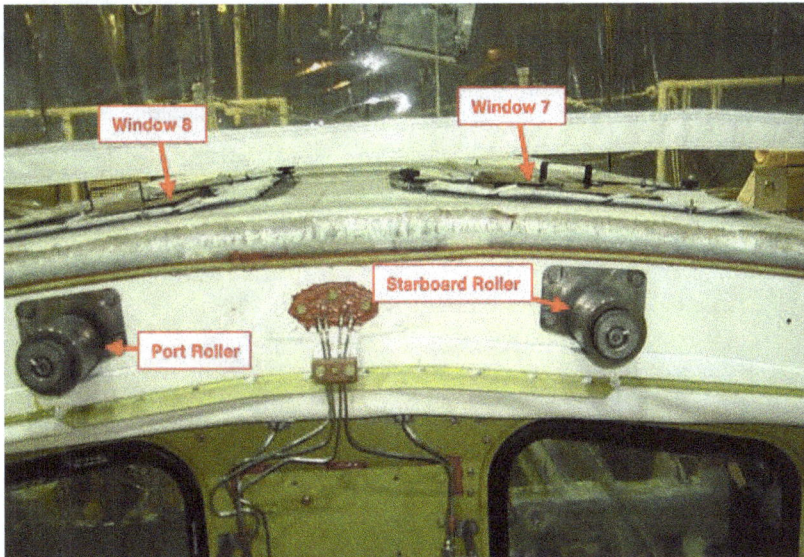

Figure 2.4-62. *Nominal configuration (*Endeavour, *OV-105) of the two inner rollers and overhead windows.*

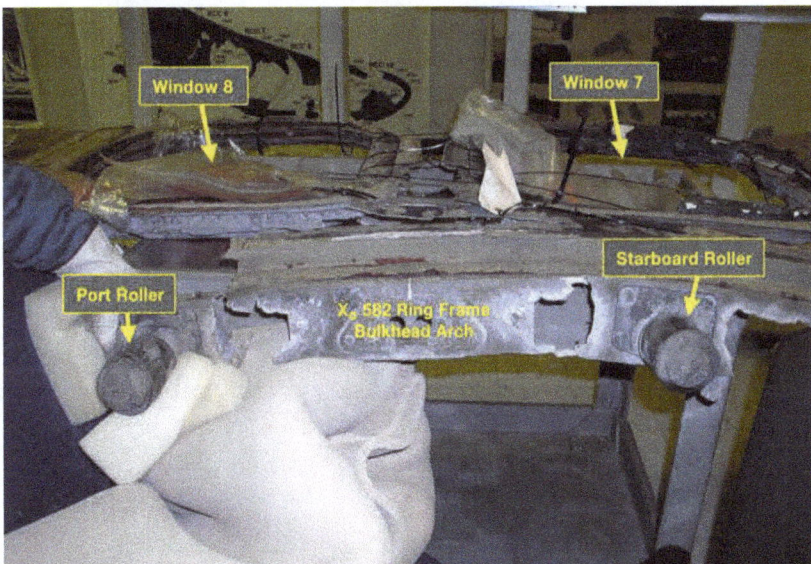

Figure 2.4-63. Columbia *debris for same location*. Note eroded rollers and eroded region of the X$_o$ 582 ring frame bulkhead arch between the rollers.

The recovered rollers for the location closest to the overhead windows both showed significant signs of erosion. While these rollers were not the only components made of titanium, they were the only ones recovered that possessed the proper material, proximity, and thermal indications and were, therefore, concluded as the source location. Given these findings, the rollers were the most probable source of titanium causing the deposition on the windows.

Since the rollers are located on the X_o 582 ring frame bulkhead arch aft and below these two windows, and protected by the PLBDs, the PLBDs must have been compromised or fully departed while the bulkhead arch was still attached to the CM. Additionally, the forebody must have been traveling aft-end forward for some period of time to have the directional thermal flow that caused the titanium to "vaporize" flow over the glass external surface (figure 2.4-64). Since no titanium was found on the internal surface of the thermal panes or on the external surface of the Windows 7 and 8 redundant panes, it can be concluded that the thermal panes were intact at the time of this event. This is consistent with other findings showing that the FF remained with the CM until breakup of both elements.

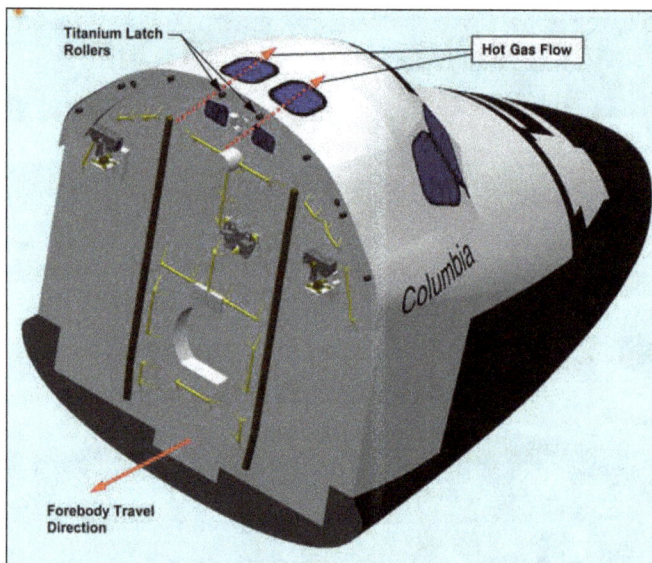

Figure 2.4-64. *Titanium deposit on windows indicates forebody traveled backwards.*

Finding. Windows 7 and 8 experienced a titanium deposition event that occurred prior to window breakup.

Finding. The most probably source for the titanium deposition on Windows 7 and 8 was the PLBD rollers. These rollers were not exposed to heat flow until after the PLBDs were compromised.

Finding. All the windows had an aluminum-rich deposition, which was consistent with a turbulent process.

A complete discussion of the thermal mechanisms that may have led to the titanium deposition is covered in Section 2.1.6.8.

2.4.5 Synopsis of forebody breakup sequence

The orbiter breakup (CE) or subsequent impacts between the CM and the FF caused small mechanical breaches on the CM skin. The FF shell stayed with the CM until the CMCE at GMT 14:00:53. At the CMCE, the FF most likely separated in two large segments, upper and lower. The departure of the FF arrowhead panel and FF Windows 1 through 6 may have triggered the departure of the rest of the FF canopy by aerodynamic forces.

Once the protective FF structure departed, the CM side skin was consumed by thermal exposure. Breakup of the CM structure occurred as a consequence of aerothermal heating and aerodynamic loading. The CM broke up with the middeck and forward bulkhead departing, most likely from starboard to port. The middeck breakup was rapid and expansive, with very little interaction between major structures.

The flight deck remained intact for some period after the middeck separated. The flight deck likely remained with the airlock and the aft bulkhead. As a relatively intact "pod" with a high ballistic number, the flight deck experienced more thermal exposure until its final breakup completed the CMCE.

The forebody was broken down to subcomponents that were too small and dispersed to see on video at GMT 14:01:10. This was described as the TD. Cascading failures and thermal damage were still occurring, but the CM no longer had any structural integrity at this time.

> **Recommendation L3-1.** Future vehicles should incorporate a design analysis for breakup to help guide design toward the most graceful degradation of the integrated vehicle systems and structure to maximize crew survival.

Chapter 3 – Occupant Protection

3.1 Crew Seats

The seats, which are the interface between the vehicle structure and the crew members, provide a source of data for the accelerations and thermal environments that the crew members experienced. This section provides background information on the design and construction of shuttle crew seats, and describes the detailed analyses performed on the *Columbia* crew seats. These analyses included review of the recovered videos recorded on orbit, review of the materials analyses performed by the *Columbia* Accident Investigation Board (CAIB)/Crew Survival Working Group (CSWG), and inspection of debris items including microscopic inspections of the inertial reels mechanisms and straps.

The following is a summary of the findings, conclusions, and recommendations for this section:

Finding. Evidence from the inertial reel straps indicates that the seats 1, 2, and 3 straps were mostly extended at the time of strap failure. The seats 4, 6, and 7 straps were extended during a material deposition period (seat 4 at least 8 in., or ~36% extended; seat 6 at least 21.25 in., or ~96% extended; and seat 7 at least 21.5 in., or ~98% extended). Medical evidence (see Section 3.4) indicates that some of the crew members received injuries consistent with insufficient upper body restraint.

> ***Conclusion L2-2.*** The seat inertial reels did not lock.

> ***Conclusion L2-3.*** Lethal injuries resulted from inadequate upper body restraint and protection during rotational motion.

> ***Recommendation L1-3/L5-1.*** Future spacecraft crew survival systems should not rely on manual activation to protect the crew.

> ***Recommendation L2-4/L3-4.*** Future spacecraft suits and seat restraints should use state-of-the-art technology in an integrated solution to minimize crew injury and maximize crew survival in off-nominal acceleration environments.

> ***Recommendation L2-8.*** The current shuttle inertial reels should be manually locked at the first sign of an off-nominal situation.

Finding. The seat 2 inertial reel strap exhibits "strap dumping" failure features. The strap failed progressively, possibly due to damage to the lateral edge of the strap from contact with the sharp edge of the strap pass-through slot.

> ***Recommendation L2-5.*** Incorporate features into the pass-through slots on the seats such that the slot will not damage the strap.

Finding. All inertial reel straps are tested with static loads at room temperature. Load testing has not been conducted to determine the loads required to fail the straps at elevated temperatures or under dynamic loads. Testing has not been conducted to determine the material properties (combustion vs. chemical degradation vs. melting) in a high-temperature/low-oxygen (O_2)/low-pressure environment.

Recommendation L2-6. Perform dynamic testing of straps and testing of straps at elevated temperatures to determine load-carrying capabilities under these conditions. Perform testing of strap materials in high-temperature/low-oxygen/low-pressure environments to determine materials properties under these conditions.

Finding. While all seat piece-parts include serial numbers, only the serial numbers of the inertial reels were recorded and tracked to a specific seat assembly. The lack of configuration management documentation hindered the process of ascribing the seat debris items to specific seat locations.

Recommendation A5. Develop equipment failure investigation marking ("fingerprinting") requirements and policies for space flight programs. Equipment fingerprinting requires three aspects to be effective: component serialization, marking, and tracking to the lowest assembly level practical.

3.1.1 Seat design and construction

Two types of seats are used on the space shuttle. The Pilot seats are used by the mission Commander (CDR) and Pilot (PLT), and Mission Specialist seats are used by all other crew members. Both types of seats provide for crew member positioning and restraint during launch, entry, and some on-orbit operations. Both types of seats are also designed to accommodate a fully suited crew member.

Seat positions are numbered 1 through 7, beginning with the CDR's position on the flight deck (seat 1) and ending with the starboard-most Mission Specialist seat on the middeck (seat 7). Seats 1 through 4 are on the flight deck (figure 3.1-1), and seats 5 through 7 are on the middeck (figure 3.1-2). Seats 1 through 5 are flown on all missions. Seats 6 and 7 are flown as required.

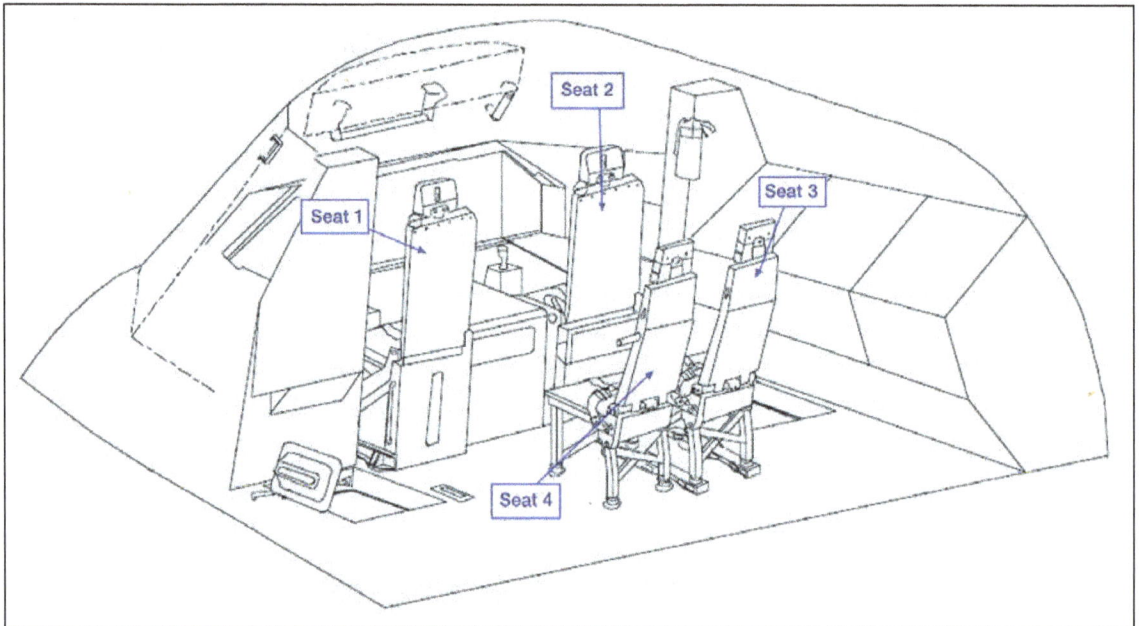

Figure 3.1-1. Depiction of the flight deck seats.

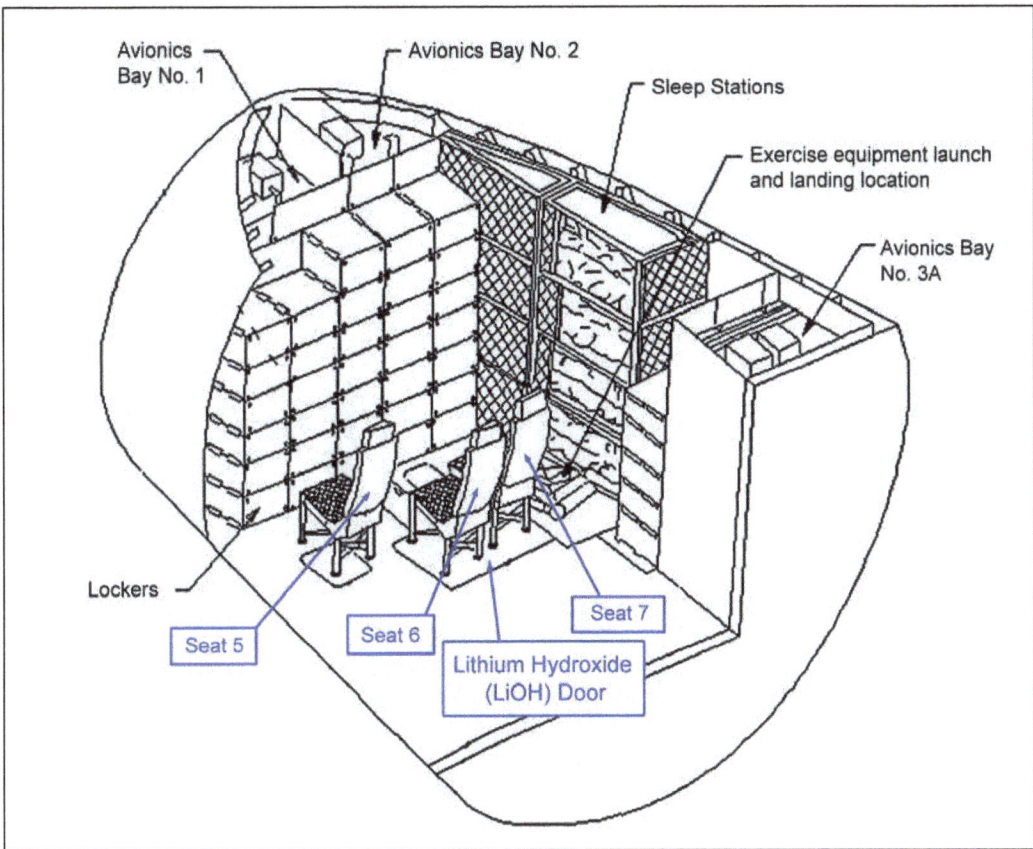

Figure 3.1-2. *Depiction of the middeck seats.* [Adapted from the Shuttle Crew Operations Manual]

The Pilot seats (figure 3.1-3) and Mission Specialist seats (figure 3.1-4) have several common design features. Both seat types have identical seatbacks, five-point restraints, headrests, and MA-8 inertial reels (a part of the restraint system).[1] The five-point harness restrains the upper torso with shoulder belts, and the lower body with lap and crotch belts. All belts connect to a rotary buckle that is permanently mounted to the crotch belt. The shoulder belts join to a single strap, the inertial reel lead-in strap, which attaches to the inertial reel mechanism mounted inside the seatback. The inertial reel will lock automatically due to accelerations pulling the strap out at 1.78 G to 2 G. A lever to manually lock and unlock the inertial reel is located on the left side of the seat pan. Both seat types accommodate the attachment of O_2 hoses and communications cables, and both have attachment brackets for cooling units used in conjunction with the crew member suits for crew member comfort.

---

[1]The MA-8 inertial reel is an off-the-shelf design used in military helicopter seats and was not designed specifically for the orbiter.

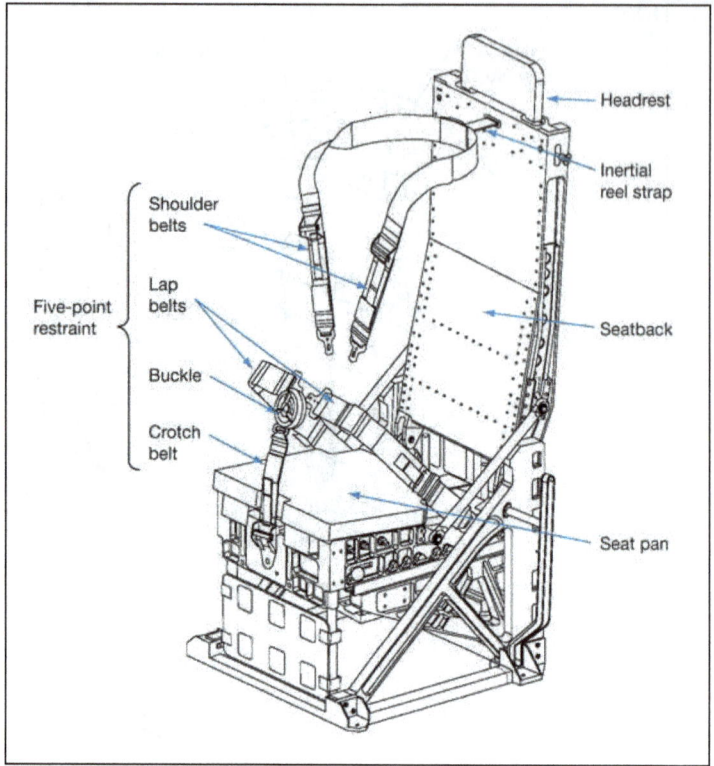

Figure 3.1-3. *Pilot seat*. [Adapted from the Space Shuttle Systems Handbook]

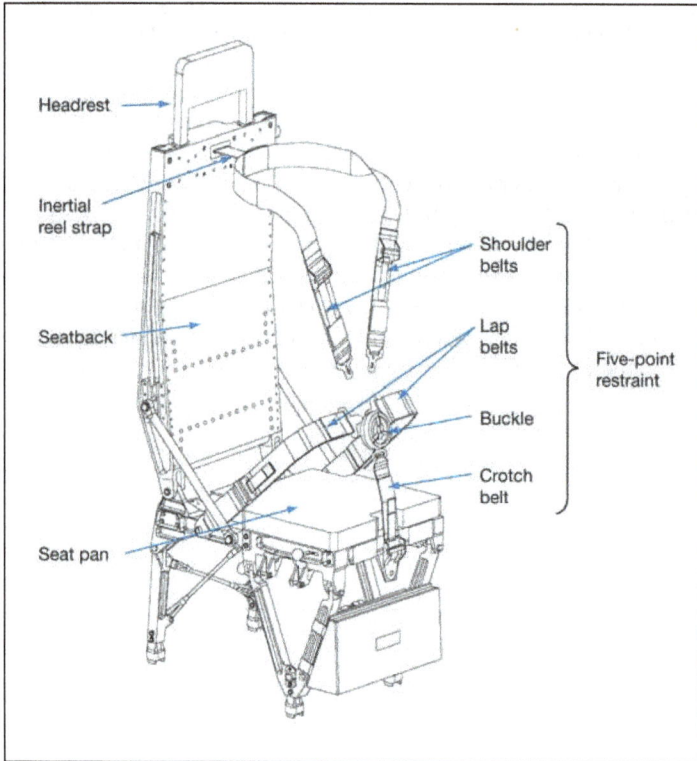

Figure 3.1-4. *Mission Specialist seat*. [Adapted from the Space Shuttle Systems Handbook]

The seat pan and base of the Pilot seats (seats 1 and 2) differ from the Mission Specialist seats. The Pilot seat base is permanently mounted to the flight deck floor and incorporates mechanisms providing up/down and forward/aft adjustability of the seat position. The Pilot seats also provide a mounting base for the rotational hand controllers (RHCs) (figure 3.1-5). RHCs are control sticks that are used by the CDR and PLT to control vehicle rotation about the roll, yaw, and pitch axes during ascent, orbit, and entry. The RHCs provide input to computers to actuate various vehicle control effectors (aerosurfaces and/or Reaction Control System (RCS) jets)).

Crew members assigned seats 3 through 7 use the Mission Specialist seats for launch and entry. Five Mission Specialist seats were flown on STS-107. Seat 3 was mounted on a special sled assembly that was unique to *Columbia*.

The Mission Specialist seats have no base, but have foldable legs attached to the seat pan. The legs are equipped with quick-disconnect fittings to allow for seat removal and stowage. Once the shuttle is on orbit, all Mission Specialist seats are detached from the floor, folded, and stowed in various locations depending on crew preference. During deorbit preparation, the crew reinstalls the Mission Specialist seats.

RHC

Figure 3.1-5. *Commander's seat and one of the rotational hand controllers, flight deck.* [Picture from a shuttle training mockup in the JSC Space Vehicle Mockup Facility, looking from starboard to port]

3.1.2 STS-107 seat configuration

The recovered middeck and flight deck videos revealed information that was related to the configuration of the crew seats and other flight crew equipment. The recovered middeck video recorded deorbit preparation (D/O PREP) activities on the middeck. Although there is no credible timestamp[2] on the video, the STS-107 crew plan[3] offers some insight into when the recording probably occurred.

The video shows that all middeck seats were installed. The seat 5 crew member is suited, and the seat 1 and seat 6 crew members are donning their suits. Based on the crew D/O PREP plan, this 30-minute video probably recorded events from approximately Greenwich Mean Time (GMT) 11:40:00 to approximately GMT 12:10:00.

The recovered flight deck video is a 13-minute, 11-second video with sound that was recorded more than 1 hour after the middeck video described above. This video recorded entry events on the flight deck from approximately GMT 13:35:34 to GMT 13:48:45. It shows that all of the flight deck seats were installed and the flight deck crew members were properly restrained. This video provides a good view of the crew members in seats 1 and 2, and fair views of the crew members in seats 3 and 4. It also shows that all of the flight deck crew members were properly secured with seatbelts to prevent floating from the seats (tight lap belts). No slack in the belts is visible in the shoulder harnesses, and the crew members are able to move their upper bodies, indicating that the shoulder harness inertial reels are not locked, which is normal at this point during the mission.

---

[2]The video has no air-to-ground audio, nor does it record any actions that can be time-verified through telemetry or onboard data recorders, so a precise time-stamp cannot be determined.
[3]Pre-mission, each shuttle crew develops a detailed deorbit preparation plan that is tailored from the formal Flight Data File procedures.

From video evidence, investigators concluded that all of the seats were installed and the flight deck crew members were properly strapped into their seats. Although no recovered video shows that the middeck crew members were strapped into their seats, medical findings and evidence in the seat debris described below confirms that two middeck crew members were fully strapped in and that one middeck crew member was at least partially restrained in the seat.

3.1.3 Seat structure

More than 68 pieces of seat structure debris were recovered. Positive assignment of the recovered seat debris to specific seat locations was difficult and required considerable analysis of subtle differences between the seats and the mounting locations because a significant portion of the lightweight seat design is common to all seven seats. Several of the recovered components were ascribable as coming from the flight deck seats (seats 1, 2, 3, and 4) and the middeck seats (seats 5, 6, and 7).

Figures 3.1-6 and 3.1-7 show seat structure pieces that were identified to each seat location. The blue items in these figures represent items that were positively identified to a seat location. The green items are those items that could be from one of two seats. Figure 3.1-8 shows the major seat structure debris pieces that could not be identified to a specific seat location (the colors in this figure distinguish the different pieces and are otherwise inconsequential). Although represented in figure 3.1-8 on just two seats, the seatback and seat restraint items could be from any of the seven seats. However, the seat pan items could only be from the Mission Specialist seats.

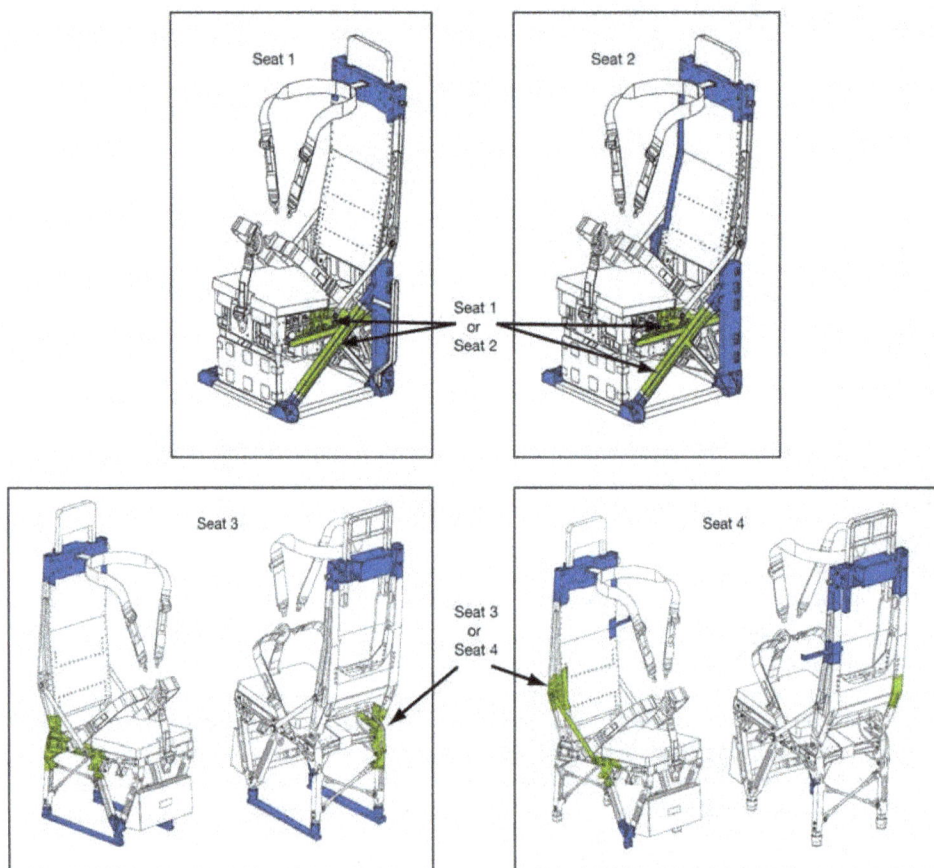

Blue items are positively identified to a specific seat;
green items could be from either one of the two indicated seats.

Figure 3.1-6. *Identified debris from the flight deck seats.* [Adapted from the Space Shuttle Systems Handbook]

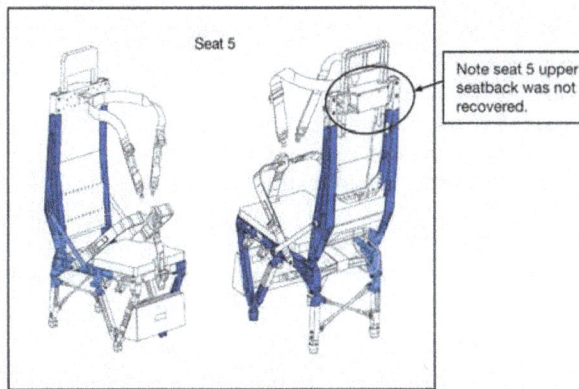

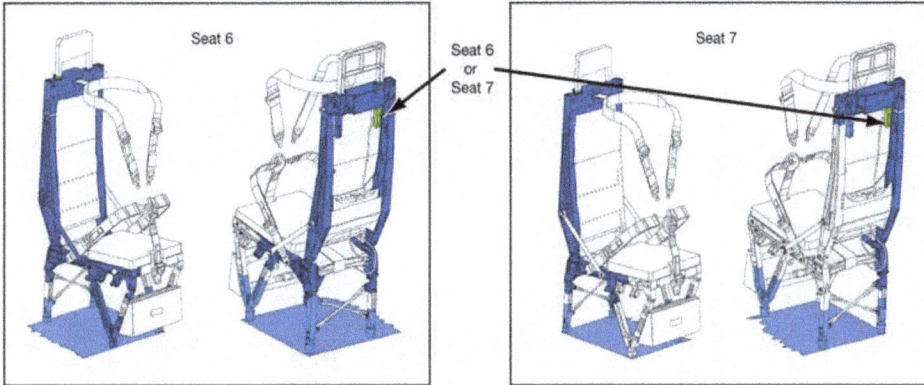

Blue items are positively identified to a specific seat;
green items could be from either one of the two indicated seats.

Figure 3.1-7. *Identified debris from the middeck seats.* [Adapted from the Space Shuttle Systems Handbook]

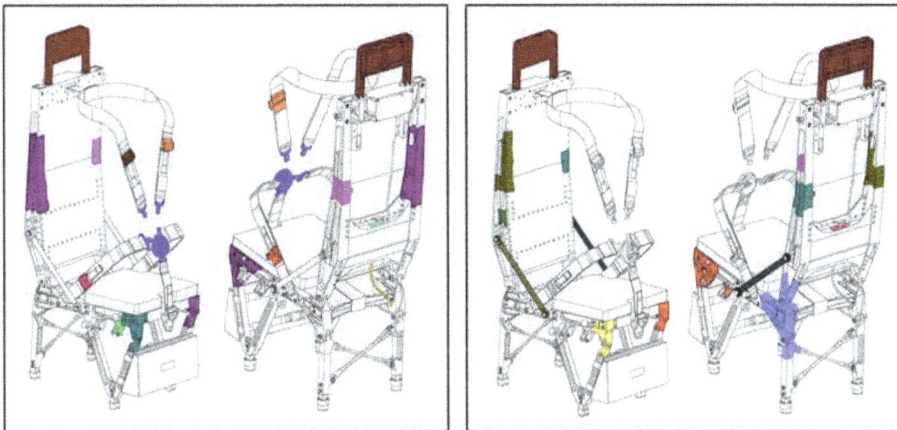

Colors are used only to distinguish different debris items.

Figure 3.1-8. *Debris from unknown seat locations.* [Adapted from the Space Shuttle Systems Handbook]

The only consistent piece of seat debris that was positively identified to six seat locations (except seat 5) was a portion of the upper seatback that included the seat restraint inertial reel mechanism and strap.

Major differences in the magnitude of thermal exposure were identified on the flight deck vs. the middeck seat locations. The flight deck seat components, as well as the flight deck floor structure, were highly melted and/or deposited with splattered aluminum on all surfaces. Close inspection of the fracture surfaces on the flight deck seat attach points revealed deformation towards the vehicle starboard direction.

Materials analysis revealed that the flight deck seats collected deposits of melted aluminum from locations throughout the cabin area. Materials that were consistent with the bulkheads and outer pressure shell (2219 aluminum), the surrounding primary and secondary structure (2219, 2024, 2124, and 7075 aluminum), and the seats (2024 and 7075 aluminum) were discovered on the upper and lower surfaces of the recovered seat debris. The CAIB Report concluded,[4] and this report concurs, that the flight deck seats remained attached to the flight deck floor panels and adjacent to the surrounding structure during a period of significant thermal exposure and material deposition.

The magnitude and distribution of heating of flight deck seat components indicates a prolonged attachment to other crew module (CM) structure during exposure to heating. This indicates that the flight deck seat components were released from the CM later in the breakup sequence than the middeck seat components. Ground plots of debris recovery locations also support this conclusion (figure 3.1-9). Figure 3.1-10 compares a portion of the seat leg from a flight deck Mission Specialist seat component with a similar component from a middeck seat.

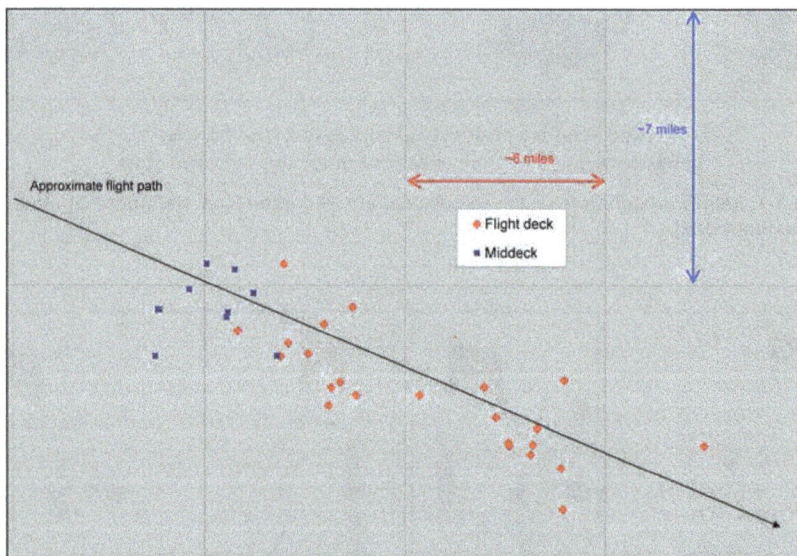

Figure 3.1-9. *Recovery locations of seat structure debris.*

---

[4]*Columbia* Accident Investigation Board Report, Volume V, Appendix G.12, Crew Survivability Report, October 2003, p. 362.

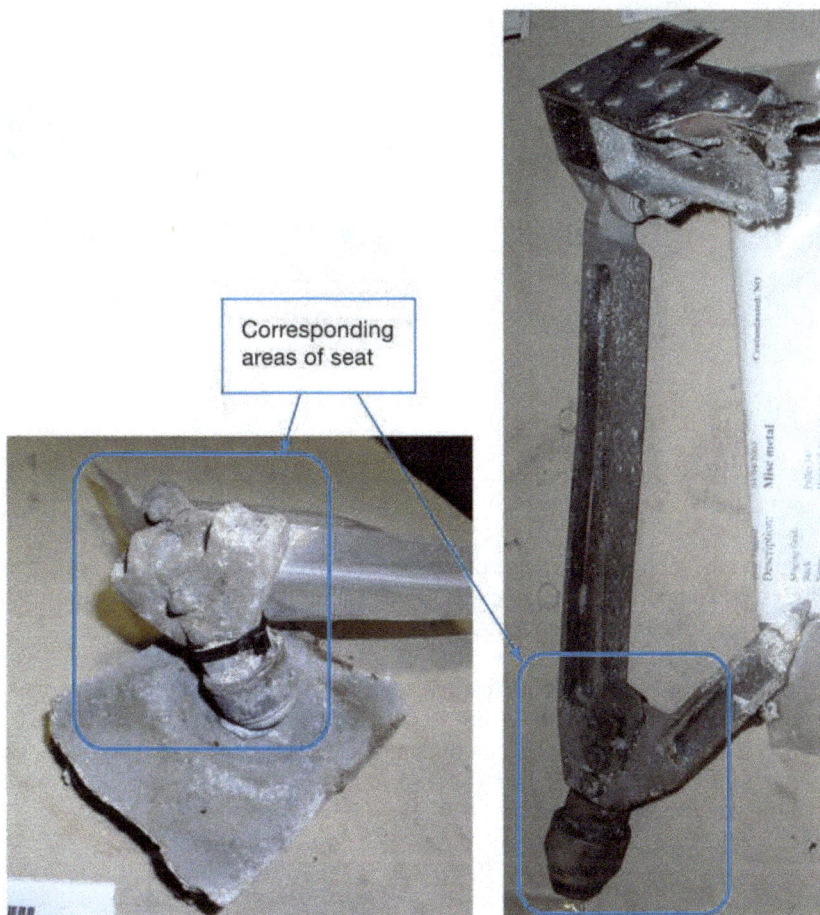

Figure 3.1-10. *Comparison of seat debris from a flight deck Mission Specialist seat (left) and a middeck Mission Specialist seat (right).*

Although all seat components experienced significant heating, the middeck seats were less eroded and fragmented when compared to the flight deck seats. Material analysis revealed that the seat 6 and 7 components collected significant deposits of melted aluminum from the lithium hydroxide (LiOH) door to which they were attached. Unlike the flight deck seat components, analysis revealed no materials consistent with the pressure shell deposited on the middeck seats. The CAIB Report concluded,[5] and this report concurs, that seats 6 and 7 remained attached only to the LiOH door during a period of significant thermal exposure and material deposition. The highly directional nature of the deposition indicates that the LiOH door/seat 6/seat 7 combination attained a stable attitude (figure 3.1-11) during thermal exposure. The close proximity of debris recovery locations indicates that the seat 6 and 7 components separated from the LiOH door shortly before ground impact.

---

[5]*Columbia* Accident Investigation Board Report, Volume V, Appendix G.12, Crew Survivability Report, October 2003, p. 362.

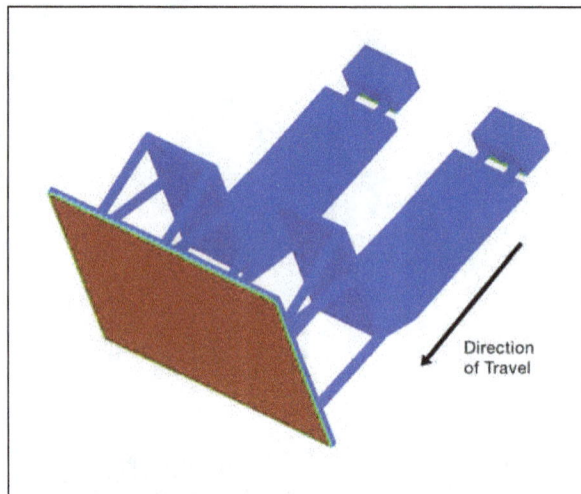

Figure 3.1-11. *Stable attitude of the lithium hydroxide door/seat 6/seat 7 combination, as indicated by deposition.*

Direction of Travel

Of the seats, seat 6 had the highest percentage of structure recovered and identified; seats 5 and 7 had slightly fewer structural components that were identified. The flight deck seats had significantly lower percentages of identified structural components.

Nearly all seat fractures occurred at minimum thermal cross-sectional areas (minimum thermal mass), away from any large heat sink locations.[6] Common seat fracture locations are shown as red lines in figure 3.1-12. It is also noteworthy that nearly all thin-sheet aluminum materials (closeout panels on the seat pan and seatback) are missing (i.e., overloaded/melted away). Additionally, with the exception of the inertial reel straps, very little belt material and seat cushion material was recovered.

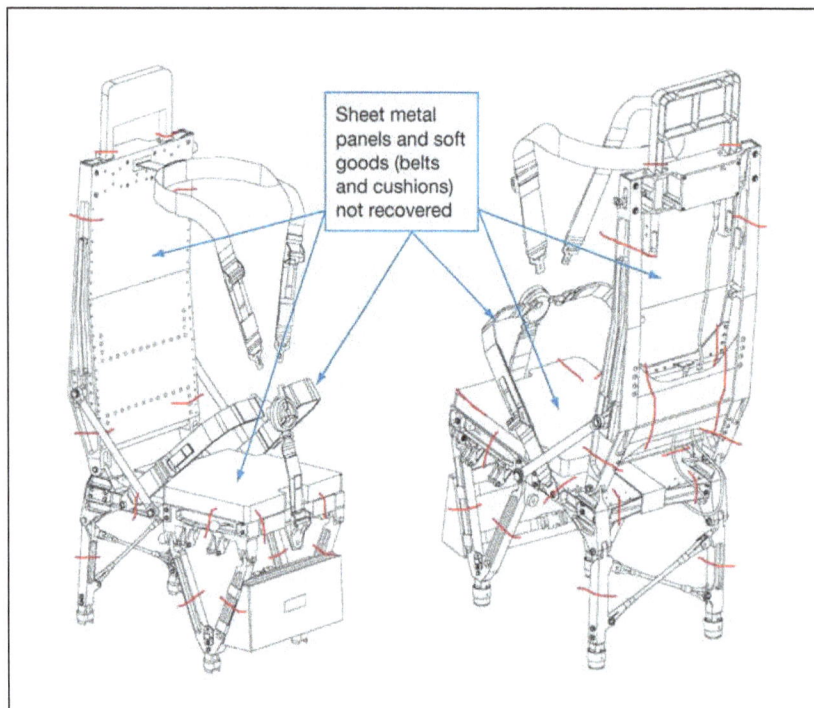

Sheet metal panels and soft goods (belts and cushions) not recovered

Figure 3.1-12. *Common seat failure locations* (shown in red). [Adapted from the Space Shuttle Systems Handbook]

Note: Mission Specialist seat shown; lower seat components on seats 1 and 2 are different.

---

[6]A heat sink is an area of the structure that has more material and takes longer to heat when exposed to elevated temperatures. Areas of minimal thermal cross-sectional area have less material and take less time to heat when exposed to elevated temperatures.

Examples of the fracture occurring at minimal thermal cross-sectional areas are evident on the seat leg failures for seats 6 and 7. Seats 6 and 7 are attached to a middeck floor panel that is the lid to the LiOH sub-floor storage compartment (figure 3.1-13). This panel, known as the "LiOH door," and the attached legs pieces are shown in figure 3.1-14.

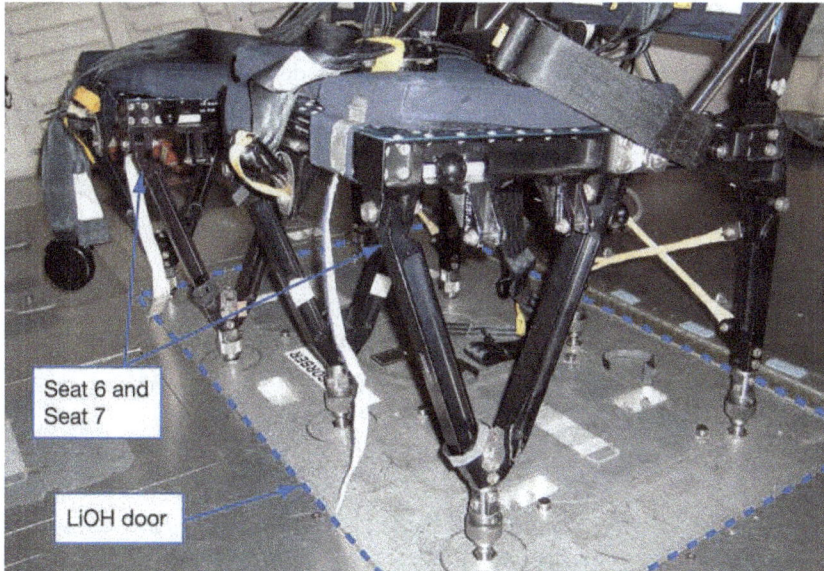

Figure 3.1-13. *Example of an intact lithium hydroxide door with seats 6 and 7 attached.* **[Picture from the Crew Compartment Trainer]**

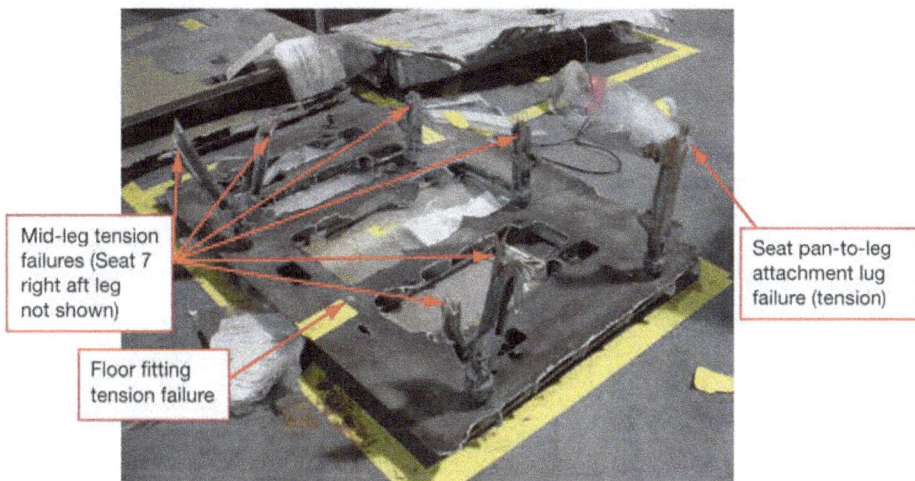

Figure 3.1-14. *Recovered lithium hydroxide door with seats 6 and 7 legs attached.*

Seats 6 and 7 leg failures occurred with the legs in tension. The lowest strength margin (and, therefore, the expected failure point) at room temperature is at the leg attachment lug at the top of the leg. Analysis indicates that the leg attachment lug should fail at around 12,000 lbs. at room temperature, but only the left aft leg of seat 6 failed at the attachment lug (figure 3.1-14). Another expected leg failure point is the seat attachment floor fitting. At room temperature, the floor fitting should fail at around 21,400 lbs., but only the right forward leg from seat 6 failed at the floor fitting.

Five of the eight seat legs failed at mid-leg locations. The room temperature failure load for mid-leg fractures of the forward leg is 45,000 lbs. and the aft leg failure load is 24,000 lbs.

The seat 7 right forward leg floor fitting is present, but the corresponding seat 7 leg was not recovered; therefore, the failure location was not the floor fitting. Because this leg is very close to the edge of the LiOH door, it is possible that the failure occurred because the locking collar[7] was thermally damaged as the LiOH door/seat 6/seat 7 complex experienced entry heating.

Structural assessments were performed on the legs (which are made of 7075 aluminum, with a melting point between 890°F (477°C) and 1,175°F (635°C)) and the floor fitting (which is made of Inconel 718, with a melting point between 2,300°F (1,260°C) and 2,440°F (1,338°C)) to evaluate materials strengths with respect to temperature. As mentioned above, a force of 45,000 lbs. is required to cause a mid-leg tension fracture of the forward leg at room temperature. The required fracturing force decreases to 33,000 lbs. (~75% of room temperature strength) at 300°F (149°C), 21,000 lbs. (~45%) at 400°F (204°C), 9,000 lbs. (~20%) at 500°F (260°C), and 4,950 lbs. (~11%) at 600°F (316°C).

A force of 21,400 lbs. is required to fracture the floor fitting at room temperature. This force decreases to 20,300 lbs. (~95% of room temperature strength) at 300°F (149°C), 20,100 lbs. (~94%) at 400°F (204°C), and 19,900 lbs. (~93%) at 600°F (316°C). These values are plotted in figure 3.1-15.

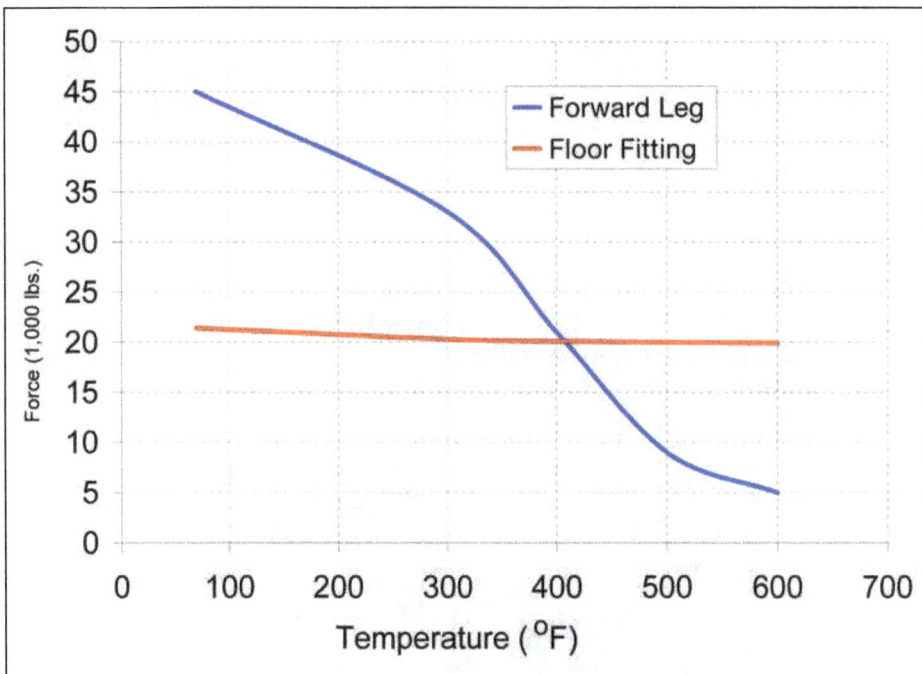

Figure 3.1-15. *Failure force of seat forward leg and floor fitting vs. temperature.*

This plot reveals that heating greater than 400°F (204°C) is needed to weaken the materials such that the leg failure would occur at the mid-leg location. Otherwise, the forward legs should have failed at the floor fitting, which did not occur.

The nylon material used for the seatbelts will lose strength as the temperature increases above 250°F (121°C), and will melt at approximately 400°F (204°C). Therefore, the same heating event that caused material properties changes in the metallic components of the seat structure also resulted in a complete loss of the nylon seat restraints.

---

[7]The locking collar is used to lock the seat leg to the floor fitting.

Inspection of the seat structure fracture surfaces revealed a "delamination" fracture pattern that is consistent throughout the seat debris items. The most drastic of the fractures makes it appear almost as if the 7075 aluminum material is constructed of a laminate material (figure 3.1-16). This phenomenon is termed a "broom-straw" fracture.

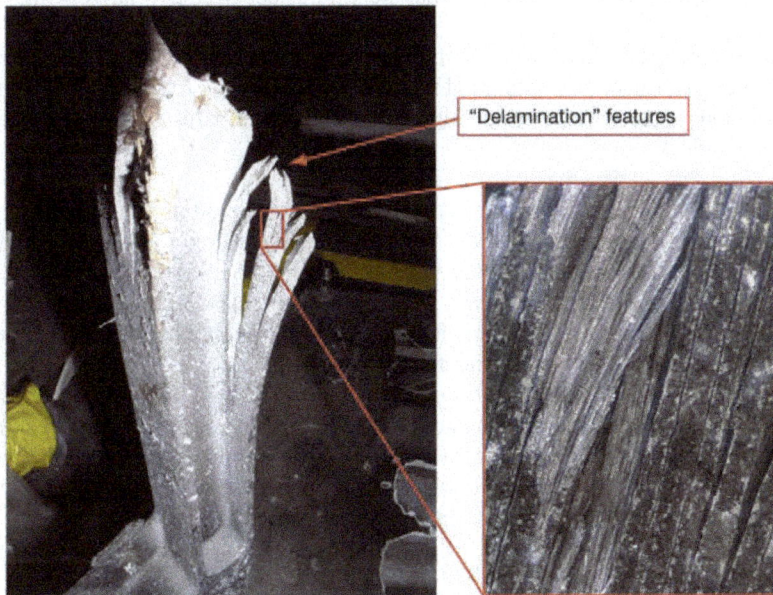

Figure 3.1-16. *Example of a "broom-straw" fracture on a Columbia seat leg.*

Scanning electron microscope analysis of the seat broom-straw fracture surface cross sections revealed fully intergranular fractures and equiaxed grain[8] microstructure. This finding is consistent with exposure to elevated temperatures and high strain rates. Equiaxed grains were discovered both along and away from the crack surfaces (figure 3.1-17).

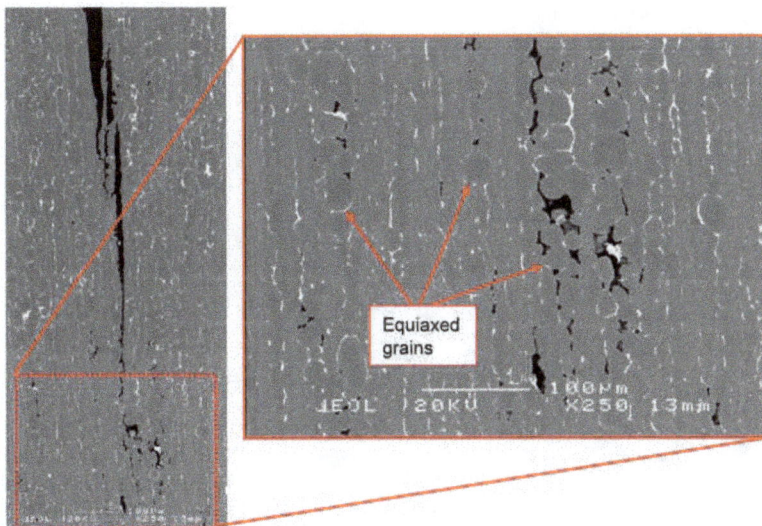

Figure 3.1-17. *Scanning electron microscope image of broom-straw fracture surface cross sections.*

---

[8]A grain of approximately the same size in all three dimensions; characteristic of a recrystallized microstructure.

Metallurgical evaluation was completed in proximity to, and away from, the crack surfaces using energy dispersive X-ray spectroscopy. This revealed heavy grain boundary precipitation that is consistent with the aluminum alloy experiencing temperatures greater than 900°F (482°C).

All of these features are consistent with material that is exposed to high temperatures. Significant lack of ductility surrounding the fracture areas indicates that the failures occurred at relatively high strain rates.

In all cases, seat failure occurred as a result of thermal exposure (resulting in material property degradation) followed by mechanical overload. The additional thermal degradation on the flight deck seat components is accounted for by a longer period of heating as a result of the higher ballistic number associated with an intact, free-flying flight deck.

Analysis of the debris led the CAIB to conclusions[9] regarding the method of seat failure (i.e., thermal exposure followed by mechanical loading), but left one question unanswered: "Why did the seats fragment so much?" As was the case with much of the *Columbia* debris, many tiny debris impact craters and/or material deposits were found on the seat debris. However, very few witness marks (dents, scrapes, divots, etc.) were found, indicating little or no impacts with debris items larger than roughly 0.25 in. Therefore, debris-debris collisions are probably not a factor in the fragmentation of the seat structure. Because the failure mechanism involved heating of the structure to temperatures exceeding 400°F (204°C) (and probably exceeding 900°F (482°C)), the nylon seat restraint material was not present as the seat structure was breaking. This led the Spacecraft Crew Survival Integrated Investigation Team (SCSIIT) to conclude that the seat was unoccupied at the time it was breaking up. No other conclusion can be made because the thermal and aerodynamic mechanisms in the high-altitude, hypersonic flight regime are not well understood.

3.1.4 Upper seatbacks/inertial reels

Six out of the seven upper seatback items (all but seat 5) were recovered. Each item was positively identified to a seat position. Each recovered upper seatback debris item contained the inertial reel/recoil mechanism and some amount of strap material, which recoiled into the inertial reel housing following strap failure. This strap material amounted to the only significant nylon webbing recovered from each of the seatbelt restraint systems. Figure 3.1-18 shows the location of the upper seatback/inertial reel.

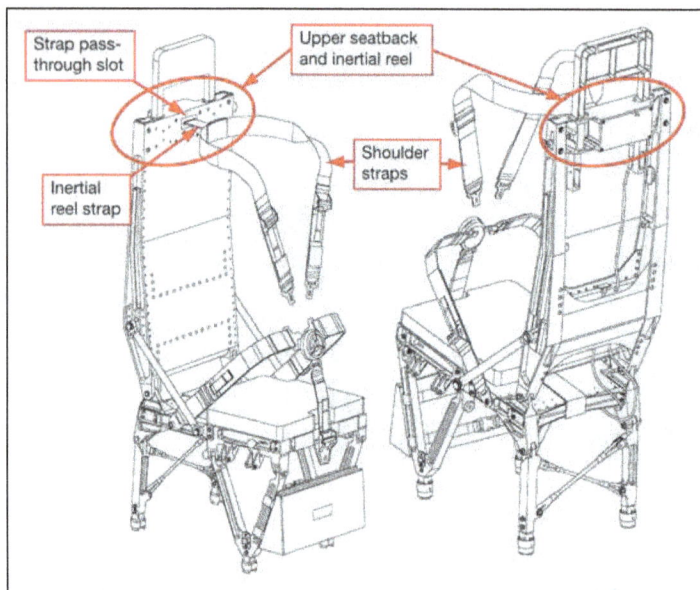

Figure 3.1-18. *Upper seatback debris item location*. [Adapted from the Space Shuttle Systems Handbook]

---

[9]*Columbia* Accident Investigation Board Report, Volume V, Appendix G.12, Crew Survivability Report, October 2003, p. 364.

As with the general seat failures that were described previously, all of the fractures at the upper seatback location occurred at the minimum thermal cross section, away from heat sinks on either side. Fracture surfaces did not exhibit melting or materials deposition, indicating that the upper seatback fractures occurred near the end of the period of heating and material deposition. Figure 3.1-19 shows all six recovered upper seatback items. Table 3.1-1 summarizes the upper seatback findings.

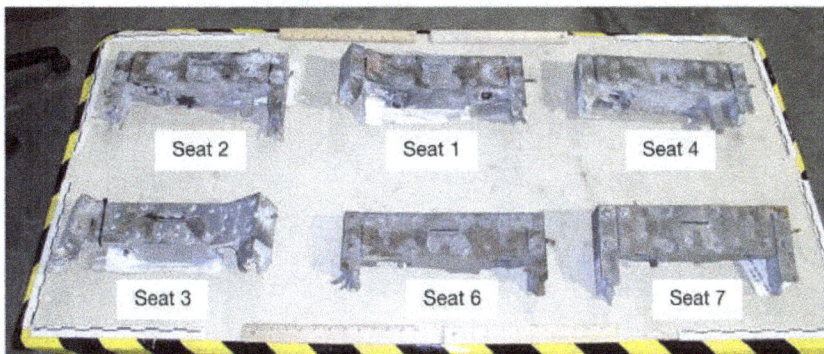

Figure 3.1-19. *Upper seatbacks* (front view).

Table 3.1-1. *Upper Seatback Comparisons*

Seat	Deformation	Fractures	Thermal Effects
Seat 1	Right frame member is bent forward, about the headrest bushing.	Tensile/bending fractures (with "broom-straw" features) are present on both the left and the right seatback frame members, just below the lower surface of the upper seatback/inertial reel assembly.	A moderate amount of splattered aluminum is present throughout, primarily deposited on the upper and lower surfaces (the upper surfaces more than the lower). No melting of fracture surfaces is noted.
Seat 2	Deformation is limited to the areas local to the fractures.	Tensile/bending fractures (with "broom-straw" features) are present on both the left and the right seatback frame members, just below the lower surface of the upper seatback/inertial reel assembly.	A moderate amount of splattered aluminum is present throughout, primarily deposited on the upper and lower surfaces (the upper surfaces more than the lower). No melting of fracture surfaces is noted.
Seat 3	Left and right frame members bent forward. Left side is bent about the headrest bushing; right side is bent just outboard of the bushing.	Tensile/bending fractures (with "broom-straw" features) are present on both the left and the right seatback frame members, just below the lower surface of the upper seatback/inertial reel assembly.	A large amount of splattered aluminum is present throughout, primarily deposited on the upper and lower surfaces. Noted the absence of deposited material on the portside bushing. No melting of fracture surfaces is noted.
Seat 4	Deformation is limited to the areas local to the fractures.	Tensile/bending fractures (with "broom-straw" features) are present on both the left and the right seatback frame members, just below the lower surface of the upper seatback/inertial reel assembly.	A large amount of splattered aluminum is present throughout, primarily deposited on the upper and lower surfaces. No melting of fracture surfaces is noted.
Seat 6	Slight deformation in the right frame member, bent about the headrest bushing.	Tensile/bending fractures (with "broom-straw" features) are present on both the left and the right seatback frame members, just below the lower surface of the upper seatback/inertial reel assembly.	A large amount of splattered aluminum material is deposited primarily on the lower surface. Noted a near complete absence of deposition on the upper surface. No melting of fracture surfaces is noted.

Table 3.1-1. *Upper Seatback Comparisons* (Continued)

Seat	Deformation	Fractures	Thermal Effects
Seat 7	Deformation is limited to the areas local to the fractures.	Tensile/bending fractures (with "broom-straw" features) are present on both the left and the right seatback frame members, just below the lower surface of the upper seatback/inertial reel assembly.	A large amount of splattered aluminum material is deposited on the lower surface. Noted a complete absence of deposition on the upper surface. No melting of fracture surfaces is noted.
Interpreta-tions		Seat fractures occur at minimum thermal cross section, away from heat sinks on either side.	Middeck items appear to have experienced significant splatter initiating from below. Flight deck items appear to have experienced significant splatter initiating from above and below. Lack of melting on the fracture surfaces indicates that the fractures occurred after the period of heating.
Conclusions	Observed deformations are affiliated with seat breakup fractures.	Fractures are consistent with general seat failure mechanism (thermal heating followed by mechanical overload).	Flight deck items experienced significant splattering initiating from above and below. Middeck items experienced significant splattering initiating from below *only*. Upper seatback fractures occurred after the period of heating.

For each seat, the inertial reel housing was removed from the upper seatback inertial reel cavity (figure 3.1-20). The cavity, housing, and mounting hardware were inspected for deformation, witness marks, debris impacts, and material deposition. Results for each seat position are generalized below.

Upper seatback (rear view).   Inertial rear cavity (rear view of intact seat).

Figure 3.1-20. *Inertial reel mounting location.*

Only seat 3 showed deformation of the inertial reel housing mounting hardware. All four mounting bolts were bent slightly. One inertial reel housing mounting lug was broken with melted metal deposits on both fracture surfaces; i.e., the fracture surface of the lug and the fracture surface of the inertial reel housing (figure 3.1-21). The inertial reel mounting hardware for the other five recovered upper seatbacks exhibited no deformations.

Mounting lug broken off

Figure 3.1-21. *Seat 3 inertial reel, looking aft.*

In all cases, melted strap material (nylon) was discovered inside the upper seatback inertial reel cavity, including on the strap rollers (figure 3.1-22). Only seat 3 had melted strap material on the external surface of the upper seatback. In this instance, the melted strap material flowed out of and away from the inertial reel strap pass-through slot, indicating that the melted material originated from inside the upper seatback inertial reel cavity. The lack of melted strap material on the outside of the seatbacks indicates that the torso restraint system failure occurred at the inertial reel strap and the remaining inertial reel strap retracted completely into the seatback, leaving no material to be melted and deposited on the external surfaces of the seats.

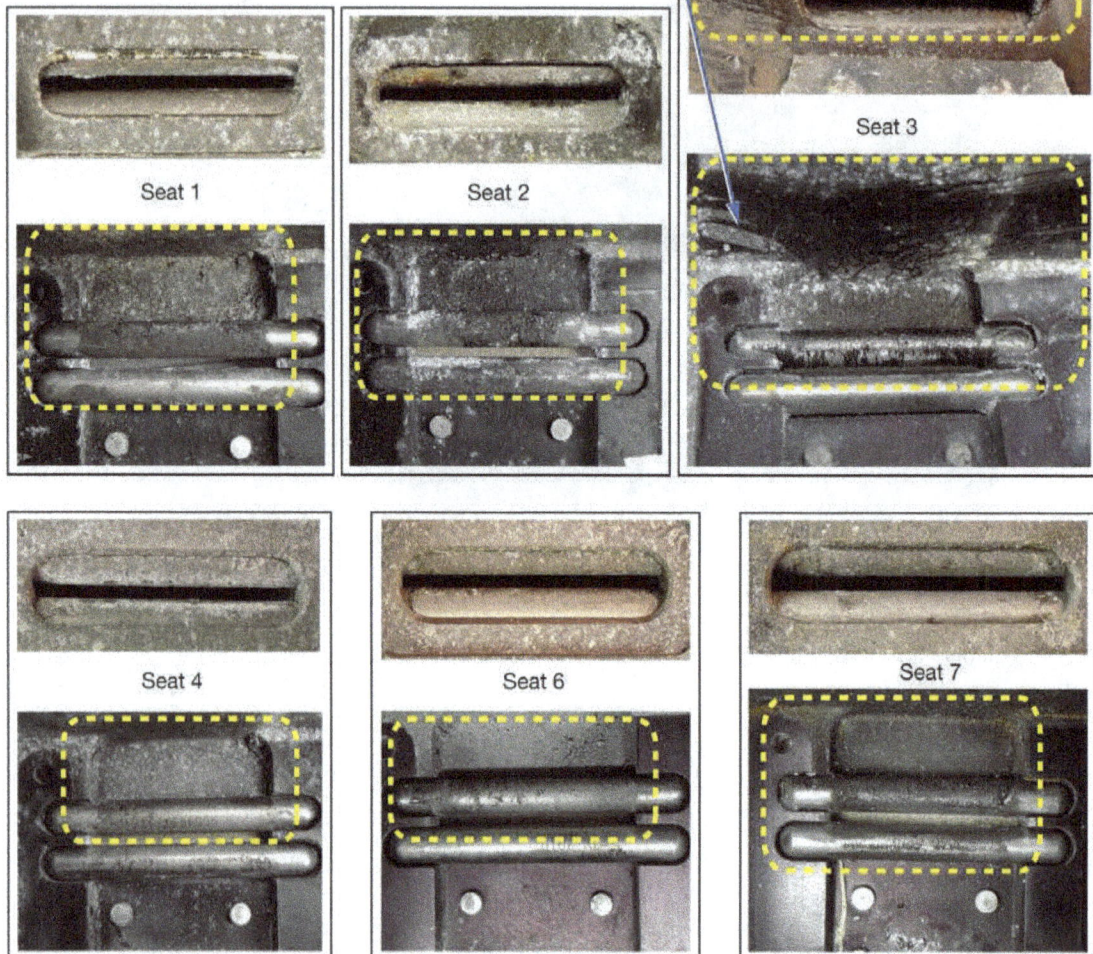

Figure 3.1-22. *Upper seatback inertial reel strap pass-through slots* – external (top)/internal (bottom); melted strap material areas outlined in yellow.

For all cases except seat 3, the melting and flow patterns of melted strap material are consistent with airflow entering the inertial reel lead-in strap pass-through slot in the upper seatback (forward-to-aft flow with respect to the seat). The melting and interior flow patterns of seat 3 are consistent with airflow entering the pass-through slot for the majority of the time that the seat was exposed to heating. Melted strap material on the exterior of the upper seatback indicates that airflow forced melted material out of the pass-through slot (aft-to-forward flow) for some period(s) of time. However, the radial flow pattern on the exterior surfaces indicates forward-to-aft airflow on the upper seatback once the melted material exited the slot. These melting and flow patterns could indicate that the upper seatback item was tumbling as it experienced heating.

In all cases, the inertial reel straps were melted only in areas that were immediately adjacent to the strap pass-through openings and along the lateral edges of the strap (figure 3.1-23). The strap material in other areas was not melted and appears normal.

(a) Front view.   (b) Side view (end cap removed).

Figure 3.1-23. *Inertial reel melted strap material.*

3.1.5 Inertial reel straps

Each recovered inertial reel mechanism was disassembled to inspect the inertial reel strap and the inertial reel locking mechanism. This inspection revealed that strap failures occurred at various locations along the strap length.[10]

Five of the six recovered strap ends terminated along a straight line. The exception was the strap for seat 2, which terminated in a jagged line. Away from, but in close proximity to, the melted areas, the residual strap material remains flexible. This is consistent with melting that occurred after the strap recoiled into the housing (the straight lines correspond to where the straps were exposed at the pass-through slot). If the whole strap was exposed to significant heating/melting, a definite thermal gradient (varying degrees of melting) and some melting and "pulling" at the broken end of the strap would be seen. However, none of the straps exhibited a thermal gradient along the strap or melting and pulling at the broken end. The demarcation between melted and non-melted areas is very distinct (figure 3.1-24), indicating that the straps were protected inside the inertial reel housings during the period of high heating.

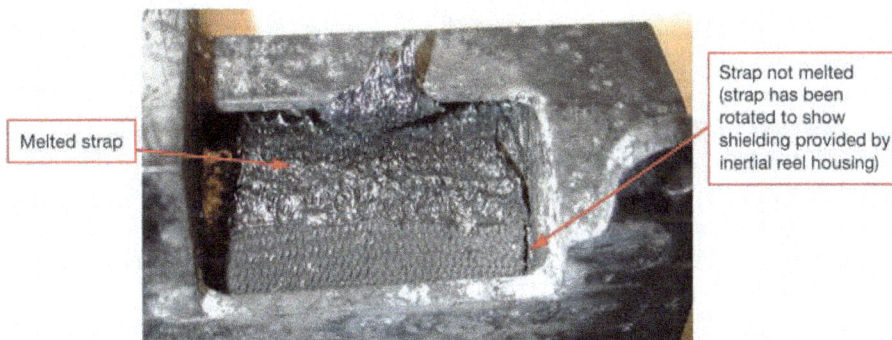

Figure 3.1-24. *Close-up of strap melt pattern.*

---

[10]The seats normally have 22 in. of inertial reel strap when measured from the reel to the shoulder harness y-split attachment.

The inertial reel straps failed primarily due to mechanical overload. Limited melting of the straps (and only in distinct areas that were unprotected by the inertial reel housing) indicates that melting was not a significant factor in the inertial reel strap failure. The mechanical overload may have been affected by elevated temperatures that weakened the straps, but these temperatures were not sufficient to cause obvious thermal damage to the strap. Significant strap melting occurred after the inertial reel straps failed and recoiled into the housing. Therefore, the inertial reel straps failed (and the *Columbia* crew members were, at most, partially restrained in their seats) prior to the end of the period of thermal exposure.

Metallic material was discovered on upper and lower strap surfaces when the straps were extended for inspection. Melted aluminum material was deposited on the straps for four of the six recovered inertial reels. The two exceptions were seats 1 and 2, both of which had strap failures at or near the inertial reel end of the straps. The deposition differed in character from the top-level depositions seen on the windows, which were more diffuse and uniform. Seat strap deposition consisted of globules or spheroids of metallic material (> 1/32 in.) that are widely scattered across the strap. Generally, there were no more than three or four deposited globules of metal per strap. Deposits on seats 4, 6, and 7 straps were close to the inertial reel end of the straps – areas of the straps that would not be exposed when the crew members were sitting upright in their seats. For material deposition to occur in these areas, the crew members had to be leaning to the side or leaning forward (or a combination of both), thereby extending the straps out of the inertial reel housing. The deposition of these material globules on the straps occurred before crew separation from the seats, yet after the CM pressure shell had been breached and the CM had been depressurized.

Evidence from the inertial reel straps indicates that the seats 1, 2, and 3 straps were mostly extended at the time of strap failure. The seats 4, 6, and 7 straps were extended during a material deposition period (seat 4 at least 8 in., or ~36% extended; seat 6 at least 21.25 in., or ~96% extended; and seat 7 at least 21.5 in., or ~98% extended). This indicates that the crew members were in seats 1 through 4 and seats 6 and 7 with at least the shoulder belts attached to the seatbelt buckle (the seat 5 upper seatback was not recovered). It is concluded that the inertial reels did not lock.[11]

Finding. Evidence from the inertial reel straps indicates that the seats 1, 2, and 3 straps were mostly extended at the time of strap failure. The seats 4, 6, and 7 straps were extended during a material deposition period (seat 4 at least 8 in., or ~36% extended; seat 6 at least 21.25 in., or ~96% extended; and seat 7 at least 21.5 in., or ~98% extended). Medical evidence (see Section 3.4) indicates that some of the crew members received injuries consistent with insufficient upper body restraint.

> *Conclusion L2-2.* The seat inertial reels did not lock.

> *Conclusion L2-3.* Lethal injuries resulted from inadequate upper body restraint and protection during rotational motion.

> *Recommendation L1-3/L5-1.* Future spacecraft crew survival systems should not rely on manual activation to protect the crew.

> *Recommendation L2-4/L3-4.* Future spacecraft suits and seat restraints should use state-of-the-art technology in an integrated solution to minimize crew injury and maximize crew survival in off-nominal acceleration environments.

---

[11]This conclusion is consistent with the entry simulation X-axis loads (discussed in Section 2.1.3 and shown in figure 2.1-16) remaining below the inertial reel auto-lock threshold of 1.78 G. Additionally, this conclusion is consistent with findings described in the Department of Defense Joint Service Specification Guide JSSG-2010-7, Crash Protection Handbook. The handbook describes a failure mode of MA-6 type inertial reels in which the reels failed to lock in crashes involving X-axis loads below the auto-lock threshold and subsequent Z-axis loads forcing the seat occupant down and forward, resulting in the occupant's torso being unrestrained during the crash dynamics. The handbook states that the "MA-6/MA-8 units were shown to be deficient in design and proven to be unreliable in survivable crash conditions." This precipitated a revision to MIL-R-8236, which is the military specification governing performance criteria for inertial reels used in military aircraft. The update to the military specification and the publication of the crash protection handbook occurred after the shuttle seats were designed.

Recommendation L2-8. The current shuttle inertial reels should be manually locked at the first sign of an off-nominal situation.

Indentations matching the linear ridges on the inertial reel spool were observed 0 to 4 in. from the strap attach location (figure 3.1-25). These linear indentations near the inertial reel, were present on all the straps (except the seat 1 strap, which failed next to the inertial reel attachment). Inertial reel straps for 11 other shuttle flight seats in inventory (not *Columbia* debris) and training seats were inspected for similar linear features. The inertial reel straps for all seats exhibited the same waves that were found on the *Columbia* inertial reel straps. Therefore, the waves are a result of strap stowage, not an indication of loading on the strap.

Linear features matching ridges on initial reel spool

Figure 3.1-25. *Example of linear waves on inertial reel straps.*

The inertial reel strap findings for each upper seatback are summarized below. Table 3.1-2 compares the inertial reel strap findings.

Table 3.1-2. *Inertial Reel Strap Comparisons*

Seat	Strap Failure Location	Extent of Melting/Condition of Residual Strap Material
Seat 1	The strap failed at approximately 2% of the extended length, in proximity to the attachment to the inertial reel (< 1/2 in. away from the inertial reel, out of the original 22 in. of strap).	Residual strap material exists only at the end of the strap still attached to the inertial reel spool.
Seat 2	The strap failed at approximately 20% of the extended length, in proximity to the attachment to the inertial reel (~4.5–5 in. away from the inertial reel, out of the original 22 in. of strap). The strap failed at an approximately 45-degree angle, over a 1- to 2-in. length.	The strap melted only at the exposed strap pass-through areas and the lateral edges of the inertial reel housing. The strap material that was shielded by the inertial reel housing did not melt. This suggests that mechanical failure occurred before significant thermal exposure.
Seat 3	The strap failed at approximately 90% of the extended length, in proximity to the attachment to the shoulder harness y-split (~20 in. away from the inertial reel, out of the original 22 in. of strap). Strap failure appears to have occurred in close proximity to the shoulder split attachment, probably at the stitch stress concentration. Two raised sections at a slight angle (~20 degrees) from perpendicular to the strap axis, observed 2 to 4 in. from the strap attach location. Marks are similar to "bird-caging" features found in dynamic failures of cable/cord. Approximately 10 small metallic bits of debris, largest approximately 1/8 in. in diameter located approximately 16.5 to 18 in. away from reel attachment.	The strap melted only at the exposed strap pass-through areas and the lateral edges of the inertial reel housing. The strap is melted completely through one layer near the upper pass-through. The strap material that was shielded by the inertial reel housing is not melted. This suggests mechanical failure before significant thermal exposure.
Seat 4	The strap failed at approximately 80% of the extended length, in proximity to the attachment at the shoulder harness y-split (~18 in. away from the inertial reel, out of the original 22 in. of strap). One very small ball (< 1/16 in.) of melted metallic material was located approximately 14 in. from the reel attachment, on the lateral edge of the upper strap surface. (Note that the debris came off during inspection.)	The strap melted only at the exposed strap pass-through areas and the lateral edges of the inertial reel housing. The strap is melted completely through one layer near the upper pass-through. The strap material that was shielded by the inertial reel housing is not melted. This suggests mechanical failure before significant thermal exposure.
Seat 6	The strap failed at approximately 90% of the extended length, in proximity to the attachment at the shoulder harness y-split (~20 in. away from the inertial reel, out of the original 22 in. of strap). Strap failure appears to have occurred in close proximity to the shoulder split attachment, probably at the stitch stress concentration. Three pieces of metallic debris were deposited on the strap surfaces: 1. A small ball of melted material (< 1/32-in. dia.) deposited on the upper surface, located approximately 16 in. from the reel attachment. (Note that described debris came off during inspection.) 2. A small fragment of melted material (< 1/32-in. dia.) deposited on the lateral strap edge, located approximately 13 in. from the reel attachment. 3. A small fragment of melted material (< 1/32-in. dia.) deposited on the lateral strap edge, located approximately 1.75 in. from the reel attachment.	The strap melted only at the exposed strap pass-through areas and the lateral edges of the inertial reel housing. The strap is melted completely through one layer near the upper pass-through. The strap material that was shielded by the inertial reel housing is not melted. This suggests that mechanical failure occurred before significant thermal exposure.

Table 3.1-2. *Inertial Reel Strap Comparisons* (Continued)

Seat	Strap Failure Location	Extent of Melting/Condition of Residual Strap Material
Seat 7	The strap failed at approximately 50% of the extended length (~11 in. away from the inertial reel, out of the original 22 in. of strap). Three pieces of metallic debris were deposited on the strap: 1. One approximately 1/10-in. diameter, roughly spherical piece of metal debris is stuck to the edge (the right side of seat) of the lower surface of the strap, approximately 1.5 in. from the inertial reel. 2. One 0.5 in. × 0.3 in. roughly triangular (~0.05-in.-thick) piece of metal debris is stuck to the lower surface of the strap, approximately 8 in. from the inertial reel. 3. Several small, roughly spherical metal debris items are stuck to the upper side of the strap fragment, along the edge, approximately 2 in. from the end.	The strap melted only at the exposed strap pass-through areas and the lateral edges of the inertial reel housing. The strap is melted completely through one layer near the upper pass-through. The strap material that was shielded by the inertial reel housing is not melted. This suggests that mechanical failure occurred before significant thermal exposure.
Interpreta-tions	Seats 1 and 2 straps failed at or near the end of the strap at the inertial reel (the belt was almost fully extended). The seat 3 strap shows evidence ("birdcage" witness marks near the inertial reel) suggesting that the belt was almost fully extended at the time of failure. The seat 4 strap shows evidence (melted material deposited approximately 14 in. from inertial reel) indicating that the belt was partially (~8 in. or ~36%) extended during the period of material deposition. The seat 6 strap shows evidence (melted material deposited approximately 1.75 in. from the inertial reel) indicating that the belt was mostly (~21.25 in. or ~96%) extended during the period of material deposition. The seat 7 strap shows evidence (melted material deposited approximately 1.5 in. from the inertial reel) indicating that the belt was mostly (~21.5 in. or ~97%) extended during the period of material deposition.	The straps exhibit areas of flexible strap material right next to areas of melted strap. The "border" between the areas is defined by the presence of the inertial reel housing shielding portions of the strap. If thermal exposure (significant strap melting) was a major factor in the failure of the strap, a thermal exposure gradient over some finite length should be present on the straps. The lack of a thermal exposure gradient, and the presence of evidence indicating drastic differences in thermal exposure, point to the strap failure mode as being primarily mechanical in nature.
Conclusions	The seats 1, 2, and 3 straps were mostly extended at the time of strap failure. All others straps failed at locations ranging from 50 to 90% of strap length (away from recoil attach point). The seats 4, 6, and 7 straps were extended during the material globule deposition period: The seat 4 strap was at least 8 in. extended; the seat 6 strap was at least 21.25 in. extended; and the seat 7 strap was at least 21.5 in. extended. The presence of melted metal globules deposited on several of the straps indicates that the mechanical overload of the straps occurred after exposure to a thermal environment resulting in globule deposition. This evidence confirms crew members were in seats 1, 2, 3, 4, 6, and 7 at the time of inertial reel strap failure, and the inertial reels did not lock.	In all cases, mechanical overload of the inertial reel strap occurred independent from significant thermal degradation. Note that the failure of the inertial reel strap may have been affected by material property deg-radation due to elevated temperatures.

Seat 1 experienced strap failure at the end of the strap next to the inertial reel, and seat 2 experienced strap failure near the end of the strap next to the inertial reel. The seat 1 strap failure (figure 3.1-26) occurred in a straight line along the strap attach point shear plane (the expected failure point if the strap is fully extended).

The seat 2 strap failed at an approximate 45-degree angle, over a 1- to 2-in. length, approximately 4.5 in. from the inertial reel (figure 3.1-26). The way in which the seat 2 belt failed is unique among the six recovered inertial reel straps. This failure is similar to a failure mode known as strap "dumping."[12] It is theorized that loading of the inertial reel strap laterally against the pass-through slot (figure 3.1-27) resulted in damage to the strap and, eventually, strap failure.

Figure 3.1-26. *Seat strap failures: seat 1* (left) *and seat 2* (right).

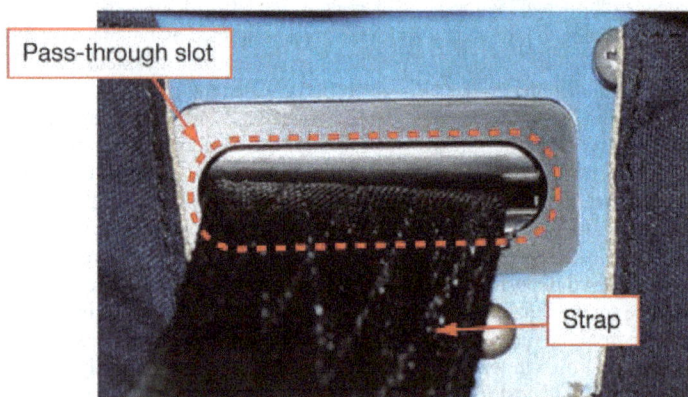

Figure 3.1-27. *Seat strap pass-through slot (intact seat shown).* [Picture from a shuttle training mockup in the JSC Space Vehicle Mockup Facility]

Finding. The seat 2 inertial reel strap exhibits "strap dumping" failure features. The strap failed progressively, possibly due to damage to the lateral edge of the strap from contact with the sharp edge of the strap pass-through slot.

> *Recommendation L2-5.* Incorporate features into the pass-through slots on the seats such that the slot will not damage the strap.

The seat 3 inertial reel strap (figure 3.1-28) failed at approximately 90% of the extended length (~20 in. away from the inertial reel). This strap failure occurred in close proximity to the shoulder harness y-split attachment, which is the expected failure point if a strap is not fully extended (the inertial reel strap will fail at around 5,200 to 5,300 lbs.). Wave features, which were observed approximately 2 to 4 in. from the strap attach location at a slight angle (~20 degrees) from perpendicular to the strap axis, were similar to the "birdcaging"[13] features found in dynamic failures of cable/cord. Because these birdcage features are close to the inertial reel end of the strap, the strap failure occurred with the strap mostly extended (retracted only 2 to 4 in.). Approximately 10 small metallic bits of debris, the larger of which are approximately 1/8 in. in diameter, were located approximately 16.5 to 18 in. away from the inertial reel attachment.

---

[12]Dumping is a strap failure mode caused by progressive strap failure that can be preceded by damage to the lateral edge of a strap.

[13]Describes the appearance of a multistranded rope or strap that has been subjected to compression or a sudden release of tension load. The outer strands are displaced outward, forming a cage-like appearance.

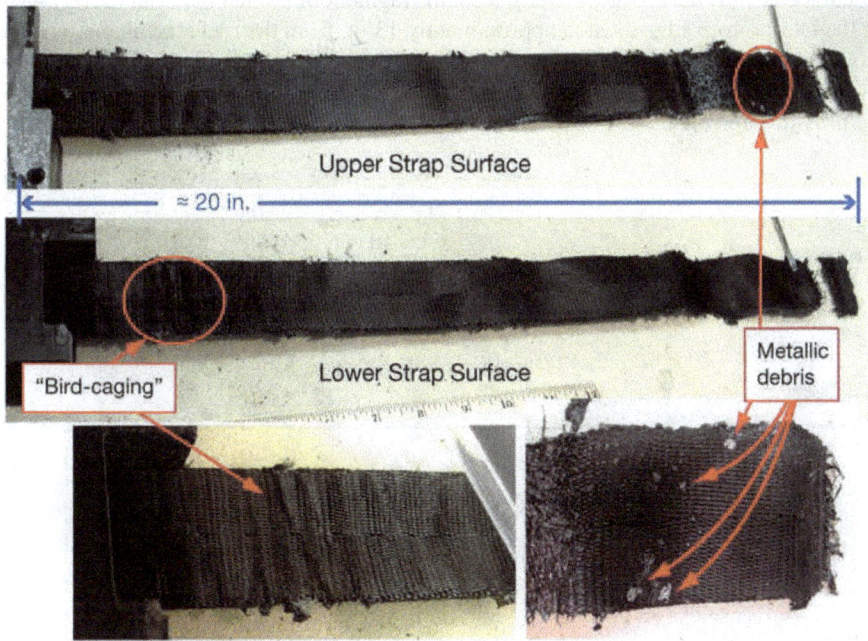

Figure 3.1-28. *Seat 3 inertial reel strap.*

The seat 4 inertial reel strap (figure 3.1-29) failed at approximately 80% of the extended length, which is in proximity to the attachment at the shoulder harness y-split (~18 in. away from the inertial reel). One very small ball of melted metallic material (< 1/16 in.) was located approximately 14 in. from the inertial reel attachment on the lateral edge of the upper strap surface (the debris came off during inspection.). No witness marks were observed along the entire strap length.

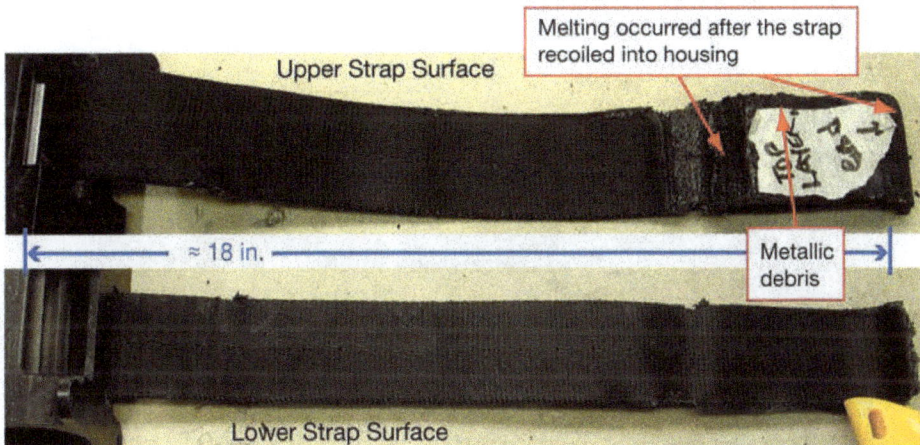

Figure 3.1-29. *Seat 4 inertial reel strap.*

The seat 6 inertial reel strap (figure 3.1-30) failed at approximately 90% of the extended length (~20 in. away from the inertial reel). Strap failure occurred in close proximity to the shoulder harness y-split attachment (the expected failure point if the strap was not fully extended). No witness marks were observed along the entire strap length. Three pieces of metallic debris were deposited on the strap surfaces: a small ball of melted material (< 1/32-in. dia.) deposited on the upper surface, approximately 16 in. from the reel attachment (debris came off during inspection); a small fragment of melted material (< 1/32-in. dia.) deposited on the lateral strap edge located approximately 13 in. from the reel attachment; and a small fragment of melted material (< 1/32 in. dia.), which was deposited on the lateral strap edge located approximately 1.75 in. from the reel attachment. The strap was melted completely through one layer near the strap pass-through opening.

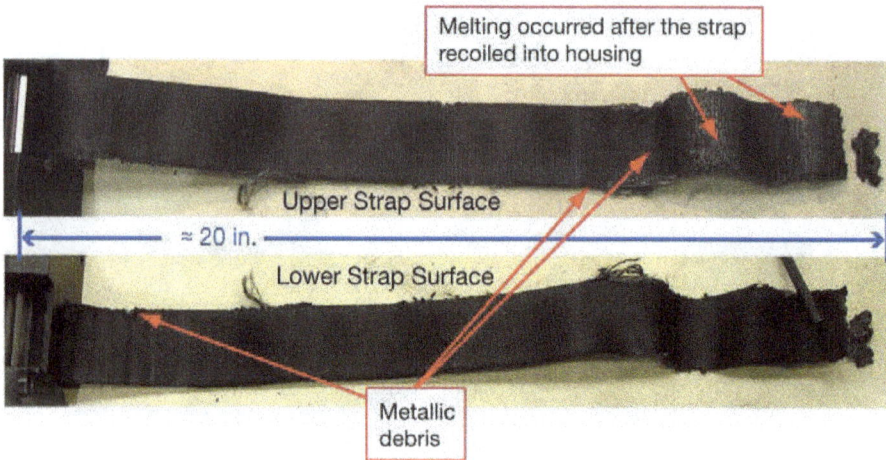

Figure 3.1-30. *Seat 6 inertial reel strap.*

The seat 7 inertial reel strap (figure 3.1-31) failed at approximately 50% of its length (~11 in. away from the inertial reel). No witness marks were observed along the entire strap length. Three areas of metallic debris deposits were noted on the strap: one approximately 1/10-in.-diameter, roughly spherical piece of metal debris stuck to the edge (the right side of seat) of the lower surface of the strap, approximately 1.5 in. from the inertial reel; one 0.5-by-0.3-in. roughly triangular (~0.05-in.-thick) piece of metal debris stuck to the lower surface of the strap, approximately 8 in. from the inertial reel; and several small, roughly spherical metal debris items stuck to the upper side of the strap fragment, along the edge, approximately 2 in. from the end. Localized melting caused the strap to separate into two pieces after the strap had recoiled into the housing.

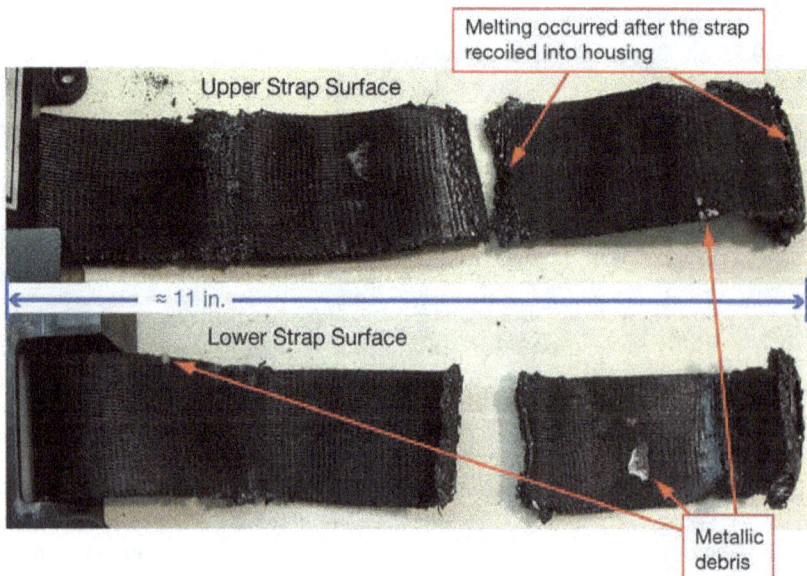

Figure 3.1-31. *Seat 7 inertial reel strap.*

The inertial reel straps are certified to sustain at least 5,000 lbs. of load (at least 2,500 lbs. of load when the strap is fully extended). Vendor testing indicates that the straps fail at approximately 5,200 lbs. (or at ~3,500 lbs. if the strap is fully extended). These values are for straps tested statically at room temperature; the loads required to fail the straps at elevated temperatures are unknown. Additionally, the material properties (combustion vs. chemical degradation vs. melting) in a high-temperature/low-O_2/low-pressure environment are not known, neither are the strap properties in highly dynamic loading situations (high loads over very short time periods). Therefore, the inertial reel straps alone could not provide sufficient evidence for determining the loads at which the straps failed.

Finding. All inertial reel straps are tested with static loads at room temperature. Load testing has not been conducted to determine the loads required to fail the straps at elevated temperatures or under dynamic loads. Testing has not been conducted to determine the material properties (combustion vs. chemical degradation vs. melting) in a high-temperature/low-O_2/low-pressure environment.

> **Recommendation L2-6.** Perform dynamic testing of straps and testing of straps at elevated temperatures to determine load-carrying capabilities under these conditions. Perform testing of strap materials in high-temperature/low-oxygen/low-pressure environments to determine materials properties under these conditions.

3.1.6 Inertial reel locking mechanisms

After inspection of the inertial reel straps was completed, each of the inertial reel/recoil mechanisms was disassembled (figure 3.1-32).

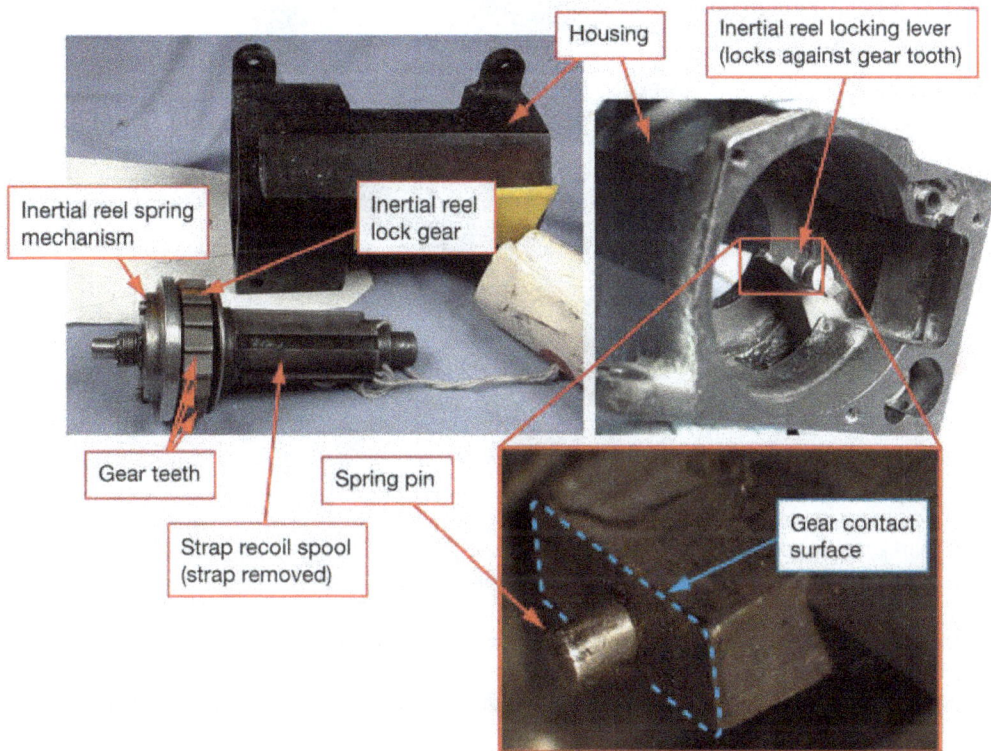

Figure 3.1-32. *Inertial reel mechanism.*

The shuttle inertial reel can be locked manually; it also has an auto-locking feature that will lock with a strap acceleration of 1.78 G to 2 G (accelerations pulling the strap out at ~57 to 64 ft/sec²). The automatic lock functions when the inertial reel spring mechanism engages the inertial reel locking lever against the corresponding inertial reel locking gear tooth surface. The strap is prevented from further extension, although recoil is possible. A spring pin is in the center of the locking lever contact surface.

All six of the recovered inertial reel mechanism gears and locking levers were inspected under stereomicroscope for evidence of mechanical loading, impact, adhesive wear, cracking, and plastic deformation. The results were consistent in all cases: the gear contact surface of the inertial reel locking lever showed no obvious witness marks; and only one tooth on *each* of the six mechanisms on the inertial reel locking gear showed evidence of witness marks (figure 3.1-33).

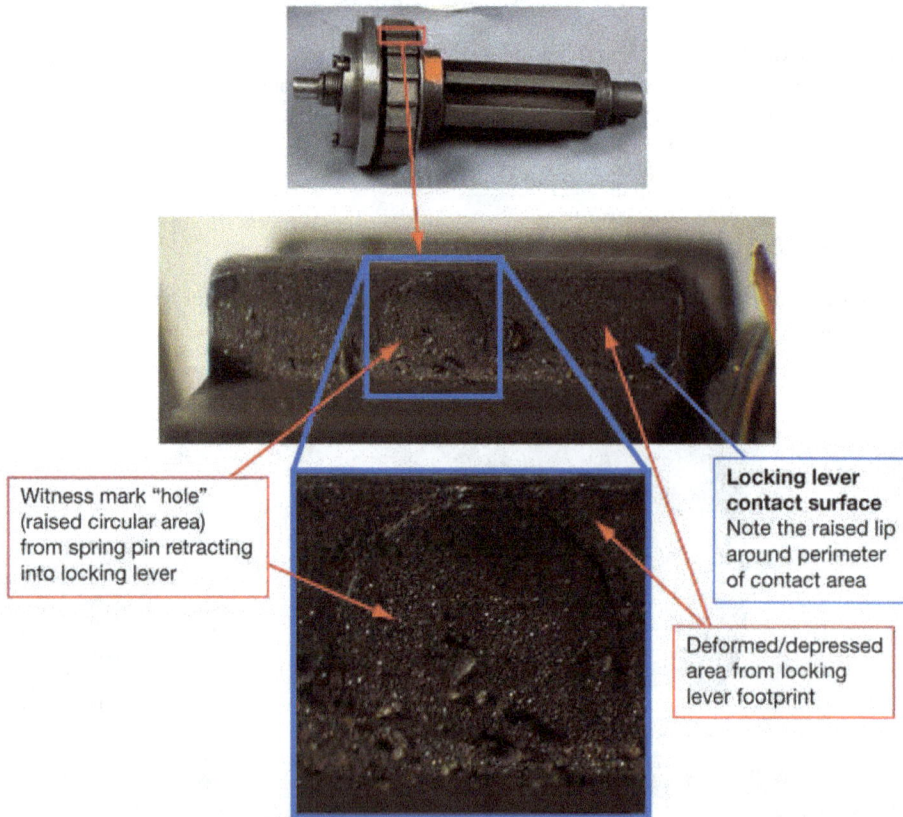

Witness mark "hole" (raised circular area) from spring pin retracting into locking lever

Locking lever contact surface
Note the raised lip around perimeter of contact area

Deformed/depressed area from locking lever footprint

Figure 3.1-33. *Inertial reel locking gear witness mark.*

The witness marks can be explained by a significant loading event causing force translation through the functional strap restraint system and the inertial reel locking lever, and finally to the corresponding locking gear tooth surface. This force caused plastic deformation of the locking gear tooth, embossing the locking lever contact area on the gear tooth surface. The raised lip around the perimeter of the contact surface (figure 3.1-33) represents the outer edge of the locking lever contact area. The raised circular feature seen in this figure is caused by the spring pin retracting into the locking lever (below the contact surface of the locking lever). This feature is an area where the locking gear is *not* deformed by the locking lever.

It was initially thought that the witness marks were an indication of inertial reel strap loading during the accident. However, extensive investigation of inertial reels from training seats and other flown shuttle seats as well as a new inertial reel revealed identical circular witness marks (figure 3.1-34).

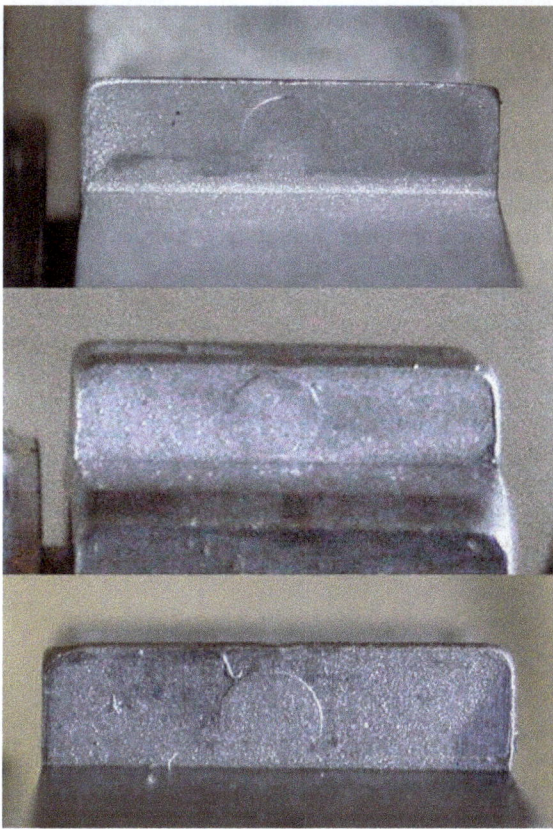

Figure 3.1-34. *Circular witness marks on inertial reels from a training seat* (top), *a flight* (non-*Columbia*) *seat* (middle), *and a new* (unused) *flight-qualified inertial reel* (bottom).

All inertial reels/straps of this type are proof-tested by the manufacturer with a 3,350-lb. static load prior to delivery to the customer. Because these inertial reels (and the *Columbia* inertial reels) were proof-tested, the investigation concluded that the witness marks on all of the inertial reels are a result of proof-testing. The absence of a second witness mark on the inertial reel led to the conclusion that the inertial reel straps failed at forces below the equivalent of a 3,350-lb. static load. However, it is probable that the straps failed under dynamic loading. As discussed above, the straps' load-carrying capabilities under dynamic loading are unknown.

A relatively large gap was discovered between the inertial locking mechanism and the lock gear on the seat 4 inertial reel (figure 3.1-35).

Figure 3.1-35. *Separation gap on the seat 4 inertial reel.*

Upon further inspection, it was noted that the locking mechanism had permanently "jumped" out of the normal track position, creating a gap between the inertial mechanism and the locking gear. Because this configuration would not function nominally, it is concluded that this "jump" occurred during the accident.

The observed gap between the inertial reel locking mechanism and the locking gear on seat 4 is consistent with information in the manufacturer's experience base and occurs in crash events resulting in accelerations above 100 G at the strap input. However, the inertial reel locking mechanism can also be moved away from the locking gear by an impact along the axis of the inertial reel spool. Therefore, the observed failure (and resulting gap) could have been caused by impacts during the breakup dynamics, ground impact, or actual loads on the inertial reel strap. Because the cause of failure cannot be determined positively, no conclusions are possible regarding seat 4 inertial reel strap loads/accelerations.

3.1.7 Belt adjusters

Each seat restraint system includes five seatbelt adjusters (figure 3.1-36). Four of the seatbelt adjusters (out of the 35 that were on *Columbia*) were recovered. All of the belts are the same width and all of the adjusters are identical, so determination of the origin (which seat or which belt) was not possible. The findings are described below.

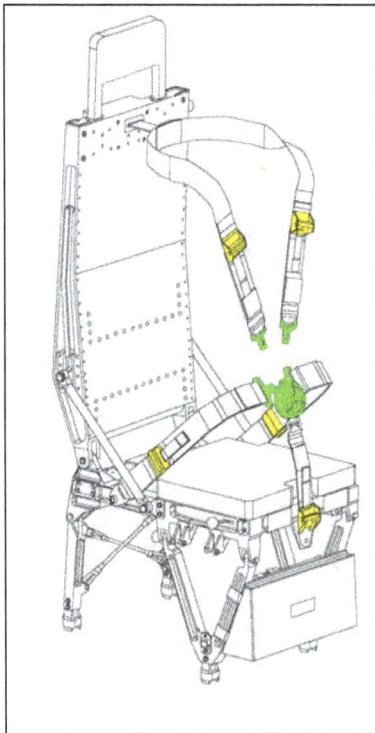

Figure 3.1-36. *Seat belt adjusters* (yellow) *and restraint buckle* (green). [Adapted from the Space Shuttle Systems Handbook]

The belt adjusters exhibited evidence of exposure to material deposition and heating. Molten metal debris impacts were observed in varying degrees (with no apparent directionality). Miniscule amounts of residual melted nylon belt material were observed on the adjusters.

Three of the four adjusters exhibited witness marks within the slider bar slot surface (figure 3.1-37). Tension on the belt will cause the slider bar to contact the slider bar slot as the bar rotates and slides within the slot. The witness marks in the adjusters are a result of significant loading on an intact restraint belt.

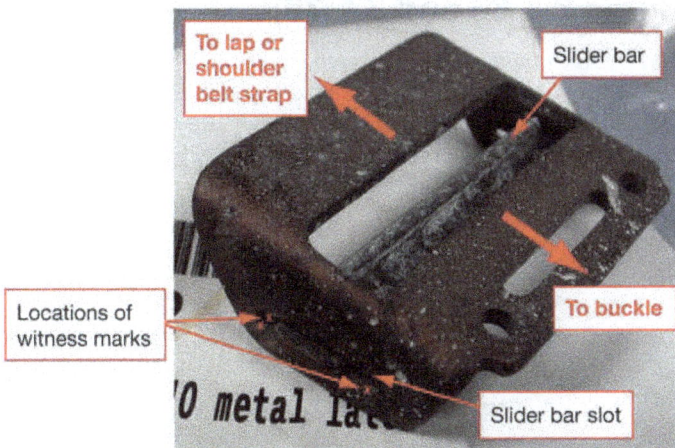

Figure 3.1-37. *Recovered belt adjuster rear view.*

The fourth adjuster was missing the slider bar, and experienced fractures on both the left and the right slider bar slots. These fractures were apparently caused by the slider bar "blowing out" away from the body surface (figure 3.1-38).

Figure 3.1-38. *Fractured belt adjuster.*

The fracture surfaces exhibited delamination fractures (the "broom-straw" fractures described in section 3.1.3), indicating material property degradation due to elevated temperature exposure combined with high strain rate loading. The heating occurred quickly and allowed material degradation of the metallic belt adjuster to occur without compromising the material properties of the nylon restraint belt to the point that the belt could not transmit forces to the adjuster.

The slider bar "blowout" on one of the four strap adjusters was the result of significant loading on an intact restraint belt. It cannot be determined positively when this strap adjuster failure occurred.

The witness marks and the slider bar "blowout" fracture are indicative of significant loading events causing force translation through the intact belt restraint system, resulting in the impact of the slider bar against the corresponding slot surface within the belt adjuster housing.

3.1.8 Restraint buckle (figures 3.1-36 and 3.1-39)

One of the seven 5-point seatbelt buckles on board *Columbia* was recovered (figure 3.1-40). Positive identification to a seat position was not possible. Although the buckle experienced significant entry heating, resulting in substantial melting of the outer plastic housing, the structure remained intact with all five belt tongues still in place.

Figure 3.1-40. *Recovered five-point seatbelt buckle (front and back).*

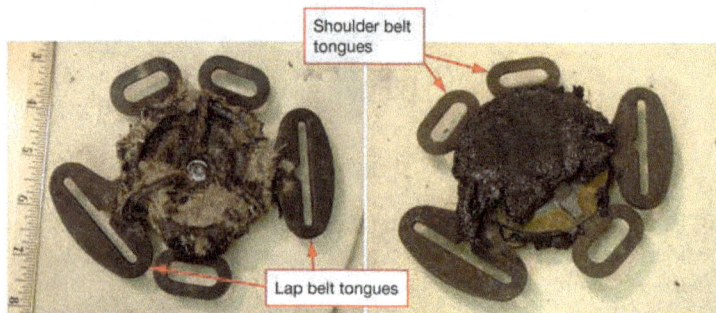

Figure 3.1-39. *Demonstration of the five-point seatbelt buckle.*

The buckle assembly was disassembled for analysis. The two shoulder belt tongues and the two lap belt tongues were bent outward slightly (i.e., away from the crew member). Both of the shoulder belt tongue latching pins had very shallow linear features corresponding to the mating surfaces of the belt tongue (figure 3.1-41). These features may be witness marks as a result of dynamic loading events that were experienced in flight by the restraint system or of forces imparted to the five-point attach buckle during disassembly by the CAIB investigation team. A definitive conclusion cannot be made based on these witness marks alone. The other latching pins did not exhibit any noticeable marks or deformations.

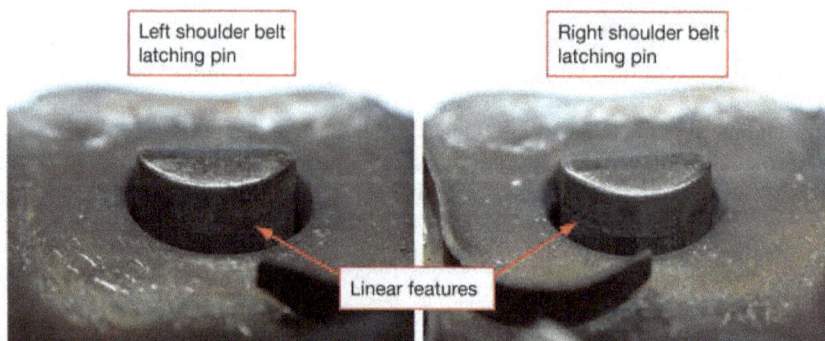

Figure 3.1-41. *Indentations on latching pins, looking from center of buckle towards outer edge.*

Shadowing[14] on the belt tongues indicates that the belt tongues were in the position shown in figure 3.1-42(b). These positions, when compared to the positions of the tongues being straight out from center of buckle (radially, figure 3.1-42(a)), are: the shoulder belt tongues are moved toward the centerline (medially), the crotch belt tongue is moved toward the right side of the seat, and the lap belt tongues are moved down. It cannot be determined positively when the deposition (and shadowing) occurred, so no conclusions can be made relative to timing.

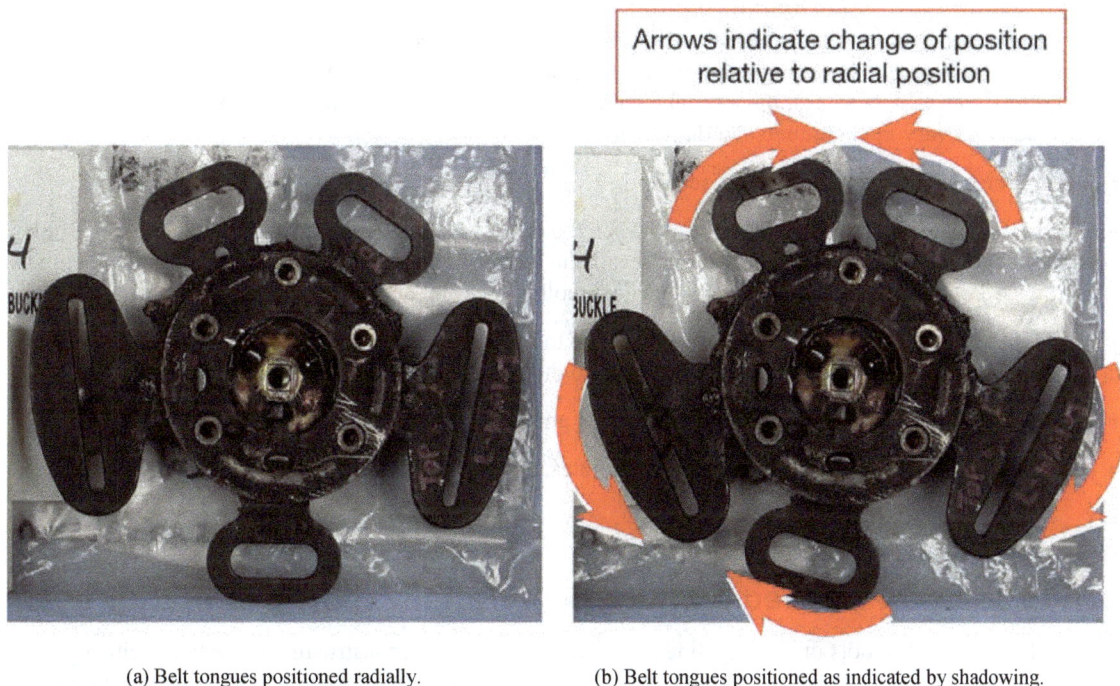

Arrows indicate change of position relative to radial position

(a) Belt tongues positioned radially.   (b) Belt tongues positioned as indicated by shadowing.

Figure 3.1-42. *Belt tongue positions.*

Because all of the seats use a common seat restraint design, none of these pieces could be positively identified as being from a specific seat. This means that the belt adjusters and the restraint buckle could be from any of the seven seats on *Columbia*. It is even possible that the items were from the same seat. However, because the items vary in condition (different amounts of heating and debris impacts) and were found in widely spread locations (> a 14-mile spread), it is unlikely that these items originated from the same seat.

Significant forces caused witness marks on the seat restraint buckle tongues and the belt adjusters. Witness marks on the belt adjusters were caused by force transmission through intact functional belts. In the case of the fractured belt adjuster, forces on the belt were sufficient to cause the adjuster slider bar to "rupture" out of the slot. Fracture surfaces indicate that this rupture occurred at an elevated temperature. However, the heating occurred quickly, allowing material property degradation of the metallic belt adjuster without compromising material properties of the nylon restraint belt to the point that the belt could not transmit forces to the adjuster.

---

[14]An area that lacks or has less material deposition when compared to an adjacent area. Shadowing indicates that another item covered the shadowed area, preventing deposition.

3.1.9 Sequence

From the evidence and conclusions described above, the SCSIIT was able to develop the following sequence of events related to the seats:

- All of the seats were installed.

- Crew members were at least partially strapped into seats 1, 2, 3, 4, 6, and 7.

- During vehicle loss of control (LOC), the dynamics caused the crew members to be pulled forward and/or side-to-side. If the component of the acceleration pulling the inertial reel out of the seat (forward) was less than 1.78 G to 2 G, the inertial reels would *not* lock and the straps could be going in and out as the loads vary in magnitude and direction.

- Cabin breach and depressurization occurred.

- During the period of material deposition (when the cabin was depressurized and molten metal globules were floating around in the vicinity of the seats), the inertial reel straps were extended (the crew members were still in their seats) and the straps received molten material deposits.

- The inertial reel straps failed (predominantly due to mechanical overload, but below the equivalent of a 3,350-lb. static load) and the straps retracted into the seatbacks.

- The remaining shoulder harness belts and crotch and lap belts eventually melted/burned away.

- The seats experienced heating and high strain rates, and broke up due to thermal and then mechanical effects.

3.1.10 Lesson learned – equipment serialization and marking

One of the most useful tools in investigating an aviation accident is reconstructing the vehicle, either physically or virtually, from the recovered debris. Being able to identify the original location within the vehicle of debris items is of utmost importance in achieving an accurate reconstruction. Identifying the origins of debris items is made possible by serializing individual piece parts and subassemblies, and keeping accurate records of the piece part/subassembly serial numbers at the assembly and, ultimately, the vehicle level. This is especially useful when there are multiple units of identical or similar components, such as crew equipment, seats, engines, or structural members.

As discussed above, there are two different types of seats – Pilot seats, which are used by the CDR and the PLT, and Mission Specialists seats. The main difference between these types of seats is the design of the seat pan and the legs. The seatbacks and seat restraints are identical in design and construction for both types of seat. When the seats were manufactured, individual seat components were ink-stamped with individual part numbers and serial numbers. Configuration management records for seat components[15] were not accurately maintained, however, so identifying component locations by any surviving piece-part serial number was futile. The exceptions to this were the components associated with the inertial reels; all six recovered upper seatbacks were identified to specific seat locations.

Initially in the *Columbia* investigation, the only seat debris pieces that could be positively identified to a specific seat location were the upper seatbacks, which contain the inertial reels, and any pieces that remained attached to identifiable floor pieces. For the remaining seat debris items, reconstruction and location identification was a time-consuming, laborious process of matching pieces with the upper seatbacks and those pieces that were attached to floor panels. Eventually, 31 pieces of seat structure debris were positively identified to specific seat locations. However, almost 60 pieces of seat structure debris remained unidentified along with numerous fragments of seat soft goods. Had the individual seat components been permanently marked with serial numbers and those serial numbers tracked to the assembled seats, reconstruction and identification would have been much easier and a higher percentage of pieces could have been identified to

---

[15]Tracking the serial numbers for seat components to the top-level seat assembly's serial number.

specific seats. Therefore, space flight programs should develop failure investigation marking ("fingerprinting") requirements and policies. Equipment fingerprinting requires three aspects to be effective: component serialization, marking, and tracking to the assembly level. Marking by electronic means, metal stamping, or etching is preferable to labels or ink stamping because labels and ink stamps are not as durable in catastrophic failure scenarios. Marking in multiple locations, and on as many piece-parts in a major assembly as practical, is recommended.

Finding. While all seat piece-parts include serial numbers, only the serial numbers of the inertial reels were recorded and tracked to a specific seat assembly. The lack of configuration management documentation hindered the process of ascribing the seat debris items to specific seat locations.

> ***Recommendation A5.*** Develop equipment failure investigation marking ("fingerprinting") requirements and policies for space flight programs. Equipment fingerprinting requires three aspects to be effective: component serialization, marking, and tracking to the lowest assembly level practical.

3.2 Crew Worn Equipment

Because the crew worn equipment is the hardware that is closest to the crew members, it provides a source of data for the mechanical and thermal environments that the crew members experienced. This section provides background information on crew worn equipment and describes its configuration on STS-107. A brief review of the different types of shuttle suits is presented. Crew worn equipment, which includes the advanced crew escape suit (ACES), the personal parachute assembly (PPA), and the parachute harness, is described. The *Columbia* crew worn configuration is addressed. In addition, some aircraft in-flight breakup case studies are also considered to draw parallels between those mishaps and that of *Columbia*. Finally, *Columbia*-specific topics are addressed: the general, thermal, and mechanical conditions of the helmets and suit neck rings; the glove disconnects; the dual suit controllers (DSCs); the boots; the Emergency Oxygen System (EOS); the Seawater Activated Release System (SEAWARS); the Telonics Satellite Uplink Beacon-A (TSUB-A) search and rescue satellite-aided tracking (SARSAT) beacon; the Army/Navy personal radio communications (A/N PRC)-112 radio; and the ground plot analysis.

The following is a summary of findings, conclusions, and recommendations from this section.

Finding. The current ACES was added after the shuttle cockpit was designed and built. In many cases, the operations that the crew must perform are difficult to perform while wearing the suit. Some crew members choose between not wearing portions of the suit (gloves) to perform nominal tasks efficiently, or wearing their gloves to protect against off-nominal atmospheric situations at the expense of nominal operations or other off-nominal situation responses needing more dexterity.

Finding. Breathing 100% O_2 results in O_2-enriched air being exhaled into the shuttle cabin. Over time, this increases the O_2 concentration in the cabin, amplifying the potential for fire. Therefore, the amount of time that crew members have their visors down and are breathing 100% O_2 is limited operationally to reduce this hazard.

Finding. One crew member did not have the helmet donned at the time of the Crew Module Catastrophic Event (CMCE). Three of the seven crew members did not complete glove donning for entry. The deorbit preparation period of shuttle missions is so busy that crew members frequently do not have enough time to complete the deorbit preparation tasks (suit donning, seat ingress, strap-in, etc.) prior to the deorbit burn.

> **Recommendation L1-2.** Future spacecraft and crew survival systems should be designed such that the equipment and procedures provided to protect the crew in emergency situations are compatible with nominal operations. Future spacecraft vehicles, equipment, and mission timelines should be designed such that a suited crew member can perform all operations without compromising the configuration of the survival suit during critical phases of flight.

Finding. Inspection of all seven recovered helmets confirmed that none of the crew members lowered and locked their visors.

> **Conclusion L5-1.** The current parachute system requires manual action by a crew member to activate the opening sequence.

Recommendation L1-3/L5-1. Future spacecraft crew survival systems should not rely on manual activation to protect the crew.

Conclusion L4-1. Although the advanced crew escape suit (ACES) system is certified to operate at a maximum altitude of 100,000 feet and to survive exposure to a maximum velocity of 560 knots equivalent air speed, the actual maximum protection environment for the ACES is not known.

Recommendation L3-5/L4-1. Evaluate crew survival suits as an integrated system that includes boots, helmet, and other elements to determine the weak points, such as thermal, pressure, windblast, or chemical exposure. Once identified, alternatives should be explored to strengthen the weak areas. Materials with low resistance to chemicals, heat, and flames should not be used on equipment that is intended to protect the wearer from such hostile environments.

Finding. The current ACES helmets are nonconformal and do not provide adequate head protection or neck restraint for dynamic loading situations.

Recommendation L2-4/L3-4. Future spacecraft suits and seat restraints should use state-of-the-art technology in an integrated solution to minimize crew injury and maximize crew survival in off-nominal acceleration environments.

Recommendation L2-7. Design suit helmets with head protection as a functional requirement, not just as a portion of the pressure garment. Suits should incorporate conformal helmets with head and neck restraint devices, similar to helmet/head restraint techniques used in professional automobile racing.

Finding. Most of the suit components and subcomponents include serial numbers that are recorded and tracked to a specific crew member. This configuration management documentation aided greatly in the process of ascribing the debris items to specific crew members.

Recommendation A5. Develop equipment failure investigation marking ("fingerprinting") requirements and policies for space flight programs. Equipment fingerprinting requires three aspects to be effective: component serialization, marking, and tracking to the lowest assembly level practical.

Finding. The ACES had no performance requirements for occupant protection from elevated temperatures or fire. The ensemble includes nylon on the parachute harness straps and the boots. The ACES may not provide adequate protection to crew members in emergency egress scenarios involving exposure to heat and flames.

Recommendation L3-5/L4-1. Evaluate crew survival suits as an integrated system that includes boots, helmet, and other elements to determine the weak points, such as thermal, pressure, windblast, or chemical exposure. Once identified, alternatives should be explored to strengthen the weak areas. Materials with low resistance to chemicals, heat, and flames should not be used on equipment that is intended to protect the wearer from such hostile environments.

Crew escape equipment (CEE) enhances the crew members' capability to escape safely from a disabled orbiter on the launch pad, in atmospheric flight, or following landing.[1] This equipment includes the crew worn equipment and the crew escape pole.[2] For in-flight bailout scenarios, the CEE is designed for use during controlled subsonic gliding flight conditions.

---

[1] Program Requirements Document for Crew Escape Equipment, NSTS-22377, Revision B, October 1994.
[2] The crew escape pole is discussed in Section 2.4.

3.2.1 Shuttle suits

The shuttle was originally designed to be operated in a shirtsleeve (bare-hands) environment for all phases of flight, including launch and landing. The U.S. Air Force SR-71 pressure suit was worn for the first four shuttle missions (STS-1 through STS-4), which were considered test flights. Following these first four missions, the shuttle was declared fully operational and shuttle crews wore standard flight suits (light-weight fabric coveralls) with a helmet and a portable air supply that was intended principally for ground egress and cabin smoke protection. After the *Challenger* accident in January 1986, NASA started to use the launch and entry suit, beginning with STS-26 in 1988, and later phased in the more capable ACES beginning in 1994. This addition of pressure suits to a vehicle that was designed for shirtsleeve operations resulted in human/machine interface incompatibilities, especially with switches and other hand-operated controls. Because of this, some crew members decide between wearing gloves for full protection in emergency scenarios or not wearing gloves to be able to perform nominal tasks (keyboard entries, manipulating displays and controls, etc.) efficiently.

Finding. The current ACES was added after the shuttle cockpit was designed and built. In many cases, the operations that the crew must perform are difficult to perform while wearing the suit. Some crew members choose between not wearing portions of the suit (gloves) to perform nominal tasks efficiently, or wearing their gloves to protect against off-nominal atmospheric situations at the expense of nominal operations or other off-nominal situation responses needing more dexterity.

> *Recommendation L1-2.* Future spacecraft and crew survival systems should be designed such that the equipment and procedures provided to protect the crew in emergency situations are compatible with nominal operations. Future spacecraft vehicles, equipment, and mission timelines should be designed such that a suited crew member can perform all operations without compromising the configuration of the survival suit during critical phases of flight.

Crew worn equipment (figure 3.2-1), which is used by shuttle crew members during launch and entry, provides the necessary protection and survival equipment to sustain the crew below 100,000 feet and ensures

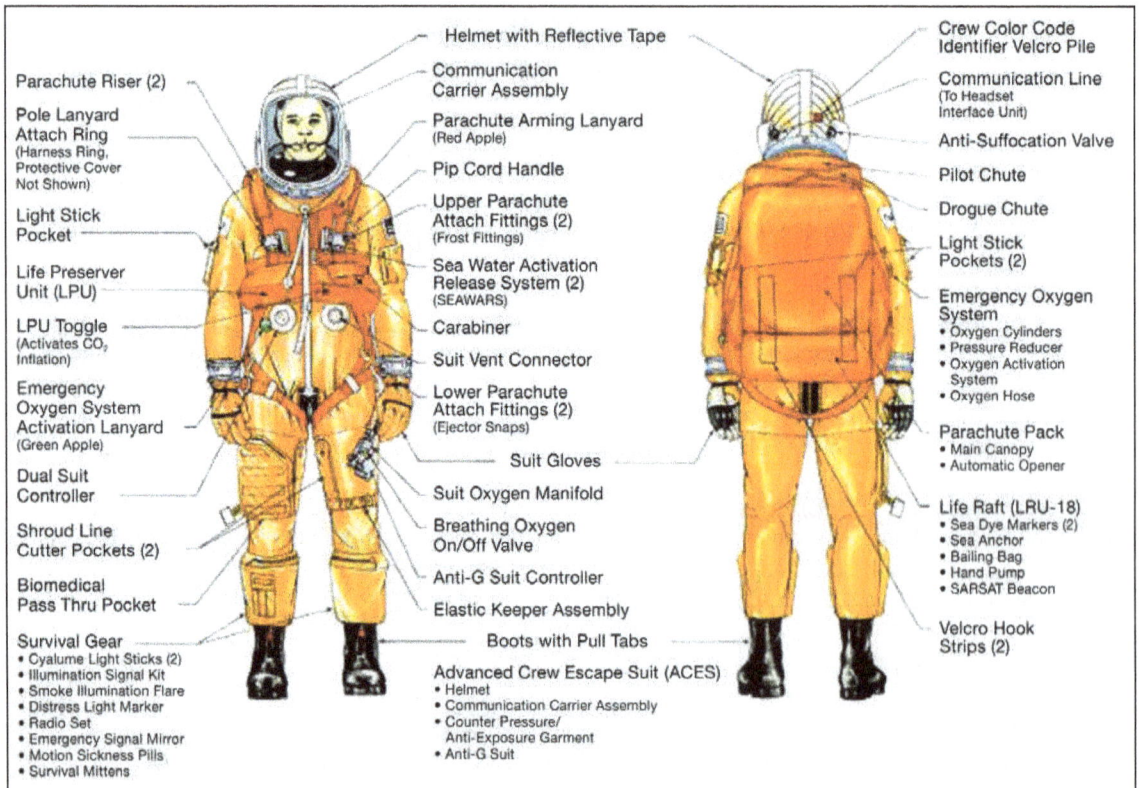

Figure 3.2-1. *Crew worn equipment: advanced crew escape suit, harness, parachute, and survival gear.*

the crew's safety after vehicle egress. The crew worn equipment consists of the ACES and g-suit, a PPA, a parachute harness, and survival gear. The equipment provides the crew with altitude and atmospheric contamination protection during normal launch/entry operations and emergency conditions.

3.2.1.1 *Advanced crew escape suit*

The ACES is a full-pressure suit that is capable of applying static pressure over the entire body. A positive-pressure regulator delivers orbiter- or EOS-supplied 100% O_2 to the helmet at a pressure that is slightly above suit pressure. Breathing 100% O_2 results in O_2-enriched air being exhaled into the cabin. Over time, this increases the O_2 concentration in the cabin, amplifying the potential for fire. Therefore, the amount of time that crew members have their visors down and are breathing 100% O_2 is limited operationally to reduce this hazard.

Finding. Breathing 100% O_2 results in O_2-enriched air being exhaled into the shuttle cabin. Over time, this increases the O_2 concentration in the cabin, amplifying the potential for fire. Therefore, the amount of time that crew members have their visors down and are breathing 100% O_2 is limited operationally to reduce this hazard (see **Recommendation L1-2**).

The outer covering of the ACES is flame-resistant orange Nomex. Just beneath the outer layer of the Nomex fabric is a woven, open-link net restraint layer made of Nomex cord that provides structural support for the suit. Under the restraint layers is a pressure bladder made of nylon laminated to GORE-TEX® to wick body moisture away when unpressurized, while holding pressure when inflated. The ACES incorporates a rear entry pressure-sealing zipper for suit donning and doffing, a neck ring (figure 3.2-2) for helmet attachment, and wrist rings for glove attachment (figure 3.2-3).

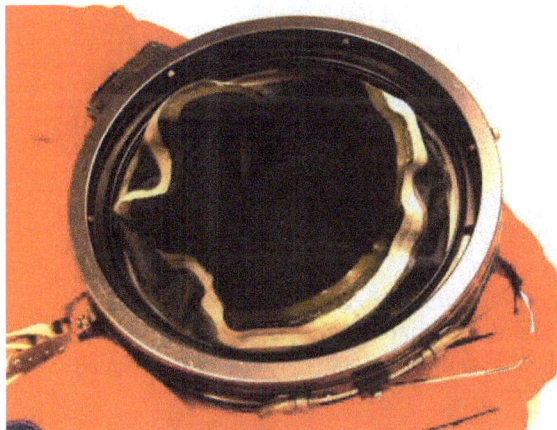

Figure 3.2-2. *Example of neck ring.*

Figure 3.2-3. *Example of wrist ring.*

The helmet (figure 3.2-4) attaches to the ACES neck ring. The neck ring has a latch (figure 3.2-5) that secures the helmet to the suit. Sliding the latch halves together moves six "latch dogs" to secure the helmet on the neck ring. Sliding them apart retracts the dogs, allowing removal of the helmet from the neck ring. Two independently rotating visors on the front of the helmet provide a dark sunshield and a clear pressure visor. The pressure visor is closed and locked by pulling the visor and the bailer bar down into the locked position. To open the pressure visor, a latch on the bailer bar lock must be pushed down and two buttons on either side of the lock must be pressed. This allows the bailer bar to unlock, after which the visor can be opened. O_2 is delivered to the helmet interior by a spray bar.[3] The shell of the helmet is made of fiberglass with a coating of reflective tape. The pressure visor is a laminate made of polycarbonate and polymethylmethacrylate.

---

[3]The spray bar is a tube with numerous small holes in it that direct O_2 towards the crew member's face.

The helmet must be attached to the neck ring and the pressure visor must be closed and locked to pressurize the suit.

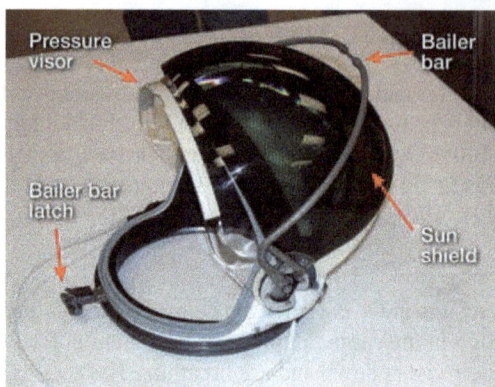

Figure 3.2-4. *Example of helmet.*

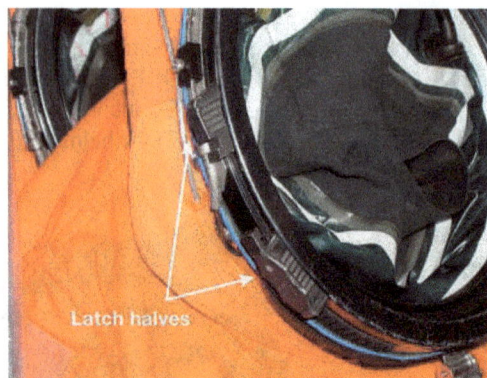

(a) Open.

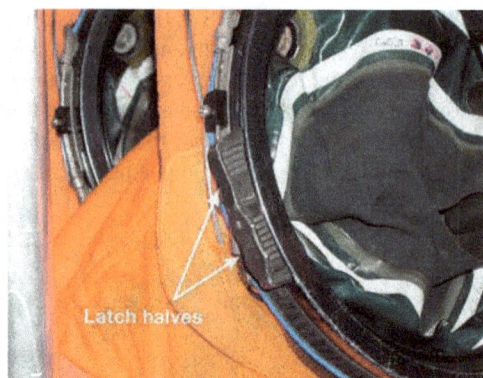

(b) Closed.

Figure 3.2-5. *Example of neck ring latch.*

The helmet provides an interface between the communication carrier assembly (CCA) and the orbiter communications system. The CCA, or "comm cap," contains microphones and earphones. The communications cable passes through the lower left side of the helmet and connects to a headset interface unit, which in turn connects to the orbiter communications system.

Detachable gloves (figure 3.2-6) attach to the ACES sleeve via mating rings and must be worn for the ACES to provide full protection. The rings provide an airtight seal and allow the gloves to swivel for improved mobility. The gloves have adjustable straps around the palm to prevent "ballooning" during suit pressurization and to allow for flexion at the palms.

Figure 3.2-6. *Example of gloves.*

The ACES is worn in conjunction with Rocky 911 commercial off-the-shelf boots (worn over the pressure bladder) that have rubber soles and leather and nylon upper sections (figure 3.2-7).

Figure 3.2-7. *Example of boots.*

Crew members wear a g-suit (figure 3.2-8) under the ACES during shuttle entry and landing. The g-suit bladders surround the abdomen, thighs, and calves and apply pressure to the crew member's lower abdomen and legs. The g-suit is used to counteract the effects of orthostatic intolerance upon return to 1-G conditions after exposure to microgravity. The g-suit is made from Nomex and nylon and has lacing to achieve a proper fit. It is pressurized with suit O_2. Pressure is controlled manually by the crew member. The g-suit connects to the ACES O_2 manifold via a quick disconnect (QD) hose.

Various garments are worn under the ACES and the g-suit for crew comfort. These garments include a liquid cooling garment (LCG), thermal underwear, wool socks, and a diaper. The LCG (figure 3.2-9) consists of thermal underwear shirt and pants with tubes sewn into the fabric on the inside. A cooling unit (thermal electric liquid cooling unit (TELCU) or individual cooling unit (ICU)) cools and pumps water through the LCG's network of tubing to cool the crew member. The water supply and return lines are fed through a plug located on the right thigh area of the ACES.

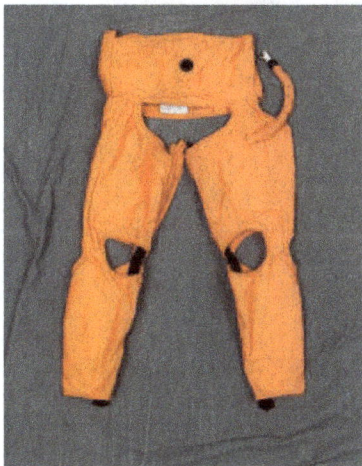

Figure 3.2-8. *Example of g-suit.*

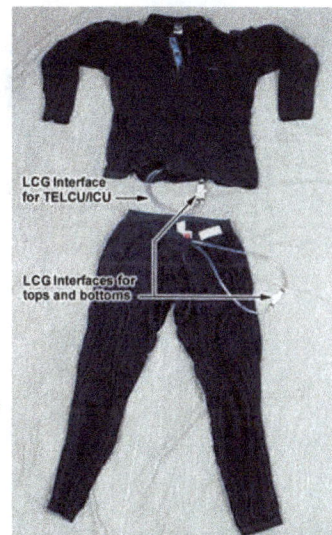

Figure 3.2-9. *Example of liquid cooling garment.*

One AN/PRC-112 radio (figure 3.2-10) is flown per crew member and is located in a survival gear pouch inside a pocket on the right shin area of the ACES. The PRC-112 radio is constructed of an outer aluminum casing, plastic external switches, a plastic external battery pack, and various internal electronic components.

Figure 3.2-10. *Example of Army/Navy personal radio communications-112 radio.*

3.2.1.2 *Parachute harness*

The parachute harness (figure 3.2-11) is a system of interwoven nylon webbing straps that provides an interface between the PPA and the crew member. The straps provide body support for crew members during bailout, emergency egress, and water rescue operations. Integrated into the harness are a carabineer, an EOS, an emergency water pack, and a life preserver unit (LPU).

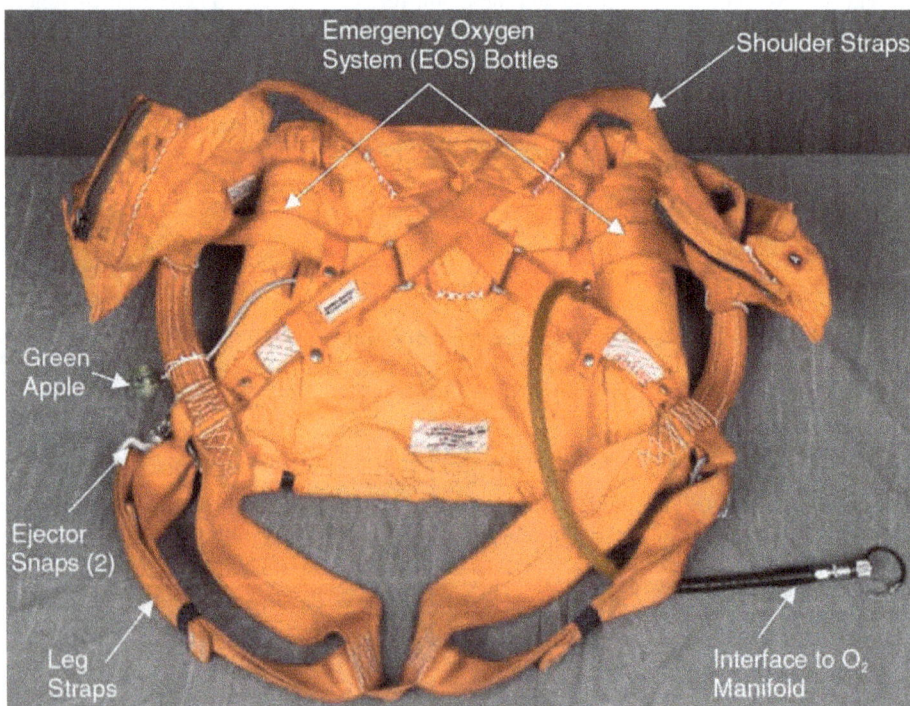

Figure 3.2-11. *Example of parachute harness.*

The EOS (figure 3.2-12), which is located within the parachute harness, consists of two bottles that are pressurized with O_2 at 3,000 pounds per square inch (psi). Each bottle has a pressure regulator that reduces the pressure down to 70 psi. A common manifold delivers the 70-psi O_2 from both bottles to the O_2 hose that connects to the ACES O_2 manifold via a QD. The system, activated by pulling the "green apple" activation knob on the right side of the harness, provides 381 liters of O_2.

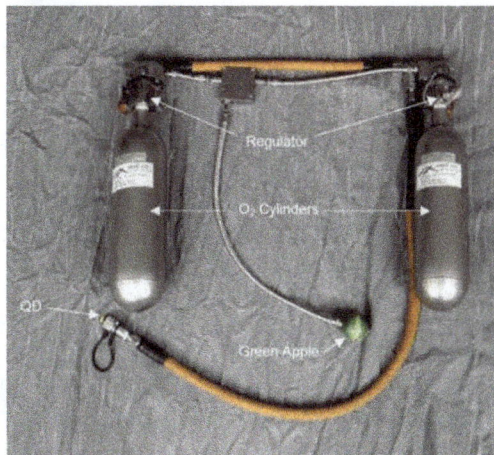

Figure 3.2-12. *Example of Emergency Oxygen System.*

3.2.1.3 *Personal parachute assembly*

All crew members wear a PPA (figure 3.2-13). It is secured to the crew member's parachute harness at four locations. The parachute risers are connected to two attachment points on the harness which are called Frost fittings, and the metal triangular rings are secured to the ejector snaps on the harness. Two SEAWARS, one on each parachute riser, are part of the fittings on the PPA risers that interface to the harness. These two SEAWARS are designed to automatically release the risers from the harness upon immersion in saltwater. The SEAWARS consist of an outer aluminum casing, plastic external components, various internal electronic components, and a small pyrotechnic device.

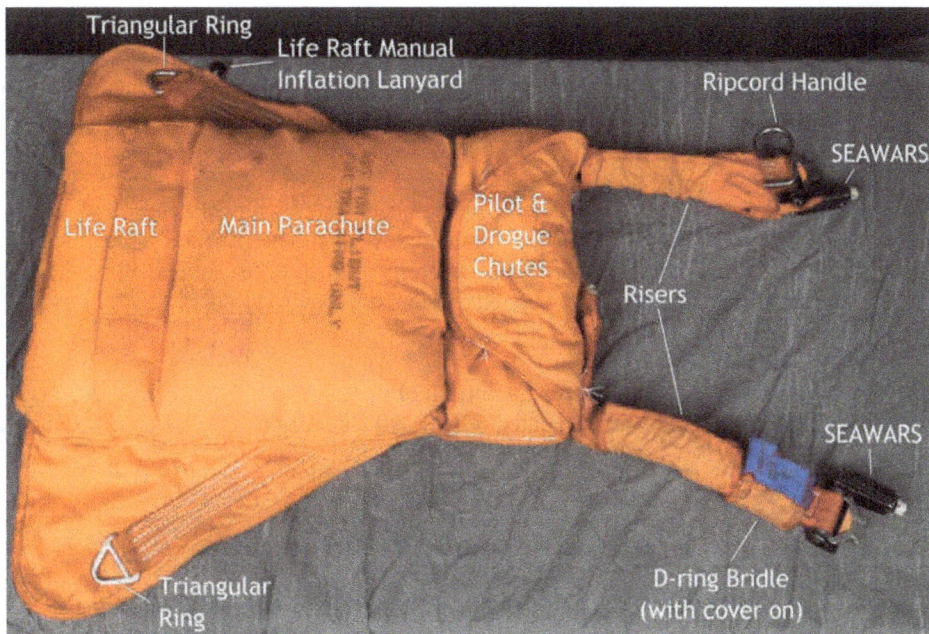

Figure 3.2-13. *Example of personal parachute assembly.*

The outer covering of the parachute pack is made of Nomex. The parachute pack has three compartments: an upper compartment that contains a pilot chute and drogue chute, a middle compartment that contains the main parachute and the automatic opening device (AOD), and a lower compartment that contains a one-person life raft. The right riser houses the D-ring bridle, which connects to a lanyard hook on the escape pole for use during in-flight bailout. The D-ring bridle and lanyard hook initiate the parachute opening sequence. The left riser has a manual rip cord, which can be used to initiate the parachute opening sequence. Both methods require some crew action (either attaching the D-ring to the escape pole lanyard or pulling the rip cord) to initiate the parachute sequence.

> **Conclusion L5-1.** The current parachute system requires manual action by a crew member to activate the opening sequence.

> **Recommendation L1-3/L5-1.** Future spacecraft crew survival systems should not rely on manual activation to protect the crew.

The main parachute is a circular canopy made from nylon with Kevlar reinforcement.

The life raft is rubber and is inflated by two carbon dioxide (CO_2) bottles. A TSUB-A SARSAT beacon (figure 3.2-14) is secured to the life raft and activates automatically upon main parachute deployment. The SARSAT beacon consists of an outer aluminum casing, plastic external switches, and various internal electronic components.

Figure 3.2-14. *Example of search and rescue satellite-aided tracking beacon.*

3.2.2 Crew worn equipment configuration

Recovered videotape from *Columbia* revealed information related to the configuration of the crew worn survival equipment, including the helmets and the ACESs. The recovered middeck video shows the seat 5 crew member suited (except for the helmet and gloves) and the seat 1 and seat 6 crew members donning their suits. The middeck video, which does not include views of other flight deck crew members donning their suits, ends prior to the seat 7 crew member donning the ACES.

Recovered flight deck video shows the flight deck crew members suited with helmets and gloves on (except one crew member, who had not completed donning gloves by the end of the video). This video also shows the helmet and helmet neck rings in close proximity to the crew members' chins. Because the helmets appeared to be restrained, investigators concluded that the crew members had the proper tension on the neck ring tie-down straps.

Although all crew members were wearing the main portion of the suit at the time of the accident, at some point the suits completely failed and separated. The SCSIIT investigated similar cases to understand the mechanism of suit failure.

3.2.3 Aircraft in-flight breakup case studies

The following civil aviation accidents provide examples of cases of passenger clothing being shed (body denuding) during in-flight breakups:

- Air India Flight 182 was flying at 31,000 feet over the Atlantic Ocean on June 23, 1985 when a terrorist bomb exploded in the baggage compartment. The Boeing 747 aircraft broke up in flight, and at least 21 of the 131 recovered bodies were denuded.
- Iran Air Flight 655 was mistakenly shot down by a U.S. Navy ship on July 3, 1988 while flying over the Persian Gulf. After missiles hit it, the Airbus A300 aircraft broke up in flight at an altitude of 13,500 feet. The denuded bodies of the passengers were recovered from the Persian Gulf waters.

- Pan Am Flight 103 was blown up by a terrorist bomb over Lockerbie, Scotland on December 21, 1988. The bomb went off when the Boeing 747 aircraft was at roughly 31,000 feet and 313 knots airspeed; numerous passengers who had separated from the aircraft prior to ground impact were denuded.
- COPA Flight 201 broke up over the jungle in Panama on June 6, 1992. The Boeing 737 aircraft broke up at approximately 13,000 feet while in a high-speed dive (the pilots entered the dive because of a faulty attitude indication that was due to a wiring problem). Many of the passengers' bodies were denuded.

These aviation accidents involved lower altitudes and slower speeds than those associated with the *Columbia* accident. However, a military accident more similar to the *Columbia* accident occurred on January 25, 1966 involving an SR-71 test flight.[4] The pilot lost control of the aircraft, and the SR-71 broke up while flying at approximately Mach 3 at over 75,000 feet (~400 knots equivalent airspeed (KEAS)). The pilot survived, but the reconnaissance systems officer was killed. Points of similarity to the *Columbia* accident include:

- The SR-71 accident occurred at high speed and relatively high altitude.
- The SR-71 aircraft breakup dynamics resulted in a fatality.
- The SR-71 aircraft breakup dynamics included crew member separation from the vehicle.
- The seat restraint straps failed.
- The SR-71 pressure suit is very similar to the shuttle ACES in design and construction.
- The dynamic pressure at the *Columbia* CMCE was roughly 405 pounds per square foot (psf) and the dynamic pressure at SR-71 aircraft breakup was roughly 398 psf, a difference of less than 2%.

However, there are also notable differences between the two accidents. These include:

- The SR-71 pilot's suit pressurized automatically (as designed) when the cockpit depressurized due to the aircraft breakup. The pilot attributed his survival to the pressurized suit, which protected him from the low-pressure/low-O_2 environment as well as the aerodynamic forces (windblast) that he experienced when he separated from the aircraft. As discussed in the sections below, the *Columbia* suits did not pressurize because the crew members did not lower visors or activate the suit O_2 system. Additionally, three crew members did not complete donning gloves, which is required for the suit to pressurize.
- While the *Columbia* crew members were exposed to a similar dynamic pressure environment as the SR-71 crew members, the thermal environment of the *Columbia* accident was much more severe than that experienced during the SR-71 breakup.
- Because of the altitude differences, the chemical environment (higher concentration of more reactive monatomic oxygen) of the *Columbia* accident differs from that of the SR-71 breakup.
- The *Columbia* suits did not remain intact, whereas the SR-71 pressure suits did remain intact.

Aerodynamic analysis indicates that the equivalent airspeed of the CM at the CMCE (GMT 14:00:53) was roughly 400 KEAS, and that it increased to 560 KEAS by the time of Total Dispersal (TD) (GMT 14:01:10). The ACES is designed to maintain structural integrity and pressure response capability when exposed to at least a 560-KEAS windblast. Since the suit is certified by NASA to meet this requirement based on its similarity to the pressure suit used by the U.S. Air Force, it was not subjected to windblast tests for certification. By contrast, the U.S. Air Force suit was tested in a certification program in 1990 during which it was exposed to a 600-KEAS windblast (the suit was worn by a manikin that was restrained in an ejection seat with the helmet visor down and locked). During the first test (suit not pressurized), the helmet sun shield separated from the helmet and a life preserver unit inflation tube separated from the life preserver unit. During the second test (the suit was pressurized to 2.99 psi), both shin pockets (survival gear storage pockets) were forced open. No other relevant anomalies were observed.

In the U.S. Air Force windblast test configuration, the helmet visors were lowered, which is notably different from the position of the *Columbia* visors. Debris evidence indicates that the *Columbia* helmet

---

[4]*Aviation Week & Space Technology*, August 8, 2005, pp. 60–62.

visors were up. With the helmet visor up, the helmet cavity presents a high drag configuration that could contribute to a mechanical failure of the suit/helmet interface, leading to suit disruption.

Standard materials testing data exist for the suit materials (GORE-TEX®, Nomex, nylon, etc.). However, the data are for tests conducted at "normal" environmental conditions (sea-level atmospheric temperature, pressure, and composition). Little laboratory test data exist on the performance of the materials in extreme environments. The lack of laboratory data presents an information gap regarding how the materials properties of the ACES are affected by exposure to the thermal and chemical environments at the altitudes and speeds experienced by *Columbia*. Although the more severe thermal and chemical environment of the *Columbia* accident may have weakened the suit materials, hastening suit disruption, the extent to which the thermal/ chemical environment contributed to suit disruption cannot be determined from the debris and because the environment's affects on suit materials is not understood.

> ***Conclusion L4-1.*** Although the advanced crew escape suit (ACES) system is certified to operate at a maximum altitude of 100,000 feet and to survive exposure to a maximum velocity of 560 knots equivalent air speed, the actual maximum protection environment for the ACES is not known.

> ***Recommendation L3-5/L4-1.*** Evaluate crew survival suits as an integrated system that includes boots, helmet, and other elements to determine the weak points, such as thermal, pressure, windblast, or chemical exposure. Once identified, alternatives should be explored to strengthen the weak areas. Materials with low resistance to chemicals, heat, and flames should not be used on equipment that is intended to protect the wearer from such hostile environments.

3.2.4 Recovered debris

Although only a small percentage of ACES fabric was recovered, many hard suit components were recovered (Table 3.2-1). There was no obvious pattern to explain why the hard components that were associated with some crew members were recovered while those associated with other crew members were not recovered. Figures 3.2-15 through 3.2-21 show the major crew worn equipment components that were recovered and ascribed to specific crew members.

Table 3.2-1. *Recovered Crew Worn Components*

Advanced Crew Escape Suit	
Helmets – seven flown	All seven helmets recovered and identified to crew members.
Suit-side helmet neck rings – seven flown	Six recovered. Seat 6 not recovered; seat 4 recovered separately from helmet; all others recovered attached to helmets.
Glove disconnect rings – 14 flown (seven right and seven left)	Nine recovered (five right and four left). • Seat 2 right side (attached to suit-side ring) • Seat 3 left side (attached to suit-side ring) • Seat 4 right side (not attached to suit-side ring) • Seat 4 left side (not attached to suit-side ring) • Seat 5 right side (attached to suit-side ring) • Seat 5 left side (attached to suit-side ring) • Seat 6 left side (not attached to suit-side ring) • Seat 6 right side (not attached to suit-side ring) • Seat 7 right side (not attached to suit-side ring)
Suit-side glove disconnect rings – 14 flown (seven right and seven left)	Seven recovered (three right and four left). • Seat 2 right side (attached to glove ring) • Seat 3 left side (attached to glove ring) • Seat 4 left side (not attached to glove ring) • Seat 5 right side (attached to glove ring) • Seat 5 left side (attached to glove ring) • Seat 6 right side (not attached to glove ring) • Seat 7 left side (not attached to glove ring)

Table 3.2-1. *Recovered Crew Worn Components* (Continued)

Advanced Crew Escape Suit	
A/N PRC-112 radios – seven flown	Four recovered. • Seat 1 • Seat 2 • Seat 4 • Seat 7
Boots – 14 flown (seven right and seven left)	Fifty-three items identified as ACES boot fragments recovered. Eight sole fragments identified from six different boots. Four matches of the sole fragments made, resulting in a total of one complete left sole, three complete right soles, a left heel fragment, and a left toe fragment.
Suit fabric	Approximately 30 fragments of suit material recovered (only one-third is cover layer material); none could be ascribed to a specific crew member.
Miscellaneous ACES components: DSC, suit breathing regulator, suit pressure relief valve, bio-instrumentation pass-through (BIP) plug (one spare BIP plug flown; it was not recovered), O_2 manifold and g-suit controller, suit vent inlet and elbow fitting, etc. – seven each flown	Four DSCs recovered. • Seat 1 • Seat 2 • Seat 3 • Seat 4 One suit breathing regulator (seat 1) recovered. Three suit pressure relief valves recovered. • Seat 3 • Seat 4 • Seat 7 Three BIP plugs recovered. • Seat 4 • Seat 6 • Seat 7 Four O_2 manifold/g-suit controllers recovered. • Seat 4 • Seat 5 (with entire EOS O_2 hose and ~34 in. of suit O_2 hose attached) • Seat 6 • Seat 7 (g-suit controller portion missing) One O_2 hose quick disconnect (seat 5) recovered. One suit vent inlet with elbow fitting (seat 6) recovered.
Parachute Harness	
EOS bottles – 14 flown (two per crew member)	Ten whole bottles and three bottle fragments (not ascribed to a specific crew member) recovered. • Seat 1 • Seat 3 • Seat 4 (two) • Seat 5 • Seat 6 (two) • Seat 7 • Two whole bottles not ascribed to a specific crew member • Three bottle fragments not ascribed to a specific crew member.
FLU-8 (life preserver unit inflation devices) – 14 flown (two per crew member)	Seven recovered. • Seat 1 • Seat 2 • Seat 3 (two) • Seat 4 • Seat 5 • Seat 6

Table 3.2-1. *Recovered Crew Worn Components* (Continued)

Parachute Pack	
SEAWARS – 14 flown (seven left, seven right)	Six recovered: four left and two right (none had parachute riser strap attached, all mated to the Frost fittings and had ~12 in. of parachute harness strap attached). • Seat 1 left • Seat 2 right • Seat 4 right • Seat 5 left • Seat 6 left • Seat 7 left
AODs – seven flown	Two recovered. • Seat 4 • Seat 7
SARSAT beacons – seven flown	Six recovered. • Seat 1 • Seat 3 • Seat 4 • Seat 5 • Seat 6 • Seat 7
Fabric	More than 180 fragments of parachute canopy, parachute cord, parachute pack, and parachute harness strap material recovered; none could be ascribed to a specific crew member.

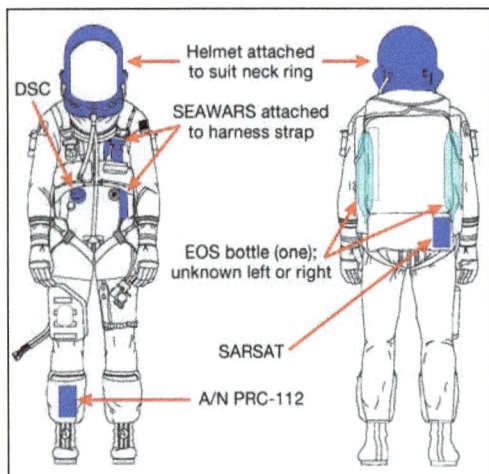

Figure 3.2-15. *Seat 1 recovered crew worn equipment.*

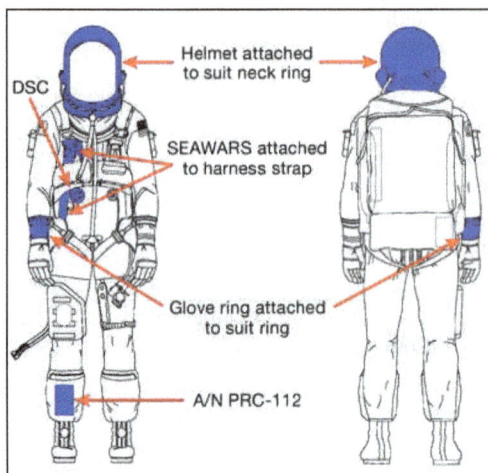

Figure 3.2-16. *Seat 2 recovered crew worn equipment.*

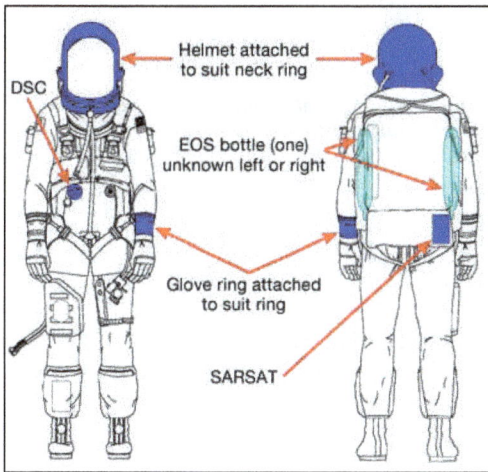

Figure 3.2-17. *Seat 3 recovered crew worn equipment.*

Helmet attached to suit neck ring

DSC

EOS bottle (one) unknown left or right

Glove ring attached to suit ring

SARSAT

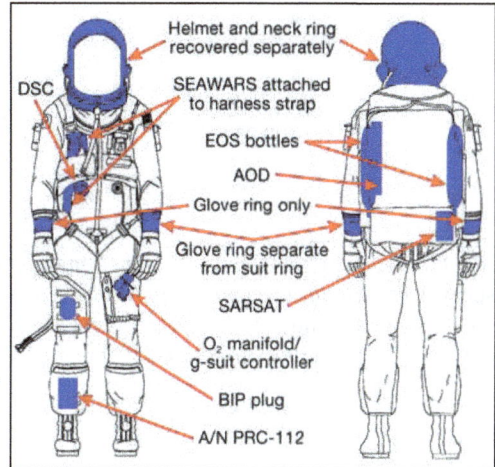

Figure 3.2-18. *Seat 4 recovered crew worn equipment.*

Helmet and neck ring recovered separately

DSC

SEAWARS attached to harness strap

EOS bottles

AOD

Glove ring only

Glove ring separate from suit ring

SARSAT

O_2 manifold/ g-suit controller

BIP plug

A/N PRC-112

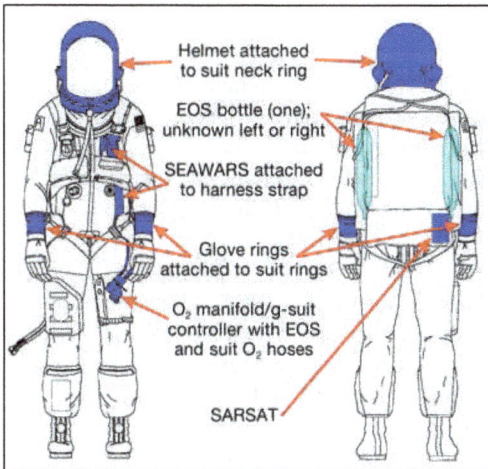

Figure 3.2-19. *Seat 5 recovered crew worn equipment.*

Helmet attached to suit neck ring

EOS bottle (one); unknown left or right

SEAWARS attached to harness strap

Glove rings attached to suit rings

O_2 manifold/g-suit controller with EOS and suit O_2 hoses

SARSAT

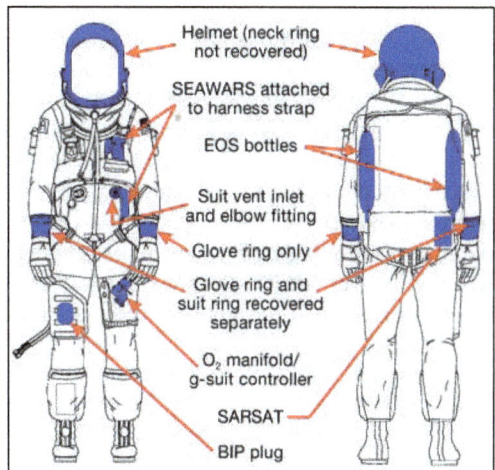

Figure 3.2-20. *Seat 6 recovered crew worn equipment.*

Helmet (neck ring not recovered)

SEAWARS attached to harness strap

EOS bottles

Suit vent inlet and elbow fitting

Glove ring only

Glove ring and suit ring recovered separately

O_2 manifold/ g-suit controller

SARSAT

BIP plug

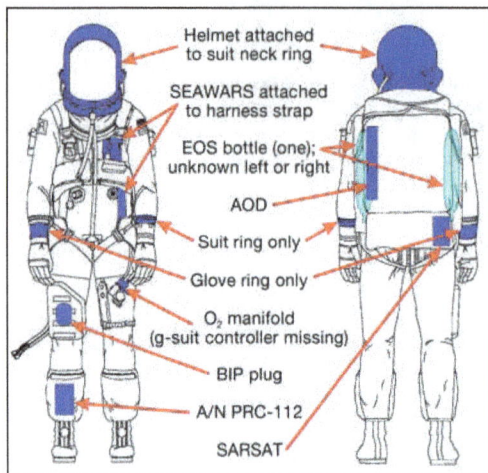

Figure 3.2-21. *Seat 7 recovered crew worn equipment.*

Labels in figure: Helmet attached to suit neck ring; SEAWARS attached to harness strap; EOS bottle (one); unknown left or right; AOD; Suit ring only; Glove ring only; O₂ manifold (g-suit controller missing); BIP plug; A/N PRC-112; SARSAT

Most of the suit components and subcomponents include serial numbers that are recorded and tracked to a specific crew member. This documentation aided greatly in the process of ascribing debris items to specific crew members. Thus, a very high percentage of recovered crew worn equipment was identified to specific crew members, aiding the post-accident analysis.

3.2.5 Helmets

Undamaged helmets and neck rings are shown in figure 3.2-4 and figure 3.2-2, respectively.

3.2.5.1 *General condition*

All seven helmets and six of the seven neck rings were recovered. Five of the seven helmets were recovered with the neck ring attached. One neck ring was recovered separate from the helmet. Inspection revealed that this neck ring had been mechanically removed from the helmet due to fracture of the latch mechanism. Detailed inspection of all helmets and helmet-to-neck-ring interfaces indicates that all crew members except one had their helmets on and latched at the time of the CMCE.[5] Helmet separation from the suit occurred between the suit-side neck ring and the suit fabric interface.

Finding. One crew member did not have the helmet donned at the time of the CMCE. Three of the seven crew members did not complete glove donning for entry. The deorbit preparation period of shuttle missions is so busy that crew members frequently do not have enough time to complete the deorbit preparation tasks (suit donning, seat ingress, strap-in, etc.) prior to the deorbit burn (see **Recommendation L1-2**).

The condition of each helmet shows effects from mechanical loading and thermal exposure. Effects from thermal exposure were generally consistent across all helmets, except for the helmet that was not donned at the time of the CMCE. This helmet had more pressure visor material remaining. The effects from mechanical loading were generally consistent across all seven helmets. The magnitude and distribution of mechanical damage was not severe, except for damage caused by ground impact.

---

[5] According to experienced astronauts, shuttle crews often struggle to complete all actions in the time allotted, giving priority to time-critical orbiter systems activities and reordering the tasks as necessary. Deorbit preparation activities frequently extend into the time after the deorbit burn and entry interface. Per the STS-107 crew's deorbit preparations plan, the crew member whose helmet was not donned was the last crew member scheduled to ingress the seat and don the helmet.

3.2.5.2 *Thermal condition*

Thermal effects were apparent throughout all helmet surfaces. Significant variations in thermal conditions were noted from helmet to helmet (both interior and exterior helmet surfaces). The reflective tape was missing from all of the helmets, and fiberglass delaminations of various sizes and depths were observed. Some white paint remained, except in the areas removed via fiberglass delamination. Residual paint on the exterior helmet surfaces shows signs of damage (pitting) that are consistent with impacts with many small debris items. Figure 3.2-22 shows examples of delaminations and pitting damage.

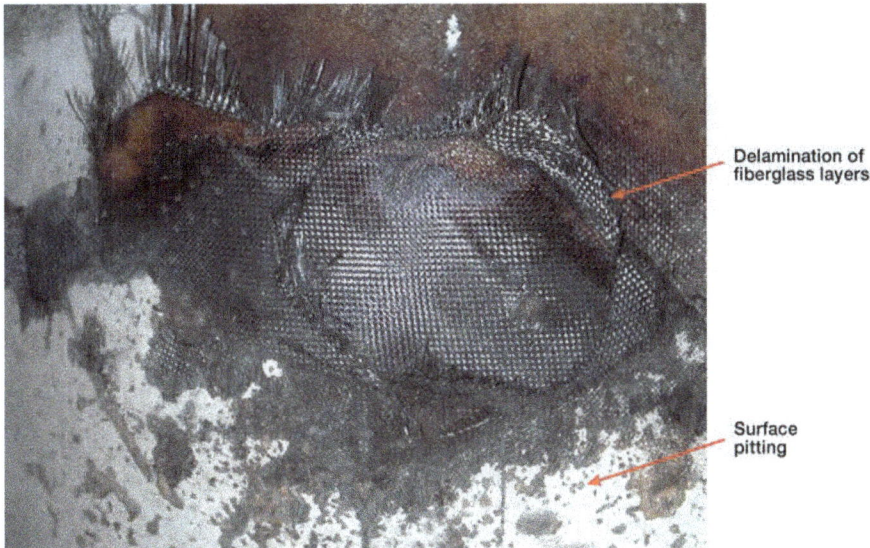

Figure 3.2-22. Delamination and pitting damage.

Small amounts of residual melted suit material were discovered, all of which were confined to the helmet/neck ring area. Melted suit bladder materials (nylon and Teflon) were observed on both sides of the helmet/neck ring interface on all helmets (except on the one helmet for which the suit-side neck ring was not recovered). Nomex material was absent from the internal and external helmet surfaces. Inspection of the suit bladder clamp interface on the neck ring yielded only nylon and Teflon (no Nomex) materials. This indicates that the Nomex material failed mechanically before the thermal decomposition temperature of 932°F (500°C) was reached. Helmet separation from the suit occurred primarily due to mechanical (aero-dynamic) forces; the helmets were not "melted off" the suit. Mechanical (aerodynamic) disruption of the suit occurred prior to completion of the heating period. Melted suit material was deposited onto the helmet and neck ring areas after mechanical separation of the neck ring (small fragments of suit material were still clamped into the neck ring upon mechanical separation).

On three of the seven helmets, the upper visor reinforcement bar was recovered with some pressure visor material still attached; no sun shield material remained on any of the helmets. The upper and lower visor bars along with visor materials on each of the other four helmets were not recovered. The visor is constructed of a laminate of polycarbonate and polymethylmethacrylate. These materials do not have a true melting point but instead have a glass transition temperature.[6] The glass transition temperature for poly-methylmethacrylate is approximately 230°F (110°C). The glass transition temperature for polycarbonate is approximately 300°F (149°C). Thermal gravimetric analysis testing was conducted to determine the temperature at which thermal decomposition (pyrolysis) in air begins. The thermal decomposition temperatures for poly-methylmethacrylate, fiberglass epoxy resin, and polycarbonate are 572°F (300°C), 735°F (391°C), and 752°F (400°C), respectively. These temperatures are for pyrolysis in air. Tests conducted in nitrogen (N_2) yielded thermal decomposition temperatures roughly 55°F (13°C) to 90°F (32°C) higher. It is unknown

---

[6]The temperature above which the mechanical properties of a material are reduced significantly and the material will flow.

whether these temperatures would be higher or lower in a low-pressure, monatomic oxygen (highly reactive) environment.

On all three helmets that have remaining pressure visor material, the polymethylmethacrylate flowed and pyrolyzed and the polycarbonate flowed in some places but did not pyrolyze (figure 3.2-23). Therefore, helmet visor materials experienced at least 300°F (149°C), which is the glass transition temperature of polycarbonate, to over 572°F (300°C), the pyrolysis temperature of polymethylmethacrylate, but certainly less than 752°F (400°C), which is the pyrolysis temperature of polycarbonate. Although there were small localized areas of fiberglass pyrolysis, in no case was there global pyrolysis of the helmet fiberglass material, indicating that the helmets did not experience temperatures globally above 735°F (391°C).

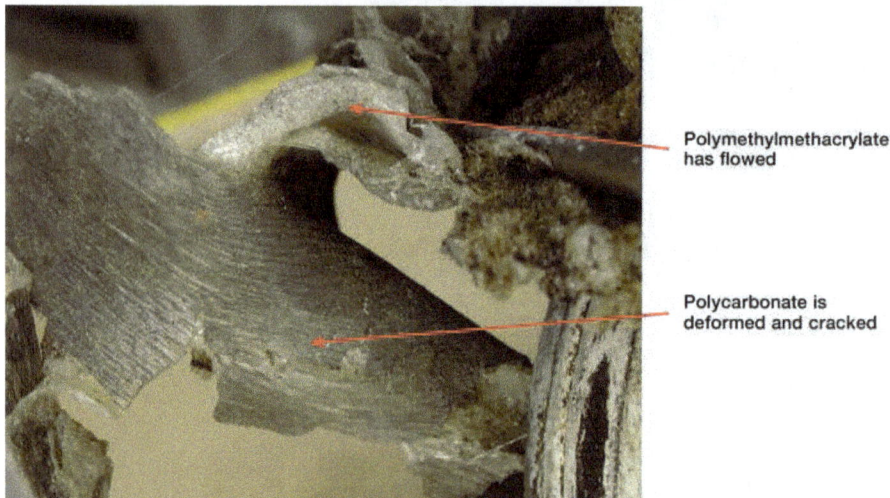

Polymethylmethacrylate has flowed

Polycarbonate is deformed and cracked

Figure 3.2-23. *Helmet pressure visor thermal effects.*

The Object Reentry Survival Analysis Tool (ORSAT) was used to predict thermal damage for helmets released at various times in the trajectory. The helmet was modeled as an 11-in.-diameter sphere weighing 6.3 lbs., with an initial temperature of 80°F (27°C). The analysis concluded that free-flying helmets released around GMT 14:01:03 would have received heating sufficient to cause damage similar to that seen on the visors of the three recovered helmets. Ballistics analysis provided helmet release times consistent within 10 seconds of the ORSAT-predicted time, confirming the approximate time of helmet separation from the CM. No significant inconsistencies were noted among ballistics analysis, ORSAT analysis, materials testing, and debris observations.

Despite the flow of the visor material, this material is notably absent from the helmet visor seal around the face opening, indicating that the visors were not in contact with the visor seal when heating occurred and were not down and locked.

Finding. Inspection of all seven recovered helmets confirmed that none of the crew members lowered and locked their visors (see **Recommendation L1-3/L5-1**).

3.2.5.3 *Mechanical condition*

In all cases, the helmet structure remained intact. The helmets experienced a range of localized mechanical damage (fractures), but did not experience massive structural damage from external impacts prior to ground impact. External helmet impacts were insignificant in size and random in distribution. Detailed inspections differentiated the sources of internal impacts.

Finding. The current ACES helmets are nonconformal and do not provide adequate head protection or neck restraint for dynamic loading situations.

> **Recommendation L2-4/L3-4.** Future spacecraft suits and seat restraints should use state-of-the-art technology in an integrated solution to minimize crew injury and maximize crew survival in off-nominal acceleration environments.

> **Recommendation L2-7.** Design suit helmets with head protection as a functional requirement, not just as a portion of the pressure garment. Suits should incorporate conformal helmets with head and neck restraint devices, similar to helmet/head restraint techniques used in professional automobile racing.

The hold-down cables on each neck ring were severed at the attach points to the cable guide tubes due to mechanical overload (figure 3.2-24). Most cable guide tubes experienced significant plastic deformation. The guide tubes display evidence of external contaminants (i.e., melted metal and suit material) and thermal effects on top of the fractures and localized deformation. This indicates that mechanical loading preceded exposure to the thermal environment. Rotation of the helmet relative to the normal forward position was observed on all neck rings varying from 90 to 180 degrees. Major cable guide tube deformation and helmet rotation indicates that a significant loading event occurred where helmets were removed via a mechanical (nonthermal) mechanism.

Figure 3.2-24. *Hold-down cable guide tube.*

One of the seven helmets was recovered with the bailer bar still attached. All other helmets had the bailer bar mechanically removed, although the bailer bar cam mechanism remained in place on the starboard and portside helmet interfaces.

The bailer bar latch mechanisms on five of the seven helmets remained attached to the helmets in good condition (figure 3.2-24). This would not be expected if the crew members had lowered and locked their visors. Mechanical separation of the bailer bar would be accompanied by fracture of the latch assembly if the visor was down and the bailer bar was locked. The other two helmets experienced latch mechanism separation due to failure of the fasteners that attach the latch mechanism to the helmet before subsequent deposition of melted suit materials. This suggests that latch separation was followed by suit melting. Neither of the two helmets shows evidence of the indentation or deformation that would be associated with forces expected if the bailer bar ripped the latch from the neck ring. Combined with the absence of melted visor material on the visor seal, this confirmed the conclusion that none of the crew members lowered and locked visors.

3.2.6 Glove disconnects

Undamaged glove disconnects are shown in figure 3.2-25.

Figure 3.2-25. *Examples of intact suit-side* (left) *and glove-side* (right) *glove disconnects.*

Twelve glove disconnect debris items were recovered, corresponding to six of the seven *Columbia* crew members (no glove disconnect rings were recovered for seat 1). Although evidence of exposure to entry heating was noted on all disconnect rings, the level of heating varied from item to item, with differences between the left and right sides of items from the same crew member. Inspection of recovered disconnect rings indicates that three crew members did not have their gloves mated to their suits for entry. Inspection indicates that three crew members had gloves mated to their suits. The recovered flight deck entry video supports these conclusions. The video also indicates that the crew member in seat 1 also had gloves mated to the suit.

Finding. One crew member did not have the helmet donned at the time of the CMCE. Three of the seven crew members did not complete glove donning for entry. The deorbit preparation period of shuttle missions is so busy that crew members frequently do not have enough time to complete the deorbit preparation tasks (suit donning, seat ingress, strap-in, etc.) prior to the deorbit burn (see **Recommendation L1-2**).

Melted aluminum deposits and/or tiny craters were observed on most of the glove and suit rings. Melted suit material (figure 3.2-26) was discovered on all recovered disconnect rings. Close inspection of the suit bladder clamp interface on the suit-side disconnect rings revealed mainly nylon and Teflon materials. In all cases, minimal amounts of Nomex remained clamped to the interface. Overall, the amount of residual melted suit material seems to correlate with the general magnitude of heating; that is, higher-magnitude heating resulted in the pyrolysis of residual suit material. As with the helmets, deposition of melted suit material on the glove disconnect areas occurred after mechanical separation. Small fragments of suit material were still clamped in the disconnect after mechanical separation. The failure modes at the disconnect ring (the ring-to-suit material interface) were similar to those observed in the suit-side helmet disconnect rings (see Section 3.2.5).

Figure 3.2-26. *Recovered suit-side* (upper left), *glove-side* (upper right), *and mated* (bottom) *glove disconnects.*

3.2.7 Dual suit controllers

An undamaged DSC is shown in figure 3.2-27.

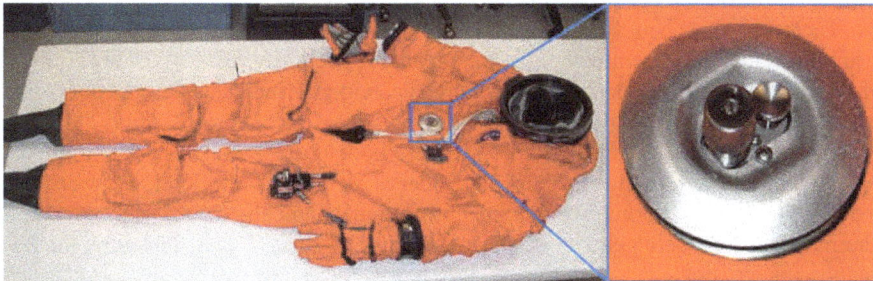

Figure 3.2-27. *Example of an intact dual suit controller.*

Four DSCs were recovered, all from flight deck crew members and all presenting similar appearances. One DSC (figure 3.2-28) was disassembled and inspected to determine the thermal environment exposure. The suit material edges were ragged with some localized melting around the edges, indicating that there was a mechanical disruption of the suit followed by thermal exposure to the suit fabric that was still attached to the DSC body. The DSC body back surface (i.e., the surface inside the suit) had many craters and one penetration (~0.5 in. × 0.25 in.). This indicates a rapid disruption of the suit, releasing the DSC while it was still in close proximity to the CM debris cloud. Some melting of the suit material to the DSC body caused the suit flange to adhere to the DSC body. All internal soft goods (O-rings, seals, diaphragms) were intact and showed no signs of mechanical or thermal damage. Both of the aneroids[7] were still hermetically sealed with the sealing solder intact. The suit cover-layer Velcro was melted to the suit restraint-layer Velcro in some

---

[7]Small, sealed metal bellows, sensitive to air pressure, that are part of the suit pressurization control mechanism.

areas. Although melting of the torn fabric edges and the Velcro® indicates thermal exposure, the complete lack of thermal damage to any of the internal soft goods suggests that the thermal exposure was limited in intensity and/or duration.

Figure 3.2-28. *Recovered dual suit controller front* (left) *and back* (right).

3.2.8 Boots

Undamaged boots are shown in figure 3.2-7.

A total of 53 possible boot fragments were recovered, including eight sole fragments from six boots and fragments of leather uppers and inserts. All fragments exhibited mechanical and thermal damage. No identifiable pieces of the shoe laces or nylon sections of the boots were recovered.

The condition of each recovered boot sole shows effects from mechanical loading and thermal exposure. Mechanical loading resulted in the removal of the boot leather uppers from five of the six recovered soles and fracture of four of the six recovered boot soles. Forensic evidence indicates that the boots were worn by the crew members during thermal exposure and that the boots failed mechanically prior to the conclusion of thermal exposure. Evidence indicates that the soles of these boots failed, with either the toe section or the complete sole being removed first followed by the remainder of the boot. The nylon lower sections of some of the boots appear to have been thermally penetrated prior to mechanical removal of the leather upper. Effects from thermal exposure were generally consistent across all soles. Edges, including the fracture edges, exhibited thermal erosion.

In an attempt to match the observed thermal damage, boot soles of flight-like boots were heated in an oven to identify the range of thermal effects with varying thermal exposure. The test samples were exposed to 750°F (399°C), 1,000°F (538°C), or 1,250°F (677°C) at normal atmospheric pressure conditions (~14.7 psi, ~20% O_2) for 15, 30, 45, or 60 seconds. The materials showed no significant changes in appearance until they combusted. This initially puzzled the team until it became clear that the presence of O_2 was affecting the results. The tests were repeated using new samples that were heated in an N_2 purge (<3% O_2). Results of the revised test protocol appeared to be similar to the recovered boot sole fragments. The test samples that most closely matched the recovered debris items were those that were exposed to 1,000°F (538°C) for 30 to 45 seconds or 1,250°F (677°C) for 15 to 30 seconds. However, no credible scenario could be envisioned in which the *Columbia* boots would be exposed to these temperatures for the length of time indicated by the tests, so the test results could not be correlated directly to the debris observations. Because the test conditions (~14.7 psi, 97% to 99% N_2, 1% to 3% O_2) did not sufficiently approximate the entry environment conditions (low ambient pressure, monatomic O_2, and possibly high dynamic pressure), they are a potential source of error in this analysis.

3.2.9 Emergency Oxygen System

An intact EOS is shown in figure 3.2-12.

Ten whole EOS bottle/reducers were recovered (eight were ascribed to specific crew members), each with no O_2 remaining. All 10 of the recovered EOS bottle/reducers have similar appearances, with some variance in the amounts of material deposition. Additionally, three fragments of bottles were also recovered (figure 3.2-29).

Figure 3.2-29. *Recovered whole Emergency Oxygen System bottle/reducers* (left and center) *and a bottle fragment* (right).

No nylon material from the parachute harness adhered to the bottle/regulator assemblies. Minor evidence of elevated temperatures, directional burn marks, and discrete external impacts are visible. As with most of the *Columbia* hardware, corrosion that occurred while the debris was on the ground is also evident. X ray revealed that all of the EOSs were activated. However, the crew members were trained not to activate them unless their visors were down. Therefore, activation of the EOS (achieved by applying tension to the activation cables) probably occurred as the bottles separated from the harnesses rather than by crew action.

The overall appearance of the 10 recovered whole EOS bottle/reducers suggests that each EOS bottle/reducer assembly experienced similar thermal and mechanical environments. Each EOS assembly was mechanically extracted from the harness as temperatures were rising; then for a short duration and nearly simultaneously, they experienced ballistic heating and some metal pellet-like impacts. This indicates a rapid disruption of the parachute harness, releasing the EOS while it was still in close proximity to the CM debris cloud.

The EOS bottle fragments exhibited irregular edges along the fracture surfaces, some outward bent edges, evidence of heating on the inner surfaces, and some deposited/flowed black material along a fracture surface.

Heating on the inner surfaces indicates that the bottle failure occurred before the end of entry heating. Neither the cause of the bottle failures nor the status of the bottles at the time of failure (pressurized or unpressurized) could be determined.

3.2.10 Seawater Activated Release System

Six SEAWARS were recovered. None had automatically ignited, and all had both of the Frost fitting male and female halves still mated. All SEAWARS had similar appearances (figure 3.2-30), consisting of the SEAWARS assembly still attached to approximately 12 in. of nylon parachute harness strap. None had any parachute riser material attached.

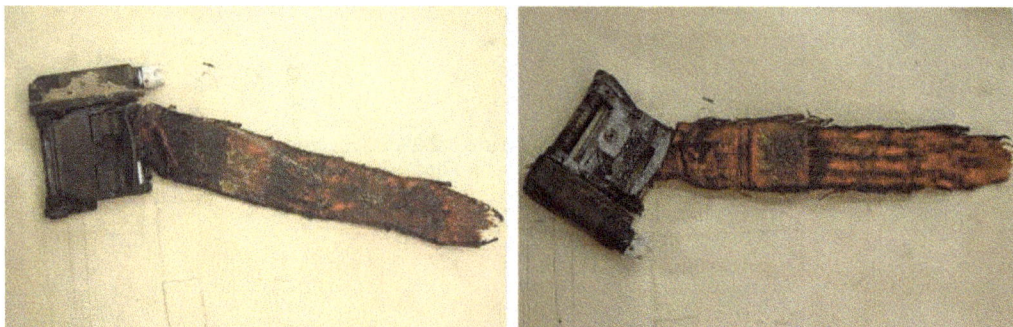

Figure 3.2-30. *Seawater Activated Release System with Frost fittings (male/female) still mated, with a segment of harness strap attached* (two different SEAWARS are shown).

Each SEAWARS/strap item shows evidence of directional melting, burning, and mechanical loading. The surviving length of parachute harness strap is consistent with a harness strap failure at the waist. While the terminating ends show evidence of melting on the top layer of fabric, these ends appear to have failed primarily due to mechanical overload, not melting. Some surface melting occurred along the length of the straps, with distinct directionality away from the SEAWARS towards the broken end of the strap, corresponding to a head-to-foot direction when the harness is on a crew member. These directional heating/melting features are present on both sides of the straps, with little difference between the front and the back of the straps. Because both sides of the straps show signs of heating, the heating and melting must have occurred after the straps had separated from the crew member. Localized heating on the metallic SEAWARS Frost fitting buckle suggests intense, short-duration heat exposure. Each SEAWARS and strap was mechanically extracted from the harness webbing before experiencing this short-duration heating. Directionality suggests that each piece of riser trimmed with the SEAWARS into the airflow (figure 3.2-31).

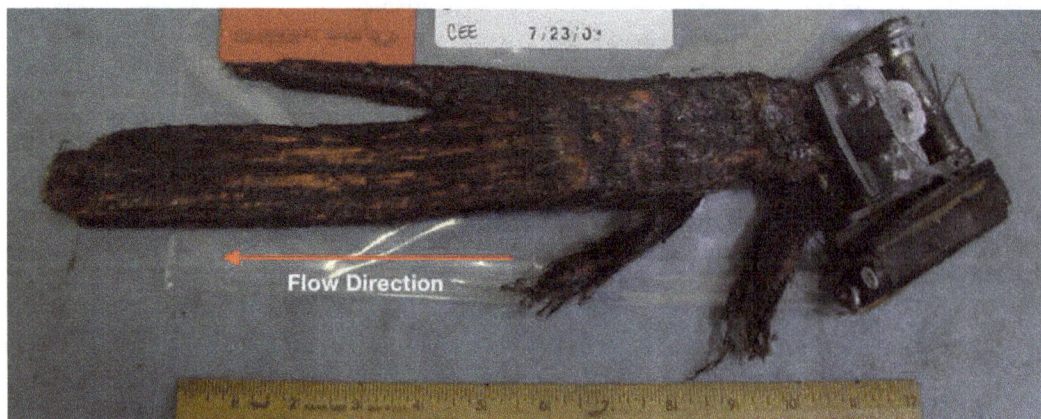

Figure 3.2-31. *Directional melting on Seawater Activated Release System/parachute harness strap.*

3.2.11 Telonics Satellite Uplink Beacon-A search and rescue satellite-aided tracking beacon

An undamaged TSUB-A SARSAT beacon is shown in figure 3.2-14.

One TSUB-A SARSAT beacon is flown per crew member; it is located in the survival raft packed in the PPA. Six SARSAT beacons were recovered; none of them were activated during the accident. All six recovered SARSAT beacons show similar thermal and mechanical damage (figure 3.2-32).

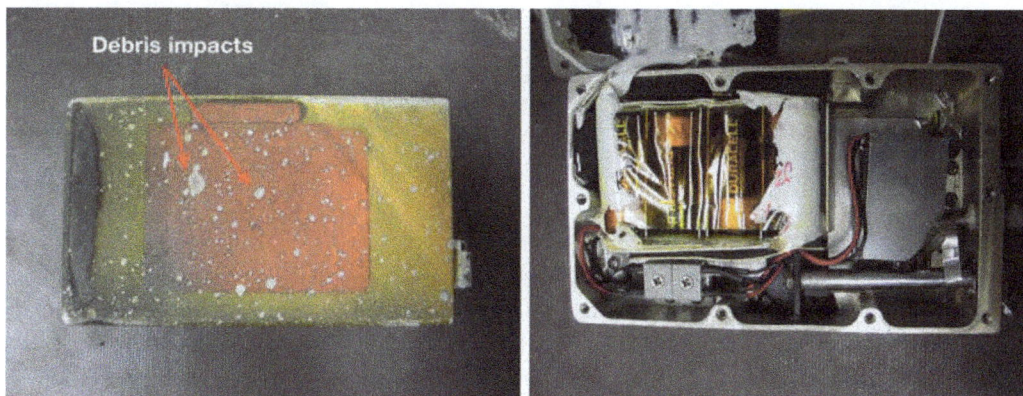

Figure 3.2-32. *External* (left) *and internal* (right) *views of a recovered search and rescue satellite-aided tracking beacon*.

No material from the raft or the PPA was adhered to the outer aluminum casing. Various amounts of paint remained, displaying evidence of impacts with small, hot metal pellets. Corrosion was also present. All external plastic was melted. Internal inspection of the beacons revealed minor solder re-flow, with most of the components mechanically and electrically intact. Two SARSAT beacons were tested and, when externally powered, functioned properly. For each of the six SARSAT beacons, the processing module was extracted and interfaced with ground support equipment to read the beacon's unique identifier, thereby allowing all six beacons to be ascribed to individual crew members. Each SARSAT beacon was mechanically extracted from the PPA as temperatures were rising; then for a short duration and nearly simultaneously, it experienced high heating and a hot metal pellet-like shower. This indicates a rapid disruption of the parachute packs and release of the SARSAT beacons while they were still in close proximity to the CM debris cloud.

3.2.12 Army/Navy personal radio communications-112 radio

An undamaged A/N PRC-112 radio is shown in figure 3.2-10.

Four A/N PRC-112 radios were recovered, all of which looked similar (figure 3.2-33). No Nomex material from either the suit pocket or the survival gear pouch was adhered to the outer aluminum casing. Various amounts of paint showed evidence of impact with small, hot metallic pellets. Corrosion was also present. All external plastic was melted or missing. Internal inspection of the radios revealed evidence of moderate heating, with the center-most components experiencing only minor solder re-flow. For each of the four radios that were recovered, the control module was extracted and interfaced with ground support equipment to read the unique identifier of the radio. Thus, all four radios were ascribed to a specific crew member. Overall appearance suggests that each A/N PRC-112 radio experienced similar thermal and mechanical environments. Each radio was mechanically extracted from the ACES pocket and the survival gear pouch as temperatures were rising and, for a short duration and nearly simultaneously, each of the radios experienced high heating and a hot metal pellet-like shower. This indicates a rapid disruption of the suit survival gear pockets, releasing the radios while they were still in close proximity to the CM debris cloud.

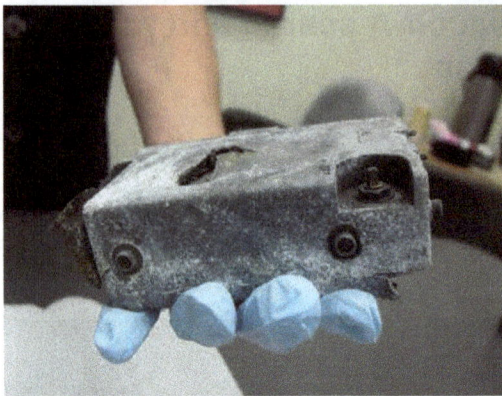

Figure 3.2-33. *External* (left) *and internal* (right) *views of a recovered Army/Navy personal radio communications-112 radio.*

3.2.13 Ground plot analysis

To help understand the breakup sequence of the CM, the recovery locations of pieces of crew equipment were analyzed to determine the order in which the crew members separated from the CM. The human body's complex geometry is difficult to model for precise ballistic analysis and can result in significant variations in trajectory. As a result, the recovery locations of the crew remains is unreliable data for determining the order in which the crew members separated from the CM. Therefore, recovery locations of the crew remains were not included in this analysis.

Items selected for analysis (helmets, SARSAT beacons, and A/N PRC-112 radios) were chosen due to the small variations in the conditions of like items (i.e., all helmets were recovered in the same general condition) and their "regular" shapes (the helmets are roughly spherical, and the SARSAT beacons and A/N PRC-112 radios are rectangular prisms), which have known aerodynamic properties and result in predictable free-flying trajectories.[8] Thus, all like items had similar flight characteristics in the fall to the ground, and can be used to determine the relative order of crew member separation from the CM. It was assumed that the items separated from each crew member in the same manner and roughly the same amount of time after each crew member separated from the CM.

As discussed above, all seven helmets were recovered and ascribed to crew members.[9] The helmets impacted the ground from west to east in the order of seat 7, seat 6, seat 5, seat 4, seat 2, seat 3, and seat 1 (figure 3.2-34).

Six SARSAT beacons were recovered (all except seat 2) and ascribed to crew members. The SARSAT beacons impacted the ground from west to east in the order of seat 6, seat 7, seat 5, seat 4, seat 1, and seat 3 (figure 3.2-34).

Four A/N PRC-112 radios were recovered and identified to crew members (seat 1, seat 2, seat 4, and seat 7). The longitude data for the seat 4 radio is highly suspect, however. These data indicate that it was recovered more than 50 miles east of all other seat 4 crew equipment debris. It is assumed that the original data point was recorded incorrectly and is off by 1 degree, so a "corrected" data point was used in the analysis. The A/N PRC-112 radios impacted the ground from west to east in the order of seat 4, seat 7, seat 1, and seat 2 (figure 3.2-34).

---

[8]Other crew equipment items were not selected for detailed ground plot analysis (EOS bottles, SEAWARS, etc.) due to their irregular shapes, which would result in lower confidence in the flight characteristics being similar.

[9]One of the middeck crew members did not have the helmet attached to the suit at the time of the breakup. It is possible that the helmet was released from the CM at a different time than it would have been if it had been attached to the suit.

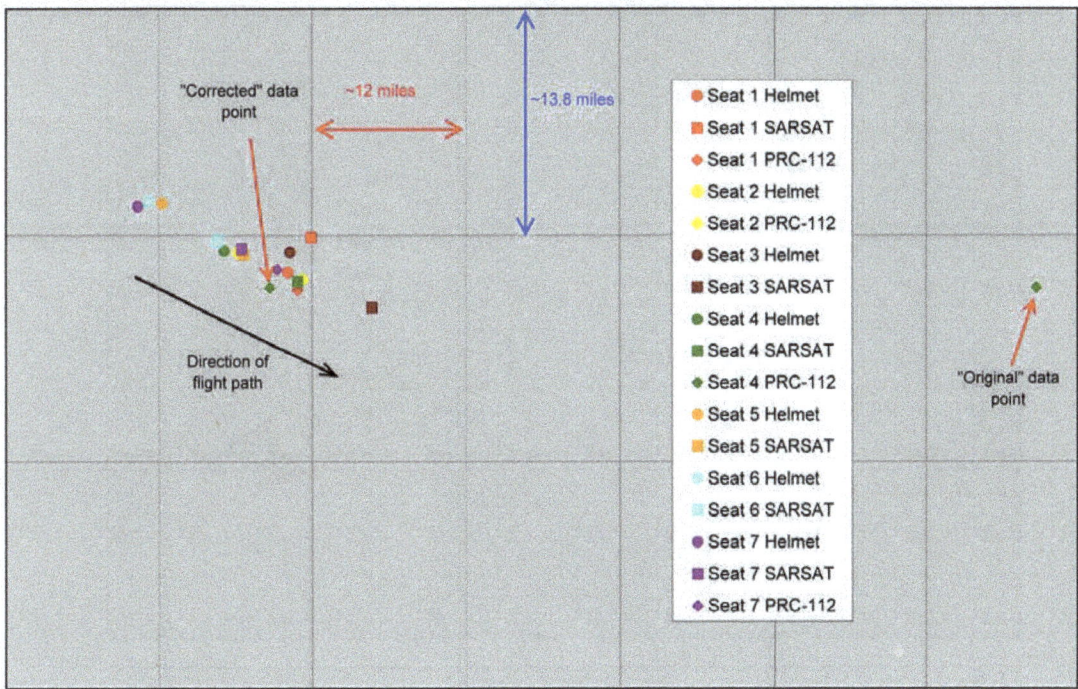

Figure 3.2-34. *Relative location of helmets, search and rescue satellite-aided tracking beacons, and Army/Navy personal radio communications-112 radios.*

The investigation team concluded that the middeck crew members separated from the CM before the flight deck crew members. Orders within the middeck/flight deck groups cannot be determined conclusively, but it appears that seat 6 and seat 7 equipment items were first out of the middeck and seat 4 equipment was first out of the flight deck. All CEE associated with specific crew members was plotted by crew member. Analysis looking at the relative centers of the areas of recovered items for each crew member supports this conclusion. Figure 3.2-35 shows the locations of the recovered items for each crew member. The upper plot shows items for the flight deck crew members; the lower plot show items for the middeck crew members. Both plots are the same scale and represent the same geographic area.

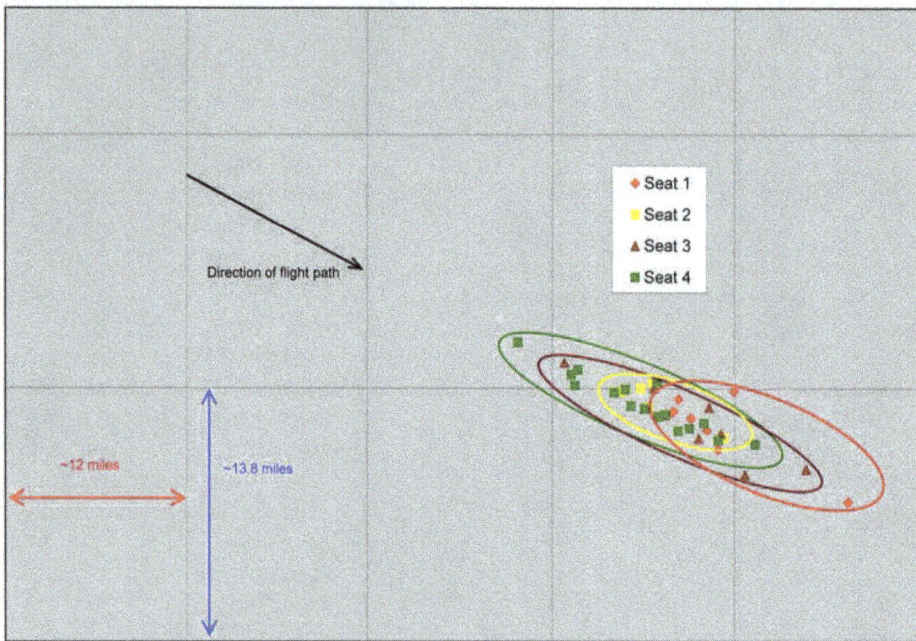

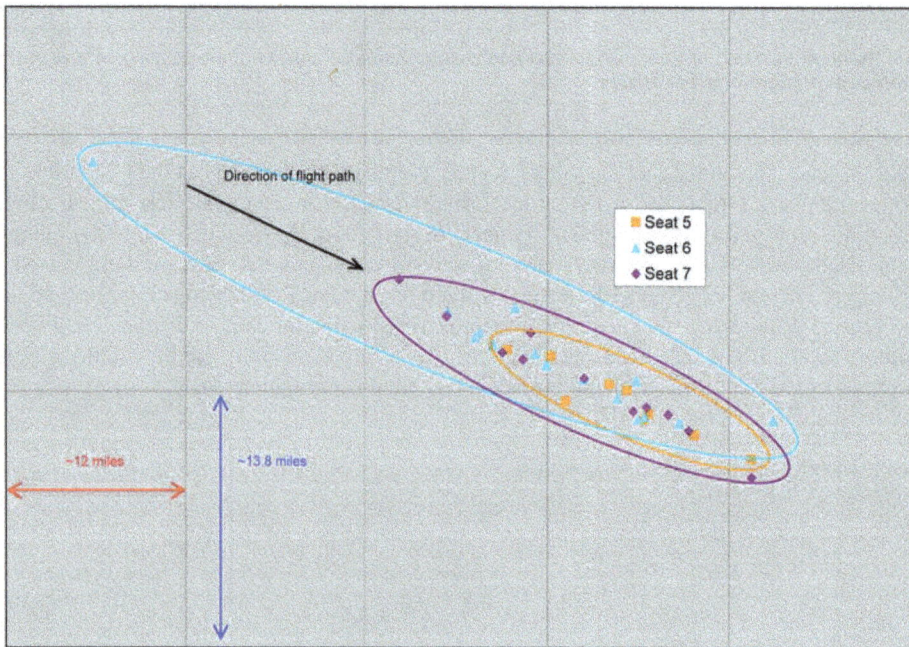

Figure 3.2-35. *Relative locations of all crew escape equipment – flight deck crew members (top) and middeck crew members (bottom).*

3.2.14 Lessons learned

3.2.14.1 *Equipment serialization and marking*

One of the most useful tools in investigating an aviation accident is physically or virtually reconstructing the vehicle from the recovered debris. Being able to identify the original location within the vehicle of debris items is of utmost importance in achieving an accurate reconstruction. Identifying the origins of debris items is made possible by serializing individual piece parts and subassemblies, and keeping accurate records of the piece part/subassembly serial numbers at the assembly and, ultimately, the vehicle levels.

This is especially useful when there are multiple units of identical or similar components, such as crew equipment, seats, engines, or structural members.

The hard components of the ACES, parachute harness, and parachute pack are serialized and tracked to the top-level assembly. Records are kept regarding which crew member is using which suit, harness, and parachute pack. Because of this meticulous recordkeeping, the recovered helmets, glove rings, SEAWARS, DSCs, ACES pressure relief valves, BIP plugs, O_2 manifolds, EOS bottles, and AODs were identified to a specific crew member. In most cases, the serial numbers are etched or physically stamped on the components, aiding identification. For the SEAWARS, the SARSAT emergency beacons, and the A/N PRC-112 survival radios, the identification labels were damaged or destroyed by entry heating. Identification was possible by disassembling the units and inspecting subcomponent serial number labels (in the case of the SEAWARS), or by reading the programmed unique transponder information in the beacon and radio electronics.

Finding. Most of the suit components and subcomponents include serial numbers that are recorded and tracked to a specific crew member. This configuration management documentation aided greatly in the process of ascribing the debris items to specific crew members.

> **Recommendation A5.** Develop equipment failure investigation marking ("fingerprinting") requirements and policies for space flight programs. Equipment fingerprinting requires three aspects to be effective: component serialization, marking, and tracking to the lowest assembly level practical.

3.2.14.2 *Suit requirements and design*

The crew escape suits (the ACESs) were designed to enable survival of crew members during egress and escape from the shuttle in emergency situations. There were specific requirements for the suit to protect crew members from contaminated atmosphere and smoke. As with other materials used in the shuttle, the suit materials were required to be nonflammable or self-extinguishing. However, the suit assembly did not have functional requirements to protect the crew members from environments involving elevated temperatures or fire, as might be present during an emergency egress due to a fire at the launch pad.

As part of the certification testing of the U.S. Air Force suit, suits were subjected to flame pit tests in which suited manikins were placed in a jet fuel fire for 3 seconds and then removed. The suits performed well, with no structural failures and no expected burns to the occupant (based on temperature sensors on the manikin). The ACES is similar to the U.S. Air Force suit, so it may be expected that the ACES would perform well in similar tests. However, the ACES ensemble has some design and materials differences from the U.S. Air Force suit. One notable difference is the use of nylon on the ACES parachute harness straps and the boots. The use of nylon presents a potential weakness in the suit if the suit is used in an environment entailing elevated temperatures or fire.

Finding. The ACES had no performance requirements for occupant protection from elevated temperatures or fire. The ensemble includes nylon on the parachute harness straps and the boots. The ACES may not provide adequate protection to crew members in emergency egress scenarios involving exposure to heat and flames.

> **Recommendation L3-5/L4-1.** Evaluate crew survival suits as an integrated system that includes boots, helmet, and other elements to determine the weak points, such as thermal, pressure, windblast, or chemical exposure. Once identified, alternatives should be explored to strengthen the weak areas. Materials with low resistance to chemicals, heat, and flames should not be used on equipment that is intended to protect the wearer from such hostile environments.

3.3   Crew Training

Crew training, while not a factor in causing the *Columbia* accident, is nonetheless an important element of this report. This section will provide an overview of generic astronaut training and examine *Columbia*-specific crew training. Finally, this section will provide an in-depth analysis and discussion of the impact of training on the actions taken by the STS-107 crew as events unfolded during vehicle entry.

The following is a summary of findings, conclusions, and recommendations for this section.

Finding. The current training regimen separates vehicle systems training from emergency egress training. Emergency egress training sessions exercise the procedures and techniques for egressing the shuttle CM without emphasizing the systems failures that caused the emergency condition. The egress training events are performed on different days from the systems training events, with little discussion of the transition between systems malfunctions and the decision to egress the vehicle. Crew members become conditioned to focus on problem resolution rather than crew survival; the training does not adequately prepare the crew to recognize impending survival situations. It is possible that the STS-107 crew members did not close and lock their visors during the vehicle LOC dynamics (before cabin depressurization) because they were more focused on solving vehicle control problems rather than on their own survival.

> **Recommendation L1-1.** Incorporate objectives in the astronaut training program that emphasize understanding the transition from recoverable systems problems to impending survival situations.

Finding. Emergency egress training for a vehicle LOC/breakup is based on extrapolated data and basic assumptions from the *Challenger* accident for aerodynamic modeling and CM dynamics. The vehicle LOC emergency egress procedures taught to shuttle crews do not address a vehicle LOC occurring during entry.

> **Recommendation L2-1.** Assemble a team of crew escape instructors, flight directors, and astronauts to assess orbiter procedures in the context of ascent, deorbit, and entry contingencies. Revise the procedures with consideration to time constraints and the interplay among the thermal environment, expected crew module dynamics, and crew and crew equipment capabilities.

> **Recommendation L2-2.** Prior to operational deployment of future crewed spacecraft, determine the vehicle dynamics, entry thermal and aerodynamic loads, and crew survival envelopes during a vehicle loss of control so that they may be adequately integrated into training programs.

> **Recommendation L2-3.** Future crewed spacecraft vehicle design should account for vehicle loss of control contingencies to maximize the probability of crew survival.

3.3.1 Overview

NASA's astronaut training program is designed to provide the systems familiarization and flight skills that are required for astronauts to operate the shuttle and carry out mission tasks effectively and efficiently. The training is structured in a building-block format, beginning with workbooks and briefings and progressing to lessons that use sophisticated trainers and simulators.

Individuals selected as astronaut candidates (ASCANs) undergo a training and evaluation period, which lasts over 1 year. This training introduces ASCANs to generic shuttle systems and flight operations, and it prepares them for more in-depth follow-on training as assigned crew members. The curriculum includes training on the Data Processing System; the Guidance, Navigation, and Control System; vehicle control and propulsion systems; the Communications and Tracking System; crew habitability; and shuttle crew escape (emergency egress) equipment and procedures. Additionally, ASCANs are given International Space Station (ISS) systems training. Upon completion of the ASCAN training, the student possesses a functional knowledge of the shuttle systems, ISS systems, and flight operations procedures.

After the initial training period, ASCANs receive advanced training that may lead to assignment to a shuttle flight crew. While awaiting flight assignment, ASCANs maintain proficiency in shuttle systems and flight operations through recurring proficiency lessons in the shuttle mission simulator (SMS). They may also elect to take single system trainer refresher lessons in various orbiter systems. In addition, ASCANs receive mission-specific courses such as Payload Deployment and Retrieval System, Mobile Servicing System, rendezvous and proximity operations, and extravehicular activity (spacewalk).

Upon assignment to a specific flight, crew members progress to flight-similar operations (ascent, orbit, and deorbit/entry) lessons and begin the appropriate mission-specific courses. Assigned crew training also includes flight-specific shuttle mockup training sessions on crew habitability and crew escape. Ascent/entry flight operations training prepares orbiter crew members for crew ingress, orbital insertion, deorbit burn, and landing.

Manual flying techniques are covered in several lessons. All CDRs and PLTs become proficient in manual skills for nominal[1] ascents, aborts, and entries. Crew coordination (space flight resource management (SFRM)) objectives are included to enhance team effectiveness and to ensure mission safety and success.

All training courses lead to integrated simulations. The integrated simulations build the team coordination between the crew and the flight control teams in the Mission Control Center (MCC) that is necessary to ensure a successful mission.

The training regimen encourages systems knowledge and problem resolution through appropriate analysis of displays and use of checklists. While the regimen incorporates scenarios that involve multiple systems failures, in general it is considered nonproductive to train scenarios from which there is no recovery and so those cases are not simulated. No simulation cases are intentionally scripted to result in explosive cabin depressurization or vehicle LOC. Unrecoverable conditions are not intentionally presented to the crew during training. There have been isolated cases in which simulations have ended in a vehicle LOC, but those instances are usually a result of an unrealistic number of simulated systems failures occurring at the same time, seemingly unrelated simulated failures interacting in unforeseen ways, failures being entered into the simulation computer system in the wrong order, crew or flight control team error or miscommunication, or an unexpected failure of the simulation computer.

Training is segregated based on the topic, the activity, and the limitations of each training facility. Sessions for systems failures, which may eventually result in the need to perform an emergency egress of the vehicle, are conducted in a facility that adequately simulates the software and hardware responses of the orbiter and provides an accurate representation of the flight deck interior only. In contrast, emergency egress procedure training is conducted in a volumetrically correct mockup of the entire CM that lacks the capacity for simulating systems malfunctions. The purpose of the emergency egress training sessions is to exercise the procedures and techniques for egressing the shuttle CM without emphasizing the systems failures that "caused" the emergency egress. The training events are performed at different times on different days with little discussion of the transition between a systems malfunction and the decision to egress the vehicle.

---

[1] Within acceptable boundaries.

Finding. The current training regimen separates vehicle systems training from emergency egress training. Emergency egress training sessions exercise the procedures and techniques for egressing the shuttle CM without emphasizing the systems failures that caused the emergency condition. The egress training events are performed on different days from the systems training events, with little discussion of the transition between systems malfunctions and the decision to egress the vehicle. Crew members become conditioned to focus on problem resolution rather than crew survival; the training does not adequately prepare the crew to recognize impending survival situations. It is possible that the STS-107 crew members did not close and lock their visors during the vehicle LOC dynamics (before cabin depressurization) because they were more focused on solving the vehicle control problems rather than on their own survival.

> **Recommendation L1-1.** Incorporate objectives in the astronaut training program that emphasize understanding the transition from recoverable systems problems to impending survival situations.

The emergency egress training program includes classroom sessions on shuttle CM egress procedures in the event of an LOC and vehicle breakup. This training is given to new astronauts as part of the ASCAN training program, and is given to flight-assigned shuttle crews in flight-assigned training just prior to launch in the escape systems refresher class. The training discusses procedures (figure 3.3-1) that are based on extrapolated data from the *Challenger* accident. The analysis of the *Challenger* data used basic assumptions for the vehicle/CM attitudes and dynamics.[2] However, this analysis had not been updated using the more sophisticated techniques available since the *Challenger* accident. Aerodynamic modeling performed for the current investigation provided estimates of CM dynamics following vehicle breakup. These dynamics differed from the assumptions that were made in the Shepherd-Foale Report, and have significant bearing on the LOC/breakup egress procedures. Additionally, the emergency egress procedures and training cover the case of a vehicle breakup during ascent,[3] which has a relatively benign thermal environment when compared to the entry trajectories. The *Columbia* accident brought to the forefront the higher aerodynamic and heat stresses that adversely affect the shuttle CM survival (and, therefore, crew survival) found in the entry environment.

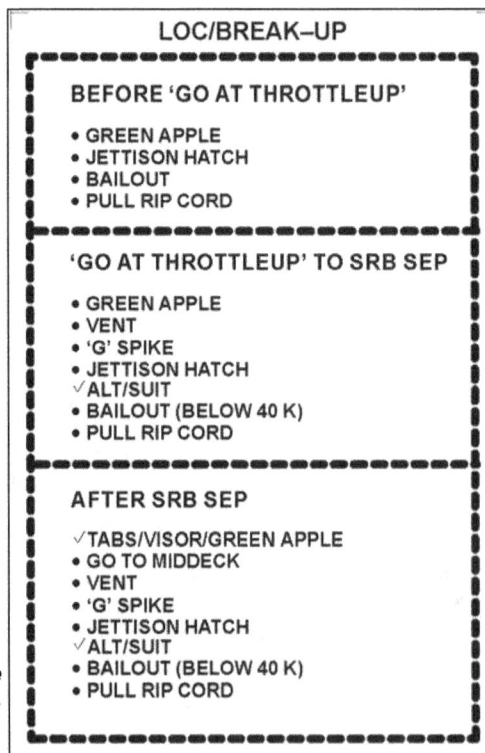

> LOC/BREAK–UP
>
> **BEFORE 'GO AT THROTTLEUP'**
> - GREEN APPLE
> - JETTISON HATCH
> - BAILOUT
> - PULL RIP CORD
>
> **'GO AT THROTTLEUP' TO SRB SEP**
> - GREEN APPLE
> - VENT
> - 'G' SPIKE
> - JETTISON HATCH
> ✓ALT/SUIT
> - BAILOUT (BELOW 40 K)
> - PULL RIP CORD
>
> **AFTER SRB SEP**
> ✓TABS/VISOR/GREEN APPLE
> - GO TO MIDDECK
> - VENT
> - 'G' SPIKE
> - JETTISON HATCH
> ✓ALT/SUIT
> - BAILOUT (BELOW 40 K)
> - PULL RIP CORD

Figure 3.3-1. *Loss of control/breakup cue card as flown on STS-107.*

---

[2]Crew Bailout Procedure for LOC/Breakup Report, B. Shepherd and M. Foale, September 25, 1989.
[3]Prior to STS-107, NASA had not experienced a vehicle breakup during entry.

Finding. Emergency egress training for a vehicle LOC/breakup is based on extrapolated data and basic assumptions from the *Challenger* accident for aerodynamic modeling and CM dynamics. The vehicle LOC emergency egress procedures taught to shuttle crews do not address a vehicle LOC occurring during entry.

>*Recommendation L2-1.* Assemble a team of crew escape instructors, flight directors, and astronauts to assess orbiter procedures in the context of ascent, deorbit, and entry contingencies. Revise the procedures with consideration to time constraints and the interplay among the thermal environment, expected crew module dynamics, and crew and crew equipment capabilities.

>*Recommendation L2-2.* Prior to operational deployment of future crewed spacecraft, determine the vehicle dynamics, entry thermal and aerodynamic loads, and crew survival envelopes during a vehicle loss of control so that they may be adequately integrated into training programs.

>*Recommendation L2-3.* Future crewed spacecraft vehicle design should account for vehicle loss of control contingencies to maximize the probability of crew survival.

3.3.2 *Columbia* crew training

All STS-107 crew members completed the applicable ASCAN, core systems refresher, and ascent/ entry flight operations training programs. The crew also completed all prescribed flight-specific training programs, including ascent and entry proficiency simulator training sessions, crew habitability, and crew escape/crew survival training as well as the prescribed ascent and entry integrated simulations prior to the launch delay in June 2002. An additional five ascent integrated simulations and three entry integrated simulations were completed prior to the January 2003 launch. The prescribed post-insertion and deorbit preparation simulations were also completed.

Launch delays, which were caused by main engine flowliner issues, resulted in the need to repeat some training. The entire crew repeated the water survival lessons (classroom and in-water sessions) in November 2002. The terminal countdown demonstration test (TCDT) was postponed several months. The crew had already completed the prelaunch ingress/egress mockup training session in June 2002, just prior to the postponement. Once the TCDT date was finalized, the CDR opted to retake the prelaunch ingress/egress lesson in November 2002. The escape systems refresher lesson was given in June 2002 but was also repeated in December 2002 at the CDR's request. The crew also performed additional ascent and entry proficiency training sessions in the simulator following the launch delay in June 2002.

Throughout their training, the STS-107 crew members displayed expert orbiter systems knowledge, correct and thorough procedure execution, and excellent SFRM techniques. The crew was very rigorous in verbalizing and verifying procedural steps and routinely took time to brief SFRM topics before each simulation. The launch delays kept the crew in training together for more than 2 years, resulting in a well-trained and finely tuned team. For example, during a simulation run, the *Columbia* crew exercised crew coordination skills by performing the entire run without verbal communication. The crew worked through the systems failures and malfunctions by knowing the systems, procedures, and each other's duties, while using nonverbal communication when appropriate.

3.3.3 Analysis and discussion

Following the loss of *Columbia*, the STS-107 training records were reviewed and the crew instructors were questioned with respect to failure scenarios that would result in the crew members closing visors. This research established that the crew had experienced numerous simulation scenarios in training with procedures requiring lowered visors (e.g., smoke/fire, cabin leaks, broken window panes, contingency aborts, systems failures resulting in an in-flight bailout, etc.). In line with the standard training regimen, very few simulations were performed with the crews wearing the ACES. When performing simulations unsuited, the STS-107 crew would verbalize the suit-specific steps of the procedures.

The CDR, Mission Specialist 2 (MS2), and MS3 were veterans of previous space flights. The CDR, PLT, MS1, MS3, and Payload Specialist 1 (PS1) were all professional military pilots, and the MS2 and MS4 were experienced civilian pilots. All members of the *Columbia* crew were highly trained and outstandingly competent. The training regimen that they underwent emphasized systems knowledge and problem resolution through appropriate analysis of displays and the use of checklists. As was previously stated, unrecoverable conditions are not presented to a shuttle crew during training. It is likely that the STS-107 crew members did not close and lock their visors before cabin depressurization because they were focused on solving the problems that had been presented to them rather than on their own survival. Upon cabin depressurization, a survival situation would be immediately apparent from a physiological perspective. The fact that they still did not close and lock their visors indicates that they were rapidly incapacitated and unable to do so.

3.3.4 Training effectiveness case study

Analysis of switch positions on recovered control panels can reveal crew actions. This analysis can provide insight into the crew members' thought processes and motives, which reveals their knowledge of vehicle systems.

Following the 1986 loss of *Challenger*, a review of the recovered equipment showed that the crew took a few actions after the breakup and prior to losing consciousness.[4] During the *Columbia* CM reconstruction, the switch panels were examined to determine whether any switches were out of the expected positions. Of the recovered flight deck panels (highlighted green in figure 3.3-2), an estimated 10% to15% of the switches were found out of position with respect to the expected positions for the entry timeframe.

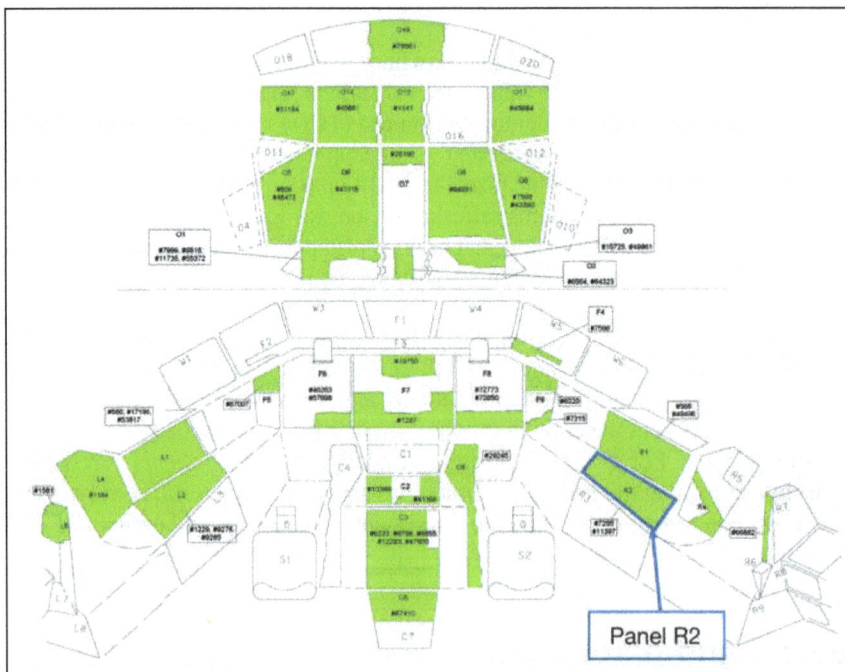

Figure 3.3-2. Columbia *recovered flight deck forward panels* (highlighted in green).

---

[4]Report from Dr. Joe Kerwin to Rear Adm Truly, http://history.nasa.gov/kerwin.html, July 28, 1986.

No further analysis was performed at that time as it was not possible to know whether the switch positions were due to crew action, the mishap, or handling during debris recovery. However, the R2 panel, to the immediate right of the PLT on the flight deck (figure 3.3-2), warranted further investigation. This panel contains the primary controls for the auxiliary power units (APUs). The APUs drive the hydraulic pumps that provide hydraulic pressure to the flight control surface actuators. During entry, these surfaces become increasingly important to vehicle control, and loss of hydraulic pressure can have catastrophic results. Reconstructed general purpose computer (RGPC)-2 data revealed that the hydraulic systems failed prior to GMT 14:00:03 while the crew was conscious and capable of taking action.

When recovered, the R2 panel was folded back in on itself (figure 3.3-3), protecting the innermost switches from manipulation during recovery operations. The switch positions were considered to be unaltered by external factors, making the switches valuable in determining crew actions.

The panel was pried open during the investigation and switch positions were reviewed. In figure 3.3-4, all out-of-position switches are outlined with a pink box and noted with a pink dot.

Figure 3.3-3. *Recovered R2 panel from* Columbia.

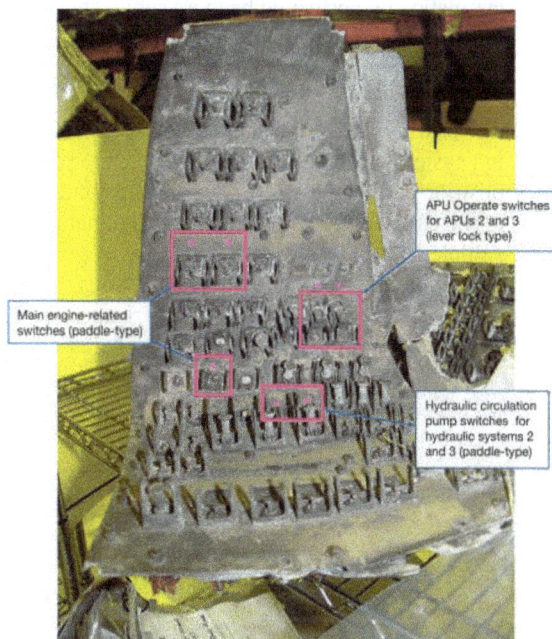

Figure 3.3-4. *Recovered R2 panel from* Columbia *(after it was unfolded). Pink dots are above the out-of-configuration switches.*

The APU 2 and APU 3 Operate switches ("lever lock" switches that require two independent actions to change the switch position) were found in the "injector cool" position as opposed to the nominal "start/run" position. These switches are used when starting and stopping the APU. The "injector cool" position is used to cool down the APU after a shutdown prior to restarting it.

The remaining out-of-position switches are paddle-type switches that require just a push to move out of position. The switches on the left side of the panel are main engine-related switches that are not used during entry.[5] The two paddle-type switches in the lower center of the panel are the circulation pump switches for hydraulic systems 2 and 3. These were found in the "on" position as opposed to the normal "off" position. The circulation pump is used for thermal conditioning of the hydraulic fluid while the orbiter is on orbit. The pump also is used to keep the hydraulic reservoir pressurized. This pump is not powerful enough to deploy the landing gear, but it can provide some hydraulic pressure if activated. The pump is neither used nominally on entry nor is it used in off-nominal procedures.

At the end of RGPC-2 (GMT 14:00:05), all three APUs were operating but the hydraulic systems pressures and quantities were zero, presumably due to a loss of hydraulic fluid from damage to the left wing. While the crew members could not know the reason for the low hydraulic pressure, they would know from training that a loss of hydraulic pressure would result in a vehicle LOC such as they were experiencing.

In response to a hydraulic system failure, the procedures require shutting down the APUs by placing the APU Operate switches in the "off" position, then moving the APU Operate switches into the "injector cool" position to cool down the APUs before attempting a restart of the APUs. RGPC-2 data indicate that the APU Operate and hydraulic circulation pump switches were in their nominal, expected positions.[6] Therefore, these switches changed position after GMT 14:00:05, 13 seconds prior to the Catastrophic Event (CE).

Because the R2 panel was recovered folded in half and the APU Operate switches were not accessible, it is concluded that these switch positions were not altered during recovery operations. While the possibility exists for the lever lock switches to move due to random debris-debris interaction, the requirement for specific physical actions to enable switch movement makes it much more probable that the PLT deliberately moved the switches in an effort to regain hydraulic pressure and control of the vehicle. The paddle switches for the circulation pumps would be more subject to movement due to debris contact. However, the switches that were out of position (for hydraulic systems 2 and 3) correspond to the same APUs (APUs 2 and 3), lending credence to the theory that the actions were deliberate.

The catastrophic events that led to the loss of *Columbia* are not simulated in training or covered by existing systems procedures. The crew's attempt to recover at least two APUs by selecting the "injector cool" position and, in the interim, providing some hydraulic pressure to the flight control surfaces through the use of the circulation pump demonstrates remarkable aplomb. Their effort to regain hydraulic pressure to recover vehicle control shows excellent knowledge of the orbiter systems and problem-resolution techniques. This also indicates that deliberate crew actions (such as manipulating specific switches) were possible for some period of time after GMT 14:00:05, indicating that the CM was still pressurized and the dynamics of the out-of-control vehicle were not incapacitating.

---

[5]These were paddle-type switches, so the positions could have been changed by any of the various methods described above. They are not relevant to entry systems, so no analysis was necessary.
[6]E-mail: from Jeff Kling, STS-107 Ascent/Entry Mechanical Maintenance and Crew Systems Officer, to Pam Melroy, November 8, 2005.

3.4 Crew Analysis

This section contains the analyses and results regarding what happened to the STS-107 crew. It encompasses the awareness that the crew had of events, crew actions in response to those events, and the events of lethal potential to which the crew was exposed. This analysis is meant to aid current and future spacecraft designers in developing vehicles and systems that incorporate the lessons learned from this accident.

The analysis is based on two types of data: objective data (e.g., medical forensic findings, on-board and downlinked vehicle instrumentation data, recovered on-board video data, and air-to-ground crew communications) and derived data (e.g., ballistics, thermal analysis, aerodynamic analyses, shock wave interactions, motion modeling, thermal injury mapping, and material testing). Although this section describes the best "data fit," it is subject to some inherent uncertainty due to the lack of data, both actual and experimental, on human exposure to conditions that are similar to the atmospheric entry environment.

Evidence indicates that the crew was aware of the LOC and was taking actions that were consistent with an attempt to recover hydraulic pressure. Once the depressurization event occurred, the crew was rendered unconscious or deceased and was unaware of the subsequent physical and thermal events. There is no evidence of crew error contributing to this accident.

[REDACTED.] Cause of death of the crew was unprotected exposure to high altitude and blunt trauma.

The first section discusses crew awareness. Next, medical findings are described by injury categories. A chronological sequence of the events with lethal potential is presented followed by a summary.

The following is a summary of the findings, conclusions, and recommendations from this section:

[REDACTED.]

> **Conclusion L1-1.** After loss of control at GMT 13:59:37 and prior to orbiter breakup at GMT 14:00:18, the *Columbia* cabin pressure was nominal and the crew was capable of conscious actions.

> **Recommendation L1-4.** Future suit design should incorporate the ability for crew members to communicate visors-down without relying on spacecraft power.

Finding. Tissue samples revealed evidence of ebullism.[1]

> **Conclusion L1-3.** The crew was exposed to a pressure altitude above 63,500 feet, indicating that the cabin depressurization event occurred above this altitude.

---

[1]Ebullism is defined as the formation of bubbles in bodily fluids under reduced environmental pressure.

Finding. The depressurization event occurred prior to the loss of circulatory function.

Finding. No conclusion could be drawn as to the rate of cabin depressurization based on medical evidence.

> **Conclusion A8-1.** Spacecraft accidents are rare, and each event adds critical knowledge and understanding to the database of experience.

> **Recommendation A8.** As was executed with *Columbia*, spacecraft accident investigation plans must include provisions for debris and data preservation and security. All debris and data should be cataloged, stored, and preserved so they will be available for future investigations or studies.

Finding. None of the six crew members wearing helmets closed their visors.

> **Conclusion L1-5.** The depressurization incapacitated the crew members so rapidly that they were not able to lower their helmet visors.

> **Recommendation L1-3/L5-1.** Future spacecraft crew survival systems should not rely on manual activation to protect the crew.

Finding. One crew member appears to have been restrained only by the shoulder harness and crotch strap.

> **Recommendation L1-2.** Future spacecraft and crew survival systems should be designed such that the equipment and procedures provided to protect the crew in emergency situations are compatible with nominal operations. Future spacecraft vehicles, equipment, and mission timelines should be designed such that a suited crew member can perform all operations without compromising the configuration of the survival suit during critical phases of flight.

Finding. Injuries were consistent with the crews' upper bodies not being securely held to the seatbacks and with evidence indicating that the inertial reel straps were extended at the time of failure.

Finding. Injuries were consistent with the crews' upper bodies not being supported during the time of dynamic motion.

> **Conclusion L2-3.** Lethal injuries resulted from inadequate upper body restraint and protection during rotational motion.

> **Recommendation L2-7.** Design suit helmets with head protection as a functional requirement, not just as a portion of the pressure garment. Suits should incorporate conformal helmets with head and neck restraint devices, similar to helmet/head restraint techniques used in professional automobile racing.

> **Recommendation L2-8.** The current shuttle inertial reels should be manually locked at the first sign of an off-nominal situation.

> **Recommendation L2-9.** The use of inertial reels in future restraint systems should be evaluated to ensure that they are capable of protecting the crew during nominal and off-nominal situations without active crew intervention.

Finding. Crew members experienced traumatic injuries in areas corresponding to the seat restraint system.

> ***Conclusion L3-4.*** The seat restraint system caused lethal-level injuries to the unconscious or deceased crew members when they separated from the seat.

> ***Recommendation L2-4/L3-4.*** Future spacecraft suits and seat restraints should use state-of-the-art technology in an integrated solution to minimize crew injury and maximize crew survival in off-nominal acceleration environments.

> ***Recommendation L3-1.*** Future vehicles should incorporate a design analysis for breakup to help guide design toward the most graceful degradation of the integrated vehicle systems and structure to maximize crew survival.

Finding. No significant levels of carbon monoxide or cyanide (combustion by-products) were identified in any of the body fluids.

Finding. There was no evidence of thermal injury to the respiratory tracts.

> ***Conclusion L1-4.*** The crew was not exposed to a cabin fire or thermal injury prior to depressurization, cessation of breathing, and loss of consciousness.

[REDACTED.]

3.4.1 Crew awareness

3.4.1.1 *Preflight*
The crew members of STS-107 were placed in protective quarantine at the Johnson Space Center astronaut crew quarters on January 9, 2003 where their health was monitored by the assigned crew surgeons; no health issues were observed.

3.4.1.2 *Launch*
As reported in the CAIB Report, at 81.9 seconds mission elapsed time, post-launch video showed a piece of insulating foam striking the left wing of the orbiter. The remainder of the ascent phase went without incident; and the crew, which was unaware of the debris impact at this time, proceeded with the mission as planned.

3.4.1.3 *Orbital operations*
The *Columbia* orbiter performed satisfactorily on orbit and the crew worked well as a team, accomplishing all scientific goals. On Flight Day 8, the crew was notified via email about the foam strike, but was told it was "not even worth mentioning other than wanting to make sure that [the crew is] not surprised by it in a question from a reporter." The capsule communicator (CAPCOM)[2] also relayed that there was "no concern for [reinforced carbon-carbon] or tile damage" and that there was "absolutely no concern for entry". A video clip of the strike was included with the e-mail.[3] No changes in the mission profile were thought necessary or recommended by the shuttle Mission Management Team, and the entry was flown as originally planned.

---

[2]The main individual with whom astronauts on a flight communicate with on the ground at the MCC in Houston.
[3]*Columbia* Accident Investigation Board Report, Volume I, August 2003, p. 36.

3.4.1.4 Deorbit preparations

The crew started the planned deorbit activities on Flight Day 16 (February 1, 2003). Per pre-mission planning, the crew began working items on the De-orbit Preparation checklist at GMT 09:15:30. At GMT 11:11:18, the flight deck crew entered a computer command (OPS 301) to initiate the *Pre-deorbit Coast* sequence. The Commander, Pilot, Mission Specialist 2, and Mission Specialist 4 seats were located on the flight deck. The Mission Specialist 1, Mission Specialist 3, and Payload Specialist 1 seats were located on the middeck. Figures 3.4-1 and 3.4-2 show the seating arrangements in the CM.

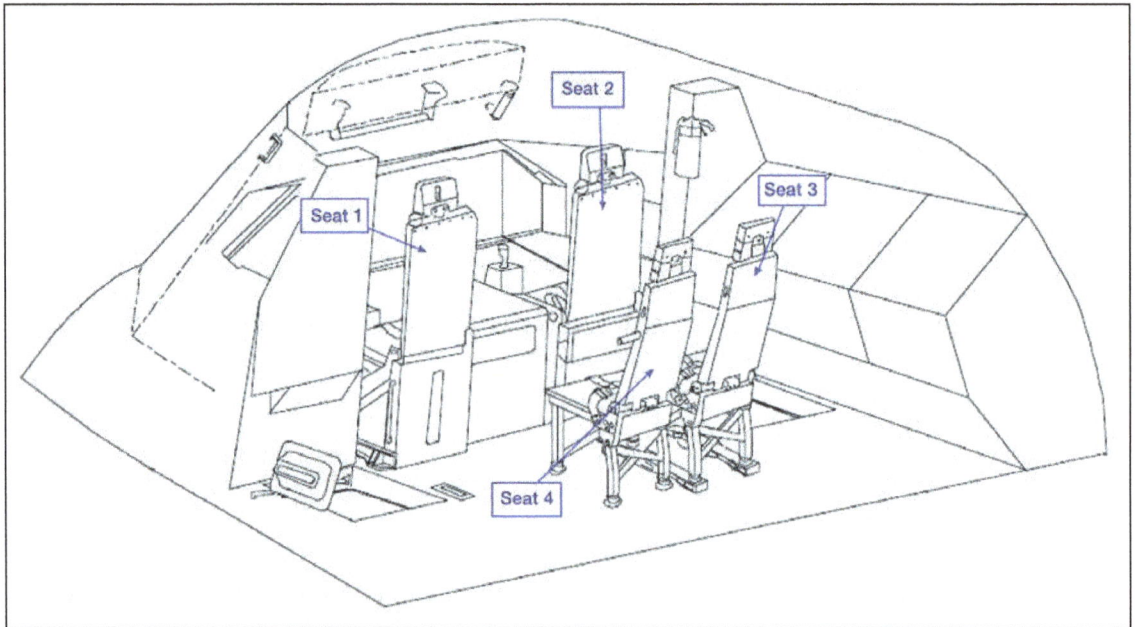

Figure 3.4-1. *Depiction of the flight deck seats.*

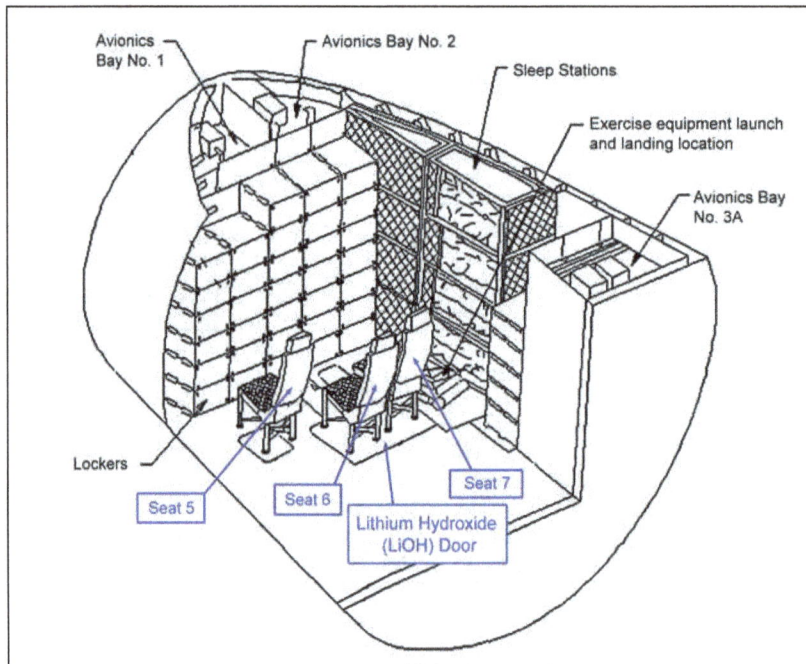

Figure 3.4-2. *Depiction of the middeck seats.* [Adapted from the Shuttle Crew Operations Manual]

Recovered middeck video, documenting events from approximately GMT 11:40:00 to GMT 12:10:00, shows the middeck crew members donning their ACESs and preparing the middeck for return. Figure 3.4-3 is a video frame-capture from the beginning of the tape showing the crew members in various states of landing preparation.

At GMT 12:10:10, the video ends with the middeck crew starting the escape pole installation procedure. Two of three crew members seen in the video were wearing their ACESs, but their gloves and helmets were not mated (typically not performed until after the crew member is strapped in their seat) and one crew member had not yet donned the ACES.

3.4.1.5 *Entry*

The initial phase of entry went without incident. A recovered flight deck video (figure 3.4-4), which runs from approximately GMT 13:35:34 to GMT 13:48:45, provides insight into the crew events taking place on the flight deck.

Figure 3.4-3. *Video frame-capture from the recovered middeck video.*

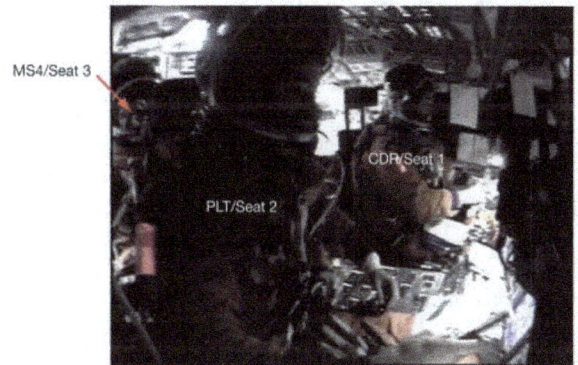

Figure 3.4-4. *Video frame-capture from the recovered flight deck video.*

The video shows that, at GMT 13:36:04, the CDR bumped the RHC accidentally (figure 3.4-5). Movement of the RHC out of the centered position caused the digital autopilot (DAP) to "downmode" from "Auto" mode to "Inertial" mode. When this occurred, a "DAP DOWNMODE RHC" caution-and-warning message was displayed, the INRTL button on the C3 panel was illuminated, and a tone, which can be heard in the recovered flight deck video, was annunciated. An immediate reactivation of the autopilot was performed by the CDR. The CAPCOM in the MCC then requested the CDR to enter "another Item 27," which is a command to fully recover the vehicle attitude from the bumped RHC. The CDR complied, and is heard on the recovered videotape commenting that he had forgotten that he needed to perform an Item 27 after a stick bump but had noticed the guidance needles "weren't really down where they needed to be." This indicates that the CDR was scanning the displays, noticing that the guidance needles were not centered after the stick bump, and was properly processing the information.

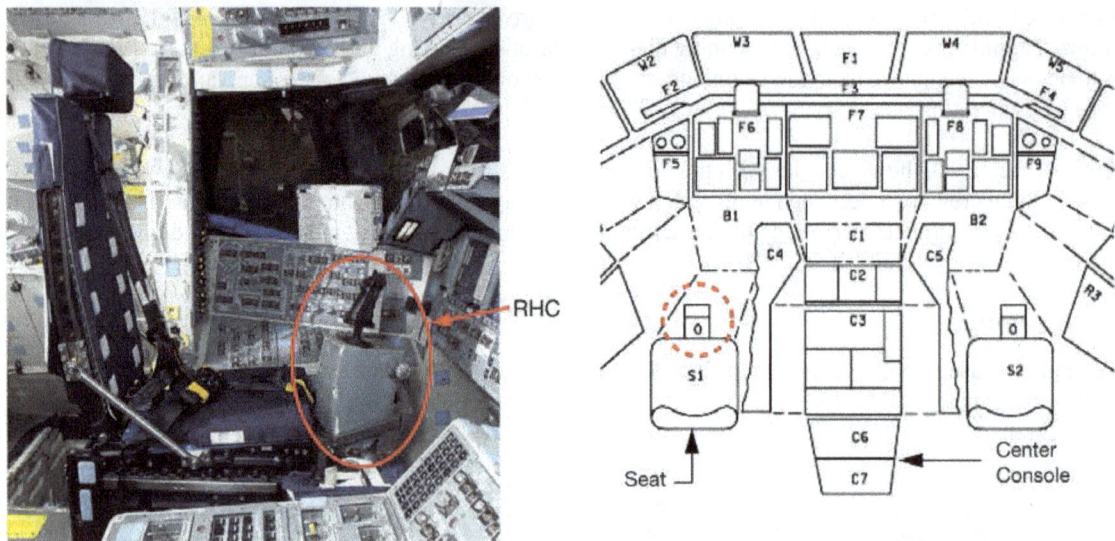

Figure 3.4-5. *Location of the Commander's seat and the rotational hand controller.* [Left picture from a shuttle training mockup in the JSC Space Vehicle Mockup Facility, looking from starboard to port; right picture adapted from the Space Shuttle Systems Handbook, looking from aft to forward]

At GMT 13:39:09, the CDR executed the OPS 304 command to load the computer software that was used to execute entry. The CDR and PLT both verified that the OPS 304 command was properly executed.

The CDR is next seen finishing a drink bag, as part of his required fluid-loading protocol,[4] and floating it back to a crew member for disposal. When the CDR went to pass the last drink bag to the crew member for disposal, that crew member requested that the CDR wait so that the crew member could finish donning the ACES gloves before the gravity levels increased. The video shows that the crew member had partially donned the gloves but did not mate the connecting rings to the ACES. Investigators, therefore, concluded that dealing with the disposal of the water bags and other loose items plus performing flight engineer duties (e.g., assisting the CDR and PLT with checklist items and throwing switches) caused the delay in configuring the ACES.

Recorded telemetry indicates that entry interface (EI) occurred at GMT 13:44:09. Upon observing the time cue for EI on the displays, the CDR states, "Just past EI." The crew then remarks on the flashes of plasma that are visible through the windows (these flashes are a normal part of entry). Shortly afterwards, the CDR requests that everyone perform suit integrity and communications checks. At GMT 13:45:24, three of the four flight deck crew members are observed performing successful communications and suit pressure integrity checks. One of the flight deck crew members could not participate in the suit pressure integrity check since that crew member's gloves were not completely donned at this time.[5] After completion of the check, a crew member asked the CDR whether the crew members were to keep their visors down after the test. The CDR replied, "No."

It should be noted that due to a limitation of the orbiter-suit system, the normal configuration for entry is with the visors up. If the crew keeps the visors down and the O_2 flowing for entry, the O_2 that is being vented from the ACES would increase the cabin O_2 concentration to a level that would violate the hazard controls for fire prevention. In addition, flown astronauts and crew trainers who were interviewed concerning this indicate that the visors also restrict the crews' field of vision and can interfere with nonverbal communication. With the visors down, inter-crew verbal communication is dependent on orbiter main power; there is no battery backup. A loss of power requires either verbal communications with the visor open or nonverbal communications, which can be hindered by having the visor down.

---

[4]Fluid loading is one of the medical countermeasures that is used to mitigate orthostatic intolerance due to the fluid shift and blood plasma level changes that are experienced during space flight and the return to a gravity environment.
[5]The gloves must be mated to the suit for the ACES to pressurize.

By the end of the recovered video at GMT 13:48:45, plasma is visible through the windows (this is normal for this phase of entry) and three of the four flight deck crew members are observed with their ACES suits and helmets on, visors open, and gloves mated and are seated with restraint harnesses on. One crew member was suited with helmet on, visor open, left glove on but not mated, and right glove off and was seated with restraint harness on. There was no indication that the crew was aware of any problems with the orbiter.

Analysis of the telemetry from the O_2 supply system at GMT 13:54:30 shows a signature that is consistent with a second suit pressure check and/or g-suit pressurization by three to five crew members. Since no video or audio was recovered from this timeframe, it is unknown which crew members performed this check. It is possible that the flight deck crew member who was not ready for the first suit and communications check participated in this one.

Between GMT 13:58:39 and GMT 13:58:56, four left tire pressure fault messages were recorded by the Backup Flight Software. These messages were annunciated on the crew displays and accompanied by an audio tone. The fault messages indicated a loss of pressure on the left main landing gear tires. These indications also were presented to the flight control team in the MCC. The CDR and PLT called up the fault page for these messages and reviewed the information. One of the failure scenarios that the crew practiced during training was a circuit breaker trip that resulted in one-half of the tire pressure sensors being disabled. A circuit breaker trip would disable some sensors for all of the tires (left main gear, right main gear, and nose gear), but the failure signature during the accident involved all of the tire pressure sensors on the left main gear only. So the indications that the crew saw would be familiar, although different from what they saw in training. At GMT 13:58:48, the crew began a call to the MCC but that call was broken and not repeated. Brief interruptions of communications often occur due to the tracking and data relay satellite antenna pointing angles changing relative to the orbiter's transceivers. This specific dropout of communication was expected.

At GMT 13:59:06, 10 seconds after the fourth of four tire pressure fault messages, telemetry indicated that the "LEFT MAIN GEAR DOWN" lock sensor transferred to "ON." Other sensors indicated that the landing gear door was still closed and the landing gear was locked in the "up" (stowed) position. These mixed signals caused the left landing gear position indicator to display a "barber pole" (figure 3.4-6), which indicates an indeterminate landing gear position. Post-accident analysis of the data and recovered debris indicates that the left landing gear was locked in the "up" position and the landing gear door was closed. The signal indicating that the gear was down was a false signal that was likely triggered by damage to the sensor system (sensor, wiring harness, etc.). Based on training experience, the crew was probably attempting to diagnose the situation given that it involved the same landing gear as the tire pressure messages and indicated a potential landing gear deployment problem.

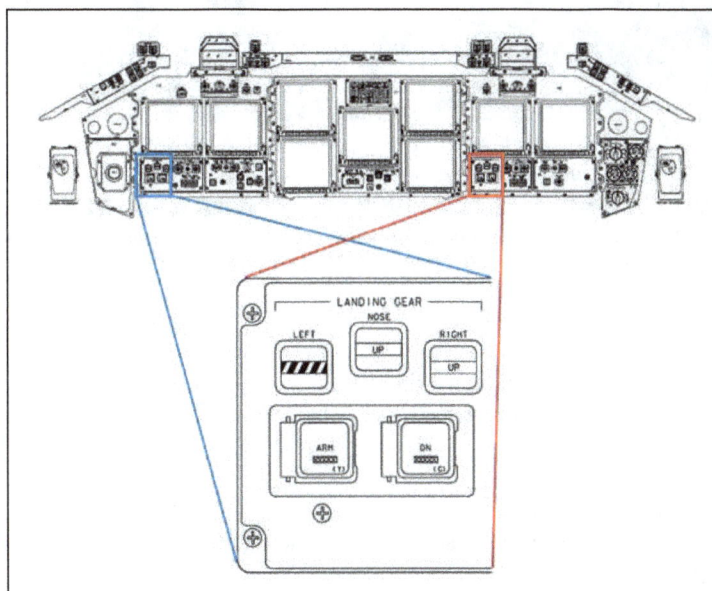

Figure 3.4-6. *Landing gear indicator panel, identical on both sides of the flight deck forward display panels.* Left indicator showing "barber pole" (indeterminate position). [Adapted from the Space Shuttle Systems Handbook]

Twenty-six seconds after the left main gear "talk-back" displayed the barber pole, the last audio transmission from the crew, "Roger, uh …," was received (GMT 13:59:32). The CAPCOM replied to the partial transmission to let the crew know that the flight controllers saw the tire pressure fault messages and did not "copy" the last transmission.

Analysis of RGPC data indicated that the Primary Avionics Software System recorded a fault message that was associated with the removal of Flight Control System (FCS) Channel 4 (CH4) from the control loop at GMT 13:59:33. This message would result in the annunciation of a Master Alarm. While there is no crew action associated with this frequently trained FCS fault message other than to perform a message reset, the crew likely called up a display to analyze the failure. Crews are trained to troubleshoot systems errors, and this crew would have been evaluating this new message along with the previous tire pressure and landing gear down-lock indications to assess whether there was a common system fault that could account for all of these messages.

3.4.1.6 *Loss of control*

Based on engineering analysis and modeling (see Section 2.1), hydraulic pressure, which is required to move the flight control surfaces, was lost at approximately GMT 13:59:37. At that time, the Master Alarm would have sounded for the loss of hydraulics and the crew would have become aware of a serious problem. It is probable that the loss of hydraulic pressure as a result of the damage to the left wing resulted in an uncontrolled pitch-up and loss of vehicle control. A visual simulation of the pitch-up associated with this LOC scenario is shown in figure 3.4-7.[6] The flight deck crew would have been the first to be aware of this owing to the changing light levels, the view of the horizon through the windows, and the information on the flight displays. Space-adapted crews are reported to be very sensitive to motions and G-loads. As the orbiter motion dynamics began to increase, all of the crew members likely would have sensed this motion and been aware of the off-nominal[7] situation. At GMT 13:59:46, a "Roll Ref" alarm message was annunciated, indicating that the orbiter had exceeded the limits of the entry drag profile. In conjunction with the hydraulics messages and the unusual motion of the orbiter, the "Roll Ref" message would have reinforced the fact that a serious problem had developed.

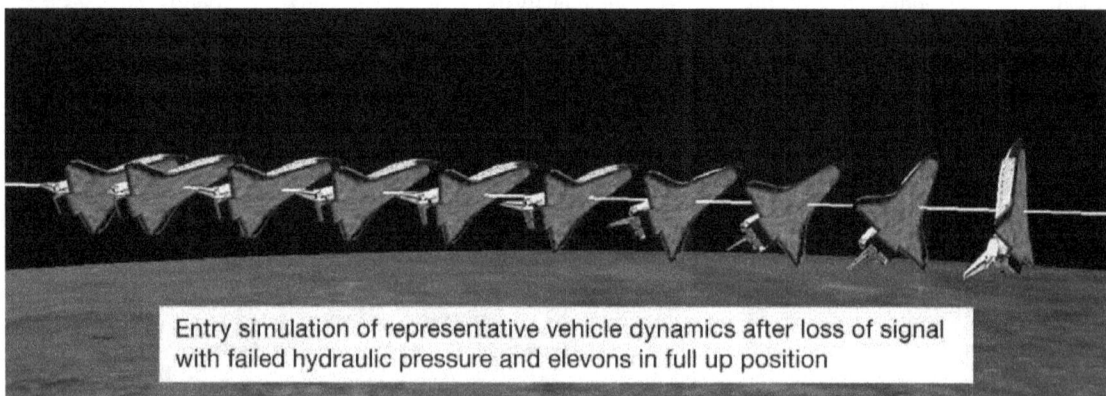

Entry simulation of representative vehicle dynamics after loss of signal with failed hydraulic pressure and elevons in full up position

Figure 3.4-7. *Sequence (1-second intervals) showing a simulation of orbiter loss of control pitch-up from GMT 13:59:37 to GMT 13:59:46.* **White line indicates vehicle trajectory relative to the ground.**

Based on the orbiter LOC entry simulation, the representative motion showed that the predominant orientation of the orbiter remained "belly-into-the-wind" with large excursions in pitch, roll, and yaw. This

---

[6]Vehicle dynamics are based on aerodynamic modeling using orbiter aerodynamic models and accelerometer data.
[7]Outside of acceptable limits.

motion can be characterized[8] as a slow (30 to 40 degrees per second), highly oscillatory spin. Analysis of the accelerations describes the overall motion of the crew as a swaying to the left and the right ($\pm Y_{Crew}$ axis, eyeballs right and left[9]) combined with a pull (deceleration) forward ($+X_{Crew}/-G_x$, eyeballs out) against the seat harness straps. Z-axis accelerations pushed the crew (vertically) down into the seat ($-Z_{Crew}/+G_z$ axis, eyeballs down). Figure 3.4-8 shows a depiction of the sign convention and the resulting motion for accelerations.

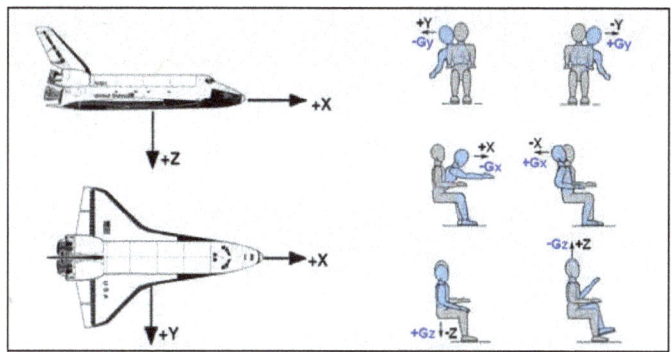

Figure 3.4-8. *Depiction of the orbiter and crew member axes conventions.* (Note: Z axis sign convention for crew is opposite from orbiter.)

Figure 3.4-9 shows representative loads based on modeling in all three axes, including the effects of increasing rotational loads. Models showed that accelerations were initially low, and peaked between 2 G and 3.5 G by the time of the CE (separation of the forebody from the midbody). The dashed black lines (upper and lower) on the chart indicate human performance limits based on NASA-STD-3000.[10] The representative loads, which are based on modeling, were well within these human performance limits.

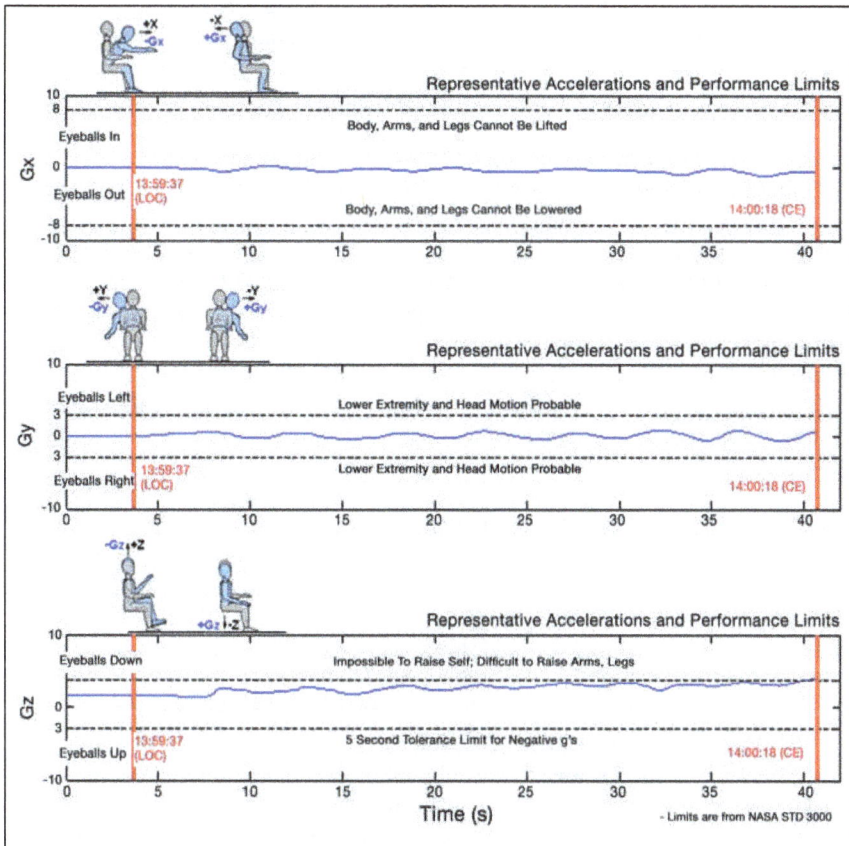

Figure 3.4-9. *Representative accelerations from modeled motion analysis from loss of control to the Catastrophic Event (see Section 2.1).* Black dashed lines show human performance limits.

[8]See descriptors in Flight Test Demonstration Requirements for Departure Resistance and Post-Departure Characteristics of Piloted Airplanes, Air Force MIL-F-83691B, Change 1, May, 31, 1996.
[9]In the crew axis convention, the physiological reaction is in the opposite direction of the acceleration vector; i.e., an acceleration in the $+X_{Crew}$ direction pushes crew members into their seat (i.e., forces the crew in the $-X$ direction, resulting in a $+Gx$ reaction, also known as "eyeballs in").
[10]NASA-STD-3000, *Man-Systems Integration Standards*, Volume I, Section 5, Revision B, 1995.

Post-flight analysis of crew equipment revealed that none of the six recovered seat restraint inertial reel mechanisms locked prior to failure. This resulted in the upper bodies of the crew members being unrestrained. Loads of these magnitudes and rates would not be expected to produce crew injuries or prevent the crew from performing most actions. However, these loads, augmented by the loose harness configuration and mass of the crew worn equipment, would require the crew members to brace themselves.

As the LOC scenario progressed, the dynamic motion environment would be expected to increase the susceptibility to motion sickness and disorientation, particularly in those who had no visual reference (i.e., those on the middeck) or who were novice space flyers.

Based on seat debris and medical analyses, one crew member was not fully restrained before loss of consciousness. Only the shoulder and crotch straps of this crew member appear to have been connected. The normal sequence for strap-in is to attach the lap belts to the crotch strap first, followed by the shoulder straps. Analysis of the seven recovered helmets indicates that this same crew member was the only one not wearing a helmet. Additionally, this crew member was tasked with post-deorbit burn duties. This suggests that this crew member was attempting to become seated and restrained when the LOC dynamics began. Given the motion of the orbiter, the lap belts hanging down between the closely spaced seats would have been difficult to locate and grasp.

RGPC data indicate that between GMT 14:00:02 and GMT 14:00:04, the vehicle was experiencing a right yaw rate of at least 21 deg/sec, which was the sensor limit, and a right roll rate of 7 deg/sec, followed by a left roll rate of 23 deg/sec that was associated with a nose-down pitch rate of 5 deg/sec (because of possible inertial measuring unit saturation, these values may be inaccurate). All available data indicate that the crew cabin environment (temperature, atmosphere) and systems (APUs, fuel cells, lighting, etc.) were still generally nominal; however, the hydraulic pressures and quantities were indicating zero.

RGPC data show that a message reset was performed by the CDR or PLT sometime between GMT 13:59:37.4 and GMT 14:00:05. This action is a normal crew response to a fault message and requires a crew member to manually acknowledge the message by keyboard entry on the center panel. RGPC-2 data indicate that the RHC was moved beyond neutral sometime between GMT 14:00:01.7 and GMT 14:00:03.6, triggering a "DAP DOWNMODE RHC" message at GMT 14:00:03.637. This message was likely due to a crew member bumping the RHC out of the null position due to the oscillatory motion of the orbiter. At GMT 14:00:03.678, the orbiter autopilot was returned to the AUTO mode. Returning the DAP to AUTO mode requires either the CDR or the PLT to press one or two buttons that were located on the glare shield. These actions indicate that the CDR or the PLT was still mentally and physically capable of processing display information and executing commands, and that the orbiter dynamics were still within human performance limitations.

Recovered debris revealed that the APU Operate switches on flight deck panel R2 were in positions that were consistent with an attempted restart of two of the three APUs[11] (figures 3.4-10 and 3.4-11). The hydraulic circulation pump paddle switches for the two hydraulic systems corresponding to the two APUs were also turned on. While turning on the hydraulic circulation pump is not in any crew emergency checklist, the pump can provide some hydraulic pressure, and this action shows good systems knowledge by the crew members as they responded to the limited information presented to them and worked to restore orbiter control. The APU Operate switches are "lever-lock" switches that require three actions to change the position. They must be (1) pulled outwards to disengage the switch lever from the lock, (2) moved to the desired position, and (3) released (figure 3.4-12). The switches are spring-loaded to hold them in the detents. The RGPC data indicate that all of these switches were in the nominal configuration up to GMT 14:00:04.826. These findings strongly suggest that despite the very dynamic vehicle motion, the PLT was still capable of taking appropriate actions to attempt a recovery of the hydraulic pressure by performing an APU restart at some time after GMT 14:00:04.826. Based on the panel R2 switch throws and the lack of visors being lowered, it is probable that the crew never realized that the vehicle LOC situation was unrecoverable and had become a survival situation.

---

[11]APUs supply the hydraulic pressure to the flight control surfaces.

Figure 3.4-10. *Location of the R2 panel.*
[Picture from the Shuttle Training Simulator]

Figure 3.4-11. *Recovered R2 panel from* Columbia *(after it was unfolded).* Pink dots are above the out-of-configuration switches.

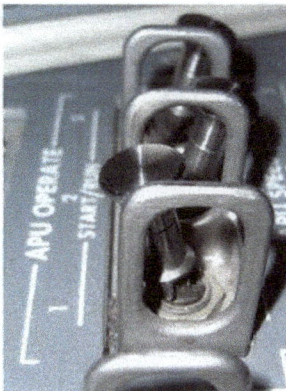

Figure 3.4-12. *Operation of the lever-lock controller switches* (side view).

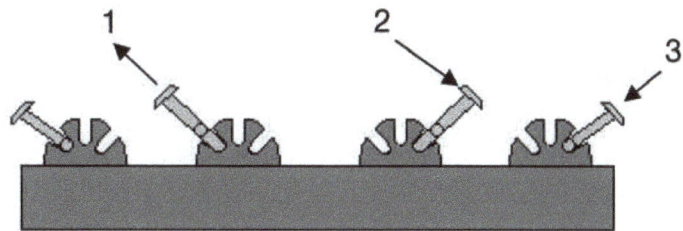

Analysis of several sources of information indicates that the forebody separated from the midbody at or shortly after GMT 14:00:18. It is unknown what accelerations occurred during separation of the orbiter forebody from the midbody; however, ballistic analysis estimates that the translational G that was experienced by the orbiter forebody at the CE decreased from approximately 3.5 G to 1 G. It is also likely that there were additional translational and rotational loads acting on the crew at this time. Analysis of structural debris supports that multiple small impacts occurred between the forward fuselage (FF) and the CM, including in the area of middeck Volume E (figures 3.4-13a and 3.3-13b).

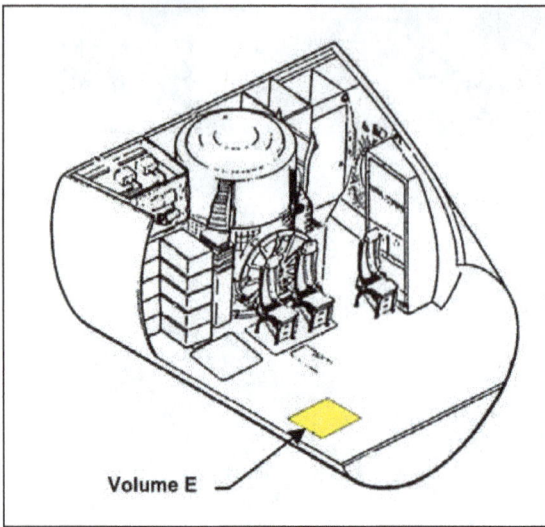

Figure 3.4-13a. *View of middeck floor and Volume E, looking aft.*

Figure 3.4-13b. *Scenario showing how the crew module pressure vessel could impact the forward fuselage, and the middeck Volume E could impact the crew module pressure vessel, with resultant damage.*

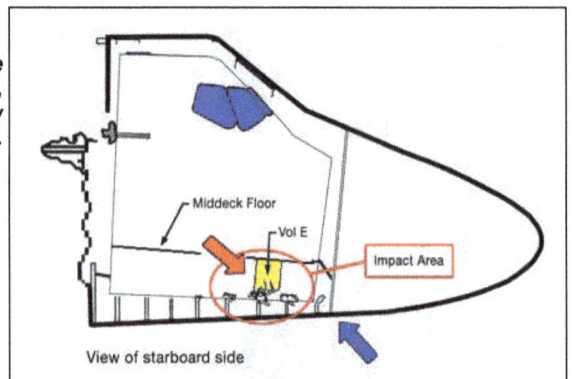

[REDACTED.]

[REDACTED. Figures 3.4-14a through 3.4-13d.]

[REDACTED.]

> **Conclusion L1-1.** After loss of control at GMT 13:59:37 and prior to orbiter breakup at GMT 14:00:18, the *Columbia* cabin pressure was nominal and the crew was capable of conscious actions.

When the vehicle forebody separated from the rest of the vehicle, all resources from the midbody were lost, including power from the fuel cells. This resulted in the loss of all powered lighting, crew displays, radio, intercom, ventilation, and main O_2 supply. The flight deck would still have had light entering the cabin from the windows as well as from the activated chemical light sticks on each arm of the ACES and positioned throughout the cabin. The middeck would have been in total darkness except for some light filtering through the two inter-deck openings and from the activated chemical light sticks. This would indicate a survival situation.

> **Recommendation L1-4.** Future suit design should incorporate the ability for crew members to communicate visors-down without relying on spacecraft power.

It was concluded that the crew was incapacitated and incapable of action at or shortly after the CE. As a survival situation, one of the first crew actions that would be expected after the CE would be for the crew to manually lower their visors and turn on their EOS. As detailed in Section 3.2, none of the crew members wearing helmets closed their visors. The accelerations derived from the representative motion modeling (figure 3.4-10) would not have prevented this action. Since they did not, it was concluded that the crew members were incapacitated due to other factors. This will be discussed in the following section.

This concludes the discussion of crew awareness. Injury classifications are discussed next.

3.4.2 Injury classifications

3.4.2.1 *Exposure to high altitude*

[REDACTED.] Given the level of tissue damage, the crew could not have regained consciousness even with re-pressurization. Survival was possible, but not likely, even with immediate and extensive medical intervention at this point. Although respiration would cease after depressurization, circulatory functions can exist for a short period of time.

[REDACTED.]

Finding. Tissue samples revealed evidence of ebullism.

> **Conclusion L1-3.** The crew was exposed to a pressure altitude above 63,500 feet, indicating that the cabin depressurization event occurred above this altitude.

[REDACTED.]

Finding. The depressurization event occurred prior to the loss of circulatory function.

There is very limited data on human exposure to space-equivalent vacuum. [REDACTED.] Although the *Soyuz 11* cabin depressurization was relatively slow (reportedly taking more than 3.5 minutes to depressurize to 0 psi), it was stated that the depressurization was fatal to the crew in roughly 30 seconds.[12] Because the exact scenario cannot be positively identified, no conclusions with respect to the rate or timing of cabin depressurization can be made from the medical findings. [REDACTED.]

Depressurization events in aviation have led to extensive studies on "time of useful consciousness (TUC)." TUC is generally based on the remaining amount of O_2 in the tissues that is permitting brain functions to continue. Various factors affect the TUC (i.e., exertion, depressurization rate, pre-exposure O_2 partial pressure, G-loads, adrenaline loading, etc.). Since the shuttle cabin uses air, the pre-exposure O_2 partial pressure was only 21% O_2 (the normal for sea-level). Based on debris and structural evidence, the most likely time for the initiation of cabin depressurization was at orbiter breakup (CE) at GMT 14:00:18. Based on video evidence, the depressurization was complete no later than GMT 14:00:59 (figure 3.4-15), and likely much earlier (see Section 2.3). This corresponded to an altitude range of 181,000 feet to approximately 140,000 feet. Traditional aviation TUC would correlate a rapid depressurization at these altitudes to a TUC of 12 seconds.[13] This would have been enough time for the crew to close their visors and initiate O_2 flow, and yet they did not (see Section 3.2).

---

[12]"A History of the Apollo-Soyuz Project, Midterm Review." http://history.nasa.gov/SP-4209/ch8-2.htm.

[13]*Joint Aerospace Physiology, Air Education and Training Command/Bureau of Medicine and Surgery*, February 1998.

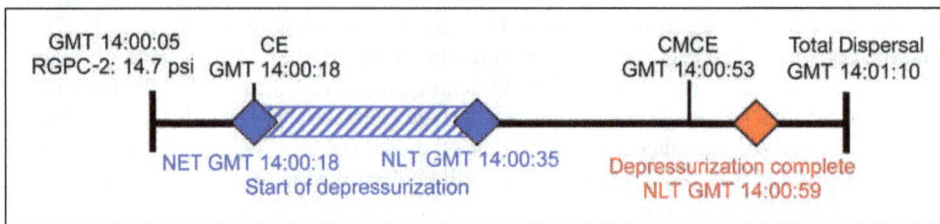

GMT 14:00:05   CE   CMCE   Total Dispersal
RGPC-2: 14.7 psi GMT 14:00:18   GMT 14:00:53   GMT 14:01:10

NET GMT 14:00:18   NLT GMT 14:00:35   Depressurization complete
Start of depressurization   NLT GMT 14:00:59

Figure 3.4-15. *Cabin depressurization timeline.*

However, additional research discussed in *Joint Aerospace Physiology, Air Education and Training Command/Bureau of Medicine and Surgery* shows that the physiological response to hypoxia during a rapid depressurization event at this extreme altitude (181,000 feet) would have reduced the conventional TUC interval by 50% (i.e., 12 seconds would have been reduced to 6 seconds). In addition to the depressurization effects, the physical exertion against the G-forces that the crew experienced at this time would further reduce the available metabolic O_2 reserves[14] and increase the CO_2 partial pressure. Also, NASA research data indicate that de-conditioned crews have a reduced tolerance to G-loads.[15] Further, anecdotal reports from accidental exposure to vacuum confirm much shorter periods of awareness as reported by survivors.[16]

The 51-L *Challenger* accident investigation showed that the *Challenger* CM remained intact and the crew was able to take some immediate actions after vehicle breakup, although the accelerations experienced were much higher as a result of the aerodynamic loads (estimated at 16 G to 21 G[17]). The *Challenger* crew became incapacitated quickly and could not complete activation of all breathing air systems, leading to the conclusion that an incapacitating cabin depressurization occurred.[18] By comparison, the *Columbia* crew experienced lower loads (~3.5 G) at the CE. The fact that none of the crew members lowered their visors[19] strongly suggests that the crew was incapacitated after the CE by a rapid depressurization.

From this time forward, the crew members would have been unconscious, totally unaware of events, and unable to brace against the loads. With the configuration of the ACES (i.e., visors up and three crew members without gloves donned), the depressurization was an event of lethal potential. Had the ACES been configured with the visors down and locked, gloves on, and EOS activated, the depressurization event by itself probably would have been survivable.

Finding. No conclusion could be drawn as to the rate of cabin depressurization based on medical evidence.

Finding. None of the six crew members wearing helmets closed their visors.

> **Conclusion L1-5.** The depressurization incapacitated the crew members so rapidly that they were not able to lower their helmet visors.

> **Recommendation L1-3/L5-1.** Future spacecraft crew survival systems should not rely on manual activation to protect the crew.

> **Conclusion A8-1.** Spacecraft accidents are rare, and each event adds critical knowledge and understanding to the database of experience.

---

[14]Naval Aviation Survival Training Program, G-Tolerance Brief: *G-Tolerance Improvement Program*, 2004.
[15]K. V. Kumar and W. T. Norfleet, "Issues on Human Acceleration Tolerance After Long-Duration Space Flights," NASA Technical Memorandum 104753, October 1992.
[16]Description of Altitude Chamber Mishap, *Roundup*, Volume 6, No. 6, Jan. 6, 1967.
[17]JSC-22175, STS-51L, JSC Visual Data Analysis Sub-Team Report, Appendix D9, June 1986.
[18]Report from Dr. Joe Kerwin to Rear Adm. Truly, http://history.nasa.gov/kerwin.html, July 28, 1986.
[19]See Section 3.2 Crew Worn Equipment.

Recommendation A8. As was executed with *Columbia*, spacecraft accident investigation plans must include provisions for debris and data preservation and security. All debris and data should be cataloged, stored, and preserved so they will be available for future investigations or studies.

3.4.2.2 *Mechanical injuries*

Mechanical injuries were isolated to the period of time at which they most likely occurred, based on engineering analyses of motions and accelerations.

Pre-Catastrophic Event
[REDACTED.]

Catastrophic Event to Crew Module Catastrophic Event

A very dynamic motion environment existed after the CE (GMT 14:00:18); this environment became more intense as the CMCE (the breakup of the forebody) approached at GMT 14:00:53. Figure 3.4-16 shows representative loads on the unconscious or deceased crew members based on aerodynamic modeling of the forebody dynamics post-CE. The black dashed lines showing human performance limits[20] are for conscious crew members. Based on the conclusion that the rapid depressurization occurred at or close to the time of the orbiter forebody separation, the crew was unconscious or deceased and unable to brace against these loads.

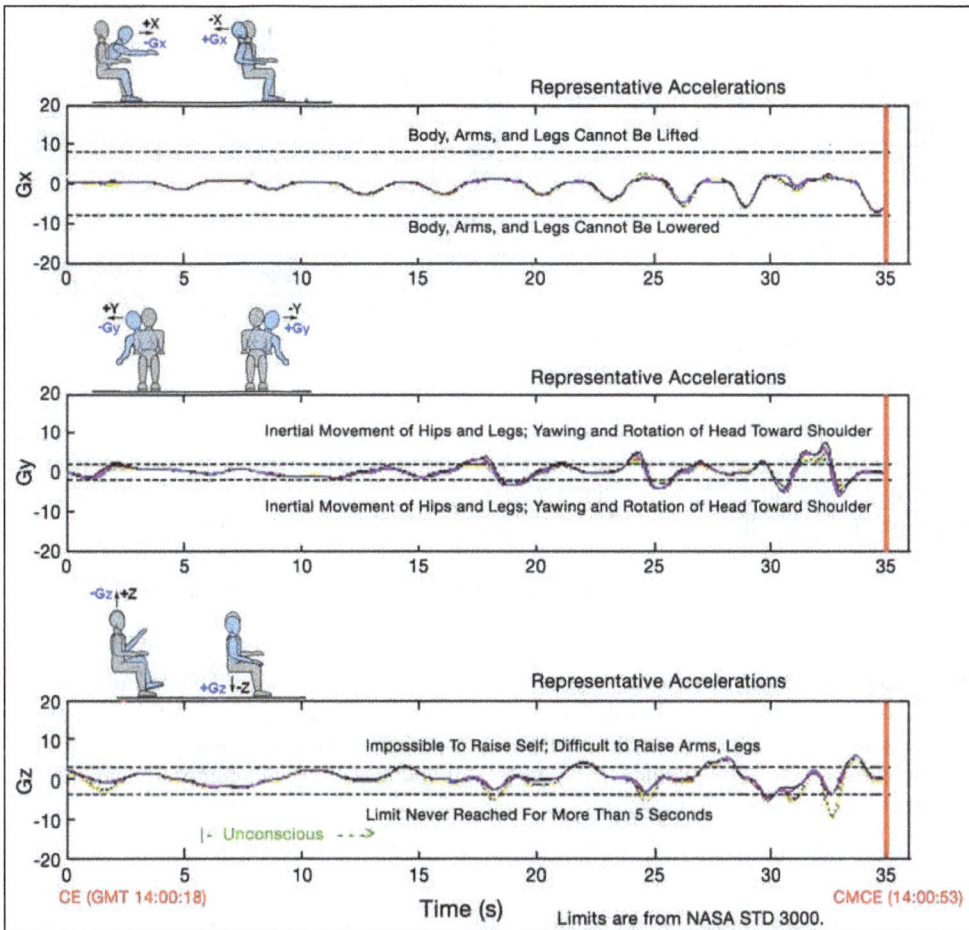

Figure 3.4-16. *Representative acceleration profiles from the orbiter breakup (Catastrophic Event) to the orbiter forebody breakup (Crew Module Catastrophic Event) based on aerodynamic modeling.* **Black dashed lines show human performance limits.**

---

[20]NASA-STD-3000, *Man-Systems Integration Standards*, Volume I, Section 5, Revision B, 1995.

For the first 15 to 20 seconds, the modeled loads would not cause serious injuries to a conscious crew member who was capable of active bracing. An unconscious or deceased crew member would have been more susceptible to injury.

The crew is normally restrained in the seats by a five-point harness system (figure 3.4-17). A lap belt secures the lower torso. A crotch strap prevents "submarining."[21] Two shoulder harnesses, which attach to an inertial reel via the inertial reel strap, secure the upper torso.

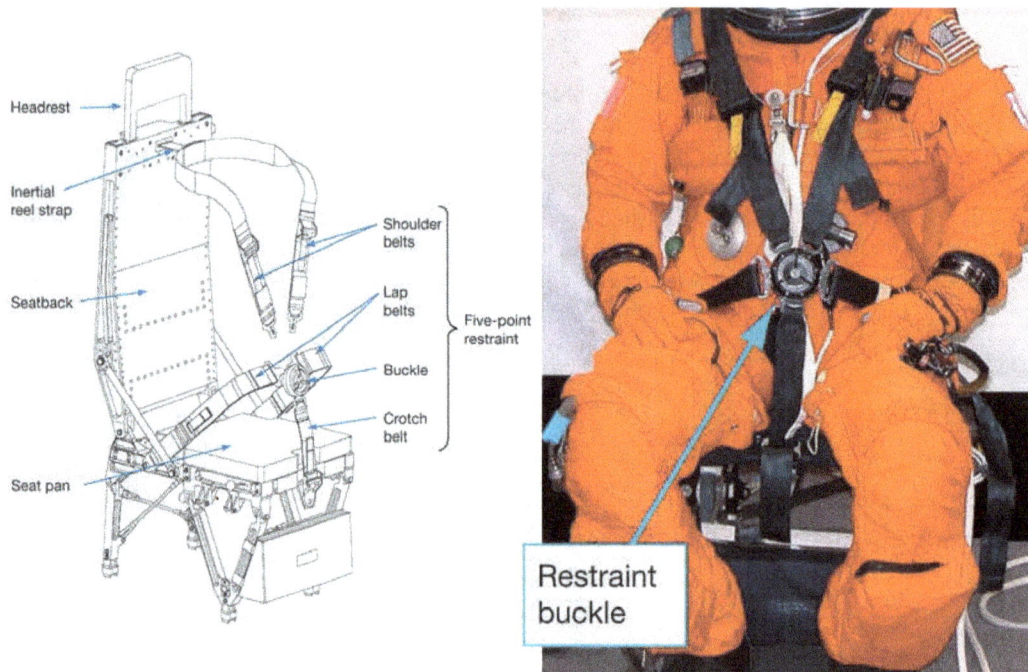

Figure 3.4-17. *Detail of the five-point harness.*

Engineering analysis of the STS-107 restraints indicates that most of the inertial reel straps were extended and did not lock or retract prior to failure of the straps. The inertial reels are normally unlocked to allow the crew to access displays and controls with a full range of motion. With the inertial reels unlocked, the crew members' upper bodies were left unrestrained during the forebody dynamics. [REDACTED.]

Finding. One crew member appears to have been restrained only by the shoulder harness and crotch strap.

> **Recommendation L1-2.** Future spacecraft and crew survival systems should be designed such that the equipment and procedures provided to protect the crew in emergency situations are compatible with nominal operations. Future spacecraft vehicles, equipment, and mission timelines should be designed such that a suited crew member can perform all operations without compromising the configuration of the survival suit during critical phases of flight.

[REDACTED.]

Figure 3.4-18 provides a demonstration of an integrated seat/suit/crew member in entry configuration.

[REDACTED.] Figure 3.4-19 shows an interior view of an intact, pristine ACES helmet demonstrating exposed hardware. [REDACTED.]

---

[21]"Submarining" is when the occupant slides forward and beneath the lap belt.

Neck ring

Figure 3.4-18. *Demonstration of an integrated seat/suit/crew member in entry configuration.*

Figure 3.4-19. *Example of an intact, pristine advanced crew escape suit nonconformal helmet (not an STS-107 helmet).*

[REDACTED. Figure 3.4-20.]

Finding. Injuries were consistent with the crew's upper bodies not being securely held to the seatbacks and with the evidence indicating that the inertial reel straps were extended at the time of failure.

Finding. Injuries were consistent with the crew's upper bodies not being supported during the time of dynamic motion.

Conclusion L2-3. Lethal injuries resulted from inadequate upper body restraint and protection during rotational motion.

Recommendation L2-4/L3-4. Future spacecraft suits and seat restraints should use state-of-the-art technology in an integrated solution to minimize crew injury and maximize crew survival in off-nominal acceleration environments.

Recommendation L2-7. Design suit helmets with head protection as a functional requirement, not just as a portion of the pressure garment. Suits should incorporate conformal helmets with head and neck restraint devices, similar to helmet/head restraint techniques used in professional automobile racing.

Recommendation L2-8. The current shuttle inertial reels should be manually locked at the first sign of an off-nominal situation.

Recommendation L2-9. The use of inertial reels in future restraint systems should be evaluated to ensure that they are capable of protecting the crew during nominal and off-nominal situations without active crew intervention.

Crew Module Catastrophic Event

[REDACTED.] Engineering and ballistic analyses of the orbiter forebody failure indicate that the middeck separated prior to the flight deck. Crew members on the middeck separated along with the middeck accommodations rack, middeck lockers, sub-floor components, and Modular Auxiliary Data System/orbiter experiment data recorder – a scenario that is supported by debris plots (see Section 2.2). Based on structural design analysis, thermal damage, and position in the debris field, the flight deck "pod" and the CM aft bulkhead stayed intact for a longer time. The location of the recovered flight crew equipment, which is plotted in figure 3.4-21, supports the middeck departing prior to the flight deck.

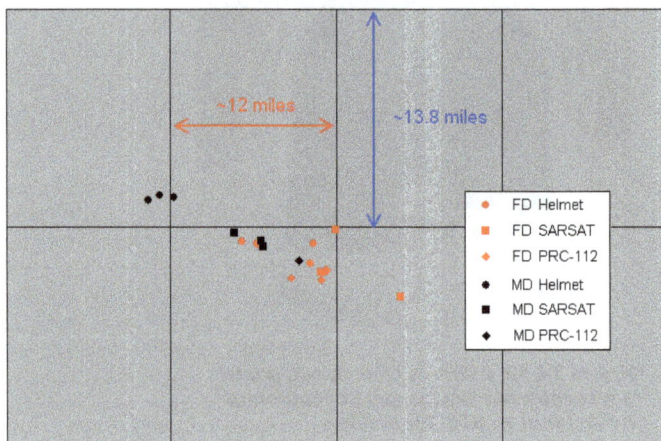

Figure 3.4-21. *Ground location of the recovered flight crew equipment.*

[REDACTED.]

[REDACTED. Figure 3.4-22.]

[REDACTED.]

Finding. Crew members experienced traumatic injuries in areas corresponding to the seat restraint system.

Conclusion L3-4. The seat restraint system caused lethal-level injuries to the unconscious or deceased crew members when they separated from the seat.

Recommendation L2-4/L3-4. Future spacecraft suits and seat restraints should use state-of-the-art technology in an integrated solution to minimize crew injury and maximize crew survival in off-nominal acceleration environments.

Recommendation L3-1. Future vehicles should incorporate a design analysis for breakup to help guide design toward the most graceful degradation of the integrated vehicle systems and structure to maximize crew survival.

3.4.2.3 *Thermal exposure*

[REDACTED.]

Finding. No significant levels of carbon monoxide or cyanide (combustion by-products) were identified in any of the body fluids.

Finding. There was no evidence of thermal injury to the respiratory tracts.

> **Conclusion L1-4.** The crew was not exposed to a cabin fire or thermal injury prior to depressurization, cessation of breathing, and loss of consciousness.

[REDACTED.]

[REDACTED. Figure 3.4-23.]

[REDACTED.]

The ambient absolute pressure condition at separation was approximately 0.03 psi.

3.4.3 Identified events with lethal potential

1. **The first event with lethal potential was depressurization of the CM, which started at or shortly after orbiter breakup.** Existing crew equipment protects for this type of lethal event, but operational practices and hardware limitations were such that the ACESs were not in a protective configuration. The current shuttle ACES relies on the crew to lower and lock the visor; therefore, complete protection from a depressurization event depends on a permissive environment. Design solutions that do not require crew action are achievable.

2. **The second event with lethal potential was unconscious or deceased crew members exposed to a dynamic rotating load environment with nonconformal helmets and a lack of upper body restraint.** Current shuttle seat and helmet design and operational practices did not protect the crew members from this lethal event. Complete strap-in with inertial reels locked would reduce the risk of injury/death; however, even in this configuration, the current seat-suit restraint system provides limited protection from dynamic G events (i.e., no lateral restraints, no control of extremity motion, and no head-neck support). Better restraint designs that include head-neck support (i.e., conformal helmets), extremity control, and spine support are achievable to reduce the risk of injury/death.

3. **The third event with lethal potential was separation from the crew module and the seats with associated forces, material interactions, and thermal consequences. This event is the least understood due to limitations in current knowledge of mechanisms at this Mach number and altitude. Seat restraints played a role in the lethality of this event.** Although the seat restraints (e.g., narrow width) played a significant role in the lethal mechanical injuries, there is currently no full range of equipment to protect for this event. The event was not survivable by any means currently known to the investigative team, with the exception of ensuring the integrity of the CM until the airspeed and altitude are within survival limits. This is not possible for the current space shuttle design; however, future vehicle designs incorporating a principle of "graceful degradation" and CM stabilization are possible.

4. **The fourth event with lethal potential was exposure to near vacuum, aerodynamic accelerations, and cold temperatures.** Although current crew survival equipment may be capable of protecting the crew, it is not certified to protect the crew above 100,000 feet. At the altitude and speeds at which the unconscious or deceased crew members departed from the CM, the environmental risks include lack of O_2, low atmospheric pressure, high thermal loads as a result of deceleration from high Mach numbers, shock wave interactions, aerodynamic accelerations, and exposure to cold temperatures. Existing shuttle

CEE is certified to protect up to 100,000 feet and 600 KEAS; however, the ACES is not designed to provide protection from high-temperature exposures. Anecdotal evidence from the survival of the pilot of an SR-71 mishap [*Aviation Week & Space Technology*, August 8, 2005, pp. 60–62] suggests that an intact, pressurized suit similar to the ACES can protect a crew member at an altitude of 78,000 feet and speeds of at least Mach 3 (~400 KEAS). More research is needed to close the survival gap. The only protection that is achievable is to ensure the integrity of the CM until the airspeed and altitude are within suit capability, which is currently not precisely determined.

5. **The final event with lethal potential was ground impact.** Existing shuttle CEE protects for ground impact with a parachute. However, the crew member must manually initiate the parachute opening sequence, or the parachute must be used in conjunction with the crew escape pole of the shuttle to initiate the parachute automatic opening sequence. Military and sport parachuting solutions exist for opening parachutes independent of crew action.

3.4.4 Synopsis of crew analysis

The crew was unaware of an impending survival situation prior to the LOC. At the time of LOC, the flight deck crew was probably troubleshooting the caution-and-warning messages that were associated with the FCS fault, left main landing gear talk-back, and tire pressure messages. One of the middeck crew members was likely attempting to become seated and restrained under the dynamic LOC conditions. Until the fore-body separated from the orbiter vehicle, the crew was conscious and had not suffered serious injuries. Cause of death was unprotected exposure to high-altitude conditions and blunt trauma.

[REDACTED.]

Chapter 4 – Investigative Methods and Processes

4   Investigative Methods and Processes

This chapter discusses the methods and processes that were used during the Spacecraft Crew Survival Integrated Investigation team (SCSIIT) investigation. The SCSIIT activity was a continuation of the Crew Survival Working Group (CSWG), which was formed during the *Columbia* Accident Investigation Board (CAIB) investigation. The SCSIIT structure, personnel, and investigative process and lessons learned from that process are presented. The remainder of the chapter documents the methods, processes, and tools used by the SCSIIT for various analyses.

The following is a summary of the findings, conclusions, and recommendations for this section. Some recommendations are targeted at improving future NASA spacecraft crew survival accident investigation processes. Some are general suggestions for other organizations that may be tasked with such investigations in the future. In some cases, certain findings and conclusions reflect existing NASA policies and practices that were considered particularly effective and are included for emphasis for future investigators.

Finding. NASA priorities put emphasis on Return to Flight recommendations, long-term recommendations, and observations, in that order. As a result, the SCSIIT effort suffered from low priority relative to other program recovery efforts. Team members had to divide their time between the investigation work and the work for their home organization. This led to delays in completing the SCSIIT work and, in some cases, significant decrease in availability or complete loss of members of the SCSIIT.

Finding. Formally trained NASA-designated accident investigation personnel were not available for inclusion on the SCSIIT due to the intensity of safety and mission assurance work related to Return to Flight activities. SCSIIT members were selected primarily based on their technical knowledge and experience as well as availability. Many SCSIIT members did not have formal accident investigation training. The team preparation training sessions did not include the lengthy accident investigation training that is normally provided to NASA-designated investigators.

>**Recommendation A1.** In the event of a future fatal human space flight mishap, NASA should place high priority on the crew survival aspects of the mishap both during the investigation as well as in its follow-up actions using dedicated individuals who are appropriately qualified in this specialized work.

>**Recommendation A4.** Due to the complexity of the operating environment, in addition to traditional accident investigation techniques, spacecraft accident investigators must evaluate multiple sources of information including ballistics, video analysis, aerodynamic trajectories, and thermal and material analyses.

Finding. It was not uncommon to find several versions of documents supporting CAIB and CSWG work.

>**Recommendation A6.** Standard templates for accident investigation data (document, presentation, data spreadsheet, etc.) should be used. All reports, presentations, spreadsheets, and other documents should include the following data on every page: title, date the file was created, date the file was updated, version (if applicable), person creating the file, and person editing the file (if different from author).

Finding. Concerns about public release of sensitive information relative to the crew creates obstacles to the performance of crew survival investigations.

> ***Recommendation A2.*** Medically sensitive and personal debris and data should always be available to designated investigators but protected from release to preserve the privacy of the victims and their families.

> ***Recommendation A3.*** Resolve issues and document policies surrounding public release of sensitive information relative to the crew during a NASA accident investigation to ensure that all levels of the agency understand how future crew survival investigations should be performed.

Finding. The unique nature of the event, closeness of investigators to the accident victims, lack of previous exposure to the results of such tragedies, and need to keep information confidential created stress on some members of the investigation team. Counseling was provided, but the follow-up could be improved.

> ***Recommendation A9.*** Post-traumatic stress debriefings and other counseling services should be available to those experiencing ongoing stress as a result of participating in the debris recovery and investigation. Designated personnel should follow up on a regular basis to ensure that individual needs are being met.

> ***Recommendation A7.*** To aid in configuration control and ensure data are properly documented, report generation must begin early in the investigation process.

Finding. CAIB/CSWG data were not cataloged. *Challenger* supporting data were mostly uncataloged and unorganized, limiting their usefulness for investigations. *Challenger* debris is unpreserved and inaccessible for analysis.

> ***Conclusion A8-1.*** Spacecraft accidents are rare, and each event adds critical knowledge and understanding to the database of experience.

> ***Recommendation A8.*** As was executed with *Columbia*, spacecraft accident investigation plans must include provisions for debris and data preservation and security. All debris and data should be cataloged, stored, and preserved so they will be available for future investigations or studies.

Finding. Brightening events were easier to correlate between videos than debris-shedding events.

Finding. Sun angle illumination impacted the visibility of debris in video recordings.

Finding. Not all videos segments within compilations were individually categorized. Not all videos were re-reviewed once a better understanding of events had been gained.

> ***Recommendation A11.*** All video segments within a compilation should be categorized and summarized. All videos should be re-reviewed once the investigation has progressed to the point that a timeline has been established to verify that all relevant video data are being used.

Finding. The lack of a single, standard data format for latitude/longitude data and the potential ambiguity associated with the need to convert data of different formats resulted in possible data errors.

> ***Recommendation A10.*** Global Positioning System receivers used for recording the latitude/longitude of recovered debris must all be calibrated the same way (i.e., using the same reference system), and the latitude/longitude data should be recorded in a standardized format.

4.1 Background

Crew Survival Working Group

The CSWG was formed to support the CAIB on February 21, 2003 by authorization of the Johnson Space Center (JSC) Director. The CSWG was co-chaired by directors of the Space Life Sciences Directorate and the Engineering Directorate. Membership included personnel from Space Life Sciences, Engineering, Mission Operations, and Flight Crew Operations. Dr. James Bagian, a former astronaut and crew survival investigator for the *Challenger* accident, was the flight surgeon advising the CAIB. Dr. Bagian, together with Lt. Col. Don White of the Air Force Safety Center (an expert in crew equipment investigations), were the primary liaisons to the CAIB for the CSWG.

The CSWG was tasked with a limited charter: first, to determine the cause of death of the crew, second, to determine the "survival gap" (what equipment or procedures might have kept the crew alive), and, third, to pass the results to the CAIB. The CSWG performed aerodynamic, thermal, and structural analyses on individual debris items and an intensive study of the crew helmets and seats. In the process, team members made several trips to Kennedy Space Center (KSC) to view the crew module (CM) and helmet and seat debris. The CSWG developed a timeline that is consistent with the official CAIB timeline to derive the sequence of crew survival events based on the data.

Following recovery in the field, the crew remains were transported to the Air Force Institute of Pathology (AFIP) at Dover Air Force Base, Maryland for forensic and deoxyribonucleic acid (DNA) identification analysis. The Office of the Armed Forces Medical Examiner is the department within the AFIP that was responsible for determining the cause and manner of death for the crew of *Columbia*. The AFIP issued a report to the CSWG on the findings of the autopsies in May 2003, near the end of the CAIB investigation. Tissue samples were sent to the Federal Bureau of Investigation (FBI) for additional medical forensic analysis. Due to the timing of the AFIP report to the CSWG, little medical information was available to be discussed at the CSWG team level, and only a preliminary effort to integrate the medical findings with the engineering findings was possible. The need to fully integrate these findings was recognized by the CSWG team at the time.

Four types of data – aerodynamics, orbiter, forensic hardware, and forensic medical – were collected by the CSWG. These data were submitted to Dr. Bagian and Lt. Col. White on June 24, 2003 to assist them in preparation of the CAIB Report. The CAIB Report, Volume I, which was published in August 2003, contained a short discussion concerning crew survival.[1] Somewhat more detail was released in Volume V, which was published in October 2003.[2]

Unlike the other elements of the CAIB investigation, there was no NASA process for the administrative and financial framework of the CSWG investigation. As a result, when the CAIB investigation concluded, there were no resources available to continue the CSWG work although it was clear more work remained. The CAIB did not make any formal recommendations (only observations) regarding crew survival. Due to the priority of the Return to Flight program, CSWG activities were discontinued in October 2003. No report was published by the CSWG. Efforts were made to locate new funding and to identify an organization that would manage the continuation of the investigation and publish a report.

The SCSIIT was formed in October 2004 to resume the work of the CSWG and perform a multidisciplinary analysis of the *Columbia* fatal mishap that focused on the crew, crew equipment, and CM. The specific products include: the establishment of a comprehensive, computer-searchable body of information, the virtual reconstruction of the mishap, and a comprehensive report that provides valuable understanding and information for the design of crewed space vehicles and crew safety equipment.

To learn how improvements to crew survival could be made in the future, the following questions needed to be answered:

---

[1] *Columbia* Accident Investigation Board Report, Volume I, Section 10.2, Crew Escape and Survival, August 2003.
[2] *Columbia* Accident Investigation Board Report, Volume V, Appendix G.12, Crew Survivability Report, October 2003.

- What events occurred that had lethal potential for the crew, even after the crew became deceased?
- How did the CM lose structural integrity?
- How did the crew equipment perform in the pressure, thermal, and acceleration environments that were experienced by the crew?
- What operational insight did the crew members have into the events that occurred, and was their training appropriate and adequate for the circumstances encountered?

4.2 Spacecraft Crew Survival Integrated Investigation Team Structure and Personnel

Because the SCSIIT investigation was mainly concerned with crew survival, the team had a "crew-centric" focus. Shuttle crew members are surrounded by layers of protection, with the crew equipment being the closest layer, the CM being the next layer, and the vehicle being the outermost layer (figure 4-1).

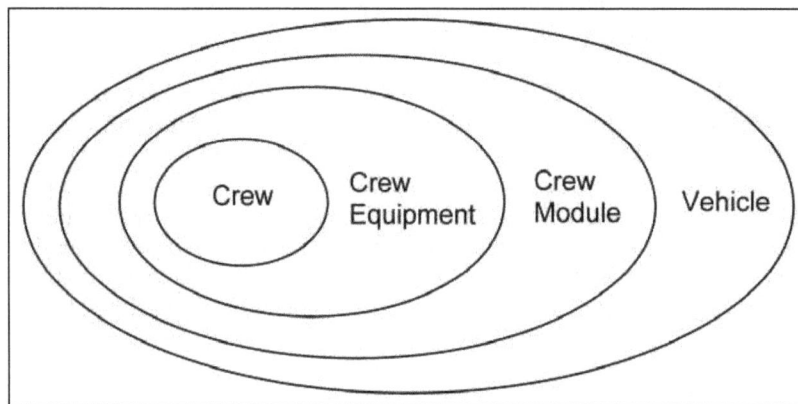

Figure 4-1. *Concentric layers of protection for space shuttle crew members.*

After it was decided that the crew and the three "layers" would be used as functional areas of focus for the SCSIIT investigation, four teams were formed. These teams were highly interdependent in the development and sharing of information.

1. The Vehicle Team was responsible for determining the dynamics of the vehicle from loss of control (LOC) until the vehicle breakup to ascertain the dynamics that the crew members experienced. This team determined the vehicle breakup sequence and the motion of the intact orbiter and the free-flying forebody, and performed all ballistics analysis on debris. Additionally, this team performed most of the thermal analyses.

2. The Crew Module Team was responsible for determining the CM environments (acceleration, thermal, and pressure) until CM breakup to ascertain the environments that the crew members experienced. This team also determined the CM breakup sequence.[3]

3. The Crew Equipment Team was responsible for determining the performance of the crew equipment (crew worn equipment, seats, etc.) to ascertain how the equipment enhanced or worsened the crew survival probabilities. Results of analysis on crew equipment were used by the Crew Module and Crew Teams to aid their analyses.

---

[3]Prior to the completion of the investigation, the Crew Module Team lead had to return to his "home" organization. The Crew Module Team was essentially dissolved and its responsibilities were spread to the other three sub-teams.

4. The Crew Team was responsible for analyzing crew awareness during the mishap and the causes of deaths of the crew members, and identifying the threats to crew survival. Analysis of crew injuries was used by the vehicle and crew equipment teams to aid in developing their conclusions.

The SCSIIT management structure is shown in figure 4-2.

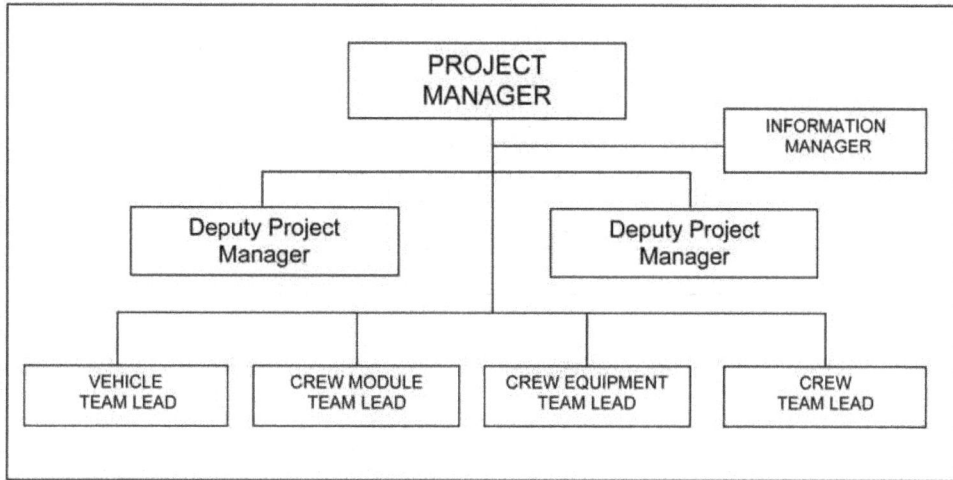

Figure 4-2. *Management and leadership structure.*

4.2.1 Team membership

Potential SCSIIT members were identified based on previous association with the CAIB and CSWG and experience in the disciplines that were necessary to conduct this investigation. Selection was made considering an individual's area of knowledge, experience, interest, and availability. The team members were expected to divide their time between the SCSIIT project tasks and their "home" organization's tasks. At times, the home organization's work had deadlines requiring work to be prioritized over SCSIIT work. This led to delays in completing the SCSIIT work and, in some cases, significant decrease in availability or complete loss of members of the SCSIIT. It is recommended that personnel on future investigation teams be temporarily "released" from their "home" organizations to be free to work full time for the investigation organization.

Finding. NASA priorities put emphasis on Return to Flight recommendations, long-term recommendations, and observations, in that order. As a result, the SCSIIT effort suffered from low priority relative to other program recovery efforts. Team members had to divide their time between the investigation work and the work for their home organization. This led to delays in completing the SCSIIT work and, in some cases, significant decrease in availability or complete loss of members of the SCSIIT.

Recommendation A1. In the event of a future fatal human space flight mishap, NASA should place high priority on the crew survival aspects of the mishap both during the investigation as well as in its follow-up actions using dedicated individuals who are appropriately qualified in this specialized work.

SCSIIT members were assigned to lead each of the four teams. The team lead was responsible for coordinating the analyses performed by that team and documenting the results. SCSIIT members were generally assigned to one team but often supported other teams when their expertise was called for. Although represented in figure 4-2 as a hierarchical team, the SCSIIT functioned as a highly integrated group with multiple interactions among the teams. This functional organization and the interactions among teams are shown in figure 4-3.

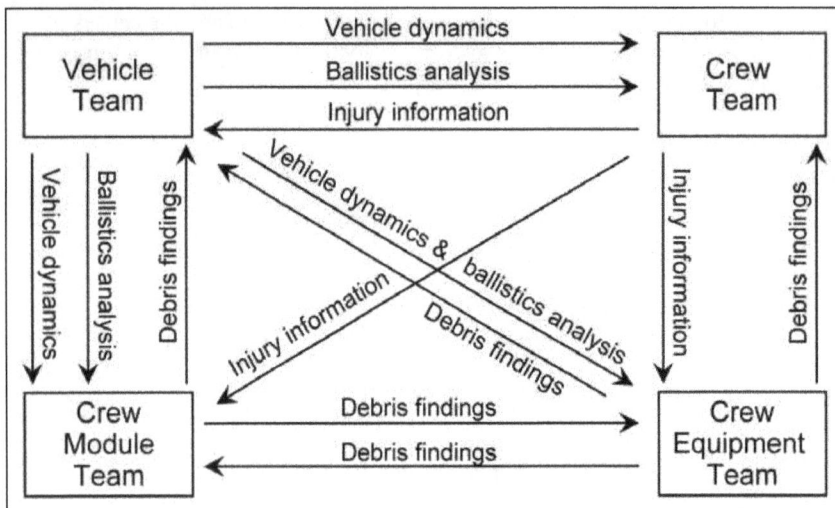

Figure 4-3. *Functional organization of the Spacecraft Crew Survival Integration Investigation Team.*

The principal SCSIIT members and their respective home organizations, backgrounds, and responsibilities are listed in Table 4-1.

Table 4-1. *Principal Spacecraft Crew Survival Integration Investigation Team Members*
(*Denotes SCSIIT members who were on the CSWG)

SCSIIT Member[4]	Organization	Background	Responsibilities
Dr. Gregory Hite	Shuttle and Exploration Division, Safety and Mission Assurance Directorate (retired 12/07)	Crew Survival	Project Manager
Dr. Nigel Packham	Safety and Mission Assurance Directorate	Safety and Life Sciences	Project Manager
Col. (ret) Pam Melroy, United States Air Force (USAF)*	Astronaut Office, Flight Crew Operations Directorate	Test Pilot, Shuttle Pilot Astronaut	Deputy Project Manager
Dr. Craig Fischer*	Space Life Sciences Directorate (retired 12/06)	Flight Surgeon and Pathology	Deputy Project Manager/ Crew Team Lead
Chrystal L. Hoelscher, Science Applications International Corporation (SAIC)	Shuttle and Exploration Division, Safety and Mission Assurance Directorate	Information/Knowledge Management and Database Administration	Information Management
Ellen Braden	Aeroscience and Flight Mechanics Division, Engineering Directorate	Aerodynamics and Flight Mechanics	Vehicle Team Lead
Mark Adams, United Space Alliance (USA)	Vehicle Integration Test Office, Flight Crew Operations Directorate	Shuttle Vehicle Integration Test Office	Crew Module Team Lead
David J. Pogue	EVA[5], Robotics and Crew Systems Operations Division, Mission Operations Directorate	Crew Systems and Crew Escape Equipment Operations Instructor and Flight Controller	Crew Equipment Team Lead
Eric Flagg, SAIC	Shuttle and Exploration Division, Safety and Mission Assurance Directorate	Military Pilot	Safety and Mission Assurance

---

[4]With the exception of Duke Tran from Boeing/Palmdale, California, all personnel were based at JSC in Texas.
[5]EVA = extravehicular activity.

Table 4-1. *Principal Spacecraft Crew Survival Integration Investigation Team Members* (Continued)
(*Denotes SCSIIT members who were on the CSWG)

SCSIIT Member[6]	Organization	Background	Responsibilities
Joe Hamilton	Habitability and Environmental Factors Division, Space Life Sciences Directorate	Military Pilot	Virtual Reconstruction / Concept Evaluation Laboratory (CEL)
Kandy Jarvis*, Lockheed Martin Space Operations/Mission Services (LMSO/LMMS)	Human Exploration Science Office, Astromaterials Research and Exploration Science Directorate	Orbital Debris and Planetary Astronomy	Video and Debris Analysis
Dennis Pate, SAIC	Shuttle and Exploration Division, Safety and Mission Assurance Directorate	Human Factors	Timeline and Crew Awareness
Duke Tran, Boeing	Boeing/Palmdale	Senior Orbiter Structures Designer	Orbiter Structural Analysis

The SCSIIT team principals called on other individuals for analysis and information. These individuals are listed in Table 4-2.

Table 4-2. *Individuals Supporting the Principal Spacecraft Crew Survival Integration Investigation Team*
(*Denotes SCSIIT members who were on the CSWG)

Name[7]	Organization
Ketan Chhipwadia*	Crew and Thermal Systems Division, Engineering Directorate
Katie Boyles Lee Bryant* Peter Cuthbert	Aeroscience and Flight Mechanics Division, Engineering Directorate
Lynda R. Estes Jeremy Jacobs* Kenneth Wong* Leslie Schaschl Brian Mayeaux	Structural Engineering Division, Engineering Directorate
Curtis Stephenson	Crew and Thermal Systems Division, Engineering Directorate
William Sarles, SAIC Paul Wilson, SAIC	Shuttle and Exploration Division, Safety and Mission Assurance Directorate
Robert Behrendsen, Barrios Technology (BAR) J. Lynn Coldiron, USA Adam Flagan, USA	EVA, Robotics, and Crew Systems Operations Division, Mission Operations Directorate
Laurie J. Bergman, Tietronix Software, Inc. Richard D. Delgado Jose Dobarco-Otero, Jacobs Technology (ESCG) William Rochelle,*[8] ESCG Ries Smith, ESCG	Space Life Sciences Directorate
Sudhakar Rajulu Kurt G. Clowers, Muniz Engineering (MEI) Sarah Margerum, LMSO Richard Morency	Habitability and Environmental Factors Division, Space Life Sciences Directorate

---

[6]With the exception of Duke Tran from Boeing/Palmdale, California, all personnel were based at JSC in Texas.
[7]With the exception of where noted, all personnel were based at JSC.
[8]Tragically, Mr. Rochelle passed away before this report was published.

Table 4-2. *Individuals Supporting the Principal Spacecraft Crew Survival Integration Investigation Team* (Continued)
(*Denotes SCSIIT members who were on the CSWG)

Name[9]	Organization
Rita Alaniz, MEI Rodney DeSoto, LMSO Chris Keller, BAR Mark Langford, LMSO Terry Mayes, USA Jeremy Reyna, Wyle Laboratories Chris Slovacek, LMSO Matt Soltis	Habitability and Environmental Factors Division/Concept Exploration Laboratory (CEL), Space Life Sciences Directorate
Danny Olivas	Astronaut Office, Flight Crew Operations Directorate (materials analysis)
Sharon Hecht, Tesseda (TES) Perry Jackson, TES Cindy Bush, TES	Information and Applications Systems Division, Information Resources Directorate
Stacey Nakamura	Safety and Mission Assurance Directorate
James Comer, USA Amy Mangiacapra, USA	*Columbia* Research and Preservation (CRP) Office, KSC
Steve McDanels M. Clara Wright	Failure Analysis and Materials Evaluation Branch, Materials Science Division, Engineering Directorate, KSC
Rick Russell	Orbiter Sustaining Engineering Office, KSC
Roy Christoffersen, SAIC	Astromaterials Research Office, Astromaterials Research and Exploration Science Directorate
Darren Cone	White Sands Test Facility
David Bretz*, LMSO Tracy Thumm*, LMSO Kathleen McBride*, LMSO Kim Willis*, LMSO	Human Exploration Science Office, Astromaterials Research and Exploration Science Directorate
Donna Shafer	Office of the Chief Counsel

Additionally, expert personnel who were external to NASA provided assistance to the SCSIIT. These persons are listed in Table 4-3.

Table 4-3. *Experts External to NASA*

Name	Organization
Dr. Robert Banks	Biodynamic Resource Corporation (BRC)
Dr. Jon Clark	National Space Biomedical Research Institute
Dr. Richard Harding	Biodynamic Research Corporation
Dr. Gregory Kovacs	Stanford University
Dr. Robert McMeekin	Previous Federal Air Surgeon
Dr. Thomas McNish	Biodynamic Resource Corporation
Dr. Charles Ruehle	Federal Aviation Administration (FAA)
Dr. Glenn Sandberg	Armed Forces Institute of Pathology
Dr. Charles Stahl	Former Chairman of the Department of Forensic Sciences at the Armed Forces Institute of Pathology
Dr. Harry Smith	Biodynamic Research Corporation

It should be noted that extensive training on accident investigation processes and procedures was not provided to the team. Many of the team members had no previous accident investigation training. Because the SCSIIT members supported the investigation on a part-time basis, taking a lengthy course in accident investigation processes and procedures was not feasible. NASA has only a small group of formally trained accident investigators across the many field centers. These investigators also have other safety duties. None

---

[9]With the exception of where noted, all personnel were based at JSC.

were available to participate in the SCSIIT investigation due to higher priority activities occurring at the time, including Return to Flight preparation.

The SCSIIT leadership felt it was preferable for NASA space flight technical experts to learn accident investigation techniques rather than to have accident investigation experts become technical experts on the space shuttle. Therefore, personnel were selected for the SCSIIT based on experience with the CAIB and CSWG and/or who were technical experts in the disciplines that were necessary to conduct the investigation. Many SCSIIT team members did not have formal accident investigation training; advice from experienced investigators was sought at various times. Accident investigation experience would have been helpful to focus the team's efforts, especially early in the investigation.

Finding. Formally trained NASA-designated accident investigation personnel were not available for inclusion on the SCSIIT due to the intensity of safety and mission assurance work related to Return to Flight activities. SCSIIT members were selected primarily based on their technical knowledge and experience as well as availability. Many SCSIIT members did not have formal accident investigation training. The team preparation training sessions did not include the lengthy accident investigation training that is normally provided to NASA-designated investigators.

> *Recommendation A1.* In the event of a future fatal human space flight mishap, NASA should place high priority on the crew survival aspects of the mishap both during the investigation as well as in its follow-up actions using dedicated individuals who are appropriately qualified in this specialized work.

4.3 Investigative Process

The team traveled to KSC to perform first-hand inspection of the *Columbia* debris. Additionally, the results of previously performed analyses were reviewed. Additional analyses were then conducted (ballistics, materials, structural, aerodynamic, etc.).[10] Individually and in small groups, the team members assessed the results and formed conclusions. These conclusions were presented at Technical Interchange Meetings (TIMs) to the entire SCSIIT to compare and integrate results and findings. The team held four TIMs (March/April 2005, June 2005, August 2005, and March 2006). This was an iterative process, with each TIM resulting in more trips, analyses, and scenario revisions. In the process, integrated products were generated, with the most important being the timeline of key events. When possible, "no earlier than" and "no later than" times were identified for key events, and sequences were built.

In many regards, this investigation presented several challenges. Space flight is a relatively new and rare experience, and, fortunately, there have been only a few fatal mishaps. Consequently, there is no integrated or widely available body of information for how to analyze spacecraft accidents for crew survival. The physics of atmospheric entry and the environment in which a human spacecraft mishap occurs are unique when compared to aviation. The SCSIIT had to break new ground in how to conduct the investigation of a singular event in such a complex environment. The team had to determine how to modify existing models and tools, which are normally used for specific nominal situations in a predictive manner, to understand the mishap environment. Multiple tools and analyses were evaluated to provide multiple sources of information to develop scenarios.

> *Recommendation A4.* Due to the complexity of the operating environment, in addition to traditional accident investigation techniques, spacecraft accident investigators must evaluate multiple sources of information including ballistics, video analysis, aerodynamic trajectories, and thermal and material analyses.

---

[10]Debris analysis and other analysis methods and tools are described in Section 4.5.

The SCSIIT collected a large amount of data from the CAIB and CSWG activities. Also, a large volume of new data was generated during the investigation. The SCSIIT Information Manager was tasked with organizing and tracking the information in a SCSIIT Project database.

The initial SCSIIT activities included review of data produced by the CSWG in support of the CAIB. It was discovered that the existing data had not been centrally cataloged or organized, making access to specific items difficult. The data that were available were dispersed among several groups, as shown in Table 4-4.

Table 4-4. *Data Types and Data Sources*

Data Type	Data Source
Medical: X rays, autopsy reports, photographs, tissue samples	Armed Forces Institute of Pathology Biodynamic Research Corporation (BRC) Space Life Sciences Directorate, JSC
Debris: Seats, suits, CM, forward fuselage (FF), windows	*Columbia* Research and Preservation Team, KSC
Technical: Ballistics, thermal, structure, and materials analyses	Engineering Directorate, JSC
Video: Recovered On-board Video, Ground Based Video	Image Science and Analysis Group (ISAG) *Columbia* Video Archives Group Concept Exploration Laboratory
Operational: Vehicle Telemetry Data	Missions Operations Directorate

Because no final report was generated by the CSWG, it was not uncommon for several versions of specific analyses or presentations to exist as scenarios were iterated. Most of the files did not include the author, date, or version of the document, making it difficult to determine the final conclusions for the subject document. Additionally, without contact information, the team members were not always able to contact the original authors of these documents to discuss the contents or obtain additional information that may have aided the investigation. Report generation must begin very early in the investigation process, using a systematic approach with established procedures. At some point in the investigation, a transition from fact-finding must give way to documenting the findings, conclusions, and recommendations.

Finding. It was not uncommon to find several versions of documents supporting CAIB and CSWG work.

Recommendation A6. Standard templates for accident investigation data (document, presentation, data spreadsheet, etc.) should be used. All reports, presentations, spreadsheets, and other documents should include the following data on every page: title, date the file was created, date the file was updated, version (if applicable), person creating the file, and person editing the file (if different from author).

Recommendation A7. To aid in configuration control and ensure data are properly documented, report generation must begin early in the investigation process.

During the SCSIIT investigation, it was discovered that *Challenger* information was cataloged with keywords and descriptions that were more oriented toward the overall mishap investigation. This may be a result of the fact that the Rogers Commission did not specifically investigate crew survival. Additionally, the data were cataloged prior to current storage techniques and re-cataloged later by different personnel well after the accident investigation was complete. It was difficult to retrieve specific documents and analyses related to the CM and to crew survival. In many cases, *Challenger* information was obtained from individuals who were involved in the original investigation. Moreover, the *Challenger* debris items are unpreserved and inaccessible for analysis as they are stored in an abandoned underground missile silo with no access or climate control provisions. The lack of debris for comparison and methods of data preservation made the *Challenger* data essentially unavailable for this investigation. It is recommended that accident investigation plans include provisions for debris and data preservation and security. All debris and data should be cataloged, stored, and preserved so they will be available for future investigations or studies.

Finding. CAIB/CSWG data were not cataloged. *Challenger* supporting data were mostly uncataloged and unorganized, limiting their usefulness for investigations. *Challenger* debris is unpreserved and inaccessible for analysis.

> *Conclusion A8-1.* Spacecraft accidents are rare, and each event adds critical knowledge and understanding to the database of experience.

> *Recommendation A8.* As was executed with *Columbia*, spacecraft accident investigation plans must include provisions for debris and data preservation and security. All debris and data should be cataloged, stored, and preserved so they will be available for future investigations or studies.

4.3.1 Public release of information

For comparison and potential application to the SCSIIT investigation, the team researched how other government agencies conduct aircraft accident investigations. SCSIIT looked at processes that are used by the National Transportation Safety Board (NTSB), FAA, U.S. Air Force, and U. S. Navy to get an overview of the agencies' crew survival investigation processes. While the investigation process used by the SCSIIT was similar to those processes used in other agencies, there were some significant differences due to the uniqueness of the spacecraft operating environment. The most notable differences regarded public release of information and the use of personnel from the affected organization for the investigation.

The NTSB and FAA must conduct their affairs publicly, while Department of Defense (DoD) investigations are considered internal matters, and documents are released at the discretion of the DoD. At NASA, the public release of sensitive information is not specifically addressed in existing accident investigation plans. As a result, there was hesitation to investigate information that was relative to what happened to the crew out of the concern that the information would result in public release, and subsequent inappropriate speculation that would be painful to both the employees and the families. A more preferable situation would be to have a pre-determined plan for what crew-related information is appropriate to release to the public, and when the information should be released.

Finding. Concerns about public release of sensitive information relative to the crew creates obstacles to the performance of crew survival investigations.

Future spacecraft accidents may result in injuries and/or fatalities. To preserve the privacy of the *Columbia* crew members and their families, access to medically sensitive data, including the crew's personal items, was provided only to those personnel who had a need to know. This practice remains in place today and must remain in place in the future for *Columbia*-related information as well as for any future aerospace incident involving human casualties. This practice will preserve the privacy of the victims and their families. Almost as important, it will ensure that future investigations can be conducted without the concern of inappropriate release of sensitive information.

> *Recommendation A2.* Medically sensitive and personal debris and data should always be available to designated investigators but protected from release to preserve the privacy of the victims and their families.

> *Recommendation A3.* Resolve issues and document policies surrounding public release of sensitive information relative to the crew during a NASA accident investigation to ensure that all levels of the agency understand how future crew survival investigations should be performed.

4.3.2 Using members of affected organizations in the investigation

In commercial or military aviation accidents, members of the organization that is affected by the accident generally are not members of the investigation teams. For example, pilots in a specific squadron would not be members of a team investigating an accident involving one of their squadron-mates. Because space flight operations are highly specialized and there are no other "external" organizations with sufficient relevant experience, it is impractical to follow this investigative practice for NASA crewed spacecraft accidents. Initially, the CSWG did not include current astronauts or crew escape operations training personnel. Astronauts were added to the CSWG and provided operational experience to the group. Crew escape operations and training joined the team during the follow-on SCSIIT portion of the investigation.

A potential downside to using accident investigation personnel who are close to the victims of the accident is the psychological impacts of the investigation on the investigators. A crew survival investigation is an emotionally charged process, causing considerable stress in the people involved in the investigation. The psychological welfare of personnel who are involved in debris recovery must be protected as part of the accident investigation process. In the aftermath of the *Columbia* accident, there were no consistent post-event stress debriefings to assist with post-traumatic stress disorder syndrome in recovery and mishap investigation personnel until later investigation phases.

Finding. The unique nature of the event, closeness of investigators to the accident victims, lack of previous exposure to the results of such tragedies, and need to keep information confidential created stress on some members of the investigation team. Counseling was provided, but the follow-up could be improved.

> **Recommendation A9.** Post-traumatic stress debriefings and other counseling services should be available to those experiencing ongoing stress as a result of participating in the debris recovery and investigation. Designated personnel should follow up on a regular basis to ensure that individual needs are being met.

4.4 Medical Process Issues

When the *Columbia* accident first occurred, the highest priority task was rescuing the crew members. When it became apparent that they had not survived, the task transitioned to recovering the crew remains. The FBI was the agency that was in charge of recovering the crew member human remains. The Bureau was assisted by members of the Environmental Protection Agency, local and state law enforcement agencies, local coroners, and members of the NASA Astronaut Office. As searchers and citizens reported possible remains, teams were dispatched to document and recover the remains. In many cases, forensic experts in the field were able to make preliminary determinations of whether the remains were human or otherwise. Initially, trained recovery personnel were used to identify human remains in the field until all principal remains were recovered. Subsequently, recovery personnel were directed to collect all remains regardless of whether or not they could positively be identified as human or other. In most cases photographs of the remains were taken in the field prior to collection to document the "as-found" condition for use during the autopsies. The quality of the photographs and the information recorded varied greatly from site to site. In some instances where human remains may have been found near spacecraft hardware, there were no established procedures for documenting these important physical relationships.

After recovery, the remains were transported first to a local morgue facility. Intake photographs were taken to document the "as-received" condition of the remains for use during the autopsies. A forensic pathologist performed a review of the remains and was able to separate out many nonhuman remains that had been collected. The human remains were then prepared and transported to Barksdale Air Force Base, and then on to the AFIP at Dover Air Force Base.

The AFIP was the government agency that was tasked with positively identifying the remains, performing the autopsies, and preparing the remains for burial. Identification of the remains relied primarily on DNA testing and dental records.

NASA expected the AFIP service to include a complete cause of death analysis in similar fashion to what the Air Accident Investigation Team had historically done and had been done for the crew of STS-51L (*Challenger*). The AFIP performed very well in the receipt of human remains, gross examination, and definitive identification, and in providing death certificates and mortuary science support. However, the gross autopsy reports were incomplete and did not contain the level of detail necessary for a thorough accident investigation. X rays that were taken were of poor quality, and no interpretation of the X rays was transmitted to NASA. The microscopic examinations included in the final autopsy reports lacked many of the specific details that were required for the investigation. Although the AFIP continues to participate in routine aircraft accident investigations, in this instance it did not have the necessary resources to integrate all of the forensic findings into a comprehensive accident investigation report due to operational commitments. Subsequently, NASA has addressed these issues with process changes.

4.5   Analysis Methods, Processes, and Tools

Various methods, processes, and tools were used by the SCSIIT sub-teams to conduct their analyses. The following sections describe these methods, processes, and tools.

4.5.1 *Columbia* debris repository

Vehicle debris is one of the most useful sources of evidence in an accident. The debris can be used to determine failure modes, fracture dynamics, and thermal exposure, helping to develop vehicle breakup sequences. The debris also can be used to support physical and virtual vehicle reconstruction. *Columbia* debris is currently stored on the 16th floor of the Vehicle Assembly Building at KSC. The CRP Office is tasked with managing the debris and the database and providing access to the debris for research purposes. The investigation would not have been possible without the careful work done by the Reconstruction Team, particularly in the area of CM reconstruction. SCSIIT members made several trips to KSC to analyze the debris.

4.5.2 Physical reconstruction

Visual and microscopic inspection of individual debris items provided details of the thermal damage to components, the directionality of melted material deposits, and the characteristics of fracture surfaces. Evidence of mechanical damage or thermal exposure on fracture edges was used to help determine the timing of the breakup. For example, significant melting of fracture surfaces would indicate that the fracture occurred first, followed by thermal exposure. In addition to the study of individual debris items, multiple debris items were studied together as reconstructions of portions of the orbiter.[11] Reconstructions were performed for the flight deck and middeck floors, CM forward (X_{cm} 200) and aft (X_o 576) bulkheads, X_o 582 ring frame bulkhead, the airlock and tunnel adapter structure, CM avionics bays, crew seats, forward Reaction Control System pod components, and selected portions of the forward fuselage. These reconstructions were useful for helping to determine breakup sequences and mechanical and thermal damage patterns across large areas.

---

[11]Physical and time constraints prohibited a full reconstruction of *Columbia*.

4.5.3 Virtual reconstruction

Using modeling, simulation, and visualization tools, members of the Concept Exploration Laboratory (CEL) at JSC integrated data from multiple disciplines to create an interactive, dynamic, 3-dimensional simulation of the orbiter. This simulation included a virtual reconstruction of the CM interior using photographs of recovered debris items. The simulation also provided a dynamic visualization of the vehicle dynamics and breakup sequence, thereby creating a virtual reconstruction not only of the vehicle but also of the accident events timeline. The CEL team also developed unique capabilities and analysis techniques to support the investigation.

Trajectory and vehicle dynamics data for the intact orbiter were entered into the simulation, as were trajectories for selected debris items. Coordinates of the recovery location for debris items were loaded into the CEL simulation, and the debris items were grouped according to their origin on the vehicle (i.e., left wing, tail structure, FF, etc.).

The simulation was used to visualize the predicted vehicle dynamics and breakup sequence. Telemetry data were used where available to represent the flight path and vehicle orientation. The reference trajectory was used to represent translation. The aerodynamic simulation of the vehicle attitude (see Section 2.1) was used between LOC and the Catastrophic Event (CE). The aerodynamic simulation of the forebody was used to represent the orientation after the CE. Figure 4-4 shows frames from the dynamic visual simulation. The image on the left represents the orbiter just before the LOC, and is viewed from a point above, behind (west), and to the left (north) of the flight path. The image on the right represents the orbiter just after the CE, and is viewed from above and behind (west) of the flight path. The three main items in this view are the payload bay doors, the forebody, and the main portion of the orbiter. The left wing has already departed from the orbiter and is not visible in this view.

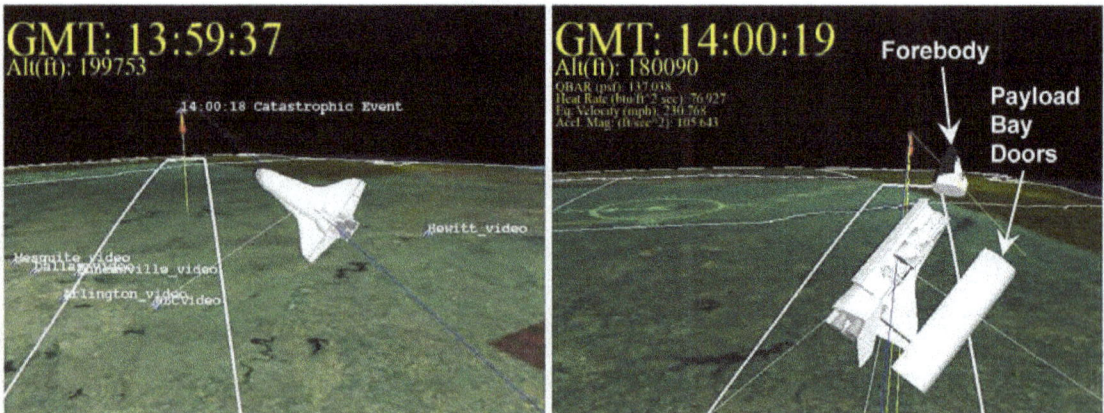

Figure 4-4. *Visual representation of the breakup sequence, looking east.*

Figure 4-5 shows the trajectories of the intact orbiter (the blue line in the upper portion of the image) and multiple debris items (different color lines) as viewed from a point in space looking down toward the Earth in a westerly direction. State borders are shown in white, and selected cities are labeled in green. The white rectangle represents the main debris search area. Almost all of Texas is visible, with east Texas shown in the center of the image. Western Louisiana is at the bottom of the image.

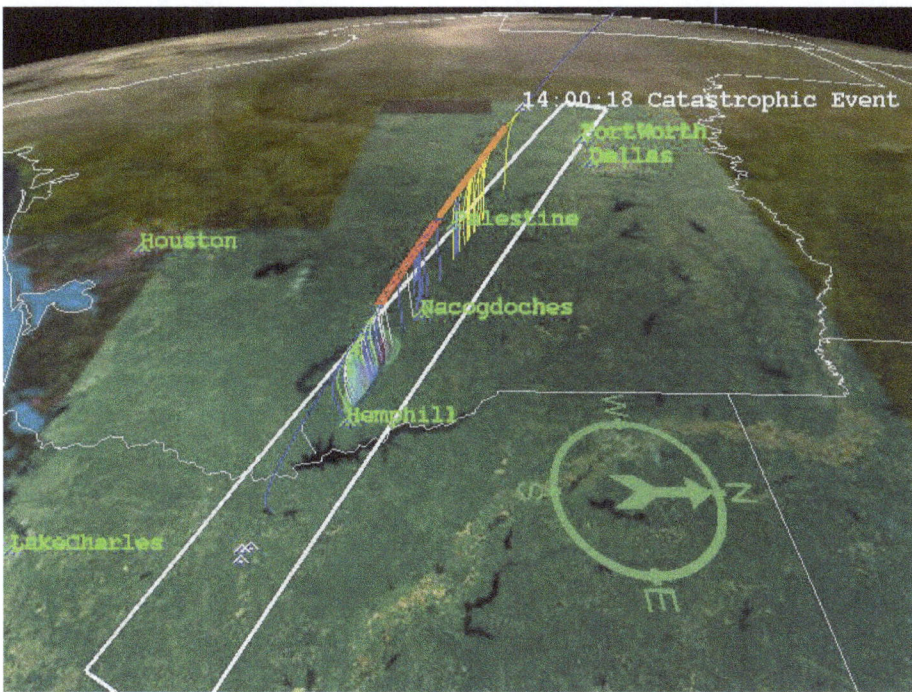

Figure 4-5. *Orbiter and debris trajectories, looking down at Texas in a westerly direction.*

The simulation could be viewed from any vantage point, such as zooming in on the CM debris spread or any of the sites from which ground-based video was recorded. Figure 4-6 shows the trajectory and CM debris groupings as viewed from a point north of the trajectory, looking south-southwest. Debris item groupings are shown as triangles and are labeled ("Airlock", Middeck floor", etc.).

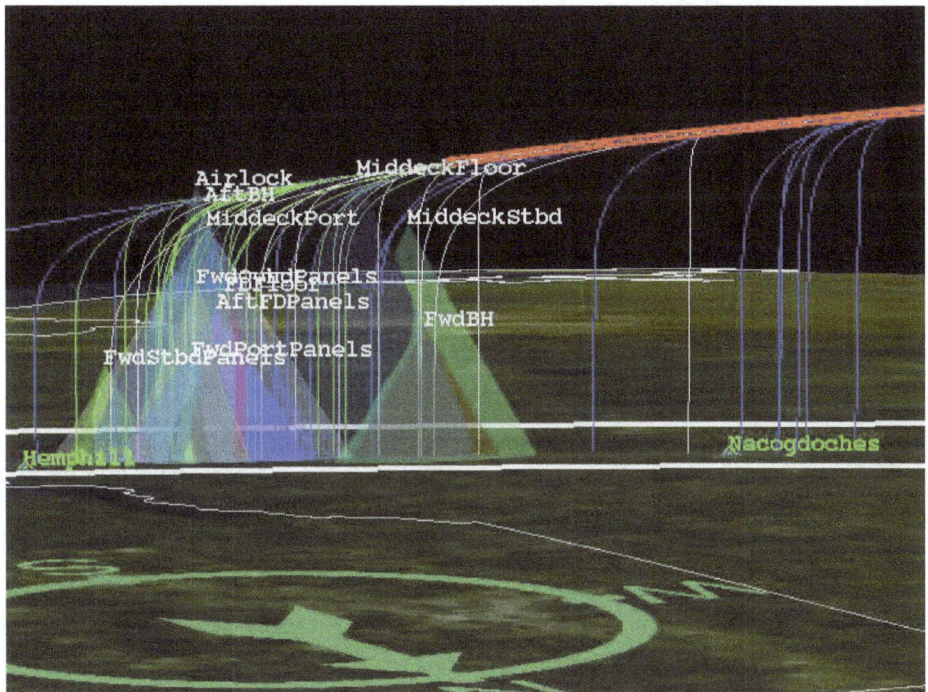

Figure 4-6. *Crew module debris groups, looking south-southwest.*[12]

---

[12]"FwdBH" = Forward Bulkhead; "AftFDPanels" = Aft Flight Deck Panels.

Portions of the TIMs were conducted in the CEL where the SCSIIT could interact with a dynamic simulation of the LOC and breakup events while simultaneously reviewing video from the various ground sites that recorded the events. This was an invaluable asset in developing and refining breakup sequences.

The CEL developed a virtual reconstruction of the cockpit by mapping photographs of selected debris items to a computer model of the crew cabin. This virtual reconstruction was used in a variety of ways to augment direct examination of the debris. Figure 4-7 shows examples of this virtual reconstruction. The image on the left shows the aft flight deck and overhead windows (looking aft) with photographs of recovered panels and window frames included. The image on the right shows several starboard side middeck floor panels and the starboard-most crew seat (the blue item in the upper right corner). This image is from a vantage point looking port and slightly aft.

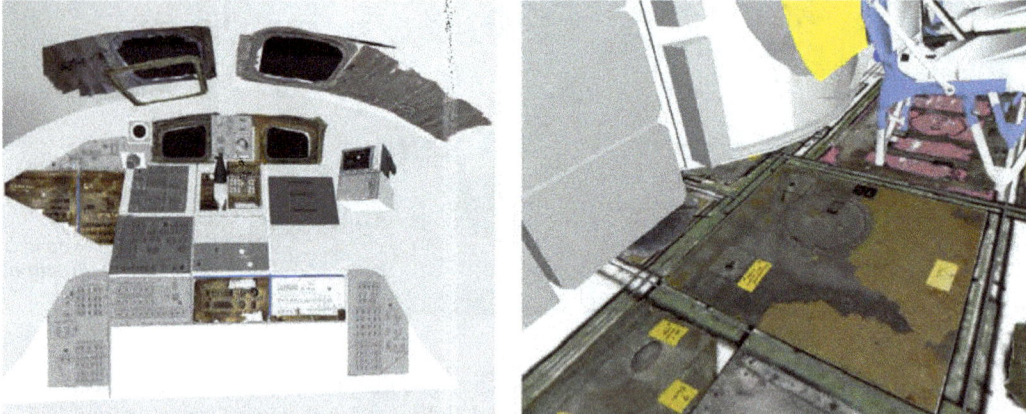

Figure 4-7. *Virtual reconstruction of the aft flight deck* (left image) *and middeck floor* (right image).

4.5.4 Motion analysis tools

The vehicle team was tasked with determining the behavior of the vehicle from LOC of the orbiter until forebody breakup. Analyses that were conducted include trajectory analysis, ballistics analysis, thermal analysis, forebody aerodynamic stability analysis, and CM survivability analysis. In addition to performing analysis on the vehicle and forebody, the Vehicle Team performed ballistics and thermal analyses on various items for the other SCSIIT sub-teams.

4.5.4.1 *Trajectory and attitude analyses*
Global Reference Atmospheric Model
The Global Reference Atmospheric Model was used to generate a representative flight day atmosphere, which is the U.S. Standard Atmosphere 1976 model. This model is a steady-state (year-round) model of the Earth's atmosphere at latitude 45N during moderate solar activity.

Intact Orbiter Simulation
The entry simulation[13] was developed to model the dynamics of the intact orbiter between loss of signal (LOS) at Greenwich Mean Time (GMT) 13:59:31 to the CE at GMT 14:00:18. This simulation provided the resulting modeled accelerations on the vehicle structure and crew during this timeframe. The simulation used the preflight predicted vehicle mass properties; downlinked general purpose computer (GPC) data including position, velocity, attitude, and any alarm/warning related data; and the Modular Auxiliary Data System/orbiter experiment recorder sensor data. The final 2-second period of reconstructed GPC data (RGPC-2) were used in an attempt to synchronize the simulation with the actual flight data. Full details about assumptions and models are described in EG-DIV-08-32, IEE Report, Appendix G – Post-LOS Analysis. The simulation used available data; therefore, there is a moderately high level of confidence in the representation of the motion and resulting accelerations.

---

[13]EG-DIV-08-32, Integrated Entry Team Report, Appendix G – Post-LOS Analysis.

Reference Trajectory

The reference trajectory is the path of the intact orbiter and forebody given assumptions about their aerodynamics properties and ballistic numbers. The trajectory provides a continuous trajectory from LOS to the Crew Module Catastrophic Event (CMCE) to main engine ground impact. The trajectory was used as a common reference for the thermal and debris ballistic analyses. The reference trajectory was generated using Simulation and Optimization of Rocket Trajectories (SORT). The reference trajectory is divided into four phases: the "Nominal Orbiter," with an average ballistic number of 108 pounds per square foot (psf); the "High-drag Orbiter," with an average ballistic number of 41.7 psf (lift generation occurs); the "No-lift Orbiter," with an average ballistic number of 41.7 psf (a 72-degree angle of attack assumed); and the "Forebody" vehicle, with an average ballistic number of 150 psf.

Simulation and Optimization of Rocket Trajectories

The SORT software program is a general-purpose, 3-degrees-of-freedom, computer-based simulation of the flight dynamics of aerospace vehicles. It was used to estimate the time of an object's release from the various configurations that the orbiter experienced during the accident. The aerodynamic forces experienced on the reference trajectories and the heating and atmospheric conditions were also generated. This program was selected because it was previously used to design the shuttle ascent trajectory and because of user familiarity.

The aerodynamic coefficients, mass, area, and ground recovery location of the debris item were entered into SORT. SORT then calculated release times of the debris item from the reference orbiter trajectory until the location where the object was found matched the calculated ground location. The calculated release time had a ±5-second error bar due to unknowns in the reference trajectory.

Shuttle Engineering Simulator

In 2003, the Shuttle Engineering Simulator (SES) provided an engineering simulation flight reconstruction of the STS-107 entry. This simulation was based on flight data that had been recorded during the descent. The ascent/entry SES was supplied with a set of data files describing the atmospheric conditions during the descent. The original task was to model the changes to vehicle aerodynamics due to the left wing damage. The results of this work are documented in the guidance, navigation, and control portion of the CAIB Report. The visual simulation was also used as part of the "crew awareness" task that was assigned to the Crew Team. Additionally, video of the three primary SES cockpit displays covering the time from GMT 13:58:19 to GMT 13:59:39 (2 seconds after the LOC) were included in the CEL visual products.

Forebody Trim Analysis

The forebody configuration was idealized in that the mass properties were held constant from the CE to the CMCE and the aerodynamic geometry was symmetric. The configuration, which was an intact FF containing the CM, ended cleanly at the X_o 576 bulkhead. At the time, the team did not yet know that parts of the X_o 582 ring frame bulkhead stayed attached to the forebody. The forebody configuration is shown in figure 4-8.

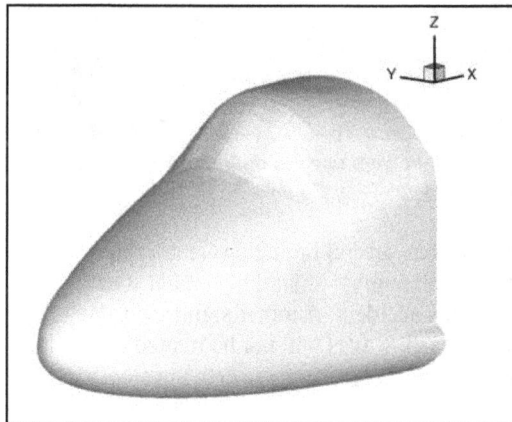

The aerodynamic coefficients and surface pressure distributions for the forebody traveling at hypersonic speeds were predicted using Snewt, an implementation of the Newtonian engineering method. The aerodynamic properties were generated at orientations through a 0-degree to 180-degree sweeps in α (angle of attack) and β (sideslip angle). The aerodynamic moments were plotted in a contour analysis to evaluate stability and were used as inputs to the 6-degree-of-freedom kinematic motion simulation.

Figure 4-8. Forward fuselage/crew module (forebody) configuration.

A sensitivity study was conducted to examine the effect of the forebody's damping moments on the motion. Because no damping moment data were available for the forebody configuration, Apollo capsule damping moments were used as the closest approximation. At the high Mach numbers that were relevant to the *Columbia* flight conditions, the effect of damping moments is low. It was found that the damping moments had little effect on the motion. As a result, subsequent analyses set the damping moments to zero.

The pitch stability of the forebody was examined by plotting the pitching moment coefficient vs. the angle of attack with the sideslip angle held at 0 degree. The roll and yaw stability was examined by plotting the rolling and yawing moment coefficients vs. the sideslip angle with the angle of attack being held constant at the pitch stable angle. This study showed that with any lateral center of gravity other than zero, neither the forebody nor the CM would achieve a stable trim attitude.

These aerodynamic data were then used as inputs to an MSC Visual Nastran simulation. The simulation, which began at approximately the CE, determined the forebody motion to approximately the CMCE. The inputs to the simulation include forebody geometry and mass properties, aerodynamic coefficients, and initial vehicle state conditions. The outputs include the motion parameters, (position, velocity, and acceleration), attitude, attitude rate, and G-loads at the seat locations. Animations of several cases (forebody, CM, initial rates, etc.) were also generated to give a better understanding of possible forebody motion. Unlike the intact orbiter analysis, no data were available for the modeled motion. Therefore, fewer conclusions can be drawn from these data. However, all analyses showed a failure to achieve a trim attitude, and video data supported this conclusion.

Snewt

Snewt is the name of a program that uses the modified Newtonian method to compute a surface pressure distribution and various aerodynamic coefficients. The program uses numerous inputs (reference area, reference length, moment reference center location, Mach number, angle of attack (α), sideslip angle (β), and the ratio of specific heat) and the vehicle outer mold line shape to compute the following coefficients (figure 4-9):

CA = axial force (force component in the body "x" direction)
CY = side force (force component in the body "y" direction)
CN = normal force (force component in the body "z" direction)
Cl = rolling moment (about the "X" body axis)
Cm = pitching moment (about the "Y" body axis)
Cn = yawing moment (about the "Z" body axis)
D = drag (force in the opposite direction of the velocity vector)
Cs = wind side force (force in the wind "y" direction)
L = lift (force in the vertical direction perpendicular to the velocity vector)

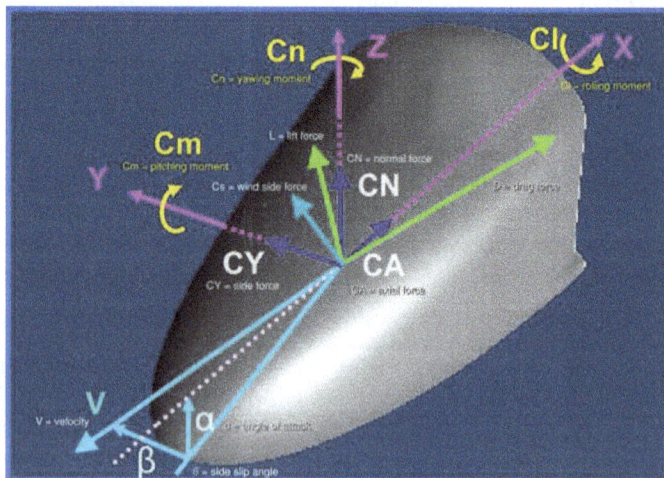

Figure 4-9. *Coefficients that were used to define the vehicle outer mode line shape.*

The modified Newtonian method is considered a simple model that has several limitations restricting its accuracy. One of the most significant limitations is that the Mach number should be greater than approximately M=5. This was appropriate for the conditions in this accident. Another significant limitation of the method is that regions of a vehicle behind multiple shock waves will not be treated correctly. An example is the portion of a wing that lies inside the bow shock.

MSC Visual Nastran Motion Working Model

The MSC Visual Nastran Motion Working Model is a conceptual design software tool that analyzes mechanical systems. It was used by the SCSIIT to simulate the motion of the forebody (both translational and rotational motion) after the CE to gain understanding into what type of accelerations and forces the forebody and crew may have experienced. This tool was chosen because many of the equations that were needed for the analysis were already embedded in the source code, thus reducing the time needed to set up, initialize, and verify the outcome. Events or additional equations could be integrated into the simulation using the built-in formula language or linking to MATLAB or Excel routines.

Objects can be imported from computer-aided drafting drawings or created within MSC Visual Nastran. An object's mass properties and initial conditions – such as position, velocity, and rotation rates – can be specified. Simulation properties can be measured and displayed in digital or graphical formats and saved to Excel files for further analysis. The MSC Visual Nastran Working Model also provides audio visual interface files of the vehicle's motion, acceleration at different locations on the vehicle, and vehicle rotational rates that can be used for additional evaluation and analysis.

4.5.4.2 *Ballistic analysis*

Ballistic analysis of the debris determined the estimated release time of individual debris items from the orbiter using the reference trajectory and the debris item's recovery location (latitude/longitude), calculated ballistic number and aerodynamic drag coefficients.

Snewt was used to calculate the average ballistic number and aerodynamic drag coefficients. To make these calculations, Snewt requires an object's mass and geometry. If these data were not available from the KSC debris database, the debris item would be measured.

SORT was used to estimate the release time of the debris piece given the reference trajectory, the average ballistic number and aerodynamic drag coefficients from Snewt, and the location of the recovered debris piece. These calculations assume that the recovered debris piece has the same mass and geometry on the ground as it had when it left the orbiter. SORT calculated the release times of the debris item from the reference trajectory until the calculated ground location matched the location where the object was found.

4.5.4.3 *Thermal analysis*

Debris thermal analysis was conducted to determine the entry heating environment for specific debris items. This analysis used the estimated release time and state of the item, the average ballistic number, and the aerodynamic drag coefficients.

Object Reentry Survival Analysis Tool

The Object Reentry Survival Analysis Tool (ORSAT) uses a 3-degrees-of-freedom trajectory model, Detra-Kemp-Riddell aero heating equations, and a 1-dimensional, finite-difference thermal conduction model to predict thermal damage to items experiencing the entry environment. ORSAT was used to predict the temperatures that specific debris items would experience after release from the orbiter. These predicted temperatures were then compared to the observed thermal damage to determine whether entry heating was responsible for the observed thermal damage. In several cases (described in Section 2.1), the observed heat damage on the recovered debris items could not be explained adequately by entry heating alone. Literature searches revealed the phenomena of shock wave impingement and shock wave interference leading to increased heating rates.[14] These phenomena and their effects on spacecraft during the hypersonic flight regime are not well understood. Due to limited resources, the team was not able to advance the understanding of these phenomena.

SINDA/FLUINT

The Systems Improved Numerical Differencing Analyzer (SINDA)/FLUINT is a general-purpose numerical thermal/fluid solver. It was used to compare heat rates for some of the thermal analyses (see

---

[14]NASA TM X-1669 "Flight Experience with Shock Impingement and Interference Heating on the X-15-2 Research Airplane", October 1968.

Section 2.1), and is maintained by C&R Technologies (www.crtech.com). The user creates a thermal model that is represented by a nodal network of capacitors and conductors, applying the relevant initial and boundary conditions. The tool solves the thermal network and returns the desired parameters (i.e., temperature, heat flow) for the desired times in the simulation.

BLIMP-K

The Boundary Layer Integral Matrix Procedure-Kinetic (BLIMP-K) code is a FORTRAN-based code that was developed during the Apollo Program by Aerotherm for NASA for use in aeroheating analyses. The SCSIIT used it to compare to ORSAT and SINDA during some thermal evaluations (see Section 2.1). This kinetic version of BLIMP allows for a finite-rate, thermo-chemical model to be used in the analysis. The BLIMP-K model, which is described in the latest User's Guide,[15] incorporates temperature-dependent catalytic models and uses a maximum of 15 nodes across the boundary layer. Subsequent modifications allow the catalytic model to be updated at each station along the streamline, as well as allowing the ability to model up to 2,500 streamline stations. Macros and scripts have been used to speed up the pre-processing of the BLIMP-K input files for large parametric studies.

Computational Fluid Dynamics Thermal Analysis of Payload Bay Door Rollers

A computational fluid dynamics (CFD) thermal analysis of the payload bay door rollers was performed by the JSC Applied Aerosciences & CFD Branch. Various Mach numbers, from M = 7.5 to M = 15, were used to determine the flow field environment and temperature at the face of a roller for an orientation with the front of the roller facing directly into the direction of travel. Figure 4-10 shows the predicted heating rate and temperature distribution along the payload bay door roller at M = 10.5. The notations q_{cw} refers to "cold wall" and q_{hw} refers to "hot wall." In this figure, the CFD solution assumes a steady-state solution, meaning that given enough time, the flow will develop into the calculated results. This is a reasonable assumption for high-speed entry because the flow field will form very quickly. The wall temperature is calculated by assuming radiation equilibrium. This means that the amount of heat that is being absorbed by convective heating is the same amount of heat that is being expelled through re-radiation. Also seen in the upper right of figure 4-10 are two photographs of payload bay door rollers. The photograph for OV-105 shows an intact roller, and the photograph for OV-102 (*Columbia*) clearly shows the erosion of the exposed titanium surface on the front face of the roller.

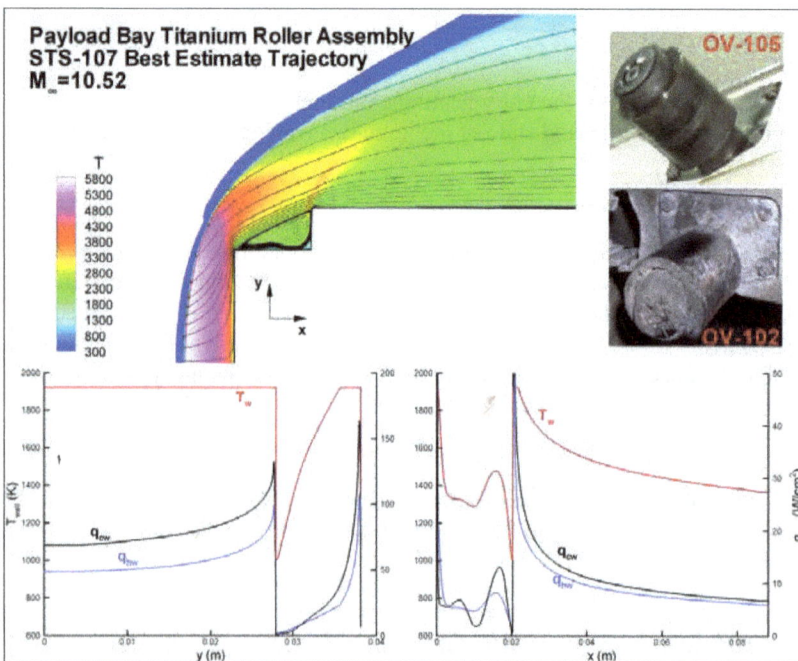

Figure 4-10. *Computational fluid dynamics analysis of heating at the tip of the payload bay door roller for orthogonal geometry into the direction of travel at M = 10.5.*

---

[15]Murray, A. L., "Further Enhancements of the BLIMP Computer Code and User's Guide," AFWAL-TR-88-3010, June 30, 1988.

Radiation equilibrium is a reasonable assumption when used for analyzing the Thermal Protection System (TPS) that was designed to have a very low thermal conductivity; e.g., high-temperature reusable surface insulation tiles or reinforced carbon-carbon panels. However, for metals that will conduct the temperature inward, there is a time factor that needs to be accounted for and a temperature distribution through the object. Thus, the CFD analysis is very good at producing accurate heating rates, and the temperature calculation can give a reasonable upper bound.

Arc jet testing

It was proposed that the erosion of both the x-links and the payload bay door rollers appeared to have some element of material selectivity. Combustion was proposed as an explanation for why some materials eroded and other materials did not and also to explain how oxide formation, which requires extremely high temperatures, occurred.

A selection of enthalpy-pressure test points was chosen based on the predicted trajectory and ballistic number of *Columbia* and the free-flying forebody. A series of tests was conducted at the Boeing St. Louis Large Core Arc Tunnel plasma arc facility. The complete results of the arc-jet testing can be found in Olivas, J. D., Mayeaux, B. M., Melroy, P. A., and Cone, D. M., "Study of Ti Alloy Combustion Susceptibility in Simulated Entry Environments," *AIAA Journal*, 2008 (submitted for publication).

4.5.5 Video analysis

4.5.5.1 *Ground-based video analysis*

The CAIB Report contains information that is related to video analysis performed during that investigation.[16] The process followed in performing video analyses is highlighted below.

When the ISAG received a video, that video was "screened." The video was watched from beginning to end, identified as an ascent or descent video, briefly summarized, entered into the ISAG database, and cross-referenced to any photographic stills or other versions of the same video.

Following the initial screening, a "detail screening" was performed on videos, noting key timing events. Key timing events may include the start and end of the data tape, the first and last appearance of the orbiter, and any visual event that was significant enough to be potentially useful in time-synchronizing with other videos. The first appearance of an object is referred to in the ISAG database as the acquisition of signal (AOS), and the last appearance of an object is referred to as the LOS.

Events appearing in multiple videos could not only time-synchronize the videos but could also provide confirmation of the nature of those events (for example, whether it was a brightness change or a debris-shedding event). Visual events that may be noted in the videos include object-brightening events, changes in the trails of the objects seen, color changes in objects and/or trails, separation and/or breakup of objects, and first and last appearance of additional objects. Generally, brightening events were easier to correlate between videos than debris-shedding events.

Finding. Brightening events were easier to correlate between videos than debris-shedding events.

The AOS of an object is usually well identified. An object can have multiple AOSs and LOSs if it passes in and out of the field of view (FOV) due to tracking of the camera or changes in the magnification factor of the video. Visual separation of objects is impacted by the zoom factor of the video, the resolution of the recording device, focus, image stability, the amount of saturation of the pixels of the recording device due to the brightness of the objects in the FOV, and the viewing geometry of the viewer in relation to the

---

[16]*Columbia* Accident Investigation Board Report, Volume III, Appendix E.4 *Columbia* Early Sighting Assessment Team Final Report, June 13, 2003, details a combination of video, photograph, radar, and ground debris information and details data submission handling. *Columbia* Accident Investigation Board Report, Volume III, Appendix E.2, STS-107 Image Analysis Team Final Report, details the investigation of the imagery of the STS-107 entry (as well as imagery of the launch and suspect foam strike).

objects and the sun. The time of visual separation of objects in an FOV will always be later than the actual time of physical separation, with delays ranging from a few video frames (1 frame ≈ 1/30 second) to several seconds.

The LOS time for an object generally indicates that the object left the FOV rather than indicating the loss of a detectable visual signal. Typically, the videographers zoomed the FOV to concentrate on the largest or brightest object that was visible (typically the orbiter or the aftbody/engines). If an object left the FOV, it can be inferred that the object probably had a lower ballistic number (higher drag) than the object that the videographer was tracking. On rare occasions, LOS equated to an inability to track the object due to multiple pieces seen and/or a scintillating (flashing) visual signal. There were also cases when the visual signal dropped below the sensitivity of the camera or the object disintegrated. The cause of the LOS was rarely noted in the database.

The video key events times were used to develop event sequences for the CAIB. After the CAIB event sequence timeline was completed, a Late Re-entry Working Group (LRWG) was formed in support of the CSWG. The LRWG's primary tasks were to determine when the CM separated from the rest of the vehicle and when it broke up, any visible event that could relate to loss of cabin atmosphere, and any large deceleration event (i.e., an abrupt slowing of the CM). The LRWG results fed into some CSWG results and fed extensively into the SCSIIT work. A few videos provided good views of the breakup of the aftbody, but these were not studied in great detail since they were not relevant to the SCSIIT.

The SCSIIT used the time-synchronizing of the ISAG and the LRWG when possible. The "Apache" video source has an accurate GMT time since it is based upon the Global Positioning System (GPS). All other videos from the eastern timeline were synchronized from Apache video, if possible, and a few videos had adequate timing information to allow cross-checks of GMT. The times for video events have estimated errors ranging from ±0.3 second up to ±2 seconds, although generally they were less. Errors that were associated with video event times are impacted by the magnification of the FOV, resolution, and viewing geometry. For example, an image with a high magnification for the camera FOV will see the start of the CE sooner than an image from an FOV that was not zoomed. Resolution may delay timing for similar reasons. Viewing geometry can prevent an event from being seen until later, if it is seen at all. A combination of error sources can lead to an accuracy of ±1 second for defined events within a single video or between videos, although the actual error may be better or worse by up to an additional 1 second.

Although all 51 videos depicting the orbiter over Texas were examined and contributed to understanding the events, five videos were found to be key for the LRWG's investigation (see Table 4-5).

Table 4-5. *Videos That Were Used*

Reference EOC No.	City in TX	Name	Latitude	Longitude
EOC2-4-0024	Arlington	Arlington	32.7	−97.1
EOC2-4-0209-B	Hewitt	Hewitt	31.4	−97.2
EOC2-4-0221-4	Mesquite	WFAA4/Mesquite	32.8	−96.6
EOC2-4-0221-3	Fairpark	WFAA3	32.8	−96.7
MIT-DVCAM	Fort Hood	Apache	31.2	−97.6
EOC2-4-0077	Burleson	NBC*	32.5	−97.3

*NBC = National Broadcasting Corporation

Late in the SCSIIT investigation, a previously unanalyzed video was discovered. Some videos received from television affiliates were compilations of video that had been collected by television camera operators and those that were submitted by the public. These compilations were not noted as "ascent" or "descent" videos if they were a mix of videos from both phases of the mission. Additionally, the original review of videos focused on identifying pre-CE videos, and this video was misclassified as recording events after the CE. It had been reviewed within the first 5 days of the video investigation related to the CAIB, when the importance of the video evidence was not fully realized. Because the video compilation did not have a "descent" annotation, both the LRWG and the SCSIIT were unaware of its existence. Additionally, because no team had initially identified it as a critical video, the original was not requested in a timely fashion. A copy of part of the original video from the television station was obtained (in addition to a copy of the full video in the ISAG archives), but the original video no longer exists. The copy provides adequate resolution for data interpretation. This video, which is designated "NBC" (EOC2-4-0076-B), added several seconds of good-quality imagery to the eastern timeline.

Finding. Not all videos segments within compilations were individually categorized. Not all videos were re-reviewed once a better understanding of events had been gained.

> *Recommendation A11.* All video segments within a compilation should be categorized and summarized. All videos should be re-reviewed once the investigation has progressed to the point that a timeline has been established to verify that all relevant video data are being used.

It was originally assumed that the sun would have little impact on the brightness of the debris pieces. If a piece of debris was generating a trail, it was expected that the debris was self-illuminating due to the thermal effects of entry heating. Objects that were tumbling were not expected to vary in brightness solely due to the sun. However, review and comparison of videos showed that the sun did contribute to the variations in illumination of objects, even those generating trails. Illumination by the sun impacted the visibility of all debris.

Finding. Sun angle illumination impacted the visibility of debris in video recordings.

In establishing the sequence of events as seen from video and photographic imagery, there are brightening events and debris-shedding events. The appearance of these events was impacted by illumination angle, viewing angle, viewing geometry, and timestamp accuracy. Additionally, the accuracy of the timeline and related data was dependent on information from the videographers. A standard information sheet was used that helped to ensure that most, if not all, key information was available.

4.5.5.2 *Forebody triangulation*

The CEL created a custom application to evaluate the relative motion between the forebody and the aft portion of the vehicle. The application used ground-based videos that were recorded from multiple locations and triangulation to determine the relative distance between the forebody and the aftbody. The original intent of the analysis was to identify the initial conditions of forebody separation (forebody rotation rates, separation rates, etc.). The time between the CE (GMT 14:00:18) and the first point where the forebody was visible in two videos (~GMT 14:00:30) made it difficult to draw conclusions about the forebody separation conditions. However, as described in Section 2.1.5, this triangulation analysis yielded information regarding the forebody motion after the CE. The analysis led to the conclusion that the forebody was rotating in all three axes at approximately 0.1 rev/sec.

4.5.6 Debris mapping/plots

Debris field analysis was conducted to investigate where components of the orbiter impacted the ground compared to where the debris items originated from on the orbiter. This analysis involved plotting the recovery locations of debris items, identifying debris groups based on their location on the orbiter (i.e., left wing, payload bay, FF, etc.), and comparing the relative positions of the groups. The debris groups were referred to as "clusters." Cluster analysis is a technique that uses the assumption that when a large number of debris items from the same structural zone (e.g., tail, wings, payload bay, CM) are considered, different clusters will have a similar range of ballistic numbers. With a similar range of ballistic numbers, the centroids of the clusters can be evaluated *relative* to each other to approximate sequencing of key events.

Ballistic analysis for selected debris items within the various clusters provided calculated release times, and the release times were used to develop ballistics-generated breakup sequences. These sequences yielded similar sequences to the cluster analysis-generated sequences, providing confidence in the breakup sequences that are described throughout this report.

Latitude and longitude errors

Performing these analyses required accurate latitude and longitude data in a format that was usable by the plotting application. The majority of the items in the KSC debris database include latitude and longitude data of the recovery location. In some cases, no latitude/longitude information exists because debris were recovered and submitted by local residents. In other cases, the latitude/longitude data may have been recorded in the field but not imported into the KSC debris database. The data that do exist in the KSC debris database were assumed to be accurate unless there were obvious factors that indicated otherwise, such as the resultant plot indicating that an object was recovered beyond the reasonable bounds of the debris field.

Much of the data that are in the KSC debris database are recorded in degrees and decimal degrees (DD.ddddd), for example: 31.31063° N, 93.87701° W. However, the database contained several other formats, including 31.21.26.3; 31 21′ 26.3″; 31 21 26.3; 31 21.263; and 31.21.263.

The first three formats were assumed to indicate degrees, minutes, seconds, and decimal seconds (DD MM SS.sss), unless information indicated otherwise. The last two formats were assumed to indicate degrees, minutes, and decimal minutes (DD MM.mmm), unless information indicated otherwise.

To be usable by the plotting application, the data were converted into the degrees and decimal degrees format (DD.ddddd). In many cases, there were indications that the data would not need to be converted, despite the placement of decimal points or spaces. If the "minutes" or "seconds" numbers were greater than 59 in either the latitude or longitude, the data were assumed to represent degrees and decimal degrees, regardless of the placement of extra decimal points. For example, although "31.87.70.1" appears to be in the DD.MM.SS.s format, it cannot be because 87 minutes and 70.1 seconds are not valid coordinates. Therefore, this latitude, and its corresponding longitude, would be used as 31.87701° (DD.ddddd) without converting the numbers. In the absence of conflicting cues, the latitude/longitude data would be converted according to the assumptions described above.

The lack of a single, standard data format for latitude/longitude data and the potential ambiguity that is associated with the need to convert data of different formats resulted in possible data errors. One possible error type is due to the numbers being entered in the DD.ddddd format when they actually represent another format. This results in a *lack of conversion*. For example, 31° 23′ 41.410″ was entered in the database as 31.2341410°. The format implies that no conversion to DD.ddddd is necessary, but the original numbers actually were in the DD MM SS.sss format and should have been entered in that format (making the need for conversion obvious). Another error type is due to the numbers being entered in a format other than the DD.ddddd format when they actually represent the DD.ddddd numbers. This results in an *unnecessary conversion* being performed on the numbers. For example, 31.31063° was entered into the database as 31.31.06.3 and erroneously converted to 31.51842°.

Finding. The lack of a single, standard data format for latitude/longitude data and the potential ambiguity that is associated with the need to convert data of different formats resulted in possible data errors.

Figure 4-11 shows a plot for a pair of number strings (31 39439 and 94 53462) that is interpreted for three different latitude/longitude formats. Point A shows the numbers interpreted as the DD.ddddd format (31.39439° N, 94.53462° W). Point B shows the numbers as a DD MM SS.sss format (31° 39′ 43.9″ N, 94° 53′ 46.2″ W). Point C shows the numbers as a DD MM.mmm format (31° 39.439′ N, 94° 53.462′ W).

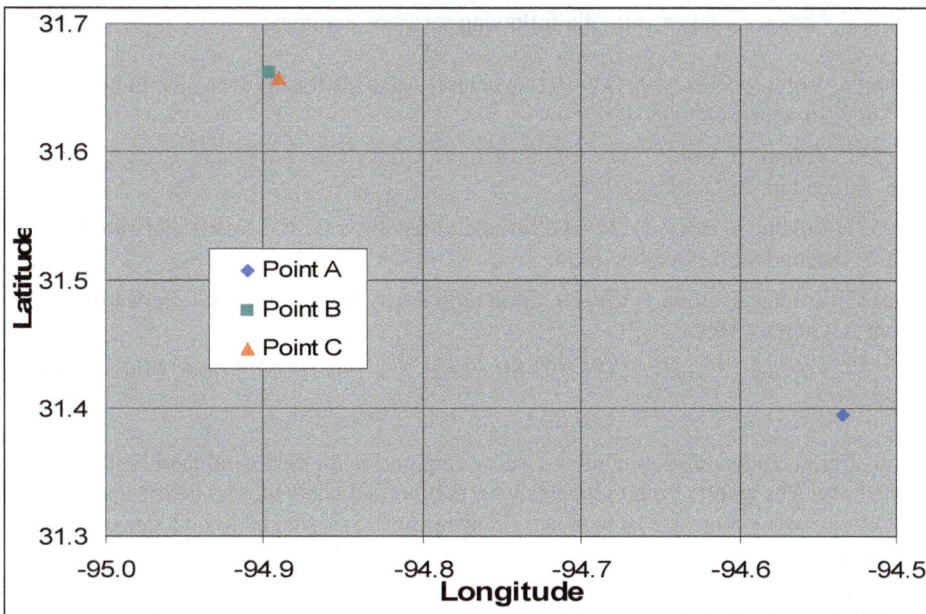

Figure 4-11. *Plot for a pair of number strings that was interpreted for three different latitude/longitude formats.*

In this example, Point A is approximately 28 miles from Points B and C; while Points B and C are less than six-tenths of a mile apart. This example demonstrates the magnitude of error that can occur if the wrong data format is assumed and a necessary conversion does not occur (or an *un*necessary conversion is performed).

To avert these errors and increase the confidence in the debris-plot-related analyses results, all forebody debris of critical importance that had suspect latitude and longitude data were researched to confirm or correct the latitude and longitude data recorded in the KSC debris database (or to provide missing data). Correcting the data involved researching the original field data sheets and the field photographs taken with GPS receivers in view, and, in some cases, using the field descriptions of the area ("100 feet north of county road X, one-quarter of a mile east of the intersection of county road X and county road Y") and GPS/mapping software to establish corrected latitude and longitude data. This was a labor-intensive task, but the need for accurate debris recovery locations warranted the effort. Accurate debris recovery locations resulted in a higher degree of confidence in the debris plotting analyses and the ballistics-related analyses.

> **Recommendation A10.** Global Positioning System receivers used for recording the latitude/longitude of recovered debris must all be calibrated the same way (i.e., using the same reference system), and the latitude/longitude data should be recorded in a standardized format.[17]

4.5.7 Structural analysis

The recovered components of the CM and FF were studied in depth to provide information relating to failure mechanisms and timing. Areas of primary interest were the identified debris from the primary load bearing structures, windows, and hatches.

Figures were taken from drawings in the Orbiter Structures CATIA library, Version 4, which was developed by Boeing Engineer Daren Cokin.

---

[17]STS-107 *Columbia* Reconstruction Report, NSTS-60501, June 30, 2003, p. 142.

OV-102 structural loads were taken from the following series of volumes:

- SSD96D0095, Volume 5, Book 1, OV-102 Structural Analysis for Performance Enhancement, Crew Module-Shell Structure, October 1997.

- STS 89-0537, Volume 5, Book 1, OV-102 Structural Analysis for 6.0 Loads, Crew Module-Shell Structure, Addendum A, October 1995.

- STS 89-0537, Volume 4, Book 1, OV-102 Structural Analysis for 6.0 Loads, Forward Fuselage Upper/RCS, Addendum A, October 1995.

- STS 89-0537, Volume 3, Book 1, OV-102 Structural Analysis for 6.0 Loads, Forward Fuselage Lower, Addendum A, October 1995.

- STS 89-0537, Volume 15, Book 1, OV-102 Structural Analysis for 6.0 Loads, Mid Fuselage, December 1992.

Estimated loads from aerodynamic simulations were compared to the structural load limits from the volumes cited above. The results from video analysis, debris field analysis, and debris inspection/analysis were used to determine the sequence of breakup. Engineering judgment and knowledge of the CM structure were used to assess the breakup.

4.5.7.1 *Structural capability*

When the orbiter was designed, structural analysis was performed to ensure that the CM and/or vehicle would maintain integrity under nominal conditions and a few defined off-nominal cases such as a crash landing. However, little work was done at that time to characterize failure modes in an effort to understand what might happen in a catastrophic situation. During this investigation, analysis was performed to determine the structural capabilities of the orbiter and the CM, and to predict failure modes and failure locations assuming certain load cases derived from the accident scenario.

The primary tools that were used to perform the assessments were structural certification documents, NASTRAN finite element structural analysis software code (using detailed finite element models of local structure), and a medium-fidelity, global-finite element model of the orbiter. With these tools, investigations were completed to assess the vehicle's flight recorder loads, CM skin stress, CM attachment linkages and fittings forces, window thermal shock, window frame distortion, crew seat structural failures, and CM floor loading.

The objective of these assessments was to gain additional insight into CM structural performance and CM structural failures consistent with the accident debris. Load comparisons, conventional structural analysis approaches based on known conditions, and parametric analyses were the primary approaches that were used.

Conventional structural analyses were performed when primary inputs permitting this type of analysis were known, or when reasonable assumptions could be made regarding structural configuration, structure temperature, and loading. These analyses were performed using the flight data accelerations prior to LOS, while the vehicle was essentially intact.

Parametric assessments were required when less information was available. Since structural certification documents are based on nominal design conditions, the parametric assessments provided information about the structural response outside the design envelope. Exposure of the orbiter vehicle structure to high heating and changes in structural configuration required a parametric approach because, in many cases, the effects of heating shifted structural failure modes and failure locations away from areas that were documented as minimum capability zones in the certification reports. Parametric analyses were used to predict trends for structural failure with increasing temperature, stress contours for combined levels of pressure and inertial loading, and failure propagation as load paths changed.

4.5.7.2 *Crew seats*

In an attempt to determine individual crew load profiles, the team set out to identify seat debris items to specific seat locations. Configuration management records (tracking the serial numbers for seat components to the top-level seat assembly's serial number) were not accurately maintained, so identifying the locations of components by any surviving piece-part serial number was futile. The exceptions to this were components that were associated with the inertial reels, and all six recovered upper seatbacks were identified to specific seat locations.

Some seat debris items were identified to seat locations based on being attached to identifiable floor panels. One item was identified due to its unique application to one seat location. Most other seat components were identified to seat locations based on matching them to identified pieces (upper seatback components or pieces that were attached to floor panels). This process of matching pieces was very time-consuming and laborious. Eventually, slightly less than half of the recovered seat structure debris items were positively identified to specific seat locations. Had the individual seat components been permanently marked with serial numbers and those serial numbers tracked to the assembled seats using rigorous configuration management and control, reconstruction and identification would have been much easier and a higher percentage of pieces could have been identified to specific seats.

Seat structure debris items were studied to determine seat failure modes with the intent of determining the forces (and, therefore, CM dynamics) and the thermal profile. The analyses included gross and microscopic inspection of fracture surfaces, scanning electron microscopy of selected sectioned items, energy dispersive X-ray spectroscopy, and comparison to structural analysis of the seat design. These analyses provided information on the temperatures that were experienced by the seat components. Additionally, the identified failure modes revealed the loading conditions, directions, and dynamics.

Inertial reel mechanisms and straps were examined to obtain information on the seat restraint loading history (and, therefore, loads on the crew members), the thermal history, CM breach timeline, and crew separation timeline. Analyses included gross and microscopic inspection of the inertial reel mechanisms, straps, inertial reel housings, and upper seatbacks. Emphasis was placed on looking for evidence of loads (deformation or witness marks on the inertial reel components, failure modes and locations of the straps, deformation of inertial reel mounting hardware), and material deposits (metal deposits on the straps, melted strap material deposits on the inertial reel housings and seats).

4.5.8 Materials analysis

Materials analyses were performed by the JSC Materials and Processing Office, the JSC Astromaterials Research Office, the KSC Failure Analysis and Materials Evaluation Branch, and the White Sands Test Facility to characterize the deposition (char) on the window panes. Additionally, Langley Research Center performed materials testing on the seat fragments.

Analyses included performing microscopy of debris items to determine fracture dynamics and mechanisms, determining the compositions of materials deposited on the debris, testing materials to determine materials properties, and heating pristine materials samples in an attempt to match debris observations.

The following techniques were used for materials analysis:

- Optical microscopy including polarized light.
- Electron microscopy including scanning electron microscopy, transmission electron microscopy, and scanning tunneling electron microscopy.
- Energy dispersive X-ray spectroscopy.
- X-ray fluorescence spectroscopy.
- Fourier transform infrared spectroscopy.
- Focused ion beam milling.
- Selected area electron diffraction.

- X-ray diffraction.
- Backscattered Kikuchi diffraction pattern.
- Differential scanning calorimetry.
- Thermal gravimetric analysis.

4.5.8.1 *Crew seats*

Analyses were performed on the material deposited on the seat components to yield information relating to the CM breakup sequence and the thermal profile. These analyses included determining deposition patterns (portions of the seats containing deposits and directionality of the depositions) and analysis of the deposited material combined with investigation of vehicle materials to determine possible sources of the deposits.

Microscopy (optical and electron) and spectroscopy techniques were performed on the seat structures to determine failure mechanisms and thermal exposure.

4.5.8.2 *Boots materials thermal testing*

Thermal testing was performed on suit boot soles in an attempt to match the observed thermal damage. Boot soles of flight-like boots were heated in an oven to identify the range of thermal effects with varying thermal exposure. The test samples were exposed to 750°F (399°C), 1,000°F (538°C), or 1,250°F (677°C) at normal atmospheric pressure conditions (~14.7 pounds per square inch (psi), ~20% oxygen (O_2)) for 15, 30, 45, or 60 seconds. The materials showed no significant change in appearance until they combusted. This puzzled the team initially until it became clear that the presence of O_2 was affecting the results. The tests were repeated using new samples that were heated in a nitrogen (N_2) purge (<3% O_2). The samples were then compared to the recovered boot sole fragments. The results of the revised test protocol appeared similar to the recovered boot soles. The test samples that most closely matched the recovered debris items were those that were exposed to 1,000°F (538°C) for 30 to 45 seconds or 1,250°F (677°C) for 15 to 30 seconds. However, there is no credible scenario in which the *Columbia* boots would be exposed to these temperatures for the length of time indicated by the tests, so the test results could not be correlated directly to the debris observations. Because the test conditions (~14.7 psi, 97% to 99% N_2, 1% to 3% O_2) did not accurately approximate the entry environment conditions (low ambient pressure, monatomic oxygen, and possibly high dynamic pressure), they are a potential source of error in this analysis.

4.5.8.3 *Helmets*

Thermal gravimetric analysis was performed on helmet materials to determine the temperatures at which thermal decomposition (pyrolysis) begins. Determining these temperatures allowed the team to determine the thermal exposure of the helmets, which was used to predict helmet release times using ORSAT.

4.5.9 Medical processes

The SCSIIT Crew Team performed an extensive examination of all available medical evidence, performed additional studies including motion, thermal, and ballistic analyses, and correlated all of this information with the results of the other SCSIIT teams to develop a detailed crew event timeline sequence, cause(s) of death, threat matrix, and survival gap. NASA used an independent consulting firm specializing in injury analysis (BRC, San Antonio, Texas) to independently identify the threats with lethal potential that were faced by the STS-107 crew members and any mitigations that were in place at the time of the accident vs. those that are possible (i.e., the survival gap).

4.5.9.1 *Crew event timeline*

The crew event timeline development began with a review of the recorded air-to-ground (A/G) audio to identify any relevant information, such as timing of crew actions and the state of crew event awareness. The recovered videos, showing middeck and flight deck activities, were also reviewed and correlated with the A/G audio and recorded telemetry.

Forensic temporal markers were used to determine the sequence of observed injuries. The medical forensic evidence was then correlated with the audio and video evidence, recorded telemetry, and analysis of the recovered flight crew equipment. These correlations helped establish the configuration of the crew escape equipment at the time of the accident. All of this information was then used to develop the detailed crew event timeline.

4.5.9.2 *Cause of death determination*

The cause of death, blunt force trauma and hypoxia[18], was originally determined at the time of the autopsies, which were conducted by the AFIP.[19] During the CAIB investigation, a team that included members of the AFIP and FBI ensured definitive identification of the bodies, performed autopsies on the crew, provided photographic and X-ray documentation of the human remains, collected tissue specimens, provided sub-specialty interpretation of the microscopic slides, and performed toxicological analyses and spectral analysis of skin deposits. The data generated by the initial examining team were used by the SCSIIT as a starting point for determining what the crew experienced, the temporal sequence of events, and how these events related to the cause of death.

The standard AFIP autopsy protocol that was followed for determining the cause of death was subsequently found to be inadequate to address some of the unique aspects of a hypersonic, high-altitude-entry accident and did not include the collection of unique evidence that would have helped to better understand the sequence of events that the crew experienced and the unique injuries that were incurred from exposure to such extreme events and conditions. Although somewhat limited, the information that was derived from the autopsy reports was invaluable and was used to develop a matrix of the injuries inflicted on each crew member. This basic injury matrix was then expanded, based on additional studies and data reviews conducted by the SCSIIT Crew Team, and integrated with information derived from X rays, photographs, and additional reviews of the histology material. The expanded injury matrix was then used to identify any common injury patterns among the crew members. This also facilitated the identification of injury patterns based on crew location (i.e., flight deck vs. middeck, starboard vs. port).

All of the X rays and autopsy photographs were reviewed. A subset of the X rays and photographs, which was identified as having potentially useful information, was digitally enhanced and re-examined. This effort identified additional information that was not previously noted.

All of the histology slides were also re-examined to aid in confirming the cause of death and determine the temporal sequencing of injuries.

The Anthropometry Biomechanics Facility in the Habitability and Environmental Factors Division of the JSC Space Life Science Directorate was used to conduct medical analysis. A Vitus 3-dimensional scanner[20] was used to scan a SCSIIT member for use as a baseline in the suited and unsuited (i.e., minimally clothed) configuration. The anthropometry of the subject was recorded and compared to the STS-107 crew anthropometry data. The baseline subject's legs, torso, arm lengths, and shoulder and hip breadths were scaled using a percentage of the baseline subject's lengths to the STS-107 crew member's lengths to match each crew member's unique anthropometry, thus providing an anthropometrically correct model for each STS-107 crew member. The end results were scaled 3-dimensional (3-D) models representing the crew members. A simple 3-D skeleton model was also scaled to match the anthropometric measurements of each crew member. Each model was converted into a 3-D mesh and imported into a modeling tool called 3DStudioMax. Using the modeling software, the scaled suited mesh was attached to an underlying scaled skeleton that allowed all the linked scans to move in concert. After the individual crew member models were developed, they were incorporated into a 3-D computer model of the *Columbia* cockpit (developed by the CEL). Autopsy reports, dermatopathology data, and a review of the field intake and autopsy photographs were used.

---

[18]*Columbia* Accident Investigation Board Report, Volume I, August 2003, p. 77.
[19]The AFIP is the U.S. government agency that is authorized to conduct autopsies on astronauts flying on U.S. spacecraft who die in the line of duty.
[20]Vitus is a 3-dimensional scanner that records a digitized image of the body using a laser to capture a surface image of an object. This surface image, which is a volumetric representation of the subject, can then be saved digitally and used to take anthropometric data.

The resulting models were then reviewed by the Crew Team. Patterns were analyzed to determine sequencing and potential injury sources.

All of this information was then integrated with the results of other SCSIIT sub-teams to form an integrated medical scenario and generate augmented autopsy reports that included all updated information and a final determination of the injury sequence leading up to death.

4.5.10 Other methods and processes

4.5.10.1 *Hygiene/drink package depressurization testing*

Intact hygiene bottles and empty (used) drink bags were recovered in the *Columbia* debris field. The fact that these items were not ruptured may indicate a depressurization rate that the CM experienced. Therefore, rapid depressurization tests were conducted in an attempt to determine the maximum depressurization rate that these packages can sustain without rupturing, determining an upper bound for the *Columbia* cabin depressurization rate.

Each trial consisted of testing shuttle-type shampoo bottles and drink bags. Testing was conducted at approximately 14.5 psi/second, approximately 18 psi/second, and approximately 31.5 psi/second. No packages ruptured at any of these rates. Because none of the shampoo bottles or drink bags ruptured during the depressurization tests indicates that the items are capable of withstanding depressurization rates greater than approximately 31.5 psi/second (which would represent an instantaneous, explosive depressurization of the CM – a scenario that is not supported by debris evidence). Therefore, no conclusions could be made regarding the rate of depressurization of the *Columbia* CM based on these tests.

4.5.10.2 *Advanced crew escape suits and crew worn equipment*

The recovered components of the crew worn equipment (advanced crew escape suit (ACES)), survival gear, parachute harness, and parachute pack) were studied in depth to provide information relating to crew injuries, CM breakup sequence, crew separation sequence, and suit disruption mechanism and timing.

Almost all of the approximately 75 recovered non-fabric components were identified to specific crew members. This identification was possible because most of the suit components and subcomponents were marked with serial numbers, and the serial numbers were recorded and tracked to the suit assembly, which was tracked to a specific crew member. In many cases, the serial numbers were stamped or etched on the hardware items and, therefore, survived the entry environment. In some cases, damage to the exterior of the component precluded reading the serial number, so other means were used to identify the hardware items. For the search and rescue satellite aided tracking beacons and the Army/Navy personal radio communications (A/N PRC) 112 radios, the processing modules were extracted and interfaced with ground support equipment to "read" the unique identifier codes of the specific unit. For the Sea Water Activated Release System (SEAWARS), the devices were disassembled and unique markings on the electronics packages were discovered. The vendor was contacted and the team was able to trace the unique markings to specific serial numbers for the SEAWARS units.

The ability to ascribe recovered crew worn items to specific crew members was critical to being able to draw conclusions based on the crew worn equipment.

4.5.10.3 *Helmets*

Visual inspection was performed on the helmets to qualify the helmet structural damage. Several nondestructive evaluation techniques were used in an attempt to determine the extent of thermal and mechanical damage beyond what the visual inspections revealed, with an emphasis on sub-surface damage. Computer tomography scans, real-time X-ray, thermographic (pulse flash and through transmission), and ultrasound testing were used to determine the structural condition of the fiberglass shell. These techniques validated damage sites that were seen in visual inspections and revealed some additional damage areas, but did not produce significant additional findings.

4.5.11 Cabin depressurization modeling

The Killer Press Model is a Microsoft Excel spreadsheet that was developed and used by NASA flight controllers to calculate pressure equalization times for two or three volumes connected in series. The user inputs the starting volume, pressure, and temperature of the air in the volumes in question and parameters for the venting path between the volumes. In all cases, the model assumes a circular vent path between volumes. The model outputs pressure with respect to time for the volumes. The cabin depressurization curves in Section 2.3 were generated using version 2.06 of Killer Press. These plots show the differential pressures between three volumes: the flight deck and middeck volume of the CM; the lower equipment bay of the CM; and the ambient atmosphere.

The following parameters were used as inputs: flight deck/middeck volume = 2,163 feet^3, pressure = 14.7 psi, temperature = 70°F (21°C); lower equipment bay volume = 337 feet^3, pressure = 14.7 psi, temperature = 70°F (21°C); vent path between lower equipment bay and middeck is ~50 in^2 (8-in.-diameter hole); ambient atmosphere volume = 1 × 10^{20} feet^3, pressure = 0.022 psi, and temperature = –67°F (–55°C).

Future Work

Future Work

Chapters 2, 3, and 4 outline the details of the analyses and assessments that were performed to understand the events with lethal potential that occurred. Many findings, conclusions, and recommendations were documented. However, it is also important to discuss the work that was not performed.

The Spacecraft Crew Survival Integrated Investigation Team (SCSIIT) was intentionally established as a small team. This allowed the team to maintain subject confidentiality, which was very important prior to publication of the report. However, the small size of the team, the resources that were available given Return to Flight and other very important programs, and the fact that most of the team members also had to continue their "regular" job assignments resulted in limitations to the investigation. Some items of interest were not accomplished due to lack of resources and schedule. A selection of future projects that may be of interest to complete in the future is listed below.

Although the list of testing and analyses below did not have significant influence on the SCSIIT conclusions and recommendations, the team believes that the data gained from these test and analyses will increase the knowledge base of the accident dynamics, spacecraft hardware, and materials commonly used in space vehicles.

Spacecraft Accident Investigation Database. To develop a database for future investigators that will be cross-referenced across all spacecraft accidents to collect information in a single location.

Suit Destruction Testing. To determine the forces that are required to disrupt the suit (remove helmet shell from helmet neck ring, remove suit-side neck ring from the suit, remove wrist rings from the suit), and to perform an analysis to determine the aerodynamic loads/windblast (Qbar or knots equivalent air speed (KEAS)) that is required to attain the above-mentioned forces with the end goal of determining the suit's actual windblast capability. This evaluation should be done for visor down and visor up configurations.

Materials Analysis/Testing. To positively determine the sources of the materials that were found deposited on various debris items. Also, testing should be done on suit materials (Nomex, GORE-TEX®, nylon), seat straps, and boot materials in a low ambient pressure (near vacuum), high heat, and atomic oxygen environment to simulate the environment at the Crew Module Catastrophic Event (CMCE) to understand how nonmetallic materials react in the thermal and chemical environments of spacecraft entry.

Static and Dynamic Testing on Seat Restraints. To determine the forces that are required to fail the inertial reel strap with the strap fully extended and the strap partially retracted, and the forces that are required to fail the straps when at elevated temperature (200°F (93°C), 400°F (204°C), 600°F (316°C)).

Helmet Destruction Testing. To determine the impact forces that are required to cause physical damage to the helmet.

Emergency Oxygen System (EOS) Bottle Testing. Investigate using microscopic analysis, materials analysis, etc. to determine the failure mode of the EOS bottle fragments and analyze the fracture surfaces to see whether the bottles were pressurized when they were fragmented.

Window Glass Analysis. To evaluate which window glass fragments collected in the field are from *Columbia* and which are non-orbiter.

Shock Wave Analysis. To better understand the effects of shock waves on aerodynamic heating in the hypersonic flight regime.

Titanium Combustion Analysis. To fully characterize the behavioral properties of titanium in an entry environment, including those that can lead to combustion.

Challenger *Analysis.* Complete an analysis on the *Challenger* debris to compare and contrast with the *Columbia* findings.

Appendix – Ballistic Tutorial

Appendix – Ballistic Tutorial

This is intended to give a reader a general understanding of ballistic number and ballistic trajectories. The objective is to give an intuitive understanding of the concepts rather than a rigorous derivation.

When the trajectory of an object is described as being ballistic, that means that the object has no control over its trajectory. As an example, think about a cannonball fired out of a cannon. Once the cannonball leaves the barrel of the cannon, the forces determining its trajectory are its own momentum (mass multiplied by velocity), gravity, drag, and winds.

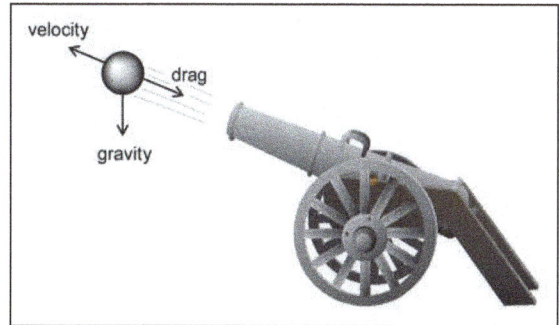

Drag is the force that slows an object down as it travels through the atmosphere, and it is always opposite the velocity vector.

Drag is a function of thickness of the atmosphere, the velocity, reference area, and the coefficient of drag, C_D. The equation for the drag acceleration is

$$Drag\ acceleration = -\frac{1}{2}C_D \frac{Area}{Mass}\rho v^2$$

where C_D = coefficient of drag
    Area = reference area
    ρ = atmospheric density (thickness)
    v = velocity

The minus sign indicates that drag acts opposite to the direction of travel. The coefficient of drag, C_D, is a function of the shape, orientation, and velocity of an object.

The ballistic number (BN) of an object can be thought as a measure of how far downrange the object will travel. The equation for the BN is

$$BN = \frac{W}{C_D * Area}$$

where   W = weight

The units for BN are pounds per square foot (psf) (C_D does not have units associated with it).

Objects with large BNs will generally travel farther downrange than objects with smaller BNs. The plot below shows the trajectories for two objects with different BNs, 20 and 2 psf. The red line is the trajectory of the 2-psf objects and the blue line is the trajectory of the 20-psf object.

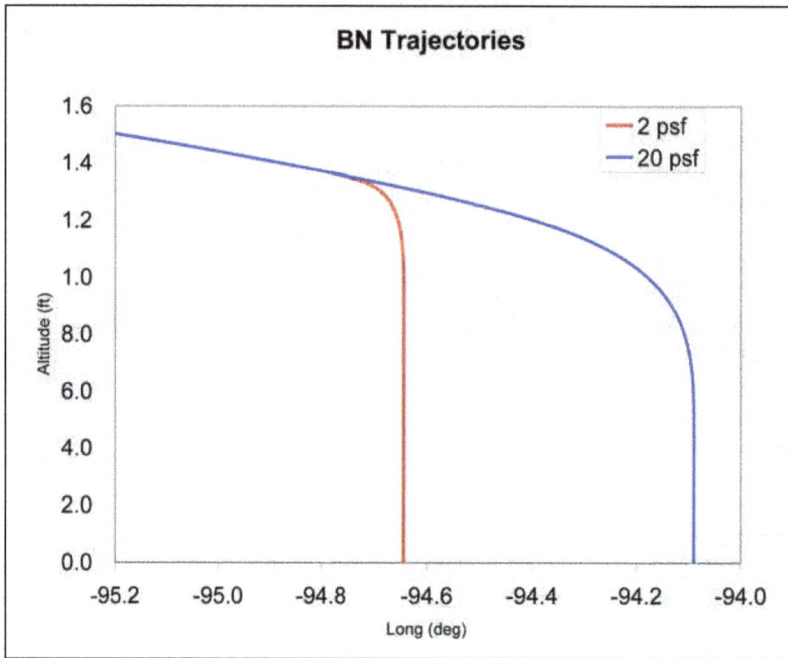

Let's look at how each of the parameters in the BN equation, weight, C_D, and area, affect the BN. First we'll look at weight.

How Weight Affects BN
Two spheres, one is made of rubber and the other is made of aluminum. Both spheres have the same area and C_D.

Area = 0.087 ft^2 (a radius of 2 in.)
$C_D = 1$

Rubber
W = 1.3 lbs.
BN = 14.9 psf

Aluminum
W = 3.3 lbs.
BN = 37.9 psf

The sphere with the larger weight, the aluminum sphere, will have a larger BN. If both spheres were released at the same altitude and velocity, the aluminum sphere would travel farther downrange than the rubber sphere because of its larger BN.

How Area Affects BN

For this example, there are two boxes, both made of aluminum. They have the same weight and C_D. The only difference is the area.

Weight = 5 lbs.
$C_D = 1$

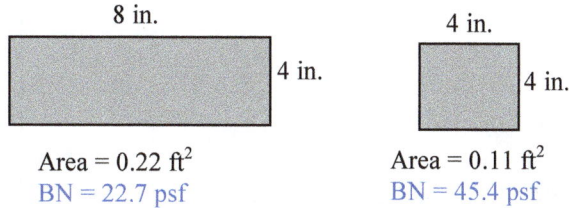

8 in.

4 in.

4 in.

4 in.

Area = 0.22 ft²
BN = 22.7 psf

Area = 0.11 ft²
BN = 45.4 psf

The box with the smaller area will have a larger BN. If both boxes were released at the same altitude and velocity, the box with the smaller area would travel farther downrange because of its larger BN.

How C_D Affects BN

Two spheres, both made of aluminum with the same weight and area. The only difference is that they are released at different velocities that give them different C_D's.

Area = 0.087 ft²
Weight = 3.3 lbs.

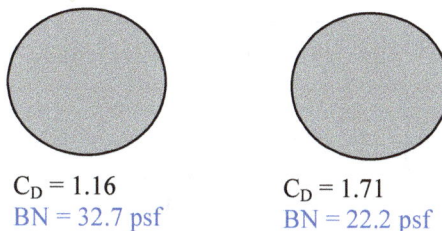

$C_D = 1.16$
BN = 32.7 psf

$C_D = 1.71$
BN = 22.2 psf

The sphere with the smaller C_D will have a larger BN. But in this particular example, the sphere with the larger BN may not have the larger downrange distance because the initial velocities were different. The next section looks at how the initial velocity affects a ballistic trajectory.

How Initial Velocity Affects Ballistic Trajectories

Two spheres have the same BN, which is equal to 20 psf, and are released at the same altitude. But they are released at different initial velocities, Mach 12.6 and Mach 1.3. The plot at the top of the following page shows the trajectories of each sphere. The red line shows the trajectory of the sphere that is released at Mach 1.3. The blue line shows the trajectory of the sphere that is released at Mach 12.6.

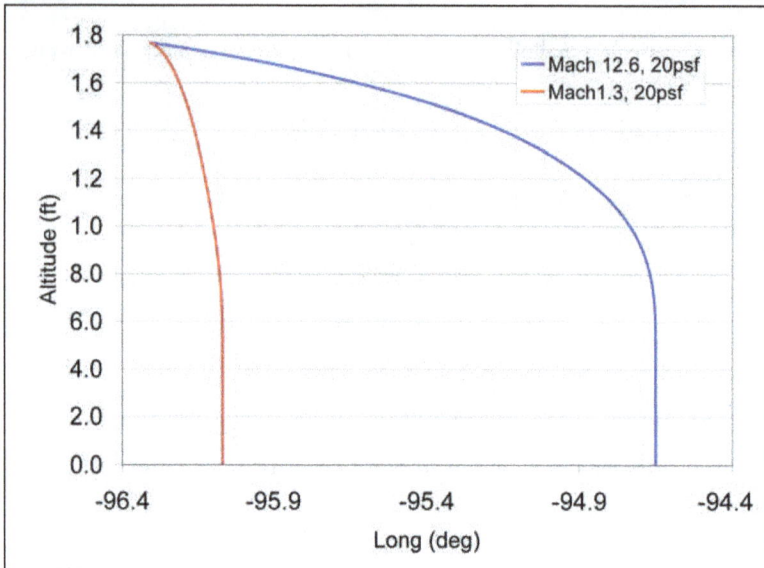

Even though both spheres have the same BN, the sphere that is released at the higher velocity will travel farther downrange. The initial conditions of the ballistic trajectory play a role in determining the downrange distance of an object.

G-loads

Another topic that is related to BN is G-load. G-loads are the accelerations acting on an object divided by Earth's gravitational acceleration. When an object with a small BN is released from an object with a larger BN, the object with the smaller BN will experience a sudden change in the accelerations acting on it. This sudden change in the accelerations is called a G-load spike. As an example, let's go back to the 20 psf and 2 psf objects that we looked at in the beginning. Both of these objects are released from a larger 150-psf object. The plot below shows the sudden change in acceleration that each object experiences. The black line before the G-load spikes is the 150-psf object, the red line is the 20-psf object after release, and the blue line is the 2-psf object after release.

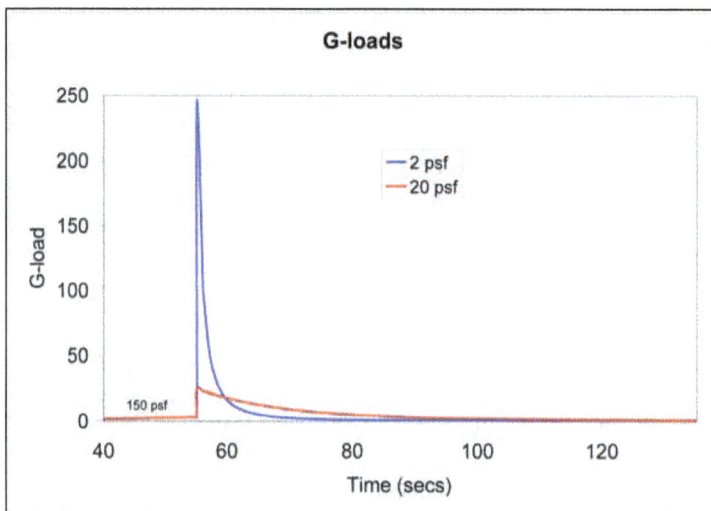

The 2-psf object experiences a much greater change in acceleration after being released than the 20-psf object. The greater the difference between the BNs of two objects, the greater the G-spike.

In a previous example, we examined the effect of initial velocity on two objects with the same BN. What would their G-loads look like? The G-loads for the Mach 12.6 and Mach 1.3 spheres are shown in the plot below. The red line is the Mach 1.3 object, and the blue line is the Mach 12.6 object.

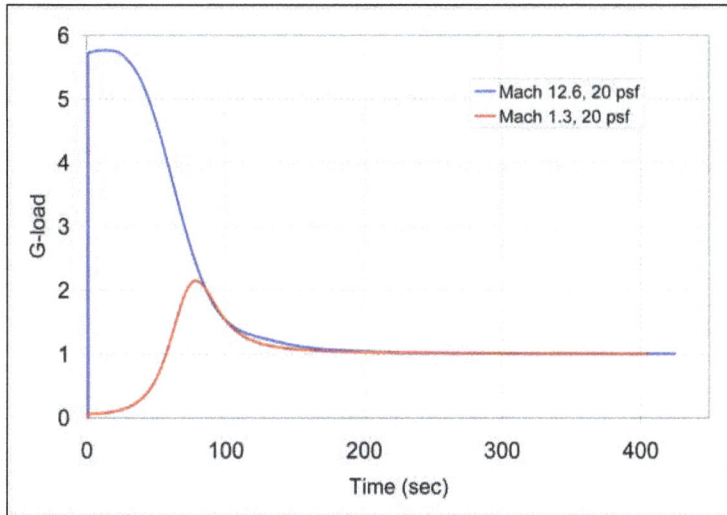

The object with the higher release velocity had a larger G-load spike. The reason for this is found in the drag acceleration equation. It is a function of velocity squared. So, the higher-velocity object will experience a greater drag acceleration than the lower-velocity object.

$$Drag\ acceleration = -\frac{1}{2}C_D\frac{Area}{Mass}\rho v^2$$

But why did the G-load of the object at the lower initial velocity ramp up? That object's velocity actually increased after release, increasing the drag acceleration, which increased the G-load.

To sum it up, a ballistic object has no control over its trajectory. The BN is a general measure of how far downrange an object will travel. The larger the BN, the farther downrange it will travel. G-load is a measure of the accelerations acting on an object. The larger the difference in BN between two objects, the larger the initial G-load, or G-load spike, will be at release. Initial conditions are important in ballistic trajectories. Different initial velocities for the same BN will generate different downrange distances and G-load profiles.

www.ingramcontent.com/pod-product-compliance
Lightning Source LLC
Chambersburg PA
CBHW081043220326
41598CB00038B/6966